NONFICTION (ADULT)   c.1

Oversize 917.8 F538

Fisher, Vardis, 1895-1968
Gold rushes and mining camps of the
 early American West /

8000026011

W9-BSX-627

Oversize

Fisher, V   917.8
                                   F538
Gold rushes and mining camps   c.1
   of the early American West

WITHDRAWN

Gold Rushes and Mining Camps
of the Early American West

Books by
Vardis Fisher

THE TESTAMENT OF MAN
Darkness and the Deep
The Golden Rooms
Intimations of Eve
Adam and the Serpent
The Divine Passion
The Valley of Vision
The Island of the Innocent
Jesus Came Again
A Goat for Azazel
Peace Like a River
My Holy Satan
Orphans in Gethsemane

THE VRIDAR HUNTER TETRALOGY:
In Tragic Life
Passions Spin the Plot
We are Betrayed
No Villian Need Be

OTHER NOVELS
Children of God
Dark Bridwell
April
Toilers of the Hills
Forgive Us Our Virtues
City of Illusion
The Mothers
Pemmican
Tale of Valor
Mountain Man

SHORT STORIES
Love and Death

NON-FICTION
God or Caesar?
The Neurotic Nightingale
Thomas Wolfe and Other Essays
Suicide or Murder?
Gold Rushes and Mining Camps of
the Early American West (with his wife)

GOLD RUSHES AND MINING

CAMPS OF THE EARLY AMERICAN WEST

BY
VARDIS FISHER
OPAL LAUREL HOLMES

ILLUSTRATED WITH
PHOTOGRAPHS

The CAXTON PRINTERS, Ltd.

CALDWELL, IDAHO   83605

1979

First printing May, 1968
Second printing August, 1968
Third printing December, 1970
Fourth printing February, 1979

© 1968 BY
VARDIS FISHER AND OPAL LAUREL HOLMES

Library of Congress Catalog Card No. 68-15028

International Standard Book No. 0-87004-043-X

Lithographed and bound in the United States of America by
The CAXTON PRINTERS, Ltd.
Caldwell, Idaho 83605
132959

FOND DU LAC PUBLIC LIBRARY

917.8
F538
c.1

To Jim and Gordon Gipson,
president and vice-president
of the institution which their
father founded and built,
we inscribe this book.

17.21

4-22-81

BT

A Note to the Reader

This book has been written not for the scholar and professional historian but for the general reader with an interest in the American West. We have tried to give a broad picture, based, as far as possible, on primary sources, and to eschew the exaggerations in popular books on the subject. Because so many books and articles in this field have played up the bonanzas and fabulous riches, and had so little to say about the fact that most of the miners never found more than enough gold or silver to buy cheap food and clothes and keep them going, we let Peter Sapp (in the camps there was a man named Sap and another named Littlebury Shoot) stand for the tens of thousands who followed the rainbow from camp to camp and at last to the Yukon, but found no pot of gold. Solitary and unloved they persevered from rush to rush, to die at last, alone, far from their birthplaces and their people. While the Floods are building their fantastically vulgar mansions, and the Mrs. Mackays are shocking all of Europe with their lack of taste, the reader will now and then be reminded that Peter and all those whom he stands for are still down in the holes, or in their dugouts and shacks. Our last glimpse of Peter and his kind, now long forgotten, is in the line of those climbing the Chilkoot Pass, still driven forward in their old age by a dream of great riches that they would never find. The book closes with the midnight sun as their epitaph.

Table of Contents

Acknowledgments

We are indebted to a great many persons, historical societies, museums, and libraries, not only for most of the photographs that appear here (see names and addresses in the back), but for extraordinary assistance in our research. We are glad to acknowledge our debt to Caroline Bancroft; James Davis of the Denver Public Library; Miss Mary K. Dempsey, Montana Historical Society librarian; Mrs. William Fray, of Colorado Springs; Mrs. Eva Gillhouse, of Las Vegas, Nevada; E. W. Harris, of Reno; Mary Hill, Division of Mines and Geology, San Francisco; James D. Horan; Anne Hyder of the Huntington Library rare book room; Ellis Lucia; Myrtle Myles of the Nevada Historical Society; Margaret Pead of the Caxton Printers; Lois Robinson, reference librarian of the Boise Public Library; J. V. Root, curator of photos, Idaho Historical Society; Louise Shadduck, secretary of the Idaho Department of Commerce and Development; Margaret D. Shepherd, curator of photos, Utah Historical Society; Miss Irene Simpson, director of the History Room, Wells Fargo Bank, San Francisco; Mrs. Gertrude Stoughton of the rare books department, Pasadena Public Library; Mrs. Enid T. Thompson of the Colorado Historical Society; Dr. John Barr Tompkins of the Bancroft; Jane Weddle, San Clemente, California; and Marion Welliver, editor of the Nevada historical quarterly and assistant director of the Society, who gave generously of knowledge and time. Because my wife, in research, has shared so many thousands of hours with me in the stacks of great libraries all over the country, not only for this book but for many of my other books, her name appropriately appears at last with mine on a title page.

VARDIS FISHER

Hagerman, Idaho
February 12, 1968

PART I

The Gold Rushes

Fact or Fiction: the Blend of History and Legend

When we said to a California librarian that Hubert Howe Bancroft often spun legends instead of writing history she replied sharply that she did not agree. Stefansson in his book *The Standardization of Error* says that errors widely and generally accepted are more convenient for most people than truth, because, as the late Professor Wendell Johnson, distinguished semanticist, put it, error "agreed upon and firmly fixed in legend and in law, is something one can count upon from day to day, even from century to century." As so many people still count upon Bancroft. It is simply a fact of life that most persons prefer the mythmakers and legend makers to the scholars; indeed, most of them have never heard of scholars and would think them very dull if they tried to read them. Professor Kent Ladd Steckmesser in his excellent *The Hero in History and Legend* says the serious historians lost the Western field by default, meaning that when they ignored "such presumably adolescent subjects as cowboys and Indians" they left a vacuum that the glamorizers have filled. Prof. David H. Stratton reviewing Helen Addison Howard's *Northwest Trail Blazers* in the Spring, 1965 issue of "Montana: The Magazine of Western History" points to the

reluctance to give up old myths and legends. It is still possible to read in popular histories of the Northwest that Sacajawea was responsible for the success of the Lewis and Clark expedition, Marcus Whitman saved Oregon for the United States, and Chief Joseph was a "Red Napoleon." On the other hand, scholarly works of history are often too stodgy for the general reader and get no further than the college library shelf and the reading list for advanced history courses.

It seems to us that Steckmesser and Stratton haven't put their case very well. It's not so much that scholarship is stodgy, that is, thick, heavy, stuffed, and lumpish, or that historians lost the Western field when they failed to write tales for adolescents about Indians and cowboys, scouts and marshals; it is that most of those who read books about the past, or the present for that matter, don't want to know what the facts are, if knowing about them will disturb their standardized errors. With no love of knowledge for its own sake they have no wish to discover whether their beliefs are false. If you tell them that Wild Bill was not the great civilizer of the West, and as a pistoleer was only an average shot, that Wyatt Earp was largely an impostor and braggart, and that Kit Carson, a small man, was not a Hercules with muscles of steel plate and a perfect command of English, they will want to run you out of town.

If a person wishing to know the history of the American West were to read a few hundred books and articles about it, chosen at random, he would come out of it with his head well crammed but there might be few historical facts in it. Even some professors and professional historians have a most unscholarly credulity, not to mention encyclopedias, for the *Britannica* as late as 1961 was telling its readers about the "McCanles gang." Too typical of Hall's four-volume history of Colorado is his account of Billy the Kid, in which, says Ramon Adams, there is not a single fact. Professor Grace Hebard's childlike approach to the demigods is revealed in her breathless words about Wild Bill: "No desperado that disputed his authority lived to repent it. He was the terror of evildoers. Members of the McCandlass [sic] gang once leagued together to put him out of the way when he was a station guard on the mail route, and at one time a roomful attacked Wild Bill alone. When the smoke had cleared away it was found that ten men had been killed, and Wild Bill had received three bullets, several buckshot, and numerous knife cuts." There is not a single historical fact in that. When a professor for whom the

1

history of the West was a major interest can go off into the wild blue of legend without knowing it we need not be surprised to find ex-President Eisenhower telling the people that he was "raised in a little town of which many of you may never have heard. But out in the west it is a famous place. It is called Abilene, Kansas. We had as our Marshal, a man named Wild Bill Hickok."* If Eisenhower grew up in ignorance of Hickok's true nature and his deeds we must wonder how completely the legend makers have taken over our public school system.

In no other field in which we have read widely have we been so frustrated as in this field of our Western history. So, now, the reader is warned. He is to suppose that there are many errors in this book, for the simple reason that in many areas we don't know, and no person yet knows, what is fiction and what is fact. In an effort to dodge the errors we have based a great deal of this book on primary rather than on secondary sources. Let us admit at once that some things probably can never be determined, such as the number of men Hickok killed (though this can be of little interest to anyone except those who make heroes of killers), or whether Murieta actually lived, or Calamity Jane was a psychopathic liar. It will take a lot of research to settle the innumerable discrepancies in the thousands of books already before us. It would, for instance, be a sizable task to determine which of two competent writers is correct, or if neither is, in their conflicting statements about Stratton and Cripple Creek. Except in the physical sciences when we enter books about the West we are in the realm of legend, and if we are not to get lost in it we must be alert on every page. Some authors have tried to justify their indifference to the historical facts by quoting J. Frank Dobie: "If this story isn't true to facts, it is true to life"; or Mark Twain: "This tale may never have happened, but it could have happened." Some writers include in their bibliographies books that are worthless as history—such as, to give absurd instances, a life of John Wesley Hardin by John Wesley Hardin or of Polk Wells by Polk Wells. It is never-never land when writers accept as fact the lives of criminals written by the criminals themselves. In this book we have tried to be guided by such statements as Dims-

dale's: after giving a calendar of crimes that led to the vigilance committees in Montana he said his task had not been pleasant but "the historian must either tell the truth for the instruction of mankind, or sink to the level of a mercenary pander, who writes, not to inform the people, but to enrich himself." His words may sound harsh to this self-indulgent age in which the accumulation of money is generally assumed to be the chief reason for living.

In this introductory essay we shall try to suggest how valueless, as history, most of the writing about the West is, not by citing instances over the whole broad field of it, though this one could easily do, but by looking at a few persons who have had hundreds of books written about them. The reader who wants a longer look at this matter can find it in *Burs Under the Saddle,* by Ramon F. Adams, in which six hundred double-column pages expose a few thousand errors in four hundred and twenty-four books. Though historian Bloom finds Mr. Adams sometimes chewing on trivia with considerable relish he admits that he writes "with great authority." The reader who wishes merely to sample this huge exposure of errors could turn to page 113 to see what happens to only one of the many books on Calamity Jane. Another excellent book in exposing the vast areas of nonsense has already been cited here, Steckmesser's *The Hero in History and Legend;* this is a fascinating study of what the mythmakers have done with Kit Carson, Billy the Kid, Wild Bill, and General Custer.

There are books, literally hundreds of them, that present some of the badmen of the West as Robin Hoods. Of Jesse James, for instance, who became a national idol, James Horan asks the question, What was he like? and says: "The answer is simply that he was a cold-blooded killer and a thief. There is no credible evidence that he ever gave one cent to a widow, or to anyone else in need, or took up arms for the helpless or the downtrodden." His teacher had been Charles Quantrill, the " 'bloodiest man in American history.' A slim, handsome man, former teacher and a superintendent of a small Bible school, he was Jesse James' teacher in the art of murder, horse-stealing, arson, and butchery." George D. Hendricks in his *The Bad Man of the West* undertook

---

* Steckmesser, 158.

to delineate their traits and characteristics but like most people with a thesis he pulled leather and rode hard. He classified Black Bart (whom see) as a poet, on no evidence except a few scraps of doggerel, and approved the opinion of an unnamed person that this road bandit was "a great reader of classics and Bible." Wyatt Earp he presented as a man over six feet tall with a "powerful frame," yet who weighed only 155 pounds. If his book is to be judged by what he says about Murieta, the legendary Mexican Robin Hood, it is worthless; and this is sad, for in his Introduction he says he is a Doctor of Philosophy. Mr. Adams gives ten columns to exposing the errors by this doctor of philosophy who majored in history, and says of his book that it "has been kept in print for over twenty years and seems to have been a popular seller, which makes it all the more regrettable that it is loaded with errors." Typical of his indifference to facts is his statement that "a mob hanged Jack McCall for shooting Wild Bill Hickok."

We are aware of the need of heroes in most persons—today they are called images—of the Kit Carsons, Wild Bills, and John F. Kennedys—and on the other side the Klondike Kates and Baby Does and Elizabeth Taylors; but it is going pretty far when Kit Carson, who was a small man, is compared to Hercules and other giants of myth, with enormous shoulders, a "voice like a roused lion," and the habit of creeping through thickets with his heavy rifle barrel held by his teeth; when a man who could barely read faced a bully and said, "Shunan, before you stands the humblest specimen of an American in this band of trappers, among whom there are, to my certain knowledge, men who could easily chastise you; but being peaceably disposed, they keep aloof from you. At any rate, I assume the responsibility of ordering you to cease your threats, or I will be under the necessity of killing you." Such preposterous bombast was seized on by biographers and anthologists and repeated word for word. No wonder that when Colonel Inman showed Kit the cover of a journal that had him protecting a trembling woman while dead Indians littered the acres all around him, he said, after studying the picture, "That thar might be true but I hain't got no reckerlection of it."

Another national hero of Homeric proportions was described by one who knew him as a "short,

THE KID

slender, beardless young man. The marked peculiarity of his face was a pointed chin and a short upper lip which exposed the large front teeth and gave a chronic grin to his expression." Joseph Henry Jackson called it sob-sistering—the outpouring of tears and prayers and hysteria and the almost frenetic idealizing of various brutal killers in the old West. What touched off the worship of the homely little brute known as Billy the Kid?—for the Bonney legend, says Steckmesser, "has penetrated every level of American culture" and Billy the Kid country "draws thousands of hero-worshipers each year and is the subject of articles in national travel magazines. . . . Buildings have been restored, markers erected, and an annual Billy the Kid pageant is staged in August." What explains it?—repressed fears and hostilities, hatred of authority, chronic anxieties?—for there he is, enthroned in "one of the great All-American legends"—a youngster who was "a tough little thug, a coward, a thief, and a cold-blooded murderer." That is one view and it is far closer to

cowardice and brutality of it. Walter Noble Burns, a specialist in stardust for those spurning the earth, endowed Billy with all the virtues stored up in the saints' legends, Hollywood's producers pitched in to help, other writers eager for profit and acclaim took up the cry, and lo and behold, Johnny Mack Brown, All-American football player, becomes Billy the Kid for millions. The his-

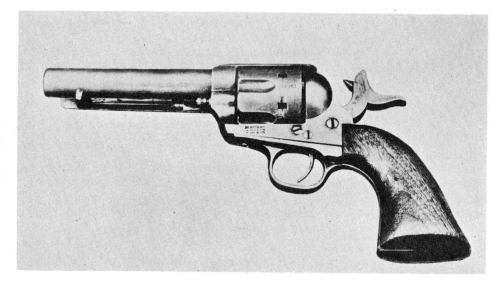

This .44 single-action Colt was taken from the Kid by Pat Garrett December 21, 1880. Some writers have said he used only a double-action but Eugene Cunningham says this pistol explodes that legend.

what he was than the view that he was a brave and gallant crusader for right and justice.

The people who knew him are certainly better judges of what he was than those who have nothing better to do than to create folk heroes for the uneducated. On his death, "The general reaction of the press was one of relief." One newspaper said that those unfamiliar with William Bonney's criminal record "cannot comprehend the gladness that pervades the whole of New Mexico and especially this county. He was the worst of criminals." Another said that despite the glamorizing of him "by sensation writers, the fact is he was a low down vulgar cut-throat, with probably not one redeeming quality." And so for a while after his death he was "an out-and-out villain, a cold-hearted wretch who giggles while his victims writhe in their death agonies." For a quarter of a century that's all he was, but then the sob sisters got busy and the satanic Billy became the saintly Billy, and the officer he shot down without giving him any chance at all became "a snarling, foulmouthed bully," though at the time of his death every newspaper that wrote about the killing was horrified and shocked by the

torians with their stodgy pages and devotion to truth can't make any headway against that—against a little squirrel-toothed runt who has become a giant who avenges insult to his mother and then is persecuted and hounded by all the human monsters of the Southwest, but manages to kill a few hundred of them and ride off into the sunset with the little girl who stuck to him through thick and thin. Today, in August, down in New Mexico you can see thousands of the graduates of our public school system going solemnly to the shrine where rest the bones of one of the most vicious and merciless killers in American history. In an earlier time, in California, the good and virtuous ladies took "fresh flowers, delicate viands, fine wines" to the brutal murderer Tiburcio Vasquez when he was in jail, and, in Nevada, to the equally brutal strangler of Julia Bulette.

"Today's student," says Jackson, "will wonder what the good ladies of San Jose saw in the little man (he stood barely five feet seven in his boots), with his retreating forehead, his sullen look, and the thick mane of coarse black hair." Jackson obviously had little knowledge of the stuff heroes are

POCKET Library

Copyrighted, 1884, by BEADLE AND ADAMS. Entered at the Post Office at New York, N. Y., as Second Class Mail Matter. May 28, 1884.

Vol. II. $2.50 a Year.   Published Weekly by Beadle and Adams, No. 98 WILLIAM ST., NEW YORK.   Price, Five Cents.   No. 20.

ROARING RALPH ROCKWOOD, The Reckless Ranger.

By Harry St. George.

CHAPTER I.

THE PRAIRIE PESTS.

THE sharp, whip-like crack of a rifle awoke a thousand echoes among the distant foothills, and broke the stilly silence of the night. Following the shot came the shrill neigh of a horse and then once more stillness.

A person with remarkably keen ears might have caught the low but hearty curse, that

ALF DIME Library

$2.50 a year.   Entered at the Post Office at New York, N. Y., at Second Class Mail Rates.   Copyright, 1885, by BEADLE AND ADAMS.   June 2, 1885.

Vol. XVI.   **Single Number.**   PUBLISHED WEEKLY BY BEADLE AND ADAMS, No. 98 WILLIAM STREET, NEW YORK.   **Price, 5 Cents.**   No. 410.

Deadwood Dick's Diamonds; or, The Mystery of Joan Porter.

BY EDWARD L. WHEELER,

AUTHOR OF "DEADWOOD DICK" NOVELS, "ROSEBUD ROB" NOVELS, ETC., ETC.

"BACK!" THE RUFFIAN CRIED, TO DEADWOOD DICK. "ADVANCE ANOTHER STEP, AND I'LL BLOW THE GAL'S BRAINS OUT!"

made of: Kit Carson is said by some to have been only five feet six, and the Kid was shorter than that—but so was Napoleon, who fills countless volumes. It's not only the good ladies who idolize the repulsive creatures: when the judge sentenced Vasquez he said to him that he had "one unbroken record of lawlessness and outrage, a career of pillage and murder," yet one of the idealizers of badmen has blandly written, "Even the judge who sentenced Vasquez stated that he did not believe him guilty." When the facts of history get in the way of those dedicated to the apotheosis of criminals they don't stand much chance. Look again for a moment at the James brothers. Popular opinion from the beginning was on their side and became so strong that it "was impossible for peace officers to enter Clay County. There was no need for Jesse and his gang to hide out now; they had their own 'iron curtain' of sympathizers, frightened sheriffs who were often in their pay, and kinfolk." When a harassed governor posted a reward of ten thousand dollars for Jesse, dead or alive, and for the reward a coward shot him in the back of the head, the kinfolk and sympathizers would have torn Bob Ford limb from limb if they could have laid hands on him. As for Jesse, he was buried in such style that one wag expressed the opinion that on resurrection day he would be mistaken for a banker. The governor who had dared to offer a reward had his political career ruined. That you can understand if you have before you the facts that on an occasion when Frank James was acquitted of the charges against him "the crowd went wild trying to get to shake his hand"; that invariably when a James or a Younger was jailed long petitions were presented which begged for their parole; and that Jesse's death made front-page news over the nation. Among the words on the white shaft above his grave were these: "Murdered by a coward whose name is not worthy to appear here." But our favorite of all the sob-sister incidents is told by Dimsdale. When Charles Forbes, a Plummer bandit without a single virtue, was acquitted of a murder he had boasted about, "Judge Smith, bursting into tears, fell on his neck and kissed him, exclaiming, 'My boy! My boy! Hundreds pressed round him, shaking hands and cheering. . . .'" It's a fact of history that a substantial part of the people is always on the side of the more spectacular criminals, and that the idealizing of them takes every possible form, including paintings of them, such as the one of Murieta in this book. As Jackson says, it didn't even pretend to be a likeness of this bandit, if in fact he existed, but was nevertheless accepted "by devout believers as an authentic portrait."

Edward Zane Carroll Judson, who called himself Ned Buntline, was a "frisky fellow of many pursuits, . . . drunkard, temperance lecturer, beggar, promoter and jailbird," who decided to write what became known as dime novels. Between 1869 and 1933 seventeen hundred such novels were written about Buffalo Bill alone, and thousands more about other idols. Among the common people there seems to be a "voracious, wolfish hunger" for heroes. This Bill, a killer of helpless beasts, a show-off and braggart, was blown up into a figure of almost cosmic proportions. Jackson points out that magazines devoted to the West, such as the *Overland Monthly,* the *Argonaut,* and *Sunset Magazine,* "doubtless with the best of intentions, printed almost entirely the inventions of rewrite experts and sensational feature-writers," who based their unhistorical moonshine on such sources as the *California Police Gazette,* which in turn often copied from pamphlets or booklets in which it is hard to find a single fact. As Jackson says in regard to the heroes, "almost nothing that has been written since 1900 . . . contains more than a bare speck or two of truth. . . . What it comes down to, then, is that you find the truth . . . only by going to original sources and then exercising your wits as nearly in the manner of Mr. Sherlock Holmes as you are able. When you have done all that you can, you are still painfully aware that there are holes in your account." Jackson points out the significant fact that some of the sheriffs performed their dangerous tasks with great courage but are seldom mentioned in books about the West, much less transformed into national idols, as so many have been who were merely spectacular outlaws. As an instance there is Thomas J. Smith, marshal in Abilene, who probably had ten times the courage of Wild Bill or Wyatt Earp, yet has aroused the admiration of practically no one. Jackson hoped that some day writers would give such men their due. That is a lot to hope for; they were on the wrong side of the law. Of Tom Pollock, one of the bravest marshals a hell town ever had, Zamonski and Keller observe that had his "career coincided with the westward push of a railroad or with the great cattle drives or with the better publicized cowtowns, Pollock might have joined that select and legendary few who seem to have blazed all the trails, dug all the gold, punched all the cattle, tamed all the towns, and fought all the Indians all by themselves."

When errors are corrected they still persist, not only year after year but decade after decade. If they are the kind that appeal to the unenlightened and the unthinking they are likely to persist until the last cow comes home. Though we published a novel about the Lewis and Clark expedition to the ocean, and made it plain in the introduction and in the story that the Indian girl did not act as their guide; and though other writers have pointed out the same obvious fact, she is still acclaimed in programs over the air, and will be in countless books yet to be written, as the bird-woman who led the men all the way from Mandan to the ocean—this ignorant child who could hardly have been more than thirteen or fourteen and whose notions of geography would have made the worst of the old maps of the West look divinely inspired. Though we gave most of two years to the research for and the writing of a book on the death of Meriwether Lewis, and pretty well exploded the myth that he was an alcoholic and a lunatic who killed himself, as late as March 12, 1965, *Time's* book editors told their readers that Lewis "died in alcoholic ruin." Such perversions of fact in the interest of sensationalism and best sellers could be cited by the thousands.

In the libraries of nonsense that make of small men the giants of legend nothing is more unrealistic than the accounts of their shooting. For our autopsy in this matter we choose Wild Bill, the most deadly shooter since William Tell shot the apple off the head of his son, and "monumental figure in the tall-tale mythology of the West." Though Frank North, Hickok's friend and fellow scout, has said that he never saw Hickok handle a gun with his left hand, James Butler nevertheless shoots with unerring accuracy and simultaneously with both hands in both directions. In exposing the fictions in this legend Ramon Adams cites a hundred and seventy-nine page references, and could, of course, have cited ten thousand more. It seems very unlikely that as a shot this man was in the same company with Phoebe Anne Oakley Mozee, known as Annie Oakley, the girl

James Butler "Wild Bill" Hickok. Tinhorn gambler or great civilizer?

who got her gun, yet he is gravely presented in book after book as the great civilizer of the West who did his magnificent job with two forty-fours. The hypnotized writer in the *Encyclopedia Americana* tells the world that Hickok "never killed but in self-defense or line of duty"; and even Stewart Holbrook took leave of his senses when faced with this legend and set it down as imperishable fact that Wild Bill was a "quiet and courteous" gentleman who "seldom swore, drank even less, and was a little quicker with Mr. Colt's Patent Revolving Firearm than any of the many ugly customers he found it necessary to shoot in the process of civilizing the frontier."*

He was in fact a tinhorn gambler, an ardent patron of the soiled doves, a chronic drinker, and a coward in more gunfights than one; but as Steckmesser says, you have only to make him a dead shot and fill him full of social purpose and you're on your way to the image of the great civilizer. You have only to take the word of J. W. Buel who, in 1880, published a thing he called a biography. He said it: "He was essentially a civilizer.... Wild

_____
* *Esquire*, May, 1950.

Bill played his part in the reformation of pioneer society more effectively than any character in the annals of American history." Seventy years later Stewart Holbrook solemnly repeated it, though a more preposterous falsehood it would be hard to find in all the writings about the West.

"It wasn't love at first sight. It was just a case of goggle-eyed, moon-struck hero worship. Colonel George Ward Nichols, an Eastern galoot, couldn't help himself." He looked into the eyes of James Butler Hickok, "eyes as gentle as a woman's," so gentle indeed that "you would not believe that you were looking into eyes that have pointed the way to death for hundreds of men." Whether Mr. Horan is right in thinking that the Colonel was goggle-eyed we do not know but there can be no doubt that he set the legend on its sturdy legs and gave it a push toward the twentieth century. Richard O'Connor thinks that Nichols had a case of "girlish enthusiasm" when he told the world about Hickok in *Harper's Monthly* in February, 1867, but O'Connor himself, whose book on Wild Bill Adams thinks one of the best, is not entirely free of the girlish attitude when he writes such words as these: "nearly blind, Wild Bill outdrew and shot gambler Phil Coe . . . a jigger of whiskey jostled out of his hand, and four men hit the sawdust." There is no evidence that Hickok was nearly blind at any time, and the four men dead in the sawdust are only a brilliant facet of the legend; but it must be admitted that by the time Mr. O'Connor tackled the job a few years ago it took a mighty man to wrestle with this legend. Ramon Adams has shot it full of holes, and Professor Steckmesser has given it a very bad time, but it must have at least a hundred years of life left in it. This we sense when we find O'Connor apparently accepting the oft-told feat of "having a tomato can thrown into the air and riddling it with twelve bullets before it fell back to the ground." With two six-shooters fired by both hands. He doesn't make clear whether he believed the story that Hickok could drive "a cork through the neck of a whiskey bottle at about twenty paces." Men who accept such nonsense have surely never fired a gun, or even discovered the nature of tin and glass. His eyes full of stardust O'Connor does gravely tell us that Hickok was "often a walking arsenal. In addition to a pair of .44s strapped to his thighs he would carry a brace of .41 Derringers in his side pockets, a Bowie knife

in his belt, a shotgun or repeating rifle crooked in his arm." If there had been a seven-inch breech-loading howitzer around no doubt he would have also carried that. What we need to be told is why a man who could put twelve bullets in a small falling can in two or three seconds needed all those weapons to protect himself.

How Nichols ever got to be a colonel is also a mystery, for he must have been a timid man forever on the run if he had need of such a ferocious hero. It was he who put in circulation the outlandish story that Hickok killed McCanles and five other desperadoes with six bullets, and (as Horan puts it) "cut four more to death in hand-to-hand combat, and walked away from the carnage in fair health except for eleven buckshot holes and thirteen stab wounds." There were only three unarmed men, and Hickok, hiding behind a curtain, shot them. Nichols told his readers that he asked the great civilizer, Have you ever been afraid? and that James, who by that time had become Wild Bill, said yes, he allowed as how he had been. It was in the Wilson Creek fight. "I had fired more than fifty cartridges and, I think, fetched my man every time." That makes him as big a liar as Wyatt Earp. No less a person than Henry Morton Stanley confided to the *New York Herald* that Hickok had told him, "I suppose I have killed considerably over a hundred." O. W. Coursey, whom Horan calls "one of the more open-mouthed biographers," brought his quivering emotions under control while he told of this civilizing episode: "Wild Bill, facing the desperate character who entered the front door, had shot him with a revolver in his left hand, while with his right hand he had thrown the gun over his left shoulder and shot the man coming in from the rear. History does not record a more daredevil act, a more astute piece of gun work." Astute means sagacious; under sagacious the dictionary cites this instance: "He was observant and thoughtful." Sagacious shooting is pretty fancy shooting, in any land or clime, but Adams spoils it by saying "there is not one shred of evidence" to support this over-the-shoulder sagacity.

Other instances of astute shooting are, as Adams summarizes them, "cutting a chicken's throat with a derringer shot without breaking its neck, shooting a hole through a silver dime from fifty paces, tossing a wooden block in a stream, flipping it into the air with a bullet striking underneath it, and then riddling the block before it could fall back into the water." Riegel's *America Moves West,* written for the schools, includes the chicken neck, dime, and telegraph pole stories, and introduces an apple. He tells the children that Hickok could cut the stem of an apple with the gun in his left hand, and hit the falling apple with the gun in his right hand. Checking school texts against the known facts about Western history might show that many of the books ought to be given to libraries in those countries where our libraries are most likely to be burned. In Appendix C of our book on the death of Meriwether Lewis we cite a few errors, most of them by nationally known professors. We quote from that Appendix:

Professor Albert Bushnell Hart in his *School History of the United States:* "The party of forty-five men, for which Congress appropriated only $2500 . . . guided by the 'Bird Woman'." There we have three errors in seventeen words. Professors Bourne and Benton tell students that "only one Indian had been killed," an error found also in Hall, Smither, and Ousley, *The Student's History of Our Country,* though Lewis wrote as plain as day in his journal that two Indians were killed. Professor Fite says that "forty-five members were included in the party" which reached the Three Forks the "next spring." Both statements are false. Casner and Gabriel say the party wintered in "the lodges of the Mandan Indians" and were "guided by a French trapper and his young wife." Both statements are false. Sometimes the errors are simply fantastic. Boyle, Shires, Price and Carman in *Quest of a Hemisphere* say that "many times" the Indian girl "saved the white men when their lives were threatened by hostile Indians." In not a single known instance did she save their lives. Thwaites, Kendall, and Paxson in their *History of the United States* (for the schools) say that "After many thrilling experiences with fierce currents, inclement weather, grizzly bears, they spent a long, bitterly cold, and almost starving winter not far from the present Bismarck, North Dakota." Thwaites at least should have known that the fierce currents and the grizzlies came after the party left its winter headquarters. Professor Frederic L. Paxson (the same Paxson) in his students' edition of his *History of the American Frontier* says that when Lewis wanted to converse with Indians "he was forced to rely upon his mulatto body servant, who by chance spoke French." Lewis had no servant of any kind. Paxson goes on to commit another whopper when he says that on "May 14, 1804, he led his band of thirty-two across the Mississippi and up the Missouri." It remained for S. P. Lee, with Louise Manly, in their *School History of the United States* to misspell both

names (Merriwether, Clarke) and to say: "Up the Missouri and its great branches and through the wild mountain ranges of the northwest they pushed their way to the Pacific slope. Here they found the two rivers which today bear their names. Down these they went, until the two rolled together into the Columbia." Students who use this text and have a map of the Northwest before them must conclude that it was Lewis and Clark who rolled together into the Columbia.

If famous professors can't do any better than that shall we expect accuracy from the biographers of James Butler Hickok and Wyatt Earp? No wonder Wild Bill sickened on it after Buffalo Bill lured him into his circus. The exterminator of the "McCandlass gang" had to murmur these tender romantic words: "Fear not, fair maid; by heavens you are safe at last with Wild Bill, who is ever ready to risk his life and die if need be in defense of weak and helpless womanhood." Though the applause may have been deafening that was about the last thing on earth that the lover of the crib girls was ready to die for. When he could no longer gag it down he left the show.

The deadly shooting by the deadeye Dicks doesn't seem very deadly after you have read some descriptions of it by trustworthy witnesses. Dimsdale saw a duel between Ives and Carrhart, two of Montana's deadliest pistol toters: "Carrhart stood still till Ives turned, watching him closely. The instant Ives saw him he swore an oath, and raising his pistol, let drive, but missed him. . . . Carrhart's first shot was a misfire, and a second shot from Ives struck the ground. Carrhart's second shot flashed right in Ives' face but did no damage. . . . Carrhart jumped into the house, and reaching his hand out, fired at his opponent." The two then blazed away at one another until the Ives gun was empty and Carrhart had one shot left. Eleven shots fired and neither man was touched. Then: "As Ives walked off to make his escape Carrhart shot him in the back." It is Dimsdale who also tells about the duel between two gamblers named Banefield and Sap. When one of them called the other a cheat the two men drew their guns and fired at close range, and though neither was hit, a dog named Toodles, lying under the table, was shot three times and killed.

If objection is made by the admirers of Western gunmen that we are citing only second raters, let's move on to the very top, noting, on the way,

Otero's statement that when he was in Las Vegas, New Mexico, in 1880, Wyatt Earp stopped there to settle an old score with a man named White. In a saloon at close quarters they fired many shots at one another and the only result was a bullet scratch across White's spine. But let's see Earp in action, as he himself told it. After his brother Morgan was shot through a window, at night, probably by a coward named Frank Stilwell, and Stilwell was ambushed and slain, probably by Earp and his friends, a warrant was got out for Earp, and while dodging it, he and his dead-shot buddies headed for Iron Springs, unaware that an outlaw known as Curly Bill and his cattle-rustling gang were there. This Bill, according to some deathless prose, was the greatest outlaw in the Southwest in those years and about as deadly with a gun as anyone you ever heard of. The only story of what happened at the Springs—if in fact anything at all happened—was told by Wyatt Earp.

Wyatt Earp. Braggart and liar or superlative frontier marshal?

After a ride of thirty-five miles he and his men were weary and off guard, and so were within gunshot of Curly's gang before they sensed danger. Thinking they had been ambushed all but

Earp fled, who at that moment was walking and leading his horse, a shotgun in his hand. Only a few yards ahead of him was Curly Bill, and he had a shotgun.

I can see Curly Bill's left eye squinted shut, and his right eye sighting over that shotgun at me to this day, and I remember thinking, as I felt my coat jerk with his fire, "He missed me; I can't miss him, but I'll give him both barrels to make sure." I saw the Wells, Fargo plate on the gun Curly Bill was using and I saw the ivory butts of Jim Hume's pet six-guns in Hume's fancy holsters at Curly Bill's waist as clearly as could be. I recognized Pony Deal, and as seven others broke for the bottomwoods, I named each one as he ran. . . .

That amusing nonsense has been solemnly accepted as fact by nearly all the writers on Earp and Tombstone whom we have read. Anyone who has used guns knows that if Earp was close enough to see that Bill's left eye was shut, and the plate on the gun, and in turn to identify eight men as they ran, he was close enough for Bill to hit him with a shotgun. Earp says he emptied both barrels in Bill's chest and practically blew him to pieces. But what was Bill doing while Earp observed the shut eye, the stolen guns, the holsters, and the eight men running away? The fact is that in his old age Earp talked to a bewitched Stuart Lake and really filled him full, and that most of the writers on Earp have accepted Lake as gospel. Adams takes forty-three columns and fourteen thousand words to demolish the thing that Lake called a biography. Frank Waters, one of the best of the writers about the West, spent a lot of time digging out the facts. He says:

The truth about Wyatt Earp lies, not in his fictional exploits on a legendary frontier, but in his lifelong exhibitionism and his strange relationships with Doc Holliday and his three wives. It will not add to his posthumous fame and the almost psychopathic interest in him manifested by the general public. But it will be a healthy sign if we can now face up to how this pathetic figure —an itinerant saloonkeeper, cardsharp, gunman, bigamist, church deacon, policeman, bunco artist, and supreme confidence man—has conned us into believing him America's most famous exponent of frontier law and justice.

And along comes Ed Bartholomew with two volumes, so that now, as Joseph W. Snell of the Kansas State Historical Society puts it, "the image of Earp's fantastic career lies broken and trampled. He was not, it turns out, the man he said he was." Bartholomew shows us, says Snell, "that Earp was once arrested for horse stealing in the Indian country, nearly killed himself by foolishly loading the sixth chamber of his pistol, protected his prostitute sister-in-law with his Wichita policeman's badge, and in many cases was nowhere near the action he claimed to have been involved in." Nevertheless, persons back East were prepared to believe everything. After the gunfight at the O.K. Corral, proudly placed by some writers at the head of the list of all the most ferocious and frightening gun battles of the old West, it was reported in Eastern papers that Earp was found with nineteen bullets in his body, yet was still breathing. The editor of the Tucson *Star* took up his quill:

Pshaw, that is not half nor a twentieth part. The Marshal had fifty-seven bullets extracted and it is believed there is about a peck yet in his body. Only a short time ago a cowboy had a Henry rifle rammed down his throat and then broken off; he spit the gun barrel out with the loss of only a tooth. It is stated as a fact that more than four-fifths of the inhabitants of the district carry one or more bullets in their body. One instance is of record where birth was given to an infant who came forth armed with two bowie knives. . . . Every house has portholes from which the cowboys are shot down. It is a great place for suicides; if a fellow wants to die with his boots on, he just steps out on the street and yells "you're another," and immediately he is pumped through from all sides. Sports play at cards with a knife in the left hand and a six-shooter in the right. It is no uncommon occurrence to see twenty men dumped out of a card room in the morning, and pitched down some mining shaft where the ore has petered out. This is but a faint picture of the situation. Our New York exchanges had better try and get the facts.

One fairly aches with sympathy for the poor Tucson editor, for it was impossible to make the matter look ridiculous by trying to exaggerate it, since no exaggeration could be more fantastic than what was accepted as fact.

In a book before us a full professor is saying that Wild Bill "could probably draw faster and shoot more rapidly and more accurately than any other man then living. He killed many men, most likely twenty-seven, although nobody cared to ask him." Such innocence in full professors is pretty painful; many persons asked Hickok how many persons he had killed and he was tickled silly every time. The same professor goes on to say: "He was always within the law and justified . . . quiet mannered, absolutely fearless, and always in control of himself." The Great Civilizer, that is, who single-handed opened the West to the wagon trains and the pioneers.

CHAPTER 2

Gold: What It Is and Where It Is Found

Fitzhugh reminds us of an obvious fact: "Had it not been for the gold and jewels of the Orient, Columbus would never have sailed westward." For those of us in the West that should answer Virgil's question,

> O cursed lust for gold,
> To what dost thou not drive the hearts of men?

From Byron the loveliest of all the metals brought this judgment:

> A thirst for gold,
> The beggar's vice, which can but overwhelm
> The meanest hearts.

And from Shelley this cry:

> Gold is a living god, and rules in scorn
> All earthly things but virtue.

Is it not strange that writers and moralists have placed the odium on the metal, instead of on the passions in man of avarice and greed and lusting for power? Beautiful metal, what a curse ignorance and cowardice have forced it to bear!

Bret Harte was looking at the greed when he wrote, "The ways of a man with a maid be strange, yet simple and tame to the ways of a man with a mine when buying or selling the same." As we shall see in the pages to come. Quiett was looking at the greed when he told his readers the story of "two men in the cool green forests of British Columbia," known to us today only as McDonald and Adams, who on Fraser River struck it rich. They set out for a trading post to buy supplies, and on the way it was greed, not gold, that compelled McDonald to murder his partner.

> Gold! Gold! Gold! Gold!
> Bright and yellow, hard and cold,

cried Thomas Hood, apparently a little beside himself. But gold is neither cold nor hard. It has long been the way of people to find scapegoats to take the guilt of their basest passions.

For how many centuries has the Phrygian King Midas been an object of pity, while the metal which should never have touched any but Cellini hands has borne the curse? The reader will recall that legend of a stupid and greedy king who was granted his wish, that everything would turn to gold that he touched. At once he began to turn apples, corn, clods of earth into gold, as well as rivers of water. When he takes up a goblet of rare wine and tries to pour it down his throat he finds himself choking to death on gold, gold, hard and cold, and would have suffered the death he deserved but for Bacchus. In the *Edda,* in the Nibelungenlied, in countless myths and legends and literary works gold has been the major theme or force, the villain and the curse, and indeed today it is the worry of politicians and governments who don't understand it. We are going to take a less hostile, and we hope more enlightened, view of it in this book.

With an astonished glance in the direction of Senator Ingalls, "It is the most cowardly and treacherous of all metals. It makes no treaty that it does not break; it makes no friend whom it does not sooner or later betray," we shall present a few facts about its history and then tell you what it is not, what it is, where it is found, and how to find it. Gold appears in all the mythologies; in innumerable legends, superstitions, fables, parables, and sermons; in the religions and philosophies, the arts and literatures; and in enough moralizing to have deadened the sensibilities of mankind ages ago. It has been associated with and even identified with the sun; it has been believed by some peoples to beget offspring, and by others to grow as a tree grows. Indeed, the ancients had only fantastic notions of its origin. As late as 1540 an essay was written on "The Generation of Metals" which said that "in some places of Hungarie at certain times of the year, pure gold springeth out of the earth in the likeness of small herbs, wreathed and twined like small stalks of hops, about the bigness of a pack thread and four fingers in length." Even much later it was believed "that the tailings of abandoned mines became so enriched as to be workable at a profit, the amount of gold thus obtained being proportioned to the interval of time elapsed, which meant the measure of the opportunity for the gold to grow."

In their fables that used gold, the ancients didn't usually make sense. In the famous Atalanta's race the picking up of apples of solid gold and running with them seems pretty foolish, when you bear in mind the fact that gold is so heavy that a sheet of it an inch thick and twelve inches square weighs almost a hundred pounds. The fair Hesperian tree laden with blooming gold must have had branches of steel, and the girls equipped with golden girdles could hardly have felt like dancing. One bewildered writer says it is impossible to divine the intention of Vulcan who made golden shoes to speed the feet of the gods. The ancient Hebrews had enough trouble with gold to have put them all out of their wits—as Keats seems to have been when he burst forth with, God of the golden bow, and of the golden lyre, and of the golden hair, and of the golden fire. Though golden lads and girls all must, like those of lead all come to dust, and though a word fitly spoken is like apples of gold in pictures of silver, the belief that every splendor and every wickedness in life are to be found in the hue of this metal seems to be pushing the matter pretty far. We have the golden tongued, the golden number, the golden mean, the golden section, the golden moment, the golden rule. And one writer is convinced that "we thrill with an old pleasure when we recall the cave where Jim Hawkins saw displayed great heaps of gold and quadrilaterals built of bars of gold. . . . English, French, Spanish, Portuguese, Georges, and Louises, doubloons and double guineas and moidores and sequins. . . ." Books are filled with the golden singsong—Aesop's golden goose, Portia's golden casket, the Hebrews' golden calf, Jason's golden fleece, the leprechaun's crock of gold—there is literally no end of it.

What *is* gold?—or perhaps we should ask first, What is it not? Let Bernewitz tell us in his handbook for prospectors:

Certain common and often valueless minerals are frequently mistaken for gold; probably the most frequent error is that of pyrite, pyrrhotite, chalcopyrite, or mica. . . . Pyrite is a sylphide of iron, brass-yellow in color, with a metallic appearance, and occurs either as distinct cubical crystals or in massive crystalline forms. It is slightly harder than steel and thus cannot be scratched with a knife. When finely powdered it has a greenish-black color. . . . Pyrrhotite is also a sulphide of iron but is usually found in massive form. It is of a bronze color and metallic luster, though when finely powdered it is grayish black. Unlike gold it crumbles into powder under the knife, while gold cuts more like lead. . . . Chalcopyrite is a sulphide of iron and copper combined and contains, when pure, nearly 35 percent copper. It is a golden-yellow mineral with a metallic appearance and is usually found in massive form. It is slightly softer than pyrrhotite and may be easily scratched with a knife. . . . These three minerals are brittle and emit sulphur when heated.

Because for persons with no knowledge of geology it is not easy to grasp the origin and nature of gold, we may be a bit repetitious here. We'll let Bernewitz open the subject:

It is important to know how and in what kind of rocks the different metals occur. For instance, the tungsten, and molybdenum are almost always found either in or near coarsely crystalline igneous rocks; on the other hand, platinum, chromium, and nickel are usually in or near close-grained igneous rocks. It should be clearly understood what is meant by a metal's being found in or near a certain kind of rock. Generally speaking, ores were not formed at the time that the surrounding rock was but were introduced later by molten rock which hardened or crystallized to form the different grades of igneous rocks, such as granite, pegmatite, gabbro, and other similar rocks. The ores themselves may be carried from the molten rock by escaping steam and fluids through cracks and crevices into surrounding rock.

The molten rock masses usually contain large quantities of water and steam. These may soak out into solid rock next to the cooling melted rock and will then change the nature of the solid rock. Sometimes the water and steam are barren of valuable ores while at other times the hot waters, soaking in, bring ores of gold, silver, copper, tungsten, or other metals.

And now for the formation of placers:

As water for centuries washes over a gold vein on a mountainside it is worn away, and the gold is carried down by some stream. As gold is heavier than ordinary sand it will be dropped in any little hollow in the stream bed, together with more or less sand and gravel. . . . Any heavy mineral that does not weather or powder easily may be found in placers. The valuable minerals so found are chromite, cassiterite (tin), monazite, gold, platinum, and some gems.

Since most of the gold rushes in this book were to placers, we need clearly to understand how they were formed:

streams carry the boulders, cobbles, and fine gravel with the released gold down to the valleys, where it settles and sometimes partly solidifies. The gravel and the gold are deposited along and in the beds of streams. In other places gravel deposited in rivers was covered with lava (basalt) and later, as the result of upheaval of various earth movements, these rivers were raised to higher levels. This is called drift-gravel, and such deposits in California have yielded large quantities of gold. . . . Placer deposits sometimes have a false bed-rock, but this is difficult to determine. At Bulolo, New Guinea, two dredges had worked about 21 months and dug 8 million cubic yards of gravel from a clay bottom at a depth of 22 feet, when drilling revealed pay gravel to a depth of 75 feet or 53 feet below the false bottom.

A fact that ought to suggest that vast quantities of gold remain to be found.*

Gold in its native state in placers is sometimes found in patterns and figurations of surpassing loveliness. Dame Shirley, who was at some of the California placers, in her twenty-third letter tells of seeing a piece worth only $2.50 by weight for which persons offered as much as one hundred dollars.

It is of unmixed gold. . . . Your first idea in looking at it, is, of an exquisite little basket. There is the graceful cover, with its rounded nub at the top, the three finely carved sides—it is tri-formed—the little stand, upon which it sets, and the tiny clasp which fastens it. In detail, it is still more beautiful. On one side you see a perfect W, each finely shaded bar of which, is fashioned with the nicest exactness. The second surface presents to view a Grecian profile, whose delicately cut features remind you of the serene beauty of an antique gem. It is surprising how much expression this face contains, which is enriched by an oval setting of delicate beading. A plain, triangular space of burnished gold, surrounded with bead-work, similar to that which outlines the profile, seems left on purpose for a name.

All fashioned by weather and water, and rolling downstream in gravel and sand!

The specific gravity and weight of clay are 1.9 and 120 pounds to the cubic foot; of quartz, 2.65 and 162; of garnet, 3.97 and 247; while gold is 19 and 1185, and platinum is 20.3 and 1268. You can get an idea of the amazing weight of gold by cutting from a sheet of paper a piece five inches square, and standing a six-inch stick vertically on the center of the sheet. Your block of gold 5x5x6 inches will weigh approximately a hundred pounds.

The word in many languages besides German and English means simply yellow. This metal is unique because of its lovely color, its luster, its high specific gravity, and its freedom from rust or tarnish when exposed to air. Among the metals it is the most ductile and malleable, and has a tenacity equal to that of silver. It will stand beating into leaves so thin that they will transmit a greenish light. It stands next to silver and copper as a conductor of heat and electricity. Though not attacked by any of the ordinary acids, it readily dissolves in a mixture of hydrochloric and nitric.

There is, properly speaking, no gold ore, this metal, so far as is known, never being mineralized by sulphur or oxygen. Though in its native state it is usually so fine as to be invisible, nearly all the gold that has been taken from the earth has been washed out of sand and gravel or picked up as nuggets or extracted from quartz in which it was imbedded. Native gold is an alloy with silver, often with traces of copper and iron. No gold entirely free of silver has ever been found: that in California was from 10 to 12 percent silver; in Australia, less than half that; while in Queensland it had only a trace, being 99.7 percent pure.

The content of gold in alloys ranges from 37.5 to 91.6, and the amount of it is indicated in carats.

The whole weight of the piece of metal is considered as being divided into twenty-four parts with this name, and the number of twenty-fourths that are pure gold is the carat number of the alloy —22-carat being, of course, 22 parts gold and 2 parts alloy. The chief alloy used is copper, which

_____

* In the Kolar fields in India some of the mines are now over eight thousand feet deep.

GOLD COINS

1. Electrum stater of Lydia (600 B.C.). 2. Persian daric (500 B.C.). 3. Stater of Philip II of Macedon (350 B.C.). 4. Roman aureus of Tiberius. 5. Florin. 6. Venetian ducat. 7. Gold penny of Henry III. 8. Noble of Edward III. 9. Guinea of Charles II.

hardens gold but gives it a reddish color. The coloring by copper is rendered invisible by adding such metals as zinc, so that it is possible to get a long-wearing wedding ring that is 9-carat. Since 9-carat is only 37.5% gold, and 25 percent silver will make the alloy white, it takes "no mean metallurgical feat" to give low-carat gold products the appearance of being all gold. Of late there has arisen a demand for different colors—green gold is an alloy with silver; mock gold is an alloy with zinc, platinum, and other metals; red gold is an alloy with copper; mosaic gold is an alloy with copper and zinc; and white gold is an alloy with silver in which it may predominate by as much as 5 to 1. Pure gold has rarely been used in the arts or for coinage. Nearly all gold ornaments and coins are alloys with copper, or with copper and silver. The coins in England were 11 to 1, in France and the U.S., 9 to 1.

The gold coinage of the ancient world was at first confined mainly to Asia, European Greece preferring silver. The earliest issues of the precious metal were not of pure gold but of electrum, a natural mixture of gold and silver containing, roughly, a fifth of the latter metal; this was the "white" gold of the Greeks. Globules of this metal with a variety of stamps on them, which were found at Ephesus and date from the eighth century B.C., are among the earliest known coins. These are regularly struck pieces bearing stamps on one side and the marks of the anvil on the other.

In European Greece Philip II of Macedon issued huge quantities of gold staters, known as philips; and his son issued even vaster quantities. Diocletian's solidus, which had one-sixtieth pound of gold, reduced to one-seventy-second by Constantine, "passed into most of the early gold coins of Europe." The Arabs called it a dinar, which for five centuries was the basis of coinage in the Muslim world. As gold increased in value, it was dropped from European coins and, for centuries about the only coin was the silver denier or penny. In 1252 Italy issued the gold florin, with 54 grains; Henry VII issued the first sovereign, with 240 grains. Henry VIII debased its gold and reduced its weight and issued the gold crown. Since the Middle Ages the coinage of the East has been chiefly silver.

In the manufacture of gold today, the faces of the backing metal and of the gold alloy are put together, sometimes with a sheet of silver solder between, and welded in a temperature of about eight hundred degrees centigrade. The two are then rolled out to the required thickness. Since the sheet of gold and the backing metal are rolled out together, "the proportion of gold to base metal remains practically constant." Of this rolled gold many tons are made each year for the production of the cheaper kinds of jewelry, including watchcases, spectacle frames, and so on. The backing metal, usually of a high-copper bronze, and the gold can be of any low carat desired, and the rolling can produce any thickness of coating down to a fifty-thousandths of an inch.

The basis of mechanical gilding is gold leaf, the gold being beaten to such thinness that an ounce will cover 250 square feet, with a thickness of four millionths of an inch. The beating is done by hand. It is "refreshing to find in these times a field in which machinery cannot compete successfully with manual labour." The metal is rolled to a thickness of about a thousandth of an inch and then beaten, "first between vellum with a heavy hammer and then between 'gold-beaters'-skin' with a lighter one . . . the leaf is then cut into pieces about 3¼ inches square, and these are placed between the pages of a thin paper book which holds twenty-five. The metal is too thin to be touched by hand without damage, and throughout the manufacture and use it is handled with wooden pincers."

Though in the ancient world most rulers sought to seize and accumulate gold, believing that it endowed them with superior virtues, and had much of it buried with them, thinking that in the afterlife it would confer divinity on them; and though some rulers accumulated enormous hoards of gold and precious stones, only in recent times has the lust for it possessed whole peoples and devastated whole areas. The superstition toward it of the ancients has carried over to modern times in those people who feel that their paper money is secure if there is gold under it. The Egyptians were possibly the first ancients to make extensive use of it, and those people around them who could find no gold in their own lands sought to acquire it by trade or war. Throughout the Near East gold became the badge of culture.

Over most of Europe men sought it in the mountains and then took to the seas, like Henry the Navigator; and the journal of Columbus reveals that gold was in his thoughts. The gold of Montezuma was a piddling poke, but the Incas had incredible piles of it, for which at the hands

of Christians they suffered dreadful massacre. Temples were plundered, graves looted, and golden images and artwork by the ton were taken to the Old World. So many men of solid gold were seen that there came into being El hombre dorado, the man who was to become El Dorado. So much silver was found that at one time four or five hundred thousand pounds of it was shipped annually to Spain. The amount of gold hoarded by the civilizations that placed it at the apex of all

Lonely prospector

good things—the Incas, Aztecs, and Chibchas of New Granada—simply staggers belief.

The lonely prospector and the gold rush came into being much as they appeared later in the northern hemisphere, in Brazil in the sixteenth century. Some of Peter Sapp's cousins were there, staring wistfully at those enriching themselves in rich diggings with slave labor. The drinking houses were there, the brothels, the soiled doves, and the hordes of psychopathic criminals who prey on those who have what they covet. The gourmet feasts were there, the finest of silks for the harlots, extravagant living for the lucky and the ruthless; and the Peters watched it all and grew old and weary with seeking and envy, and died with only the rags on their bones.

The Peters were in Siberia later, when the diggings spread down the western flanks of the Urals, and the shacks and tents of placer workers lined the streams. He was farther east when in the vast mountain masses more silver was dug out than in any other area of the Old World. His dull honest mind would never grasp the idea that if there was enough gold to gild countless statues with gold leaf, or to make angels of solid gold, there was enough for thieves. For over three centuries, as the cry of gold sent the thousands rushing to the mountains and rivers, from the silver mines at Zacatecas in 1546 to the Sutter mill on the American River in 1848 and the Klondike in the nineties, Peter Sapp and his cousins would be among them, but never close enough to the James Floods and Mrs. John W. Mackays to see more than the dazzling signs of their immense wealth.

But more than most he would see what the lust for gold did to men. He was at the burial scene which Marryat described. The grave had been dug, men were kneeling round it, and a "powerful preacher" was telling the listening world what a fine fellow he had been, when one of the men kneeling began, out of habit, to examine the earth he was kneeling on. He found it "thick with gold" and the wonderful—perhaps we should say the golden—news went from man to man around the circle. The preacher, who also had the itch, gazed thunderstruck and shouted, "Gold! You're all dismissed!" And falling to his knees he began, with the others, to paw and dig at the earth. The corpse was dragged out of the grave and buried in a spot where the soil showed no colors.

Edward F. Fitzhugh, Jr., a geologist, gives a fascinating picture of the way gold and other precious metals came into being. "When we realize the molten lavas penetrate the surrounding rocks many miles beneath the surface and overcome the weight of the great masses of rock above them, the stupendous pressures which they must exert are almost inconceivable to man . . . the activities of these deeply buried lavas have been very largely responsible for our accumulation of valuable minerals." In the strange production of gold deep in the earth there are what geologists call families —deep-seated, of which the Morro Velho lodes in Brazil are typical; the intermediate, such as the Mother Lode in California, where veins "have been followed as much as a mile vertically below the surface"; and the shallow-seated, such as, in Nevada, the Comstock, Tonopah, and Goldfield. From the latter camp a single carload of ore contained six hundred thousand dollars, but this type

of ore does not "go down," whereas in the deep-seated it may go far deeper than man can go.

Professor Emmons says:

It is believed that most gold deposits are formed by precipitation from fluids given off by cooling granitic bodies which are called batholiths. . . . Since most of the deposits that have been discovered are located at very definite positions with respect to parts of the batholiths and to roof structures, it is reasonable to suppose that undiscovered deposits also will be found at places where similar controlling features exist.

A batholith

is a great, irregular, deep-seated igneous mass that broadens downward. . . . Unlike laccoliths, batholiths have no visible floors but extend downward beyond the range of observation. They are acidic, ranging from quartz-bearing diorite through quartz monzonite and granodiorite to granite. The molten rock or magma which, cooling, forms the batholith rises by thrusting the invaded rocks aside, by melting or assimilating them, and, according to Daly, by stoping or prying off pieces of the invaded rocks which sink to depths where probably some are melted.

All this leads to the question: Are the great mines worked out? Very probably not. Many ore ledges are broken by faults and as a consequence "there are millions of tons of ores left in mines that have been abandoned on account of faulting." The cost of rediscovering the lode after a fault has severed it is often prohibitive.

As a land surface with gold ore

is worn away in regions where conditions are unfavorable for solution, gold tends gradually to become concentrated at the surface. Some of it may remain practically in place. . . . As erosion goes on, the gold-bearing mantle rock of a deposit on a slope will gradually settle downhill, constituting an alluvial deposit. As erosion is continued, however, the gold finds lodgment in streams together with sand and gravel. Such accumulations constitute the principal placer deposits.

Professor Whitney of the California State Geological Survey gives another aspect of it:

Extremely few metalliferous veins are equally rich for any considerable distance, either lengthwise or up and down; the valuable portions of the ore are concentrated in masses which are frequently very limited in extent, compared with the mass of the vein, in which they are contained. It is a fact, that indications of valuable ores on the surface do not always, nor once in a hundred times, lead to masses of ore beneath the surface of a sufficient extent and purity to be worked with profit. There are, literally and truly, thousands of places in New England where ores of the metals, including silver, copper, tin, lead, zinc, cobalt and nickel, have been observed; many of these have given rise to mining excitements, and have been taken up, worked for a time, abandoned, taken up again, abandoned again, off and on for the last fifty or even a hundred years, and always with partial, and usually with total, loss of the money invested.

Even on Lake Superior, that region which is commonly appealed to as made up of solid copper, there have been many hundreds of companies formed, and at least a hundred mines opened and worked more or less extensively; but for ten years after mining had begun to be be actively carried on there, only two of the mines had paid back to the stockholders one cent of dividend.

In England the Keeper of Mining Records "who has devoted many years to the investigation of the statistics of this branch of the Nation's wealth" has said that with the exception of iron, mining was "not on the whole remunerative." So possibly we'd do well to leave our fancy new metal-finders in storage until we've found out how many Peter Sapps there are. Still, the nuggets are just ahead of us.

No matter how hard to find, it is known that there is an almost inconceivable amount of gold still undisturbed by human greed, distributed all over the world, usually in very tiny particles, called gold dust, but also in lodes and in nuggets, and in the oceans. There are traces of gold in the ashes of plants and animals and in the metallic sulphides. But it is the estimated amount in the oceans that paralyzes the imagination—at least ten billion tons of it! Streams have washed it in but much of it has come from erosion of mountain masses in the ocean depths. Because in salt water it is seldom more than 267 parts in a hundred million, no way has been found to extract it.

We'll now look briefly at quartz, at placer, at how to find gold, at the more spectacular nuggets, and at the total amount of recovered gold in the world.

The great rush in California was to placer mines; few of those climbing the hills and streams there had ever heard of gold in rock. Indeed, when gold was found in quartz in Grass Valley and men with hammers and picks knocked stones to pieces and convinced themselves that the gold-

en stuff actually was gold, their notions of mining were horribly upset. But of the one billion eight hundred and fifty million of gold found in California by the time Emmons wrote, nearly one third came from lodes. "Much the greater part is from the Sierra Nevada, particularly from a belt that extends along the west slope of the range for 200 miles." Gravels mined underground produced about $300,000,000, running from $1 to $30 per cubic yard. The largest deposits of the Sierra Nevada batholith were in the Nevada City and Grass Valley areas. "The Mother Lode vein system extends from Georgetown to Mariposa, a distance of 120 miles. It is about a mile wide and includes numerous veins."

In California, Colorado, and a few other areas gold was found in veins buried in quartz, and the quartz was usually hard, white, and brittle. In Gold Run Gulch in the Black Hills gold was found in "blowups," which were quartz deposits on the surface, that looked as if they had been exploded out of the earth. In the Grass Valley area, according to Bowles, the Massachusetts Hill was a vein only twelve or fourteen inches in width, yet it produced over seven million dollars in gold; the Allison Ranch was a vein averaging about ten inches, that produced about $850,000.

Float is a piece of rock bearing precious minerals that has been eroded or broken off an outcropping of lode. When Bob Womack in Colorado's Cripple Creek area picked up a piece of what he thought was float and had it assayed, and found that it had $200 of gold to the ton, he still had the big job ahead of him. He knew that the outcrop was always uphill from the float, and that during the centuries it might have been buried by upheavals or landslides. He knew that a long time ago water had flowed down the mountainside where he had found the ore, and so he studied the conformation of the area to determine, if he could, where the old streambeds were. His job was to follow the float-lead up the mountain as far as he could. He was a gambler and a whiskey-drinker and not a very zealous prospector, but for three years he worked on that float-lead, off and on, and traced it up the mountain for two miles. A friend staked a claim at the highest point on Bob's trail and sank a shaft; and after missing by only three feet a three-million dollar vein, he gave up. Womack, the discoverer of the Cripple Creek goldfield that was to make Winfield Scott

Stratton a fabulously rich man, did not give up.

"Placer" is from the Spanish and means, of course, contented, satisfied. In a court case placers have been defined as "superficial deposits of gold which occupy the beds of ancient rivers." Says Emmons:

Because of its properties gold is more suitable than most other metals for accumulation in gravels. It is highly malleable and therefore is easily mashed or rounded, but a particle of gold is not readily divided. It remains intact when brittle minerals are broken. . . . It will remain solid when other minerals are dissolved and carried away. . . . Even under favorable conditions gold is dissolved less readily than copper or silver and is precipitated more readily than either. Consequently its enriched ores are likely to be found near the surface.

Says Fitzhugh:

[As] weather, water, and wind break up and wear away a gold-bearing outcrop, particles of gold are included in the disintegrated mass that was once rock. In a hilly region this mass gradually moves down the hillsides to the bottoms of gulches and valleys. Heavy rains may hasten this movement. Where there is little vegetation, with few roots to retard the progress, this creep of the gravel, sand, and soil may be so rapid that we can measure it from year to year. [If a mineral is to survive in streams] it must offer great resistance to the corrosive powers of water. . . . Gold, platinum, the mineral cassiterite (oxide of tin), and diamonds are the most important of these resistant minerals that we mine in sands and gravels.

Placer mining

Bowles says placer mines are of three classes: "the first and richest being the 'creek diggings,' comprising the bed of the creek and its low banks; the next and less productive . . . includes the higher bank; the third consists of hill diggings beyond. . . ." He was speaking there of the Boise Basin area. As Pefley (and many) have pointed out, gold sinks to the lowest possible level, being so much heavier than the materials around it, and bedrock is therefore "always the richest part of a placer mine." On the sandy beach at Nome, about 200 feet in width and 25 miles long, a million dollars was taken from the sands in two months, with the aid of every contrivance from tin pans to rockers. As late as 1907 a spot of these sands gave up $40,000 in three days of sluicing.

We have in part answered the question of where gold is found. Rickard tells of an old grizzled Cornishman who, when asked where gold comes from, scratched his pate and said, "Well, Sorr, wheer it is, it is, and wheer it ain't, theer be Oi."

Most of the California placer gold

lay on the bedrock, to which natural bottom it had settled through the gravel on account of its gravity, its descent being aided, during the geologic aeon, by such constant vibrations of the ground as is exhibited in its major form by earthquakes. Sometimes a layer of clay was mistaken for the base of the deposit . . . below which, on true bedrock, more gold was to be found. Usually the precious metal had settled into the cracks and crannies of the rock, and in such places were discovered the pockets from which gold was gathered in handfuls. Rich deposits occurred in what seemed to be the most unlikely places; even high on the hillside, and . . . on bedrocks that had a reverse slope. These were the remains of older stream-bottoms.

An experienced prospector

will pan the gravels of gulches that drain a region that contains gold lodes and will seek gold lodes in a region that contains gold-bearing gravels. It is common practice to follow up a gully, panning it at intervals, and to scrutinize the country where "colors" cease to appear. This method has proved effective in prospecting hill slopes for gold lodes in regions that have not been glaciated. The mantle rock is panned at intervals and a line may be found above which gold is absent. Trenching across such a line may disclose a lode. Not all lodes have associated placers.

The method he describes here is, of course, the one Womack used.

Another expert puts it this way:

One of the most common ways of prospecting for gold ore is to follow placer gold upstream. . . . When he comes to where these traces terminate in the channel, he tries the debris from the hillsides. By following the "colors" uphill he eventually comes to the source of the gold. This, however, is not always gold-bearing rock. The trail may lead him to a low gap in a divide that marks the channel of an ancient river.

Which is to say that the earth spread before him must be thought of in terms of millions of years.

As Rickard says, many of the tales of discovery of gold "are almost preposterous in their inconsequence"—such as that of the man who picked a wild orchid for his girl and saw gold glistening on the roots; or the weary and frustrated prospector who in anger drives his pick into a rock and opens a hunk of rich quartz. One such prospector struck a nugget, the celebrated Welcome Stranger, that weighed 2,280 ounces and was worth $42,000. A man stubbed his toe in Grass Valley and uncovered a lode; another digging a ditch to water his garden uncovered a nugget worth $1,058. A drunken sailor clambering up a hillside started a landslide and decided to sleep there. In the morning he was sitting on a deposit of gold from which he took $70,000 in two weeks.

Finding gold in Alaska was a problem because the earth was frozen deep. "The surface is covered with moss, named tundra by the Russians; beneath this comes a blanket of frozen mould, which when thawed becomes a black mud. Under this is the gravel, also frozen solid, because it has contained water. This extends to bedrock, where lies the gold. . . . It was impossible to reach bedrock by aid of pick and shovel, nor would ordinary explosives prove of much use." The miner had to thaw the moss and mould and gravel as he went down, foot by foot; but unlike southern miners he did not have to build expensive rigging in tunnels, since the frozen earth was secure above and all around him.

Moses Manuel tells us how he found the famous Homestake:

Toward spring . . . four of us found some rich quartz. We looked for the lode, but the snow was deep and we could not find it. When the snow began to melt I wanted to go and hunt it up again, but my three partners wouldn't look for it, as they did not think it was worth anything.

I kept looking every day for nearly a week, and finally . . . I saw some quartz in the bottom and some water running over it. I took a pick and tried to get some out and found it was very solid, but I got some out and took it to camp and pounded it up and panned it and found it very rich. The next day Hank Harney consented to come and locate what we called the Homestake mine, the 9th of April, 1876.

In his excellent chapter on prospecting, Emmons reminds us of an important fact, that "Most gold ores carry an iron sulphide, and this by oxidation will stain the rock red or yellow. This outcrop, generally colored by iron oxide, is the gossan or 'iron hat'. Quartz, which resists solution more than most common minerals, generally is prominent in outcrops. A common type of outcrop is iron-stained quartz." Quartz, the dictionary says, is "a form of silica in hexagonal crystals or crystalline masses. Except water, it is the commonest mineral, occurring in granite, sandstone, etc. and may be colorless and transparent, or colored." Schist is in "any metamorphic crystalline rock having a foliated structure and is readily split into slabs or sheets." Gold-bearing lodes are "likely to be overlooked" if they are low in iron sulphide, that is, without much iron stain, especially in an area of quartz schist.

Fitzhugh reminds the prospector that few metals are found uncombined with others. Gold, copper, and silver are most commonly found in "solitary form. Most of the metals in nature are in chemical union with sulphur, arsenic, antimony, oxygen, or some other substance. Some of our minerals are simple combinations of two elements, others are quite complex. . . ."

For the amateur prospector there are two common tests for gold. Its color varies from a pale to a deep yellow, but no matter how you look at it or from what angle it always looks the same. Gold is malleable and is the only yellow metal that will not break when you bend, flatten, hammer, and shape it. In the heat of a blowpipe, ore containing gold will yield yellow buttons; Cripple Creek became famous for its lustrous tellurides, both of gold and silver.

The easiest way to find gold, many a tale would suggest to you, is to own a burro and let it get loose. One of the most famous of these tales concerns the Bunker Hill lode in northern Idaho, discovered by Noah Kellogg in 1885. Without the usual embellishment, all that can be said is that Noah's donkey got away from him and he followed it and saw an outcrop. Another famous one is about Jim Butler and his donkey in the Tonopah area. His donkey or donkeys wandered away and after tracking them he found them sheltered from a storm behind a huge outcrop of black rock. Waiting for the storm to subside, he chipped off a few pieces and persuaded a friend to have them assayed. According to Stoll, Noah's donkey came to an ignominious end. Because this creature had a habit of braying night and day, its owner sold it to Kellogg for three dollars. After the mine was found and Kellogg was too busy with lawsuits to take care of his jackass, that unhappy creature, back in camp, took up his braying again, until the camp folk decided that something had to be done. Stoll says that several sticks of dynamite were tied to the beast and "once again I see him galloping toward the outskirts, and can picture the pell-mell rush of miners to escape from his immediate vicinity, and hear the explosion." There probably wasn't even a piece of hoof left of a jackass who was undoubtedly on his way to another undiscovered lode.

If you don't own a jackass or a burro, there are prospectors who would advise you to get a piece of the famous golden bough, the mistletoe; or if there is none around, a piece of willow or alder or almost any shrub will do. No man knows how many prospectors have spent a lifetime with witching stick or divining rod. Mabel Barbee, in her book about the Cripple Creek area, tells how her father, an impractical dreamer, stalked over the hills with his dowser stick "dipping and trembling dramatically" in his grasp. His argument, a common one, was that a man well-stocked with electricity could find water or silver or gold deep in the earth, when the electric current poured from his body into the stick and from the stick into the earth, where "the yellow stuff singing in the earth" called to the current in his body. There are still men traipsing all over the West to hell and gone, looking for water or gold; that they never find any seems not to bother them at all. Frank Waters makes the statement as though he believed it: "Another wizard, John Barbee, located the vein of the El Paso with a forked stick." That's more than his daughter was willing to accept.

If you don't have a dowser or a burro, the best way may be simply to stumble onto it. Or you can pan out the sites of old gambling halls. It is said that after Oro, Colorado, was deserted, a log building that had served as a gambling hall was torn down, and "$2,000 worth of gold was panned out of the dirt floor." But we prefer nuggets. Millions of years ago gold was sometimes eroded free of the veins in which it had been deposited by volcanic upheavals, and sometimes this free gold is found in pockets or cups or vugs—there are various names for it. In a Cripple Creek mine a man named Roelofs was snooping around in deep tunnels when by chance he found one. He showed it to friends. Sprague thinks they were stunned as children are stunned on seeing their first Christmas tree—"millions of gold crystals—sylvanite and calaverite." Scattered over the walls were flakes of pure gold. This pocket was about 40 feet by 20 by 15, a pretty large geode, from which were taken, 1,400 sacks of flakes and crystals which were sold for $378,000. The total production from this geode was $1,200,000 in four weeks.

We can't all be that lucky; we'll do well if we find a nugget as large as the largest in the history room of the Wells Fargo Bank in San Francisco. In this marvelous display of native gold from various California mines, chiefly placer, there must be forty or fifty tiny pans each holding a tablespoonful of gold that varies in size from barely visible particles to nuggets as large as apple seeds. Gold particles by scale of sizes are as follows: the very fine passes through a 40-mesh screen; the fine, through a 20-mesh; the medium (these are the smallest nuggets), through a 10-mesh (1/32 of an inch); coarse nuggets remain on a 10-mesh screen (1/16). Boericke presents this table from Hoffman: the medium has an average number of colors per ounce of 2,200 and at $35 per ounce is worth 1.5 cents; the fine has 12,000 and is worth one third of a cent; the very fine has 40,000 and is worth one tenth. He says that flour gold, which is found in most streams, is so fine that it may take 1,500 colors for a cent of value.

According to Quiett the first discovery of sizable nuggets was in North Carolina in 1799. Picked up in a creek bed by a lad fishing on Sunday, it was used by his family as a doorstop for years, though "almost solid gold." When a jeweler paid $350 for it, a gold rush was on in that area. Patterson spent years in the California placers and came away with positive notions. "A nugget weighing fifty-two ounces was found in Hudson's Gulch,

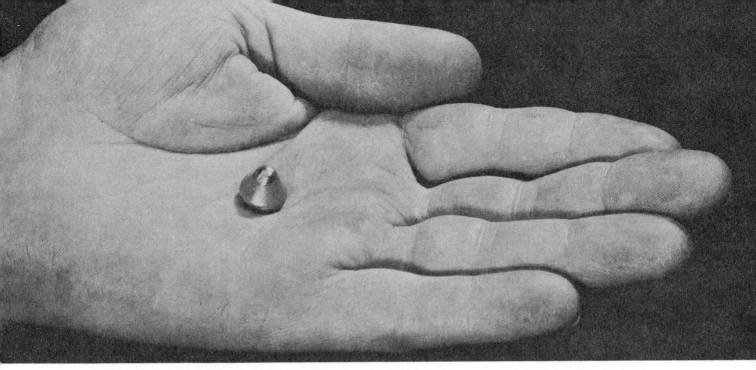

Three hundredths of an ounce of gold taken from the famous Homestake. This is all that is taken from a ton of ore.

Largest nuggets ever found in Alaska

Largest nugget here weighs 1½ pounds

shaped like a solid sphere, with a neck to it, that, in my judgment, could never have been formed by water. A smooth nugget, which weighed thirteen pounds, was found on a side-hill back of American Bar, Middle Fork. This was not washed there by water, as the river never reached that point." He thought such pieces had been shaped "while cooling from a state of fusion."

In Sonora, just south of Arizona, Yaqui Indians in 1736 found large lumps of silver. "No such masses of pure silver had ever been seen before at the surface." This is geologist Rickard speaking, so we had best take him seriously. One lump of 2,700 pounds, and several others weighing from 200 to 400, were found, "besides numerous smaller nuggets." Arizona's Silver King mine had huge crystalline clusters, and the ore was so rich that much of it assayed $1,000 to the ton. "It is said that $100,000 in pure silver was picked up on the surface and just underneath" on the western slope of the Apache Mountains. Wandering over the Dragoon Mountains, John Pearce took a second look at a huge outcropping of quartz that for years had been used as a landmark. Persons had used hunks of the quartz to make camp fireplaces, never suspecting how rich it was in silver and gold. Sometimes claims were almost nugget-rich. On the Yuba the Oriental took $734,000 from a claim only fourteen feet by twenty-four feet. The Red Star took $80,000 from a pocket.

Wells Drury, famous Comstock editor, says he found a nugget in the craw of a chicken that assayed $4.85, and had it made into cuff links. Lucius Beebe says a man named Strain found a hunk of gold near Columbia, California, that weighed thirty pounds and was appraised at $7,438.50. Early in the California rush Buffum saw a 25-pound chunk worth $5,000 (gold was only $14 to $16 an ounce then), and a piece of gold weighing 27½ ounces that was "a beautiful specimen, about six inches in length, the gold being inlaid in a reddish stone." John B. Farrish says that as a boy he saw two miners dig out of the bottom of a placer shaft a piece of almost pure gold that weighed ninety-seven pounds and was sold for $25,000 to his father's firm, A. T. Farrish and Co., San Francisco.

In the California fields a nugget that is said to have assayed $38,916 was taken from Carson Hill only fifteen feet below the surface. Glasscock, who spent a lot of time over old records, says that

GOLD DISPLAY
U.S. National Bank of Oregon
BAKER, ORE. BRANCH

the largest gold nugget ever found "on this continent" was in the shape of a giant crab, with the spread veins in the rock looking like legs. Some of these were 7½ inches broad and an inch thick and were "nothing but pure gold." He tells of having seen in a private collection in California such loveliness in gold as "no human goldsmith, unless it were Benvenuto Cellini, could have formed . . . ferns and sprays and flowers of gold, delicate, gleaming, thrilling, maddening gold. Molded in their matrices of quartz perhaps ten million years ago. . . ." He says that in the Mariposa area in the 1850s two men found a quartz boulder from which $7,600 was taken. One of the men thereupon made a hobby of boulder-hunting; he found "several" that were worth more than a thousand dollars each. "As late as 1858 one resident of the town found a two-thousand dollar rock in his back yard."

We don't know how many nuggets have been called by one writer or another the world's largest.

One of them, turned up at Montana's Cable City, is said to have been worth $19,000. Ferguson, who left California for the Australia diggings, says that a nugget weighing seventy-two pounds was found there, and that at Ballarat as many as seventy pounds of gold "have been washed out of one tub of wash dirt of six common buckets." Some of the stories of nuggets smell like tall tales. Webb Todd says that at Breckenridge, Colorado, there was turned up a "lump of almost pure gold valued at $200,000." That, if true, would be a nugget weighing more than 800 pounds, at $15 to the ounce.

There have been fabulous nuggets in legends. Mexicans called them chispas. It is said that leaf gold has sometimes been found "in shiny droplets and beads and hearts, and strings of gold crystals set in pure white quartz better than any jeweler could mount them." Whether fact or fiction, or a bit of both, one of the most fabulous nuggets found in American mines was a slab "peeled off the wall of a drift, like you'd strip the bark off a redwood tree." It is said to have been twenty-seven feet long, six feet wide, and no thicker than an inch in any part of it. A nugget found in the Urals weighed nearly a hundred pounds. An encyclopedia says the largest authentic gold nugget was found in Australia—184 pounds assaying 99.2 pure, worth $46,625 at the time is was found.

The copper Indians used for jewelry in the Ohio and Mississippi valleys had been torn loose by glaciers and carried down. Rickard tells of "Two masses of virgin copper, weighing 1,900 pounds," that were found in 1754. In 1845 there was found in the Lake Superior region a "mountain of solid iron ore, 150 feet high" that "looked as bright as a bar of iron just broken." As for diamonds, the same geologist tells us that in 1905 "a diamond weighing 3,025 carats, or 1.33 pounds avoirdupois" was found in the Premier deposit; the "stone was 4 by 2½ by 2 inches, but even at that it was only the major portion of a broken crystal. It was named the Cullinan and was presented to King Edward VII." In Amsterdam it was cut and divided into nine stones, the largest of which, of 516½ carats, is known as the Great Star of Africa and is in the British royal scepter.

So much for nuggets. How much gold has been found? In 1936 Quiett wrote: "The world's total gold production from the year Columbus landed in America to 1932 has been conservatively estimated at 1,108,000,000 ounces, which at present prices would be worth $38,780,000,000. More than half of it has been mined since 1900, and production now continues at the rate of about 20,000,000 ounces a year. Only about one-half of the gold produced since 1492 has gone into currency or bullion, however; the rest has been used in the arts or is hoarded in India, China, or elsewhere. John Hays Hammond has estimated that all the

Gold ore nuggets in the Wells Fargo History Room, San Francisco.

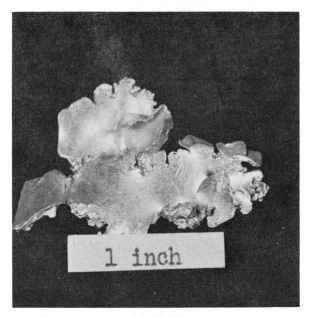

Gold Leaf

gold in the world, if melted into one lump, would make a cube 38½ feet square." Put that way, there doesn't seem to be a great deal of it. Emmons estimated it as a cube of forty-one feet, and the weight as 74,682,076 pounds.

Basing his figures on Emmons and the U.S. Bureau of Mines, Sprague lists the eighteen largest mining areas of the world:

Year of discovery	Name	Estimate in dollars	
1884	Witwatersrand, S. Africa	9,714,900,000	
1910	Porcupine, Ontario, Canada	981,195,000	
1911	Kirkland Lake, Ontario	551,547,000	
1876	Homestake, South Dakota	494,113,000	a single mine
1890	Cripple Creek, Colorado	432,974,000	475 mines
1888	Kalgoorlie, West Australia	425,000,000	
1851	Bendigo, Victoria, Australia	425,000,000	
1846	Lena, Siberia	412,620,000	
1859	Comstock, Nevada	380,000,000	2/3 silver
1882	Yenisei, USSR	371,120,000	to 1923
1882	Kolar, India	300,000,000	one lode
1849	Mother Lode, California	270,000,000	
1873	San Juan, Colorado	200,000,000	
1896	Klondike	186,000,000	
1859	Central City, Colorado	170,000,000	
1902	Goldfield, Nevada	100,000,000	
1900	Fairbanks, Alaska	85,000,000	
1898	Nome, Alaska	80,000,000	

It is estimated that in the decade 1849-59 as much gold was found as in all the preceding centuries. The Homestake at Lead (pronounced Leed) has become the greatest gold mine in the western hemisphere, and as the table shows one of the four greatest so far. In 1964 it mined and milled 5,800 tons of ore daily, or two million annually, and the average of gold *to the ton* of ore is shown by the droplet on the palm of a hand. It weighs only 31/100 of an ounce! The deepest level in the Homestake as of January, 1965, was 6,800 feet.

With the possible exception of the Homestake, says Glasscock, the Empire in Grass Valley is the most spectacular mine in the United States. It has been in continuous operation for almost a hundred years, and when Glasscock was there in 1933 had 190 miles of underground tunnels, one of which went straight down for 4,000 feet. At that time ten tons of water were lifted out of the depths every minute around the clock, and 12,000 cubic feet of air were compressed every minute to drive the drills that used annually fifty-five tons of high carbon steel, driven by twelve hundred

thousand kilowatt hours a year. In 1933 the Transvaal was producing 51% of the world's gold; the U.S. was third, Russia fourth, and Mexico fifth.

Hogg tells the dramatic story of the smuggling, transportation, and storage of British gold to Canada, at the height of the German blitz—tens of thousands of ingots weighing about twenty-seven pounds each, fifty thousand sacks of gold coin, including hundreds of thousands of Napoleons, sovereigns, and half-sovereigns from many reigns, rare guineas and many other kinds—six hundred and thirty-seven million pounds sterling which safely dared the U-boats and unloaded in Canada, with only a handful of people ever knowing about it. It surely was the greatest smuggling feat in human history.

Finally, is prospecting worthwhile today? More than ever, say some of those who write about it. In 1932 Hinckley wrote in his book, *Gold: Where to Go, How to Get it:* "The whole Northwest is rich in placer gold, particularly in such regions as the Clearwater district of Idaho, the Alder Gulch district of Montana, the Blue Hills Basin, the John Day regions of eastern Oregon, and the Swauk region along the Blewett Pass highway in Washington. Streams flowing both slopes of the Rockies have shown a high placer content, but of more recent discovery is the richness of the region on both sides of the Cascade range." He solemnly states that "More gold locations have been discovered in the western states this year than in all preceding years." Boericke, a mining engineer, reminds us that "large arid areas in parts of Arizona, Nevada and California are known to have considerable promise" as placers but need water for development.

In conclusion we may as well let Colton take a crack at the experts:

The volcanoes did their work by no uniform geological law; they burst out at random and scattered their gold in wanton caprice. . . . We have a young geologist here who can unroll the whole earth, layer by layer, from surface to centre, and tell the properties of each, and how it came to be deposited there, who unsuspectingly walked over a bank of gold, which a poor Indian afterwards stirred out with a stick. I have seen this *savan* camp down and snore soundly through the night, with a half-pound piece of gold within a few inches of his nose; and then rise at peep of day to push his learned theory into some ledge of rocks, where not a particle of the yellow ore ever existed.

CHAPTER 3

The California Gold Rush

Bell, book, and candle shall not drive me back, when gold and silver becks me to come on. That seems to have been true in every gold rush on this continent—California in 1849, Pikes Peak ten years later, the British Columbia, Idaho and Montana in the sixties, the Black Hills in 1876, Cripple Creek, the Klondike and Alaska in the nineties, and, after the turn of the century, Tonopah and Goldfield. The great California rush has been written about more than any other, and especially by persons who were there; and so we shall make it stand for them all.

The year 1849, says Harlow, "was one of the three or four great milestones in our history. Epochs ended and began with it. The golden fruitage of California's hills changed us in ten or fifteen years' time from a weak, infantile, economically dependent nation to what might almost be called a world power." Rickard puts it on a broader canvas: "The discovery of gold in California by Marshall in 1848 was the most portentous event in the history of modern mining because it gave an immediate stimulus to worldwide migration, it induced an enormous expansion of international trade, and it caused scientific industry to invade the waste places of the earth."

The story of Marshall's discovery has been told so many times that one must feel apologetic who proposes to tell it again. Marshall and the men under him had been digging a millrace for Sut-

Marshall standing before Sutter's mill

ter, and he had gone out that beautiful morning to look at the debris that had been washed down during the night. He had no thought of gold in mind. But according to his biographer his eye caught the glitter of something in a crevice, six inches under water; he picked it up and stood there, looking at it in his hand, and trying to recall the little he knew about minerals. If it were mica it would be much lighter; if it were a sulphuret of gold, it would be brittle; but if it were gold, it would be malleable. He laid it on a stone and smote it with another stone, and there must have been sensations over his scalp and down his spine when he saw that it had flattened without cracking or crumbling. The story of his gathering an ounce or two of it and rushing to Sutter has been told a thousand times.

As Zollinger tells it, there was a terrific cloudburst and drenching rain, and Marshall, soaking wet, "arrived at the fort and wildly demanded the whereabouts of Sutter. Always a mysterious man, the effect of a reckless chase of fifty miles on horseback through a slashing rainstorm, his picturesque buckskin dress, his wide-brimmed hat and Mexican *serape*, all this gave him the appearance of a madman. Dripping with water, bespattered with red mud, gasping, palpitating, he entered Sutter's office. . . ." The largest piece of those he had gathered was about the size of a melon seed; he had dented it with his teeth and had flattened it, and going to his men is reported to have said, "Boys, by God, I think I've found gold!" A Mrs. Wimmer was making soap nearby and he had her boil the largest piece in strong lye. It did not discolor. This piece Sutter eventually sent to the Smithsonian Institution, which described it as follows: "Under the microscope it shows numerous white particles imbedded in it, which are apparently of quartz. Two small thin films of quartz are still attached to it. . . . There are further many minute black points of no appreciable thickness, which are evidently iron or manganese oxide." Rickard says the quartz became imbedded when he pounded it, and the black spots are from an anvil on which it was hammered.

Sutter must have stared a few moments at the rain-drenched man who had just ridden fifty miles without a pause. Then he read an article on gold in an old encyclopedia; weighed the few pieces immersed in water; compared their gravity with that of some silver coins; tested a piece with nitric

James Wilson Marshall

acid; and "his decision spoken in that frontier store at the foot of the Sierra Nevada went forth to all the world." What a day that was to be for him!

Foreseeing that a rush to the area might ruin him, Sutter pledged Marshall and the other men to secrecy; but the secret went out as such secrets always will. Nevertheless the frenzy took a while to gather. March 15, 1848, the *Californian* told its readers, "Gold Mine Found. In the newly made race-way of the saw-mill recently erected by Captain Sutter, on the American fork, gold has been found in considerable quantities. One person brought thirty dollars worth . . . gathered there in a short time." Nobody paid much attention to that. In Sam Brannan's paper the editor snorted at the news: " all sham—a supurb [*sic*] take in, as was ever got up to guzzle the gullible." A superb take-in to guzzle the gullible the Pikes Peak rush would be, but this was no guzzle. Marshall picked up the gold January 23; two months later the California *Star* was calling the discovery a sham, but by that time men were riding into San Francisco with gold in tin cans, in buckskin pouches, in glass bottles, and the excitement began to seize all the people in the area. It was not until August 19 that the New York *Herald* announced the discovery to the world, but the news of it did not officially reach Washington until almost a year after Marshall rushed in on Sutter. Captain Mason, the military boss out that way, wrote his superiors that he had been to the American Fork diggings and had found two hundred men sweating over picks and gold pans. At Sutter's fort flour was "$36 a barrel and will soon be $50." The captain was a good prophet. On December 5 President Polk gave the news to the Congress, and before long the eastern ports were swarming with men trying to book passage to San Francisco. In the news going East was a statement to the War Department that there was more gold in the area drained by the Sacramento and San Joaquin rivers than the cost of a hundred wars like the late one with Mexico. An impostor under the name of Simpson rushed out a book called *Three Weeks in the Gold Mines* in which he said that in ten days he picked up large nuggets worth fifty thousand. It didn't take many lies like that to start the stampede.

Before looking at that fantastic phenomenon we should glance at San Francisco, from which most of it started. This Street is Impassable—Not Even Jackassable was literally true. A resident of the time has left this description of the main street: "The first portion was constructed of Chilean flour in one hundred pound sacks, which, in some places, had been pushed down nearly out of sight in the soft mud. Then followed a long row of large cooking stoves over which it was necessary to pick your way carefully, . . . Beyond these was a double row of boxes of tobacco of large size. . . . Chilean flour, cooking stoves, tobacco, pianos were the cheapest materials to be found." According to horrified persons who were there, beasts such as donkeys sometimes were thrown in their struggles and floundered in the deep mud and sank out of sight and died; and it has been conjectured that at least a few unwary persons, such as drunks, actually disappeared into it, when after dark they were taking their way across a walk of flour and stoves and pianos from one saloon to another. Indeed, Quiett says that in the rainy season the streets "were seas of mud, so deep that people had been known to drown in them, . . . Loads of brush and trees were dumped into the mud, but these were ineffective. . . . To travel at night without a lantern was to court death. . . . As for lumber, it was more expensive than bales of merchandise—almost unprocurable at $600 a thousand feet."

This mud camp was the base for the gold rush. It seems likely that the frenzy of a people in such a situation cannot be imagined but must be seen. Fortunately, we have the records left by some who were there. After A. J. Pritchard appeared in Spokane in August, 1883, with four pounds of gold in a buckskin pouch, "whiskey stood untouched upon a score of bars, faro and poker lost their hold, and within twenty minutes of the arrival of the two prospectors a mob of at least two hundred milled and banked about them." In the area where gold was found "ploughs rusted in their furrows. Herds shifted for themselves. . . . Every stagecoach brought in its quota. Every Northern Pacific train discharged a small army of men." Multiply that small frenzy by a hundred or a thousand and you have the California rush. Buffum, who was there from the beginning, says that soon after the secret was out,

one thousand people were on their way to the gold region. The more staid and sensible citizens affected to view it as an illusion. . . . Yet many a man who boldly one day pronounced the discovery a humbug, and the gold-hunters little better than maniacs, was seen on the morrow stealthily wending his way, with a tin pan and shovel concealed beneath his cloak. . . . Before the middle of July the whole lower country was depopulated . . . ere the 1st of August, the principal towns were entirely deserted.

In San Francisco there were "but seven male inhabitants, and but one store open. In the meantime the most extravagant stories were in circulation. Hundreds, and sometimes even thousands, of dollars were spoken of as the reward of a day's labour."

The *Star* was saying: "The whole country from San Francisco to Los Angeles and from the seashore to the base of the Sierra Nevada resounds to the sordid cry of Gold! GOLD! GOLD! The fields are left half-planted, the houses half-built. Everything is neglected but the manufacture of shovels and pickaxes and the means of transportation to Captain Sutter's Valley." It is said that as soon as "the worthy editor wrote and published that statement he shut up shop, paid a high price for a shovel, and started for the same place." The Reverend Walter Colton was alcalde of Monterey when in May he heard that gold had been found. He was skeptical but the "sibyls were less skeptical: they said the moon had appeared, for several nights, not more than a cable's length from the earth; that a white raven had been seen playing with an infant; and that an owl had rung the church bells." A week later belief in the discovery was only "the flash of a firefly at night" but a good old lady had been dreaming about gold. The next day, "troubled with the golden dream almost as much as the good lady," he sent a messenger to the gold area, and six days later a man came back with a piece of gold. By July 15 "The gold fever has reached every servant in Monterey; none are to be trusted in their engagement beyond a week." They all fled to the mines, and their masters with them. August 12 Colton's messenger came in with two thousand in gold. "Bob, while in my employ, required me to pay him every Saturday night, in gold, which he put into a little leather bag and sewed into the lining of his coat, after taking out just twelve and a half cents, his weekly allowance for tobacco. But now he took rooms and began to branch out; he had the best horses, the richest viands, and the choicest wines in the place." By August 16 Colton is writing of Monterey men who employed thirty Indians, and after seven weeks and three days "divided seventy-six thousand and eight hundred and forty-

The placer miner's tools

four dollars. He knows of another man who worked on the Yuba 64 days and took out $5,356; another who worked on the North Fork 57 days took $4,534; a boy of 14 who worked on the Mokelumne 54 days and brought home $3,467; a woman who worked 46 days and brought back $2,125." It is obvious that credulity in Monterey was having a very vigorous life.

But its ordinary life had gone to pot: "The master has become his own servant, and the servant his own lord. The millionaire is obliged to groom his own horse . . . the hidalgo—in whose veins flows the blood of all the Cortes—to clean his own boots! . . . And here am I, who have been a man of some note in my day, loafing on the hospitality of the good citizens, and grateful for a meal, though in an Indian's wigwam." By September 9 he had met a Scot bent double in pain, because he had seen a chunk of gold as large as his two fists and it had "given him the cholic." The remedy, Colton surmised, might be an *aurum-similia—similibus—curantur*. September 16 he met a man "from the mines in patched buckskin, rough as a badger from his hole, but who had fifteen thousand dollars in yellow dust, swung at his back."

By September 20 Colton and some friends, including the son of an ex-minister to Russia, were loading a wagon in (we may imagine) considerable haste. Some of them rode on ahead and first thing they did was to get lost in a pitch-black night, without water, food, overcoat, or blanket, with storm overhead and wolves all around. He recalled how in Monterey "six little boys and girls knelt around the chair of their father, repeating the Lord's prayer. . . . What are gold mines to this?" Nevertheless, the next morning he was pushing forward. Bands of wild horses swept past him, "rushing down the plain like a foaming torrent to the sea"; they boiled wild geese, "but neither fire nor steam could make an impression on their sinewy forms"; passed the grave of a gold rusher who had drowned, and drew a moral from it; met a group on their way back, and "a more forlorn looking group never knocked at the gate of a pauper asylum"; but pushed on and on, with the thousands, until they came to the mines, when, "I jumped from my horse, took a pick, and in five minutes found a piece of gold large enough to make a signet-ring." But come Sunday he is idle: "He must be a callous soul, who, with the

Degrees of fortune in the California Gold-diggings

hope of heaven in his dreams, can wantonly profane its spirit." But the morning after his arrival he is out with the gold diggers "tearing up the bogs" and up to his knees in mud. Weary of the mud, he splits ledges asunder and finds "particles of gold, resembling in shape the small and delicate scales of a fish." Seventy persons were working in the gulch and averaged an ounce of gold a day. "They who get less are discontented, and they who get more are not satisfied." A woman, finding at the bottom of her pan gold worth only fifty cents, hurled it back into the stream; for such "is human nature; and a miserable thing it is, too, when touched with the gold fever."

Colton bought a pick without a handle, hafted it with an ash limb, and set to work; but before long he lay under an oak and went to sleep and dreamed. He dreamed of where gold could be found and on waking struck his pick and fetched up a piece weighing three ounces. He makes the finding of gold seem simple and easy; he saw a man digging furiously on a hillside, from which he took two pounds "in the shape of watermelon seeds." He saw a gigantic fellow six feet eight in his shoes, whose "beard and hair flowed in tangled confusion to his waist . . . tearing up bog after bog" until he picked up a piece of gold that weighed an ounce; saw a German make a hole in the earth for a tent pole and pick out of it a piece weighing "about three ounces"; but in the richest deposits the average man was getting only about a half-ounce daily. When a lump of gold weighing twenty-three pounds was found, its "discovery shook the whole mines." In other ravines or on mud bars the men shoveled down through several feet of mud to bedrock, and found "gold in grains about the size of wheat kernels." Colton perceived that a successful gold hunter was "like the leader of hounds in the chase—the whole pack comes sweeping after, and are sure to be in at the death. No doubling hill, no covert, or stream throws them upon a false scent." He guessed there were fifty thousand persons "drifting up and down these slopes of the great Sierra, of every hue, language, and clime, tumultuous and confused," and ready in an instant to take off with pick and pack if they heard of a richer strike elsewhere. Colton was impressed:

I have walked on the roaring verge of Niagara, through the grumbling parks of London, on the laughing boulevards of Paris, among the majestic ruins of Rome, in the torch-lit galleries of Herculaneum, around the flaming crater of Vesuvius, through the wave-reflected palaces of Venice, among the monumental remains of Athens, and beneath the barbaric splendors of Constantinople: but none of these, nor all combined, have left in my memory a page graven with more significant and indellible [*sic*] than the gold *diggins* of California.

But he regretted that so many thousands were coming overland and by sea, for in his opinion "Not one in twenty will bring back a fortune, and not more than one in ten secure the means of defraying the expenses of his return." He understated it.

As Shinn figured it, in 1845 there were about 500 Americans in all of California, and when Marshall discovered gold only about 2,000; but by July of the next year there were 15,000, by December, 53,000, and by the spring of 1853, 300,000. At the close of '49, 60 vessels had sailed from Atlantic ports, carrying 8,000 men, and 70 vessels were up for passage. Bayard Taylor thought that the rush by land "more than equalled the great military expeditions of the Middle Ages, in magnitude, peril and adventure." Hittell, California historian who lived at the time, wrote: "From Maine to Texas there was one universal frenzy. About 1835 a lone Russian sailing vessel entered what was to be the San Francisco harbor, looked around, saw one solitary hide-droger at his labors, and sailed away. The summer of 1849 saw no less than 549 vessels in the harbor, and by August, "four hundred large ships were idly swinging at anchor, destitute of crews."

Morrell says that a person on the scene estimated that two thirds of all the white men in Oregon mounted a beast, and with a pick, bedroll, and lunch pail, headed south. Hardly believing their eyes, those who arrived at the first diggings spread north, south, and everywhere around them. Greed kept them moving. Greed made them wonder day and night if the gold was richer over that

A Punch cartoon as adapted in *L'Illustration* Paris, 1849. The English poked a lot of fun at the California diggins, and a lot of books by English visitors to California were written by them.

hill north, or that hill south. The ten men in Gould Buffum's party took $150 the first day, $1,000 the first week, but itching and feverish pushed on to the Middle Fork, where they took $416 the first day and $400 the second. All day miners were digging furiously; all night they were tossing and dreaming dreams. Peter was there, his eyes bugging, his Adam's apple shuttling up and down as he gawked and dug and listened, a lad of twenty whose whole being was fixed on gold.

And what did they do to the peaceful domain of John Augustus Sutter? Perhaps nothing more vividly and shockingly reveals the obsessive stampede of the thousands than what they did to this man's property. He had written to his people in the old country that his holdings were so huge that he hadn't yet seen all of them. The center of his domain was a stockade-fort on the Sacramento that also served as his office; around it he had ten acres of fruit and nut trees, and even two acres of roses. His living quarters had a fireplace, California laurel furniture, and even a candelabrum. Colton, whom we have followed to the diggings, said that in California the people never spoke "of acres, or even miles; they deal only in leagues." Sutter's farm or domain was sixty miles long. "Two thousand horses, fifteen thousand head of cattle, and twenty thousand sheep, are only what a thrifty farmer should have. . . . A sheep has two lambs a year; and if twins, four; and one litter of pigs follows another so fast that the squealers and grunters are often confounded. . . . It is singular how the Californians reckon distances. They will speak of a place as only a short gallop off, when it is fifty or a hundred miles distant. They think nothing of riding a hundred and forty miles in a day, and breaking down three or four horses in doing it. . . ."

Shinn puts it this way:

Captain Sutter, at that important era in early 1848, swung his malacca cane, pulled his trim military mustache, gave crisp orders to his Indian alcalde, his commander of the troops, his general superintendent, his manager of cattle, and his head farmer; paid his two hundred laborers with pieces of tin stamped with numbers representing days' work, and redeemable at his storehouses; kept open house for every traveler, explorer, or government official: he was, in a word, the most vital, sinewy, and picturesque figure in all Alta California, beside whose manliness the plotting, irascible, treacherous, dishonest governors who disgraced the closing year of Mexican rule sink into utter insignificance. His title appeared unimpeachable, his possessions secure under the American flag, his colony enterprise well managed and certain of success

—until Marshall picked up that piece of gold. Then came the rush that "ended Sutter's multifarious operations, bidding the hides rot in his tanneries, his wheat fall to the earth ungathered, his herds wander untended, and perish at the hands of outlaws." As Bowles says, thousands of acres of fine land along streams "were ruined forever. . . . There are no rights which mining respects in California. It is the one supreme interest. A farmer may have his whole estate turned to a barren waste by a flood of sand and gravel . . . if a fine orchard or garden stands in the way . . . orchard and garden must go."

As a matter of fact, forcible entry upon fenced-in property became commonplace. For some of the miners no property anywhere, not even in the heart of a city, not even a man's home, was out of bounds to their pick and shovel. In 1850 when a part of lovely Grass Valley was a mining camp, two men, with their eye on hay at eighty dollars a ton, fenced in a natural meadow. A miner sank a hole and found pay gravel, and in less than twenty-four hours the entire meadow was staked and claimed. And those who had fenced the land were denied a single foot of claim for themselves. In 1851 two miners began to sink a hole right in the business center of Nevada City. When an angry merchant asked them what they were doing, they told him and said there was no law against going to bedrock *anywhere*. He fetched a gun and drove them off. As Shinn says, "Nothing was sacred: all rights were subject to the claims of the miner. Many a case occurred where the entire town was moved to an adjacent spot, and every inch of the soil on which it had stood was sluiced away from grass roots to bedrock." In other instances miners tunneled underneath towns, and buildings fell from their foundations. Not only farms, orchards, and parts of towns were destroyed, but even public roads.

To Bancroft, Sutter afterward said, "At once, and during the night, the curse of the thing burst upon my mind. I saw from the beginning how the end will be, . . ." His fences fell, his beasts were butchered, his acres devastated, and even such buildings as the miners could use were torn down and carried away, to make long toms and rockers and sluice boxes. One who saw what was done not only on Sutter's but all over the foothills told a New York journal that it reminded him of a thousand hogs rooting up a forest for nuts. That again is an understatement, for there were fifty or a hundred thousand hogs busy there and they had long tusks.

Page says in *Wagons West* that five men ran off a bunch of Sutter's cattle,

sold them for $60,000 and departed for the States, their fortunes made . . . the stupendous robbery of his flocks and herds went on. By 1852 the settlers had taken all the livestock he owned and they were strong enough to keep the matter of his land in litigation for ten years. In the end he was granted one third of his original possessions, but the cost of obtaining this boon had been so great that he was completely ruined.

In his Diary he wrote:

left me only the sick and lame. . . . What for great Damages I had to suffer in my tannery . . . vats was left filled and a quantity of half finished leather was spoiled, likewise a large quantity of raw hide . . . even the Indians could not be keeped longer at Work. . . . The Merchants, Doctors, Lawyers, Sea Captains etc. all came up . . . all was in a great Confusion . . . all left their wives and families, . . . abandoned their houses, offered them for sale cheap. . . . People looked on my property as their own. . . . Nearly my whole Stock of Cattle has been Killed."

California for a while gave him a modest allowance, but in 1880 this man who had succored so many of the distressed on the overland trail died poor and forgotten.

The frenzy is also revealed in the fact that caste and status were swept away. Delano, who came out in 1849 and published a book about what he saw, says that it all

proved to be a leveler of pride, and everything like aristocracy of employment; indeed, the tables seemed to be turned, for those who labored hard in a business that compared with digging wells and canals at home, and fared worse than the Irish laborer, were those who made the most money in mining. It was a common thing to see a Statesman, a lawyer, a physician, a merchant, or clergyman, engaged in driving oxen and mules, cooking for his mess, at work for wages by the day, making hay, hauling wood, or filling menial offices. Yet false pride had evaporated, and if they

were making money at such avocations, they had little care for appearances. I have often seen the scholar and the scientific man, the ex-judge, the ex-member of Congress, or the would-be exquisite at home, bending over the wash-tub, practicing the homely art of the washerwoman; or, sitting on the ground with a needle, awkwardly enough repairing the huge rents in his pantaloons; or, sewing on buttons *a la tailor,* and good-humoredly responding to a jest, indicative of his present employment—thus: 'Well, Judge, what is on the docket today?' 'Humph! A trial on an action for *rents*—the parties prick anew.' 'Any rebutting testimony in the case?' 'Yes, a great deal of re-*button* evidence is to be brought in. . . .'

Social and financial inequalities between man and man were [says Shinn] "swept out of sight. Each stranger was welcomed, told to take a pan and pick and go to work for himself. The richest miner in camp was seldom able to hire a servant; . . . The veriest greenhorn was as likely to uncover the richest mine on the gulch as was the wisest of ex-professors of geology; and, on the other hand, the best claim on the river might suddenly 'give out' and never again yield a dollar. The poorest man in camp could have a handful of gold-dust for the asking from a more successful neighbor, to give him another start. . . ."

Shinn told a story that has been often repeated:

To a little camp of 1848 (so an old miner writes me) a lad of sixteen came one day, footsore, weary, hungry, and penniless. There were thirty robust and cheerful miners at work in a ravine; and the lad sat on the bank, watching them a while in silence, his face telling the sad story of his fortunes. At last one stalwart miner spoke to his fellows, saying: "Boys, I'll work an hour for that chap if you will." At the end of the hour a hundred dollars' worth of gold-dust was laid in the youth's handkerchief. The miners made out a list of tools and necessaries. "You go," they said, "and buy these, and come back. We'll have a good claim staked out for you. Then you've got to paddle for yourself."

It was not Peter Sapp who sat on the bank, or Littlebury Shoot or Rufus Roundtree,* but one much like them. Were all the miners as kind as these thirty men? Not by a thousand gold rushes. Captain Folsom wrote a long letter on September 18, 1848, that was published in the Washington *Globe* and in Edwin Bryant's book soon after-

---

* These are actual names of miners in the records.

ward. In the rush, said Folsom, "there are many runaway sailors, deserters from the army, trappers and mountaineers, who are naturally idle, dissipated and dissolute; in short, taken in the aggregate, the miners are the worst kind of laboring population." Bryant characterized one named Schwartz: "one of these eccentric human phenomena rarely met with, who, wandering from their own nation into foreign countries, forget their own language without acquiring any other. He speaks a tongue (language it cannot be called) peculiar to himself, and scarcely intelligible. It is a mixture, in about equal parts, of German, French, Spanish, and *rancheria* Indian, a compound polyglot or lingual *pi*—each syllable of a word sometimes being derived from a different language."

"For observe," wrote Mark Twain, "it was an assemblage of two hundred thousand *young* men —not simpering, dainty, kid-gloved weaklings, but stalwart, muscular, dauntless young braves, brimful of push and energy, and royally endowed with every attribute that goes to make up a peerless and magnificent manhood—the very pick and choice of the world's glorious ones." That's typical Mark Twain. The miners ranged in fact all the way from the simpering weaklings to the glorious ones. Neither Mark Twain nor Bret Harte nor any other writer we have read had a word to say about a sizable percentage of them, for whom, in this book, Peter Sapp stands as the archetype and everyman—the youngsters who flocked westward full of the most naïve hope and faith—the big rock candy mountain kids, with little education and knowledge and not much more in brains, timid and self-effacing compared to the bullies and roisterers, feeling parent-forsaken and homesick in their first years out there, but with a lot of grit and perseverance deep down in them, supported by their indestructible belief that they would strike it rich in the next gulch, over the next mountain, in the next stampede. They never struck it rich; they never did more than dig out a bare living. They were undernourished, unclean, and miserably unhappy most of the time, but at every rumor they were off again, feverish with excitement and hope; and they sometimes sat on the banks of ravines and watched the luckier ones, or peered into the saloons and gambling halls and now and then slipped silently in for a drink or two. Hundreds or thousands of them moved from rush to rush, from youth to middle

age to old age, many of them dying of malnutrition and sickness along the way, a few of them enduring all the way to the Klondike. Now and then in the pages ahead we'll catch a glimpse of Peter, a little older, grimier, seamier, but with the same light of faith in his eyes—a light indeed that will grow more intense and fixed and staring as the years pass, until in some of them the mind will be overwhelmed by it and lost in its lunacy.

Mark Twain would have been closer to the facts if he had said that the miners ranged in character from youngsters who missed their Sunday mornings in church to the blackest villains who lay in wait for the miners with their pokes of dust; and in age from boys only twelve or fourteen years old, most of whom had run away from home, to old mountain men like Caleb Greenwood, famous guide and scout, who had contempt for the consuming lusts and greeds of the gold-crazed hordes who rushed lickety split from discovery to discovery.

Bryant left a portrait of Greenwood that is worthy of its space here:

He is about six feet in height, raw-boned and spare of flesh, but muscular, and notwithstanding his old age, walks with all the erectness and elasticity of youth. His dress was of tanned buckskin, and from its appearance one would suppose its antiquity to be nearly equal to the age of its wearer. It had probably never been off his body since he first put it on. "I am," said he, "an old man— eighty-three years—it is a long time to live. . . . I have seen all the Injun varmints of the Rocky Mountains—have fout them—lived with them. I have many children—I don't know how many, they are scattered; but my wife was a Crow. The Crows are a brave nation,—the bravest of all the Injuns; they fight like the white man; they don't kill you in the dark like the Black-foot varmint, and then take your scalp and run, the cowardly reptiles. Eighty-three years last——; and yet old Greenwood could handle the rifle as well as the best on 'em, but for this infernal humor in my eyes, caught three years ago in bringing the emigrators over the *de*sart." (A circle of scarlet surrounded his weeping eyeballs.) "I can't see jist now as well as I did fifty years ago, but I can always bring the game or the slinking or skulking Injun. I have jist come over the mountains from Sweetwater with the emigrators as pilot, living upon bacon, bread, milk, and sich like mushy stuff. It don't agree with me; it will never agree with a man of my age; eighty-three last ——; that is a long time to live. I thought I would take a small hunt to get a little exercise for my old bones, and some good fresh meat. The grisly bear, fat deer, and poultry, and fish—them are such things as a man should eat. . . . Thar's beer-springs here, better than them in the Rocky Mountains; thar's a mountain of solid brimstone, and thar's mines of gold and silver, all of which I know'd many years ago, and I can show them to you if you will go with me in the morning."

One of the greatest of the mountain men.

On the Feather River, with three forks, the Yuba, with three forks, the American, with three forks, the Bear, the Consumnes, the Mokelumne, all flowing into the Sacramento, and upon the Calavaras, the Stanislaus, the Tuolumne, the Merced, all flowing into the San Joaquin, they swarmed by the tens of thousands—a hundred thousand by 1852, some say, and some put it higher. Staggering, as Borthwick says, who was there, "under their equipment of knapsacks, shovels, picks, tin wash-bowls, pistols, knives, swords, and double-barrel guns—their blankets slung over their shoulders, and their persons hung around with tin cups, frying-pans, coffee-pots, and other culinary utensils, with perhaps a hatchet and a spare pair of boots." The Chinese were even more heavy-laden, "all speaking at once, gabbling and chattering their horrid jargon, and producing a noise like that of a flock of geese. . . . horsemen galloped about, equally regardless of their own and of other men's lives." His description of several pages is one of the best left by those who were there. Of San Francisco: "The whole place swarmed with rats of an enormous size; one could hardly walk at night without treading on them. They destroyed an immense deal of property, and a good ratting terrier was worth his weight in gold dust." But the terriers became "at last so utterly disgusted with killing rats, that they ceased to consider it any sport at all, and allowed the rats to run under their noses without deigning to look at them."

Sacramento was the nearest depot of supplies. Called by some writers The Red Man's river this stream rises far up in the mountains near Shasta, and flows about 320 miles to pour its waters into the Pacific at the Golden Gate. Far up, its course is across a beautiful land of deep forests and wild flowers, with so many tributaries coming in from both sides with their cold mountain waters that Julian Dana in his book about the river estimates that "ten thousand springs and creeks and a dozen lesser rivers help to swell its broad current." The city was laid out in the spring of 1849 on the east bank, about a mile and a half west of Sutter's fort. Lots were first offered for $200 but within a year, says Delano, were as high as $30,000.

There were not a dozen wood or frame buildings in the whole city, but they were chiefly made of canvass, stretched over light supporters; or were simply tents, arranged along the streets. The stores like the dwellings, were of cloth, and property and merchandise of all kinds lay exposed night and day, by the wayside, and such a thing as robbery was scarcely known. This in fact was the case throughout the country, and is worthy of notice on account of the great and extraordinary change which occurred. There were a vast number of taverns and eating houses, and the only public building was a theatre. All these were made of canvass.

The cost of supplies like the cost of lots sky-rocketed. Quiett gives these figures, but they were by no means the highest reached: $3 for a cup of coffee, a bit of ham, and two eggs. Tin pans sold for $2.50 apiece and more. A single potato . . . cost 50 cents. A new arrival sold for $10 a dozen old New York newspapers used to fill space in his valise. Boots were $40 a pair, bread 50 cents a loaf . . . board at an ordinary house was $8 a day, rooms at hotels [and what hotels!] were $150 a month. . . . Two white shirts—$40. One fine-tooth comb—$6."

A drink of whiskey or rum cost a pinch of gold dust and barkeepers were measured by "How much can you raise in a pinch?" The famous Elephant House in Sacramento offered on its menu these items: Hash, low grade, 75¢; Hash, 18 karats, $1; Codfish balls, per pair, 75¢; Beef, plain, with one potato, fair size, $1.25; Baked beans, greased, $1; Roast grizzly, $1; A square meal, $3. The Reverend Woods, who was there, tells us that "In the spring of 1849, the single article of saleratus sold for $12 a lb.; it could be purchased in New York for 4 cents. One hundred dollars invested in this single article, deducting all expenses, would yield at the least $25,000. At the same time, building lots in Sacramento City were held at $500; in six weeks they brought $25,000." He was feeling pretty keenly about prices, for he had labored in the diggings for months and had barely fed himself.

In Marysville about 1850-51 eggs and apples

were 50¢ each, watermelons, $12. Milk at a dollar a quart "was used chiefly to dilute and disguise the water of the Yuba River, the town's only supply." The streams were so heavy with mud from diggings above that often, according to Dame Shirley, snow was hauled from the mountains, or melted there and its water was brought down. In Sacramento snow was sold for as much as $450 a load. But the Dame in the autumn of '51 rejoiced over a supper in Marysville of oysters, tomatoes, toast and coffee.

Because of heavy rains the roads, poor at best, became impassable, and people in mountain camps far from supplies faced famine. The winter of 1852-3 was so severe in Grass Valley, and the food bins so empty, that a meeting was held in Beatty's Hotel and this resolution was passed: "Whereas, when in the course of human events it becomes necessary for a people to protect themselves against want and starvation, when they are at the mercy of soulless speculators . . . we deem it right to act in self-defense. . . . Therefore we declare that in consequence of impassable roads we are short of supplies . . . there are abundant supplies . . . in San Francisco, which soulless speculators, . . . are holding for exorbitant prices, and refuse to sell. Therefore, be it, . . . 'Peaceably if we can, but forcibly if we must.' "

Severe winters, the long season of heavy rains, cold hard labor and hunger fostered both illusions and disillusionment. It is estimated that in 1849 more than 150,000 pounds of gold, worth $40,000,000, came from the diggings. In 1850 about 36,000 came by sea and 55,000 overland, and the output was much larger. One hundred thousand men were, as Morrell puts it, "in continual flux and reflux from one place to another." By 1850 Nevada City on the Yuba was the largest camp; by June, quartz was found in Grass Valley, and then came rumors of a lake somewhere that was literally a basin of gold. By 1853 many of the professional prospectors were pushing onward again, having heard of rich strikes elsewhere. By 1855 some of California's most beautiful areas had been gutted with thousands of miles of ditches, and great holes in the hillsides; and whole forests of virgin timber had been laid in waste.

And always there were false rumors and stampedes, such as that to a body of water since known as Gold Lake, which was said to have nuggets so bright that they blinded the eyes. Quiett thinks that for most of the miners "there was a great solid body of gold from which all the other gold emanated, and to find this their efforts were untiring." Some of the diggers posted signs for friends coming after them, such as that on the way to Bannack in Montana:

> Tu grass Hop Per diggins
> 30 myle
> Kepe the Trale nex the bluffe.

Some posted sarcastic jeers. But some, with more luck, wrote to the girl back home, as Jerry King did: "Perhaps I can come for you sooner than I thought, my dearest dear. They have found gold —*actually* found it—near here. With any luck I'll be home very soon with enough of it to keep you in silk dresses for always. . . . You will love this new country and we'll make it what we want it to be." Dana doesn't tell us what happened to Jerry, but a lot of Jerrys died of exposure or disease, some killed themselves, some went mad. Henry Page wrote his wife from Weaver Dry Diggins that would become Hangtown and then Placerville: "I wish that in the matter of dollars, that I now write, and may hereafter write about, that you will keep a closed mouth on the subject.—I always have and always shall, let you know everything that concerns *our* affairs but think that perhaps we had better keep such things to ourselves—I hope to go home with a goodly purse, and if I do, or do not, you will know all about it, but it is not best for others to know how much or how little I may bring. . . . " We shall see in the pages ahead that such men as Henry were not making a vice of caution.

Following the honest men who were willing to endure long hours of toil and hardship were the hordes of parasites who hoped to rob them of every ounce they found, at the gaming table or point of a gun. Johnson, who was there, took a hard look at some of them, dressed in "discolored red shirts, their ugly and dirty faces peering with cunning impudence from beneath flaming red flannel caps, which, from their shape, might be camp pudding-bags; around their waists . . . greasy leathern belts, in which revolved, at ease, a wooden-handled sheath-knife. . . . They were from New South Wales, which was already pouring its refuse population into the lovely valley of the Sacramento." The Australian Sydney Ducks were exiled professional criminals, free to go to any land on earth but their own.

What were the average earnings of the tens of thousands of miners bent over pick and shovel. On this matter there is a variety of opinion. The Johnson quoted above (*Sights in the Gold-Region*), whom Shinn seems to accept as trustworthy, thought the average earnings no more than three or four dollars a day, with not one man in a hundred making even a modest fortune. Of a hundred and twenty who went around the Horn, who were known to him, "not one had great success." Shinn thought that the five thousand miners of 1848 averaged about a thousand dollars each but "the hundred thousand miners of 1850 dug out only half as much," in a longer season. M'Collum says: "We had fallen into the very common error that prevailed with adventurers; not as to the quantity of gold in the soil of California; but as to the amount of severe labor, . . . to obtain it. Where stout able bodied men, inured to out-of-door labor, by working hard eight hours in a day, might have been pretty sure of an average earning of an ounce ($16) per day; we could not by tasking ourselves even beyond the bounds of prudence, earn half that amount." M'Collum, a medical doctor, gave up at the end of five weeks, but returned to the diggings and again barely made expenses.

A General Jesup wrote the New York *Daily Tribune* that he was reluctant to tell the whole truth lest he be scoffed at but that he believed "these to be the richest placer mines in the world." In San Francisco he was "meeting persons daily in this place who have been absent less than three months, and have returned with from $2000 to $5000 in gold dust." Some popular verses told the world that "When miners made an ounce a day in any kind of dirt, sir, and oftentimes would freely pay ten dollars for a shirt, sir. Those highly intesting times when never would a man, sir, think claims were good unless they paid ten dollars to the pan, sir. And anywhere you went to work a fortune could be made, sir, and never were afraid to trust, men paid them up so well, sir." The Peter Sapps would have liked to hang the author of those verses.

Allsop said he had seen a piece of gold that sold for $1,156 "but pieces of this size are the exceptions. . . . Still the steady miners here get ten dollars . . . some more, *much* more; many less, *much* less."

Haskins has summed it up as he saw it: "Very often the most ignorant, idle and shiftless lout would stumble by accident upon a very rich gold deposit. . . . Of the great number who prospected subsequently to '49 but a very small portion found claims that paid them for their trouble, while hundreds barely made a living." But Helper says that in his area near Marysville the gold was found "in small particles about the size of grains of sand, sometimes not half so large, sometimes much larger. . . . No lumps larger than a small pea were obtained from this bar." He labored hard for three months and had saved less than a dollar. "When I looked around me, and saw scores of dirty hungry, ragged, long-haired miners, who had toiled and labored like plantation negroes, on this and other bars, for more than two years, and who could not command as much as five dollars to save their lives, it buoyed me up. . . ."

On the other hand Buffum says that he and some companions dug round the base of a boulder weighing twenty tons, down to bedrock, "when our eyes were gladdened with the sight of gold strewn all over its surface, and intermixed with blackish sand. This we gathered up and washed in our pans, and ere night four of us had dug and washed twenty-six ounces of gold, being about four hundred and sixteen dollars." In a canyon which he crossed a number of times "without ever thinking of disturbing it" a man named Hudson took $20,000 in six weeks; a youngster of nineteen took seventy-seven ounces in one day, and ninety the next.

But the sum taken in a day or two is not an average for years. McIlhany says "the boys had made a great deal of money, averaging $150 a day. Some of the miners along the bar would go up and down the stream with a pan and find small pockets of gold that were very rich in the edge of the stream. In some instances they would wash out from $100 to $3,000 in one pan full." That is laying it on pretty thick. In contrast to that Woods says that he and his companions made two dollars each in "three weeks of hard toil," above expenses; and he was deeply affected by a young man who died near them: years would pass and his people would ask, again and again, Why doesn't he come home? Near Woods "is seated an old man of three-score and ten years. He left a wife and seven children. . . . He says when he is home-sick he can not cry, but it makes him sick at his stomach. He is an industrious old man, but has not made enough to buy his provisions. . . . Today I have weighed my little store of gold, after paying all expenses, and find it amounts, after over six weeks of hard labor, to $35." But Dame Shirley says in her third letter, "In a fortnight from that time, the two men who found the first bit of gold had each taken out six thousand dollars. Two others took out thirty-three pounds of gold in eight hours; which is the best day's work that has been done on this branch of the river; the largest amount ever taken from one panful of dirt was fifteen hundred dollars." Thirty-three pounds at $14 to the ounce would be over $7,000.

In his *Sketches* Kip says in an excellent chapter on earnings that where he and his friends "pitched our tents, the daily average to the miners did not exceed three dollars; the luckiest man among a hundred hardly averaged ten dollars, many worked hard for two, and some were plodding on in debt for their food and even for the shirts upon their backs." Pointing out that newspapers greatly exaggerated the earnings, he adds with some bitterness, "There are men who would scorn to knock a stranger down and pick his pockets, but do not hesitate to circulate such reports as draw hundreds away from their homes to die." Some miners "fabricate prodigious stories of their own luck." People who heard such boasting, and who saw the lucky ones spending freely at the bars and gaming tables sent stories home of the huge earnings, some of which got into the newspapers. Kip had heard of miners who took out ten or fifteen thousand dollars in a season "but I never saw one of those lucky individuals, or even met with others who had seen them."

Knower put the average much higher: "I have carried three hundred buckets in a day, and at twenty-five cents worth of gold in a bucket, it would amount to $75.25 to each man for his day's work, which was frequently the average. In those days all it cost for a party of three for capital to start mining was about $15." It was a large area: it depended on where the miner was. If it was Downieville, "a man could take out," Quiett says, "from $100 to $5,000 a day," though

we suspect that this writer, like some others who were never there, based his conclusions more on newspaper reports than on the books left by the diggers themselves. It is easy to imagine how impressed onlookers were who, in saloons and elsewhere, saw the shining gold dust measured out in wine glasses or water tumblers, and what glowing reports they sent to the homefolk.

So much, then, for the earnings.*

How did they take the gold out?—at Coon Holler, Mud Springs, Shingle Springs, Dogtown, Fiddletown, Dead Men's Gulch (those were some of the diggings around Hangtown); and at Jackass Flat, Murderers' Bar, Squabbletown, Growlersburg, Whiskey Hill, Poor Man's Creek, Henpeck Flat, Muletown, Sweet Revenge, Gouge Eye, Liars' Flat, Lousy Level, Git Up and Git, Mad Mule Gulch, Slapjack Bar, New Jerusalem, Bogus Thunder, Hell's Delight, Ground Hog's Glory, Last Chance. A book could be written about the names the miners gave to the spots in all the goldfields. Up in Idaho a few years later some of them seemed to be reaching for poetry—Eagle of the Light, Magnum-Bonum, Mountain Diadem, Evening Star, Seven Stars, Morning Star, Ne Plus Ultra, South Pole, North Pole, I—X—L, War Eagle and Gray Eagle, Wild Irishman, Little Joker, Queen of the West, and Pride of the Union were some of the diggings in the famous Silver City area. Furious late-comers, finding everything staked, called one spot Hog'Em in the Boise Basin; and on Tarryall Creek in Colorado they called it Grab-All. In another area it was Humbug Creek, Mad Ox Ravine, Mad Mule Canon, Skunk Flat, Death Pass, Ignis Fatuus Placer, Bloody Bend, Devil's Retreat, Hell's Half Acre, and One Horse Town.

How did they take the gold out? Not easily but simply. There was no curse of high-grading, which has been a major headache in many mines, and indeed still is. In the Kimberley mines, Engineer Rickard tells us, "one native had the nerve to swallow" nearly $4,000 worth of diamonds; another hid "21 beautiful stones weighing 348 carats in his alimentary system." A boy swallowed a diamond as big as a chestnut, weighing 152 carats. But these were pikers compared to the native who was thought by doctors to have tetanus. They opened an ugly wound in his leg and to their amazement

_____
* The reader will bear in mind that the dollar then was worth in purchasing power five or six times what it is worth today.

found inside the wound an old rag wrapped around a fortune in diamonds.

But there were, in these early placers, the fakers and deadbeats—"a race of pseudo scientists," Professor Paul calls them, "who bestowed upon themselves the title of 'professor', and talked learned nonsense about patented processes and ingenious gadgets that they had invented." None of the methods used was patented and they were anything but complex. But using them was not what Dame Shirley had thought it would be, "that one had but to saunter gracefully among romantic streamlets, on sunny afternoons, with a parasol and white kid gloves, perhaps, and to stop now and then to admire the scenery, and carelessly rinse out a panful of yellow sand (without detriment to the white kids, however, so easy did I fancy the whole process to be), in order to fill one's workbag with the most beautiful and rare specimens of the precious mineral." She swirled out a few pans and "wet my feet, tore my dress, spoilt a pair of new gloves, nearly froze my fingers, got an awful headache, took cold and lost a valuable breastpin. . . ."

Life was real and life was earnest in those diggings. As tools, sometimes a knife was enough. Buffum tells how, having dug down to bedrock, he scraped with a knife over the stone until he found a crevice that "appeared to be filled with a hard bluish clay and gravel, which I took out with my knife, and there at the bottom, strewn along the whole length of the rock, was bright, yellow gold, in little pieces about the size and shape of a grain of barley. Eureka! Oh, how my heart beat! I sat still and looked at it some minutes before I touched it, greedily drinking upon the pleasure of gazing upon gold that was in my very grasp. When my eyes were sufficiently feasted, I scooped it out with the point of my knife"—$31 from the one crevice.

Many hadn't even a pan of any kind and couldn't afford to buy one of the various contraptions and gadgets rushed to the market and shipped around the Horn, nearly all of which were worthless. Those without funds made a kind of lacework of small green branches, and while one man held it above the earth another shoveled the gravel on to it. It caught and held the larger stones, which were tossed off after each shovelful, and the part that went through the sieve was put through another screen, of finer mesh, that caught

the next largest stones. Such labor was so man-killing that it is easy to believe that those using a brush-screen gave it up, and went around pulling the sagebrush at the edges of bars and streams, to shake the stuff off the roots and examine it. G. Stuart tells how he obtained from 25 cents to $1 to the pan, from earth shaken off sagebrush roots.

In the beginning most of the miners used a pan, for, as Kip says, it had other uses: "It is no uncommon thing to see the same pan used for washing gold, washing clothes, mixing flour cakes, and feeding the mule." We'll let Buffum introduce the pan and then elaborate a little on what he says: filled with gravel the pan was "taken to the nearest water and sunk until the water overspreads the surface of the pan. The earth is then thoroughly mixed with water, and the stones taken out with the hand. A half rotary motion is given to the pan with both hands; and, as it is filled, it is lifted from the water, and the loose light dirt which rises to the surface washed out, until the bottom of the pan is nearly reached. The gold being heavier than earth sinks by its own weight to the bottom, and is there found at the close of the washing, mixed with a heavy black sand . . . the whole is dried before the fire and the sand carefully blown away."

An excellent pan was a sheet of steel or iron turned over a stout wire around its edge to make a strong rim; about 18 inches across the top, with sides sloping about 30 degrees to a bottom 10 or 12 inches across. If the depth was about three or four inches such a pan would hold 18 to 20 pounds. For cleaning black sand concentrates a copper-bottom pan with steel sides was ideal, with mercury in the bottom to catch the fine gold.

A pan was filled nearly full of gravel and sand and placed just under quiet water; one hand steadied the pan, the other stirred the contents, until all pieces of mud or of stones sticking together were disintegrated. As the water got muddy the pan was tipped to drain it off and take on fresh water. All of the larger stones were tossed out. Then the pan, still under water, was grasped with both hands and moved with a vigorous and circular motion, so that the contents were thrown from side to side. This gave the heavier gold particles a chance to sink through the swirling mass to the bottom. The pan was then taken from the water and tilted forward at about a 30-degree angle, so that lighter sands in suspension were

Panning

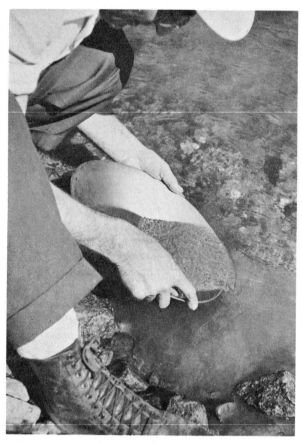

Panning

floated out. The edge was quickly dipped into water and withdrawn, to wash out all the lighter materials. It was not a bad thing to slap the pan smartly towards its bottom a number of times, to encourage the gold particles to settle. If the panning was properly done, in a few minutes there would be only a few ounces of heavy sand and the much heavier gold at the bottom. It was almost impossible to separate these by washing without losing a part of the gold. And so, as Buffum says, the general practice was to dry the sand and gold and then very carefully blow the sand off. If 7 or 8 pans figured to the cubic foot or around 200 to the yard, a man working hard and long hours could wash a ton and a half or two tons a day. If it yielded 3 cents per pan his earnings would be about $6, which seems to have been an average in the California diggings.

Naturally, men looked for a method that would be faster and less laborious. No method was found that was less laborious but some were faster—the rocker, the long tom, and the sluice box. In pan-

ning, as Caughey points out, the miner "had to stoop or squat, at the water's edge or in the water, with hands constantly in and out of the ice-cold stream, and with unflagging attention to the nature of the overflow." It was probably the experienced miners from Georgia who introduced the rocker. Since Colton was in the fields and saw and possibly used the rocker his description of it ought to be as good as any other:

The most efficient gold-washer here is the cradle, which resembles in shape the appendage of the nursery, from which it takes its name. At the end which is closed, a sheet-iron pan four inches deep and sixteen over, and perforated in the bottom with holes, is let in even with the sides of the cradle. The earth is thrown into the pan, water turned on it, and the cradle, which is on an inclined plane, set in motion. The earth and water pass through the pan, and then down the cradle, while the gold, owing to its specific gravity, is caught by cleets fastened along the bottom.

Bryant included in an appendix Governor Ma-

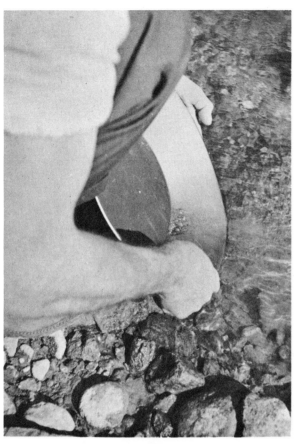

Panning

son's report to the War Department August 17, 1848:

The hill-sides were thickly strewn with canvas tents and bush arbors; a store was erected, and several boarding shanties in operation. The day was intensely hot, yet about two hundred men were at work in the full glare of the sun, washing for gold —some with tin-pans, some with close-woven Indian baskets, but the greater part had a rude machine known as the cradle. This is on rockers, six or eight feet long, open at the foot, and at its head has a coarse grate, or sieve; the bottom is rounded, with small cleats nailed across. Four men are required to work this machine; one digs the ground in the bank close by the stream, another carries it to the cradle and empties it on the grate; a third gives a violent rocking motion to the machine; whilst a fourth dashes on water from the stream itself. The sieve keeps the coarse stones from entering the cradle, the current of the water washes off the earthy matter, and the gravel is gradually carried out at the foot of the machine, leaving the gold mixed with a heavy fine black sand above the first cleats. The sand and gold mixed together are then drawn off through auger holes into a pan below, are dried in the sun, and afterwards separated by blowing off the sand. A party of four men thus employed at the lower mines averaged $100 a day.

The rocker

Buffum observed that sometimes the rocker was only a hollowed-out log, and Colton that

There are several persons among the gold-diggers here who rarely use any implement but their wooden bowls. Into these they scrape the dirt left by others, which they stir and whirl till the gold gradually works its way to the bottom. . . . This process is what they call dry washing; it is resorted to where there is no water in the vicinity, and will answer pretty well where the gold is found in coarse grains; but the finer particles, of course, escape. They rub the earth into their bowls, through their hands, detaching and throwing away all the pebbles, and then blow off the sand and dust, leaving the gold at the bottom. But on some of the streams, particularly the Yuba, the gold is too fine even for this process. It is amusing to see a group seated around a deposit, blowing the earth out of their bowls. . . . Their cheeks swell out, like the chops of a squirrel, carrying half the beech-nuts on a tree to his hole. A more provident fellow he than his two-legged superior!

The long tom was really only a larger rocker that was not rocked. It had a hopper at the upper end that emptied into a sluice trough, with riffles or cleats across the bottom. It was set in a way to slant it down, so that the water coming across the hopper and carrying earth and sand would flow down the sluice at the proper speed, with most of the gold settling downward in exactly the way it settled when carried downstream by creeks and rivers or by rain on hillsides. Even better, because able to handle more earth with the same manpower, were the longer sluice boxes, with a sizable stream, often brought along hillsides a considerable distance, flowing in at the top end, and with cleats only toward the lower end of the box. A long box with plenty of water could keep several men busy shoveling.

In some areas they had to dig many feet down to find bedrock. Since by hand it was impossible to move the earth to such a depth, over wide

The long tom

areas, the miners coyoted—that is, as Dame Shirley tells it, they drifted "coyote" holes back under the overlying earth, in search of crevices, where most of the gold was to be found. After a coyote hole had been pretty well worked out they then "cleaned it": with a knife they scraped over the entire bedrock surface, to learn whether they had overlooked any crevices.

After a few years of placer mining, quartz lodes were discovered. We shall not attempt to describe this complex mining process. Readers wishing to know how precious minerals are separated from stone can read Mark Twain's account in *Roughing It*. It is calculated to make you feel awfully tired.

Fragments of the famous California Mother Lode, Mariposa County

CHAPTER 4

The Overland Journey

What were they thinking, the tens of thousands who had come in the overland rush, as they squatted with pan by the cold streams, shoveled into rocker or long tom, or lay at night on their hard beds in tent, dugout, or under a tree? Few of them had had a clear notion of the enormous distance they would have to cover, or of what they would need in clothing, food, or transportation; they were eager, most of them, to set off the moment they read the good news, with little more than a shirt on their backs and a sandwich in their blanket roll.

The good news!—in advertisements, most of which grossly exaggerated, and in guidebooks, many of which were not guidebooks at all. All over the Eastern states and all over Europe these things were being read. An essay in the London *Times* long after the event said: "The whole community of steamship companies, editors, merchants, grocers, hotelkeepers, actors, gamblers, thieves, teamsters, drovers, and farmers were united to speed these pilgrims to El Dorado on their way. Incredible reports of nuggets discovered . . . were repeated all over Europe. But there was a determined silence regarding the appalling privations." In fairness to them it should be said that they probably knew little about the privations and the difficulties in getting there.

The New York *Herald* reported the lunacy but didn't lose its head. It told its readers that people all over the country were "rushing head over heels toward the El Dorado on the Pacific—that wonderful California, which sets the public mind almost on the highway to insanity." Men, bereft of their senses and dizzy with dreams of vast wealth, "are advertising their possessions for sale" and turning Stephen Foster's new song, "Oh! Susanna!" into "Oh! California! That's the land for me!" But later in the camps it would be, "O California! this is the land for me; a pick and

shovel and lots of bones! Who would not come the sight to see—the golden land of dross and stones! O Susannah, don't you cry for me, I'm living *dead* in Californee."

The Folsom letter had a wide influence, for he wrote back home that "I went to them in the most skeptical frame of mind, and came away a *believer*." He had no doubt that men willing to work could take out $25 to $40 a day, for he had seen some who took out from $800 to $1,000 a day. His letter appeared not only in the *Globe* but in Edwin Bryant's book and very possibly was copied in many papers. Twenty-five to forty dollars a day were a lot more than most of those in the East were making. The Louisville *Courier* published information on the route, the kind of wagon to take, the amount of food needed per person—150 pounds of flour, the same of bacon, 25 of coffee, 30 of sugar, 50 to 75 of crackers, some rice and dried peaches, salt, pepper, and a keg of lard. There was other good advice:

We would recommend emigrants who have cattle to shoe them, and we would advise, not to take them further on the route than Fort Hall (Idaho), or Great Salt Lake, but exchange them there for others, or horses. It is useless for men to start from either of these places with worn out cattle, as they never will get them to their journey's end. Thousands of cattle and horses were sacrificed, because emigrants knew no better last year. . . . We would recommend no man to overload his wagon with tools, &c., for spades, shovels, picks, &c., can be purchased cheap in California. Throw away your old yokes, chains, boxes, &c., for you will do so before you cross the desert. . . . Pack animals are the best all the time, and single men had by all means better pack. Hard bread is the best to take from either of these places; also dried beef. Clothes can be bought in Sacramento and Stockton, as cheap as in the States, and it is useless for men to overload themselves with such articles. . . . Men carrying ploughs, anvils, gold washers, stoves, or any other article of weight, may as well throw

CROSSING THE PLAINS.

V. 2 | 23

EMIGRANT TRAIN PASSING WIND RIVER MOUNTAINS.

SIOUX INDIANS.

CALIFORNIA INDIANS.

INDIANS CHASING BUFFALOES, SCOTT'S BLUFFS.

COURT HOUSE ROCK.

MOUTH OF ASH HOLLOW.

FIRST NIGHT ON THE PLAINS.

CHIMNEY ROCK.

DEVIL'S GATE.

LARAMIE PEAK.

SCENE ON THE DESERT.

CASTLE ROCK.

DRIVING STOCK ACROSS THE PLAINS.

Published by Barber & Baker, Sacramento. Copyright secured.

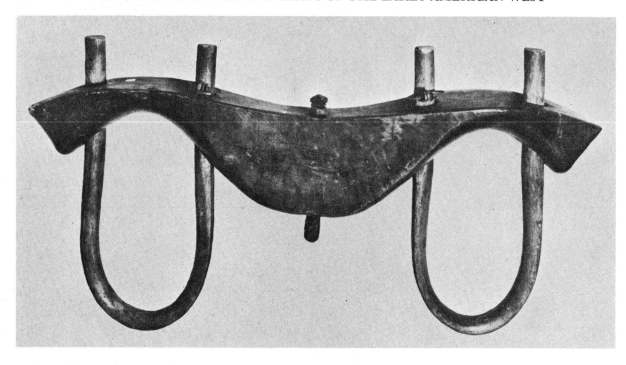

Two indispensable tools

From the motion picture of Brigham Young

them away, or dispose of them; as it is perfectly useless to wear out their cattle, and at last leave the cattle, and goods. . . .

All sound advice, except possibly about the prices in California.

Those groups were fortunate that had a cool experienced man or two to anticipate the problems and prepare to meet them. Usually such groups came to an understanding among themselves and before departure drew up an agreement. This is the pledge of the wagon train of which Page was a member:

Resolved, That we the subscribers, members of the "GREEN AND JERSEY COUNTY COMPANY" of emigrants to California, now rendezvoused at St. Joseph; in view of the long and difficult journey before us, are satisfied that our own interests require for the purpose of safety, convenience, good feeling, and what is of the utmost importance, the prevention of unnecessary delay, the adoption of strict rules and regulations to govern us during our passage: and we do by our signatures of this resolution, pledge ourselves each to the other, that we will abide by all the rules and regulations that may be made

by a vote of the majority of the company, for its regulations during our passage; that we will manfully assist and uphold any authorized officer in his exertions to strictly enforce all such rules and regulations as may be made. And further, in case any members of the company, by loss of oxen or mules, by breaking of wagon, robbery by Indians, or in fact from any cause whatever beyond their control, are deprived of the ability to proceed with the company in the usual manner, we pledge ourselves never to desert them, but from our own resources and means to support and assist them to get through to Sutter's Fort, and in fact, we pledge ourselves to stand by each other, under any justifiable circumstance to the death.

An incalculable amount of suffering and loss might have been averted if all the trains, including the Donner, had drawn up such a pledge and stuck to it. All over the East went the word, Those interested in going to California will meet at the court house at such and such a time. Hundreds of companies were formed, one exclusively of French, one of Germans, one of Methodists, one of Harvard graduates, one of Cherokee Indians; and one, founded by the matron of Sing Sing Prison, had a spinster, two widows, and fifteen

Harper's Weekly, June 12, 1869

men. Many of the men at the meetings wore "slouched hats, high boots, careless attire, and a general appearance of reckless daring and potential wealth." The journey would take that out of them. Those going by sea, according to Quiett, "had rules against strong liquor, gambling, and swearing, though on shipboard individuals might have their own private stocks."

Just about all kinds of people were in the trains or on the ships. Hulbert says there were

Farmers, lawyers, doctors, merchants, preachers, workmen; Republicans, Whigs, Federalists, Abolitionists; Baptists, Methodists, Transcendentalists, Campbellites, Millerites, Presbyterians, Mormons; white men, black men, yellow men, Germans, Russians, Poles, Chileans, Swiss, Spaniards; sailors, steamboat men, lumbermen, gamblers; the lame, squint-eyed, pockmarked, one-armed; the bearded, the beardless, the mustachioed, side-whiskered, and goateed; singing, cursing, weeping, and laughing, in their sleep; squaws in royal blankets, prostitutes in silk, brave women in knickerbockers that reach to the shoe tops, covered by knee-length skirts of similar material; the

witty, nitwits, and witless; pet cats, kittens, canaries, dogs, coons; cherished accordions, melodeons, flutes, fiddles, banjos; fortune-tellers, phrenologists, mesmerists, harlots, card sharks, ventriloquists, and evangelists from almost every state, nation, county, duchy, bishopric, island, peninsula, bay, and isthmus in all the world. . . .

That lays it on pretty thick. Some of the families rather than suffer the wrenching of separation, or leave a helpless one behind to the mercy of others, took with them persons who were too ill or too old to make the journey. Harlan in his *California* says that in his train was his grandmother, ninety years old and stone-blind.

The goods and equipment taken depended on the measure of intelligence and foresight in the leaders. There were some who actually loaded their rather frail wagons with huge pianos or organs, cast-iron cookstoves, heavy log chains, and solid oak furniture. Others set out with almost nothing. One who made the journey and returned and wrote about it came hourly upon hand-carts and foot-men slowly journeying over the undulations of the plains. Not a few of them had started out with only such clothing as they wore on their backs, and small bags containing a few pounds of corn meal and meat. We met two individuals, one fifty and the other sixty-two years old, who had left with just 20 lbs. of corn and only 1.68 dollars in money. There were innumerable poor hand-cart and foot-men hungry, in rags, shoeless, with sore and swollen feet and without shelter from the rains and chilling winds. Not a few had to meet death in its most awful form, starvation; and, what is worse, were driven by the maddening pangs of hunger to acts of *cannibalism* [he probably had the Donner party in mind]. . . . The trails were lined with cooking-stoves, clothing and even mining tools, thrown away to lighten loads; and with the rotting carcasses of horses that had perished on the way. Also with many *fresh* graves. Upon a secluded island in the Platte were the bloody remains of a little girl with a broken skull.

> For my heart is filled with grief and woe,
>   and oft I do repine
> For the days of old, the days of gold,
>   the days of Forty-nine!

The same experienced man gave excellent advice on how to treat the Indians:

The tribes likely to be met on this route are the Pawnees, the Sioux, the Cheyennes and the Arraphoes. The *less* said to the Indians, the *better*. If they are met, and manifest a friendly spirit, extend to them the usual salutation and *pass on*, manifesting *no fear*, and as little *emotion* as possible. If you make trade with an Indian and he is not satisfied, trade back *without hesitation*. The principal danger of collision arises from meeting a War Party. The *braves* are usually on horseback, and in approaching they frequently ride as though about to make a *charge*, when in reality they are only excited by curiosity. If they demand presents, they will usually leave after a gift of flour or tobacco. Should this fail, and there is danger of collision, the best weapon that can be used is an *ox-goad*, or *whip*. Let some stout man seize his whip and *thrash* away at them, and they will run much sooner and faster than if a dozen had been killed.

Except for the last statement that sounds like good advice. The Indians about whom most of the emigrants complained were the Diggers, encountered in the dreadful wastes of the Humboldt. They were horribly starved emaciated filthy creatures, naked or in rags. Now and then a party with an abundance of food would lay wagers on how much a Digger brave could eat and still walk away; and they would toss biscuits, which would disappear almost without chewing, as into the throat of a dog; and pieces of raw pork, dried fruits, rancid fat from ham, hardtack, corn, and, gulping, mouth slavering, eyes bulging, the Indian would swallow everything, in the manner of a starved dog. After a while he would make queer sounds and turn pale and act as if about to faint; he would fall back, and his bloated stomach would heave up and down, as if inside it was an animal trying to get out; but after a period of heaving and belching and groaning he would stagger up, like a drunken man, and tottering and weaving would make off toward his camp, pausing now and then to bend over, as though to rearrange the cargo of unchewed food in his stomach.

It was chiefly the Diggers who sopped arrowheads in mashed liver infused with rattlesnake poison; who crawled on their bellies toward the camps in the dead of night to steal oxen and horses or anything they could seize; and who, as famine-ridden thieves and pests, followed the wagons night and day. Page said of Indians, "we are a geting sick of them they will stand by while you are a eating and watch every mouthful you eat and you can not drive them off without giving them something. . . ."

It was also Page who in a letter to his wife told of the frenzied eagerness to get to the goldfields:

For the whole journey to this country was, a *race*; in which 30,000 men were using all their energies to outstrip each other, during the whole distance of 2000 miles—Property, by hundreds of thousands of dollars, was destroyed & throw away, cattle & horses & mules hurried along on poor feed, till they dropped down, and the topic of interest to all, was onward, & onward—By following our track one would suppose, that a large army had been flying in hot haste, only careing to save their own lives—Wagons and ox-yokes were used for fuel—huge bonfires made of bacon and pork [so the Indians wouldn't get it]—Thousands of kegs of powder blowed up—broken guns & pistols strewed the road—sacks of flour & provisions throw in large heaps upon the ground—In one day I counted forty ox chains by the side of the road, & passed fifty head of stock, dead & dying—It is supposed that one half the stock that left the States last Spring, for California died & was lost —As I look back upon this great crusade, this rush for Gold, I am lost in wonder, that so few left their bones to be a monument of this great march. . . .

Of all the comrades I had then there's
      none left now but me
And the only thing I'm fitted for is
      a senator to be;
The people cry as I pass by, There
      goes a traveling sign,
That's old Tom Moore a bummer of
      the days of Forty-nine!

The insane race for gold, that filled Page with wonder, brought out in every company the brave men and the cowards. The instances, if all gathered, would make a fascinating book. Of the cowards we'll give two instances, and of the brave men, one. A greedy and enterprising young fellow who was determined to get his hands on gold before the train, of which he was a member, could get there, rode on ahead and set fire to the prairie grass right and left, so that the beasts behind him would have nothing to eat. A posse set out in pursuit of him, overtook him, shot him from his saddle and left him where he fell. Another young man, in another company, using beads and other trinkets, lured a young Indian girl into a thicket and raped her. He was caught in the act by the girl's relatives and going to the company commander they demanded his life. No one tried to defend him. He was taken away by the Indians and shot, scalped, skinned, and cut into small pieces.

When a company was unable to get their horses and other beasts to cross a swift river a young man, who said he was a good swimmer, offered to ride his horse across, so that the other horses would follow him, once they were shoved in. Halfway across an undertow struck his beast and flipped it on its side, throwing the rider off. His right hand must have clutched the bridle rein as he went under. Watching from the shore the emigrants saw that the horse had great difficulty keeping afloat, as though something under the water were

Harper's Weekly, December 23, 1871

dragging it down. Again and again it struggled desperately, went under, came up, and struggled again; and at last reached the shore and stood there, trembling, exhausted, and very quiet. The right hand of the young man, torn off at the wrist, still clutched the rein.

Perhaps some of those toiling on the gravel bars in a downpour of rain, soaked to their skin and feeling chills, with a damp bed in a soaked tent to sleep in that night, and the signs of scurvy all over them, remembered, when looking back, those who died along the way. Sickness was as common as frenzy, frustration, and despair. The commonest disease they called cholera but it was a tent that covered a lot of ills. Page says that in the companies he saw the people "died like flies." Bancroft estimates that five thousand of them perished, or about one in every five in a certain period, for the ferry records at St. Joseph, Inde-

pendence, and Kane's Landing at Council Bluffs show that 27,000 men took the trail west that year. Some companies were fortunate enough to have a doctor. It is said that if when passing a tent, any tent, he heard sounds of distress or agony, a Dr. I. F. Morse would pause, look in, and then have the person taken to his tent or wagon, so that he could watch over him.

In that time most of the remedies were "home" remedies, and about as efficacious as a plaster of Humboldt mud. Because of the diets, poisoned water and air in certain areas, anxiety and homesickness and lack of rest, and the long hours of toil, many were crippled by rheumatism or by what they thought was rheumatism. And they used such remedies as this:

The following, taken from the writings of one of the medieval philosophers, and was believed by many to be a sure cure for rheumatism: 1st—

Brown's Hotel at Fort Laramie, 1868

The famous landmark Independence Rock on the Oregon Trail. Thousands of names were chiseled into its stone

The person must pick a handkerchief from the pocket of a maid of fifty years, who has never had a wish to change her condition; second—he must dry it on a parson's hedge that was never covetous; third—he must send it to a doctor's shop who never killed a patient; fourth—he must mark it with a lawyer's ink who never cheated a client; fifth—apply it to the part affected, and a cure will speedily follow.

Well, of course the editor had his tongue way up in his cheek but some of the remedies used by the pioneers and the emigrants overland were just as preposterous. If they died like flies it was because their notions of what they should eat were as full of superstition as their notions of medicine. Sickness in the diggings was almost if not quite the principal curse. Some of those who made the journey and were in the diggings tell us that men who seemed to be hale and well in the morning were dead and buried before sundown.

Edwin Bryant, who came in one of the overland trains and who became the first mayor of San Francisco, tells of a pathetic scene that can stand for all. Because Bryant had ministered to some of the people in his company with simple remedies and common sense a distraught and terrified father begged him to come and heal his wounded son. He found a boy

stretched out upon a bench made of planks, ready for an operation which they expected I would perform. I soon learned from the mother that the accident which caused the fracture had occurred nine days previously, and that some person professing to be a "doctor" had wrapped some linen loosely about the leg and made a sort of trough, or plank box, in which it had been confined. In this condition the child had remained, without any dressing of his wounded limb, until last night, when he called to his mother and told her that he could feel worms crawling in his leg! . . . an examination of the wound for the first time was made, and it was discovered that gangrene had set in and the limb of the child was fairly swarming with maggots.

Knowing that the lad was near death Bryant refused to amputate the leg, telling the parents that the operation would only cause severe pain and do nothing except to hasten death. The mother wept and implored, being unwilling to yield her offspring to a lonely grave in the wilderness. A Canadian Frenchman who belonged to the same emigrating party was present, and then stated that he had formerly been an assistant to a surgeon in some hospital and had seen many operations of this nature performed; he volunteered to amputate the child's limb if I declined doing it and the mother eagerly clutched at this straw of hope and frantically desired him to proceed. I could not repress an involuntary shudder when . . . I saw preparations made for an operation on that little boy. The instruments to be used were a common butcher-knife, a carpenter's handsaw, and a shoemaker's awl with which to take up the arteries. The man commenced by gashing the flesh to the bone around the calf of the leg, which was in a state of putrescence. He then made an incision just below the knee and commenced sawing; but before he had completed the amputation of the bone he concluded that the operation should be performed above the knee. During these demonstrations the boy never uttered a groan or a complaint, but I saw from the change in his countenance that he was dying. The operator, without noticing this, proceeded to sever the leg above the knee. A cord was drawn round the limb, above the spot where it was intended to sever it, so tight that it cut into the flesh. The knife and saw were then applied and the limb amputated. A few drops of blood, only, oozed from the stump; the child was dead.

Brave little fellow! If his bones are still underground somewhere on the Oregon Trail a monument should stand above them, for he is worthier of one than many in this land who sleep under them.

Perhaps worse, for many, than physical sickness was homesickness. No words can describe in many of the thousands who failed to make it and died along the Trail, the pathos of the yearning to be back home, and die there among their people, in a familiar and cherished spot. So many in the diggings, knowing that they were dying, had the same intolerable loneliness and yearning; and we can understand the heartbreak in some who would have given their chance at heaven for one pressure, one more touch, of hands three thousand miles away, and why they crawled back into thickets or into caves and died there, alone.

Says I, my dearest Sally, O Sally, fer
  yer sake I'll go to Californy and
  try to raise a stake;
Says she to me, my Peter Sapp, you
  are the chap to win; give me a
  buss to seal the deal and throw
  a dozen in!

W. H. Jackson's wagon corral at the Rock

It is easy to imagine that people on a hazardous journey would have been able to get along with one another; but such people are angels, not men. But we must bear in mind that most of them were driven day after day by a feeling of desperation, and were willing to pause only for the most desperate necessity, such as a burial. For they had all heard of the ill-fated Donner party, three years earlier, trapped by the Sierra snows; and they had read that the snows were thirty feet deep, or deeper. The thought of being driven to eat the flesh of friends and family, or of those toiling just ahead of them or just behind them, must have been in the minds of many, and drove them, man, woman, and oldest children, to a strength fivefold what they had thought they had.

And sickness and homesickness all around them only made the stronger ones all the more desperate. When they saw friends or relatives turn pale and vomit and become too weak to walk; when they saw some of them looking back, always looking back, to the far invisible land of their birth and childhood; when they saw a mother whose last of three children had died go off the road with the dead infant in her arms and sink to the sagebrush, telling them to go on and on and leave her there; and when they saw a large dog that had had enough of it refuse all offers and entreaties and bribes and turn back home, their last sight of him being a determined hound in a stiff steady trot, his nose pointed east, tempers became short and passions violent.

Andrews, in his story of the Klondike rush, says, "Many are the tales of the dividing of the out-

Coyote, perhaps the most familiar of all wild animals in the West

Fort Hall, on the Oregon Trail

fit with the trailmate, cutting the boat in halves, or splitting the Klondike stove, rather than to have it fall into the hands of the other intact. Bitter were the enmities of the lone trail to the land of gold." And so it was in all the rushes.

And if it was not the timidity or sickness or plain cussedness of trailmates it might be hunger. The records reveal that there was very little foresight in most of the emigrants, and in some, a lot of ignorance and stupidity. On the plains many of them saw the incredible and wanton waste of the buffalo hunters, who furnished meat to them —how they would open a fat beast and take only the tenderloin and a part of the kidney fat, leaving even the hams and shoulders to the wolves; and how, indifferent even to tenderloin, they used the helpless beasts for target practice. But not many of the emigrants had the good sense when flesh was abundant to jerk the choicest parts and take along a supply of dried beef, instead of heavy buffets packed with heirloom china, tall unmanageable wardrobes, or a piano for little Effie who was destined to become a great musician. By the time they were almost too weak to stagger they had thrown away the heirlooms and most of their food

—Historian Hittell says some of them poured turpentine over their sugar (and what were they doing with turpentine?), mixed their flour with salt and earth, burned their wagons, and tore their bedding to tatters, all so that neither the Indians nor emigrants coming after them could have them. It's hard to feel any sympathy for such people—even as you see them there, hungry now that their food is gone, panic-stricken, their eyes trying to see through the vast alkaline Humboldt haze to the distant Sierras. No wonder they quarreled. No wonder that in groups with members who had some education and self-discipline and an unconquerable sense of the ironies there was murder, and a brave man was driven out, destitute, to reach California or die.

Those born to the "affluent society" must find it almost impossible to grasp their problems and their torments. The want of forage for their beasts was enough to have driven them into "mental health clinics." Says Taylor:

The grass was scarce and now fast drying up in the scorching heat of midsummer. In the endeavor to hasten forward and get the first chance of pasture, many again committed the same mistake of throwing away their supplies. I was told of one man, who, with a refinement of malice and cruelty which it would be impossible to surpass, set fire to the meadows of dry grass, for the sole purpose, it was supposed, of retarding the progress of those who were behind him and might else overtake him.

This man was pursued and "shot from the saddle as he rode." In certain alkali areas the grass poisoned the beasts and turned them hairless, and the grass was also poisonous, according to Shaw in *Eldorado*, around certain springs. After their beasts died the weary people "Would take what they could carry on their backs and travel on foot. . . . Women, under the most adverse and trying circumstances, exhibited far more pati-

ence, fortitude and resignation to the inevitable than men."

Like forage for beasts, fuel for evening fires was a problem. Some of those who made the journey say that all the way across the prairies a common sight was persons on horse or afoot, far out, and women with aprons spread, looking for buffalo chips (dried dung) for their fires. So many emigrants had gone before them that all the chips along the trail had been gathered, for a distance on either side of two or three miles.

But the severest ordeal for most of them was not lack of forage and fuel, or poisoned water, or river fords or the superhuman task of taking wagons through mountain canyons, with their steep sides, fallen timber, and swift streams, or the

Brigham Young, 1864, age 63. Many emigrants were eager to see him.

Shoshone Falls on the
Trail.

Hells Canyon, deepest
in the northern hemi-
sphere.

Donner Lake

The Humboldt desolation on the California Trail.

Sierras at the end of the journey. It was the salt flats across what is now western Utah, where without water for the better part of a hundred miles, and often in dreadful heat, they had to push forward with all their strength and courage for two or three days, or nights if they traveled by night —and, right after it the hundreds of miles down the Humboldt, across what is now Nevada.

One of the most graphic accounts of the crossing is Shaw's from which we give a few details: "In preparation for what we had before us, a distance which was said to be ninety miles over the hot, burning sand, without feed or water, we gathered little bundles of bunch grass to fasten to our pack saddles for our horses . . . and also filled our canteens and two or three rubber bags. . . . Several of the boys who had clung to their yellow oilcloth overalls all the way from Independence Rock, tied up the bottoms of the legs and filled them with water." Mirages in this crossing had led to the destruction of a number of parties. Shaw and those with him saw what looked like a "small camp of emigrant wagons." This proved to be no mirage, for they found "a number of wagons with their covers quite intact and the bleached bones of human beings and animals scattered about." Thirst was made more intolerable by breezes blowing over "salty crustations, the inhaling of which parched the tongue like leather. . . . The last ten or twelve miles were a dead

level, hard and smooth as a pavement and hot as a furnace; it was every man for himself in the struggle to reach water, and we were scattered along several miles." At Pilot's Peak the exhausted emigrants camped, usually for two or three days, and then faced west into the terrifying wastes of the Humboldt.

Says Page:

The Humboldt Desert stretched a ghastly barrier across the road. Not a blade of sand-grass, not even a cactus cast a shadow. The ground was white and quick to turn to powder under the trampling of many feet, a powder that stung in nostrils and eyes, that settled in ears and hair and even penetrated between the teeth. Heat from the August sun, reflected from below, rose in visible waves, causing the distant mountains to shudder and dance, and making a hideously tantalizing mirage of water curling and breaking on the alkali flats ahead, water that retreated as the train advanced. Deeper under foot with every mile became the choking white dust, heavier with every mile grew its consistency until at the end of the second day it was dust no longer, but sand that slid under the feet of staggering oxen and men, slid and clutched again. A cruel up-grade was a climb of torture after the horrors of the flats, . . .

There is no exaggeration in that.

It is a tale of people so physicked by alkali water that they had hiccoughs and were worn down

Another view

to pale nubbins of themselves; and of women who, too sick to sit up, fell off wagons and got run over, and were then put in a wagon bound for back home, broken arm or leg, badly set, thrust out across a stinking old horse blanket. What a world of weariness and abandoned hope was in many of those begrimed and sunken faces! It is a tale of thirst-crazed beasts that staggered in pursuit of mirages, or toward the turgid poisonous waters of the river, with their owners struggling desperately to keep them away from it; of beasts that broke away and drank it and were bloated and dying, while frantic owners poured vinegar down their throats or forced them to swallow gobbets of sow-belly fat. The thirst of man and beast all the way across the desolate area was so extreme that one man swore that if

ever again he came to pure water he would drink enough to last a camel across the desert. He did, at the Truckee, and died soon after filling his stomach. The heat in late summer was so terrible that it shrank the wood in wagons until tires fell off the wheels, and sometimes the wheels fell off the wagon. The thirst was so terrible that both beasts and people went insane, and some of them, in spite of all that could be done to restrain them, went off into the desert wastes and were left there to die.

And the clouds of alkali dust kicked up by the thousands of feet and wheels were almost the worst thing of all. It seemed to some of them to be finer than flour, and it rose in huge clouds until it enveloped them and filled the sky. There were poisons in it, for it blistered the human skin

Three hubs from wagons of James Reed of tragic Donner party

on contact, and especially eyelids and lips and the inside of nostrils; and when night came people sat with mouths open, drooling, because of sores inside their mouths, which they called scurvy. The alkaline water, some of which they had to drink or die, distended their bellies and brought on violent nausea, vomiting, and headaches.

It was along these three hundred miles of the trail that people out of their minds threw themselves into the alkaline waters and died; and where the saner ones threw away almost the last of their belongings. Mark Twain said that "It would hardly be an exaggeration to say that we could have walked the forty miles and set our feet on a bone at every step! The desert was one prodigious graveyard. And the log-chains, wagon tires, and rotting wrecks of vehicles were almost as thick as the bones. I think we saw log-chains enough rusting there in the desert to reach across any state in the Union." It was there that cattle became so frightened that owners hardly dared speak to them, and women and children trudged far ahead of or behind the wagons; and in many companies the dogs were killed. It was there that crazed wives wanted to turn back, and when husbands refused to allow them, their mates set fire to the bedding.

White, in *The Forty-Niners,* does not exaggerate when he says, "Thousands were left behind, fighting starvation, disease, and the loss of cattle. Women who had lost their husbands from the deadly cholera went staggering on without food or water, leading their children. The trail was literally lined with dead animals." After the long trek down the Humboldt "had sapped their strength, they came at last to the Sink itself, with its long white fields of alkali with drifts of ashes across them, so soft that the cattle sank halfway to their bellies. . . . All but the strongest groups of pioneers seemed to break here." White admits that the picture he drew was "of the darkest aspect." Those trains with experienced and able managers went through without too much suffering; it was the greedy, the stupid, the quarreling who went insane or killed themselves or sat down and died, or in utter desperation turned around and tried to go back home, though they had seen the sign somewhere behind them:

NO ONE UNABLE TO GO FORWARD FROM HERE CAN HOPE EVER TO GET BACK

> Oh, the good time has come at last, we need no more complain, sir!
> The rich can live in luxury, and the poor can do the same, sir!
> For the good time has come at last, and we all are told, sir,
> We shall be rich as Croesus soon with California gold, sir!

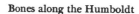

Bones along the Humboldt

Was there no lightness, no joy, nothing pleasant anywhere in that long journey? There was a little for the hardier ones, who kept their spirits up. The descent into Ash Hollow for many of them was so difficult that wagons were wrecked, beasts were killed or wounded, and even some of the emigrants suffered injury; but the lucky ones, in the right season, found there, and elsewhere along the way, a wealth of wild fruit—currants, chokecherries, gooseberries, and, above all, serviceberries.

There were the strange animals which they had never seen before, such as the buffalo and antelope and prairie dog. Horace Greeley on his journey west had refused to believe that rattlesnakes shared the burrows with the dogs until

a tremendous rain raised a creek so that it overflowed a prairie-dog town, when the general stampede of prairie dogs, owls, and rattlesnakes was a sight to behold. It is idle to attempt holding out against facts; so I have pondered this anomaly until I think I clearly comprehend it. The case is much like that of some newspaper establishments, whose proprietors, it is said, find it convenient to keep on their staff "a broth of a boy" from Tipperary, standing six feet two in his stockings and measuring a yard or more across the shoulders, who stands ready, with an illegant brogue, a twinkle in his eye, and a hickory sapling firmly grasped in his dexter fist, to respond to all choleric, peremptory customers, who call of a morning, hot with wrath and bristling with cowhide, to demand a parley with the editor. The coyote is a gentleman of an inquiring, investigating turn who is an adept at excavation, and whose fondness for prairie dog is more ardent than flattering. To dig one out and digest him would be an easy task, if he were alone in his den, or with only an owl as his partner; but when the firm is known or strongly suspected to be Prairie Dog, Rattlesnake & Co., the coyote's passion for subterranean researches is materially cooled.

Most of the pilgrims had heard of the Mormons and many of them were eager to see them.

"I'd like to be a Mormon bold, they lead such dandy lives, with all their pockets full of gold and half a hundred wives. I don't know how they manage them, for I'd like to see the man who'd dare to bring a woman home to my wife, Sarah Ann."

The Mormons only three years before had fled in mass exodus from persecution and were making their home outside the United States, in the Salt Lake Valley. Many persons going overland paused at the Mormon city to study the people, interview Brigham Young, and take notes for an essay or a book about this people. The general attitude toward the Mormons of people all over the nation was expressed in some of the mining camp newspapers. Typical is the Boise *News* of August 27, 1864, under the title "From the Ladies' Mite":

If a young woman is engaged to be married to a man and he dies before the ceremony is performed, she can still be "sealed" to him for eternity, if any good brother will act as a proxy. The children of such a marriage will belong to the dead or spiritual husband, or, in other words to the man who died before he was married. Such wives are regarded as mere secondary affairs, and are a good deal abused, both by the husband and the other wives. They are put off in one corner of the house; they and their children rationed, and only visited by the husband semi-occasionally. On one occasion such a one, with whose sad history I happen to be familiar, said to her husband that the other "women" were whipping and abusing her children and depriving her of every little comfort. He replied: "Susan, you know you are only my wife for a time, and your children belong to another; you will add nothing to my exaltation hereafter; it was only to accommodate Bro. Kimball that I "took" you, so you must get along as well as you can, and don't let me hear you complain any more. . . ."

Bro. Heber, 2nd president, better known at Salt Lake as Brigham's echo, is quite an adept in such business (arranging for a man to take more wives). Meeting Bro. T. one day he said, "Bro. T., are you doing well?" Reply: "Yes, sir." "Then you do well for the church too. How many wives have you?" "Two, sir." "That is not enough. You must take a couple more. I'll send them to you. Do you hear?" Reply: "Yes, Sir." On the following evening when the brother returned home, he found two women sitting there. His first wife said, "Ladies, this is Bro. T (all the women call their husbands brother), these are the 'sisters' Pratt. They were two widows of P. P. Pratt." One of the ladies, Sarah, then said: "Bro. T., Brother Kimball told us to call on you, and you know what for." "Yes, ladies," replied Bro. T., "but it is a very hard task for me to marry two." The other remarked: "Bro. Kimball told us you were doing a very good business, and could support more women." Sarah again took up the conversation, and enunciated the following proposition: "Well, Bro. T., I want to get married anyhow." The good brother replied, "Well, ladies, I will see what I can do and let you know," whereupon the ladies aforesaid retired. Through the influence of his bishop, Bro. T., finally compromised the matter and married Sarah, it being ascertained that Bro. Kimball wanted the house Sarah lived in, for the residence of his daughter, who was soon to be married to a son of Parley P. Pratt.

While we find some women willing to do anything which is "counseled by the authorities," however much it may outrage their sense of propriety, we find very many who would rather die than submit to such humiliation. . . . There are in Salt Lake many noble women, who despise polygamy, and deplore its terrible effects—many who can exclaim, in the language of the Apostle Paul, "I die daily." Some are on the verge of insanity, others endure daily torture in the fear that their husbands will take more wives; many suffer in silence and "Sit like Patience on a monument smiling at Grief."

After the Sink the emigrants faced the Sierras. Some of us who have driven across those great mountains many times, over Highways 40 and 50, marvel at the courage and fortitude of people who took wagons through those canyons and forests. But take them through they did. After the trek down the Humboldt they had come to Ragtown, where a hundred wagons sick unto death with thirst, heat, and alkali threw away most of what they had left; and to Wagontown, where a hun-

dred acres of desert were littered with abandoned wagons in every conceivable state of decrepitude; and to Feathertown, where pillows, thrown away, had been torn open by beasts and the winds, and their feathers scattered, till salt bushes and alkali ridges were bearded with them; and to Graveyard-town, where a train had camped and waited for its sick to die, and then had buried them in the bitter alkali wastes. And now ahead of them was Hangtown in the western foothills of the Sierras, and gold, and the fortunes they had come for. But we have seen what the average earnings were for backbreaking toil, and in the pages ahead we shall find the fortunes with the highway bandits, the professional gamblers, and a very few who combined enterprise and luck.

> They suddenly stopped on a very
> high hill, with wonder looked
> down on old Placerville;
> Ike sighed when he said, and he
> cast his eyes down, Sweet
> Betsey, my darling, we've
> got to Hangtown!

Donner monument at Donner Lake

CHAPTER 5

Other Gold Rushes

In most of their essentials the gold rushes were all alike and the story of the California rush is the story of all of them. To dwell on the things they have in common, over and over, all the way from Tombstone to the Klondike, would serve no purpose in this book. On the other hand, one important aspect or another is more prominent in this rush or in that one, such as concentration of riches and power on the Comstock, exaggeration of the discoveries in north Idaho, hardships in the Klondike, and so on. In this chapter, then, we shall proceed from one rush to another pointing out some of the features that most distinguished them.

Though in the beginning the rush in California

was to a few sites only, the Mother Lode lay north and south nearly the whole length of the Sierra foothills. From these camps there was a tremendous rush across the mountains to the Carson Valley, after silver was found there; and the concentration of riches along Mount Davidson was to have no equal in the American West, except in Cripple Creek and Homestake, much later. "I shall be candid with you," said the *Territorial Enterprise,* ". . . Humboldt County is the richest mineral region upon God's footstool. Each mountain range is gorged with the precious ore . . . almost fabulous wealth of this region—it is incredible. . . . A very common calculation is that many of our mines will yield five hundred dollars to the ton." Mark Twain put those words in his *Roughing It,* and said: "This is enough. The instant we had finished reading the above article, four of us decided to go to Humboldt . . . in Humboldt from one-fourth to nearly half the mass was silver! That is to say, every one hundred pounds of the ore had from *two hundred* dollars up to about *three hundred and fifty* in it." For those who pushed wheelbarrows across the Sierras or walked with heavy packs on their backs the semiarid almost waterless plains and mountains they came to were a formidable desolation. Mark wrote his mother that "No flowers grow here, and no green thing gladdens the eye. The birds that fly over the land carry their provisions with them."

But for those with the fortitude to endure the heat, winds, and dust the riches were there. "Not far from Mark Twain's cabin," as Quiett tells it, "a fat German prospector and two companions dug a hole 20 feet straight down in the red dirt. At that depth the two partners decided to quit, so the German worked on alone. Seven feet more straight down he dug, when he hit the dead center of a pocket containing flakes of pure gold bigger than a man's hand."

Another version says that a curious rancher filled a bag with the blue-black stuff and slipped away to Grass Valley's gold camp on the slope of the Sierras to show it to Judge James Walsh, who had it assayed. Finding that it was chiefly silver and worth several thousand dollars to the ton the judge turned sly inside and before the next morning had vanished up the mountains. The news spread, and half the people in California headed for the new diggings, with Peter Sapp pushing forward with all his energy, having nothing with him but a packet of food, a blanket, a pick with a hawthorn handle and a shovel with a broken blade.

Recorder's office in Idaho's Thunder Mountain area

The multitudes who started late were caught by an early winter, and some of them actually tried to take their mules and burros over the mountains by spreading blankets for them to walk on. The next spring the stampede was on again, with men pushing wheelbarrows, or staggering under their tools and heavy packs, or dragging small carts, while ahead of them men expensively dressed and mounted on thoroughbreds disappeared. From the west and from the east the hordes moved in—drovers with small herds of hogs or cattle, or even with chickens; sharpers with divining rods and electric "silver detectors"; farm families in rickety wagons, or afoot, with children fourteen or fifteen years old clasping the hands of smaller brothers and sisters, who clasped the hands of brothers and sisters still smaller—the young and

old, hale and crippled, rich and poor, all burning with the fever, all uttering greedy prayers, all dreaming of fabulous wealth. They did not know that they were pushing forward to a desolate, timberless God-forsaken region, above which the flying birds, Mark Twain was to write, carried their own provisions—to a land where huge fortunes would be taken but not by Peter and his pals. By the time they got there thousands of claims had been laid out, and Peter drove his stake where neither silver nor gold would ever be found. He then wiped his weary sweating brow and looked around him, as hundreds were looking round them. "My God," they asked, "where is the water?"

A smallish sandy fellow who some day would be a famous writer had this to say:

New claims were taken up daily, and it was the friendly custom to run straight to the newspaper offices, give the reporter 40 or 50 "feet," and get them to go and examine the mine and publish a notice of it. They did not care a fig what you said about the property so you said something.

Consequently we generally said a word or two to the effect that the "indications" were good. . . . If the rock was moderately promising we followed the custom of the country, used strong adjectives, and frothed at the mouth. . . . There was nothing in the shape of a mining claim that was not sale-able. . . . Every man has his pockets full of stock, and it was the actual custom of the country to part with small quantities of it to friends without the asking.

But Peter was gone from that area, his pale eyes staring north to Idaho and Montana, before Mark Twain became known for his practical jokes.

In the first twenty years 57 percent of the yield was silver, 43 percent was gold; such a combination of silver and gold with only small amounts of base metals is said to have baffled the geologists. The big lode, according to Professor Paul, could be traced for two and a half miles, running across the heads of Gold and Six-Mile canyons, on the eastern face of Mount Davidson, and underneath what would soon be the bonanza camps of Virginia City and Gold Hill. Drury says that "During the era of the Big Bonanza, for a distance of five miles reached a continuous line of roofs through Gold Hill and Virginia City, sheltering

more than 30,000 people." The famous lode was less than two miles long, and at its widest only a few hundred feet; but its blue substance, a rich silver sulphide mixed with gold, assayed as high as $4,000 to the ton. The Consolidated Virginia hoisted $50,000 worth of ore daily and became known as the Big Bonanza, a title well deserved, since in less than ten years over a hundred and five millions in silver and gold were taken from its seams. This one compact mine would have placed the four Irish bonanza kings among the world's richest men.

The Gould & Curry, says Bowles, who examined it, "has twelve hundred feet in length on the surface of the ledge, has dug down six hundred to eight hundred feet in depth, and back and forth on its line twenty or thirty times; its whole excavations foot up five millions of cubic feet, and afford some two miles of underground travel, and it has consumed more lumber to brace up the walls of its tunnels than the entire city of Virginia above ground has used for all its buildings." The visitor today who goes underground for a guided tour can get a vivid impression of the compactness of the Comstock Lode, for it is possible to

walk for hours under the small areas of Virginia City and Gold Hill.

Virginia City, says Emmons, "is the premier silver-gold camp of the United States, having produced about $380,000,000 silver and gold. The deposits are in the Comstock Lode, which occupies a fault fissure dipping 45 degrees with diorite footwall and diabase hanging wall, and in branches of the main fissure which rise toward the surface from the hanging wall. The lode was formed near the surface in Tertiary time." When George Hearst rushed in and bought a sixth interest in the Ophir, at one end of the lode, and got thirty-eight tons of ore packed by muleback across the Sierra in deep winter snows, and learned in San Francisco that the ore was worth $91,000, besides the cost of transportation and reduction, the West Coast was again swept by hysteria.

News from the Comstock swept it more than once, and not without reason, for this area produced more multimillionaires, more vulgar display of riches, and more beggars because of dishonest manipulations and stock rigging, than any other mining camp in the country; and it has been written about more than any other, with the possible exception of the Klondike, because it attracted more colorful characters and was more melodramatic.

Hordes of miners moved from the Comstock area, many of them afoot, to Oregon, Idaho, and Montana, or east to Colorado, where for many years there was one rich discovery after another. The Pikes Peak or Bust rush, in 1859, was outstanding in its frenzy, its hardships overland, and its disillusionments. A combination of the 1857 panic which "produced much unemployment and ruined many a business man," and of Missouri River towns which "vied with each other in painting the fabulous picture in the rosiest of colors and of extolling their own advantages as outfitting points," and of "guide-books galore" which said that any person who could raise $50 or $75 could easily reach the rich diggings—these, together with Horace Greeley's enthusiastic reports, sent thousands rushing westward. Greever says that "At the maximum in the spring, perhaps five thousand persons a week arrived; it is almost correct to say that for a full month there was a continuous wagon train extending from Omaha to Denver."

Greeley was duped. He was such a pompous and humorless fellow that men loved to play practical jokes on him. Determined to see for himself if the gold was there, he left his *New York Tribune* office and hustled westward. "Well, he came to Denver and was received with honors. . . . News was sent up to the Gulch that Greeley was on the way, and the boys prepared to receive him with the best they had in the cabin. Everybody knew Greeley, and they knew it wouldn't do to let him go away with a bad impression of the camp. So the boys put their heads together to see what was best to be done." And the best to be done was to salt a hole, for they figured that he would "do some panning on his own account to make sure." They showed him with a panful of ordinary dirt how to wash it, and "Mr. Greeley proceeded to put his dirt through the operation in fine shape. When he got through the bottom of his pan was covered with the bright red particles. It panned out big. He went and got some more dirt with another yield just as rich as the first." He said to them, "Gentlemen, I have examined your property with my own eyes and worked some of it with my own hands and I have no hesitation in saying that your discovery is what it is represented to be—the richest and greatest in America."

Back home he said the same thing in his newspaper, "only that he gave nearly a whole side of the paper to a glowing description and recommendation of the camp. This was circulated and devoured all over the country and it was his advice, 'Go west, young man, go west,' that caused the great Pikes Peak or Bust excitement. . . . The great mad, wild rush began from the time his articles appeared and while Horace Greeley was not far wrong, still we must admit that Colorado owed its first great boom to a salted mine." He was not far wrong if by this is meant the great strikes in Colorado in the next thirty years but he was as wrong as a man could be for that time and place. There must have been thousands of persons in the rush who would have been glad to hang him to any tree on the great plains. Greeley was wearied unto death in a fast stagecoach journey: the toiling thousands across the prairies had no shade sometimes for sixty miles but they had "alkali, cactuses, Spanish nettles, prickly-pears, dust and drought"; and they had Indians. The stupid and brutal slaughter of the red people by Colonel Chivington had "roused the Indians all over the plains to concerted attacks on the whites. They interfered with the Union Pacific Railroad then building west, tore up telegraph lines, burned stage-stations," and killed the gold seekers wherever they could find them in parties small enough to attack.

By May of 1859 "a reaction had set in, and

William Green Russell, discoverer of gold in Colorado

Greeley had heard that there was scarcely any gold, —that the diggers didn't average two shillings a day,—that Denver and Auraria were nearly deserted,—that after terrible suffering on the plains hundreds would gladly work for their board but there was no work in Denver, in short, that Pike's Peak was an exploded bubble which thousands must bitterly rue to the end of their days." Richardson saw it: "Some who started too early had hands and feet frozen. Others consumed all of their provisions before one-third of the journey was accomplished, . . . fearful suffering prevailed. The road was lined with cooking stoves, clothes and mining tools. . . . In the absence of grass, many emigrants were compelled to feed flour to their exhausted cattle. Some wandered off upon the desert . . . one . . . subsisted for several days upon

the body of his deceased brother, and when found was a raving maniac. . . ." Richardson saw one man who had turned back; he had first painted on the white canvas of his wagon, "Pike's Peak or Bust," and under it had added, in charcoal, "Busted, by thunder." The monstrous size of the bubble that would burst can be grasped in the fact that hundreds or thousands actually believed a story, widely circulated, "that the customary way of getting out gold was to slide down Pike's Peak on a heavy wooden drag provided with a bottom of iron rasps. This device ripped the gold out of the mountainside, and it would then roll down in the wake of the drag, sometimes yielding a ton of metal in a single slide. By such tales the imaginations of the optimistic adventurers were inflamed, and soon the Pike's Peak excitement became a frenzy." It became, with the possible exception of the Klondike, the most shocking and horrifying of all the gold rushes in the West.

There was, of course, a lot of gold and silver in Colorado, and in the ensuing years a great deal of it would be found. The famous California Gulch was to be its richest placer; a tourist guide says that on May 1, 1879, the whole area there

Captain E. D. Pierce, discoverer of gold in north Idaho

was a "waste of sagebrush" but four months later in Leadville there were 19 hotels, 82 saloons, 38 restaurants, 21 gambling houses, and 3 undertakers. Tabor's famous Matchless mine was there. By 1880 Leadville, one of the most hell-roaring of all western camps, was estimated to have fifteen thousand persons, and with 28 miles of streets was as large as Virginia City and Gold Hill, Nevada, combined. It was the second largest producer of silver up to its time, and its output of lead exceeded England's. Other famous Colorado camps were Central City, to which Caroline Bancroft has given an entire book, and Cripple Creek, which turned out more riches than the fabulous lode on Mount Davidson. It became, in fact, the world's fifth largest in the production of gold, and second only to the Homestake in the United States.

Professor Paul says in his *Mining Frontiers* that "Of all the new regions that opened to mining during the 1860s, Idaho and Montana came closest to making a reality of the prospector's dream of finding a new California." Credit for the first discovery, in Idaho, goes to "that intrepid trader with the Indians, Captain E. D. Pierce," who was told by an Indian "a marvelous story of a great Shining Eye that existed in the mountains. So bright was this mysterious object that the Indians had seen it at night, gleaming like a star on the wall of a mountain. . . ." Though Pierce was determined to find it, the Nez Perce Indians were hostile to white men on their lands; and when he and his party were forbidden to advance they panned some gravel on the Clearwater River and found gold. It was of such fine quality that they called their diggings Oro Fino, the Spanish for "fine gold," which became the name of the camp. A little later several prospectors tied their horses one night to tall bunchgrass and the horses pulled it up and moved away. In the roots of the grass gold was found, and the news went to Lewiston and Walla Walla and points beyond.

Oro Fino was deserted as the miners rushed to other claims, for it was thought that the placer gold that had been found indicated another mother lode. In Baboon Gulch in the Elk City area a rocker, Quiett says, cleaned $1,800 in gold dust in three hours; another miner kept three rockers going and panned out $1,000 a day; and

The famous Hangtown, later, Placerville

Florence, Idaho, ghost town

in the cabins of miners, says Bancroft, it was not an uncommon thing to see a "gold-washing pan measuring eight quarts full to the brim. . . . All manner of vessels such as oyster-cans and yeast-powder boxes or pickle bottles were in demand in which to store the precious dust." The camp at the head of the gulch was called Florence, and in the winter of 1861-62, the miners, trapped there by snow ten feet deep, wished they had never heard of gold. It is said that a sick man lived five weeks on flour, and tea made by steeping fir needles; that another lived two weeks on four pounds of flour and the inner bark of pine trees. But the news had gone all over the West and the gold rush was on. The first diggings in Idaho Territory, like those in Colorado, were to be distinguished chiefly by the grossly exaggerated stories of what had been found.

As Professor Wells says, the Portland *Oregonian* was skeptical but stoutly assured its readers that every word was true, when it published an account from a man named Mallory on the Florence diggings—"all here are now satisfied that these will prove the richest and most extensive mines yet found north of California. All claim that the center of the vast gold field has at last been found, and this is it." He goes on to tell how much this man washed out in a few hours,

and how much that one. Trimble reproduces a letter from the Portland *Advertiser,* October 29, 1861, that also appeared in the San Francisco *Daily Bulletin* of November 2, giving the excitement in Oro Fino:

On Friday morning last, when the news of the new diggings had been promulgated, the store of Miner and Arnold was literally besieged. As the news radiated—and it was not long in spreading—picks and shovels were thrown down, claims deserted and turn your eye where you would, you would see droves of people coming in "hot haste" to town, some packing one thing on their backs and one another, all intent on scaling the mountains through frost and snow, and taking up a claim in the new El Dorado. In the town there was a perfect jam—a mass of human infatuation, jostling, shoving and elbowing each other, whilst the question, "Did you hear the news about Salmon River?", "Are you going to Salmon River?", "Have you got a cayuse?", "How much grub are you going to take?", etc. were put to one another, whilst the most exaggerated statements were made relative to the claims already taken up.

The contagion spread. The *Oregonian* tossed its skepticism out the window and told its readers:

Neither California or Caribou ever presented such gold prospects as are now found on Salmon River. The facts in regard to the mineral riches there, which come to us from authentic sources, are absolutely bewildering—What are to be the results of these discoveries? Tremendous stampedes from California to the mines,—a flood of overland emigration,—a vastly increased business on the Columbia river,—the rapid advance of Portland in business, population and wealth. . . .

A man named Galbraith told an express company's agent, "I am fully satisfied that since the palmy days of California, never has any such diggings been discovered. . . . It is impossible for me, on paper, to give you any idea of things. . . . I am not excited, but I cannot do it justice." A Dr. Hawthorne sent a "highly regarded" report because of his "well known judgment and calm investigating character." But the doctor, like nearly every one else, had lost his head. A John Meldrum wrote that the excitement "has not been equaled by any since that of the children of Israel from Egypt." The Portland *Times* said it had talked to several miners in from the Salmon River area: "they pronounce the mines to be the richest ever discovered." The Salem *Oregon Argus* published reports "still more favorable" and "No one could doubt that the big rush of 1862 would be to Salmon River." It was, indeed, but all the discoveries in what is today north Idaho, before the Sunshine and its complex, were pretty small diggings compared to the Mother Lode or the Comstock.

The fever moved south, from the Salmon River to Warren and the Boise Basin, and clear to the desert of the Owyhee. The Idaho City area was an important one but nothing sets it apart from hundreds of other mining camps. The Silver City area was the most important of all the early Idaho camps and is distinguished above practically all other camps by its wars and its pitched battles. One of these small wars will be looked at on another page by Richardson, who was there. As typical of all of them, in the Silver City camp and in all other camps where they occurred, we'll look at the Golden Chariot versus the Ida Elmore, and base our summary on Professor Wells. As Wells says, during the period following the Civil War, "mining swindlers were having a hard time keeping up with all of the other frauds across the country. The scandals of reconstruction in the South and of the Grant administration nationally were only another sample of a general deluge of

dishonesty that swept the nation. . . ." In these pages we here and there point to, as other writers have pointed to, but only briefly, the dishonesties, the unscrupulous methods, the exploitation of the ignorant and gullible, that prevailed in varying degrees in all the mining camps, and no doubt prevail in all areas of life where greed and avarice are dominant motivations. The hostilities and deceptions of these two groups of miners are typical of all the camps.

"Rather than risk heavy expenses and losses through protracted litigation," the Beachy* and Fogus interests agreed to divide the property, and to have a piece of neutral ground between them to lessen the danger of conflict. The agreement was honored on both sides for a while but then the Chariot deliberately broke through into the Elmore workings, and a little later advanced to the Elmore main shaft "in an offensive marked by heavy firing." The Chariot forces threatened to destroy the Elmore underground timber supports and bury their opponents. Fogus sent for reinforcements, including a batch of toughs from Nevada, "and the Owyhee war (as this battle was known) assumed serious proportions." Though there was heavy firing from both sides only two men were killed in the first hostilities. The Chariot group seized the Fogus shaft and held it; the Fogus forces then came to the surface to lay siege. At this point the sheriff came under criticism for not marching in to stop the fighting, though "Just how the sheriff, all by himself, was going to stop a hundred armed men from firing on each other in underground workings was never proposed." He did close the saloons. The fighting so far had been down in the mines, and each side, lying for all it was worth, pretended that it was driving burglars out and protecting its property. And each side claimed to own the entire mine.

Hearing of the war deep in a mine, Governor Ballard up in Boise sent a deputy marshal and an Indian fighter named Robbins to the area, with a proclamation calling on both sides to submit to legal processes. Ballard then hastened to the scene. "By late that night, a new agreement had been reached, with formal deeds drawn. . . ." A little later in a drunken brawl a man named More was killed, and his friends were ready to lynch some of the Chariot miners, when Ballard ad-

---

* The same Beachy who tracked down the murderers of Magruder; see Vigilantes.

dressed the people of Silver City and summoned troops from Boise. Down across the desert came ninety-five soldiers with a brass cannon and the war was over.

There was nothing very unusual about the gold rushes in Montana. Quiett, who calls his book *A Panorama of Gold-Rushes,* gives most of his chapter on the Montana diggings to Plummer and his gang. Over and over again the story has been told of the six prospectors who camped on a small stream lined with alder trees, in the Tobacco Root Mountains, and were astonished when they panned some gravel to find it rich in gold. July 3, 1863, Stuart met the stampede coming from Bannack, across the mountains, to the new bonanza: "They were strung out for a quarter of a mile, some were on foot carrying a blanket and a few pounds of food on their backs, others were leading packhorses, others horseback leading pack animals." Bannack had been almost completely deserted. "People flocked to the new camp from every direction; . . . every sort of shelter was resorted to, some constructed brush wakiups; some made sug-outs, some utilized a convenient sheltering rock, and by placing brush and blankets around it constructed a living place; others spread their blankets under a pine tree. . . ."

That was very commonplace gold-rushing. Of more interest are some observations by prospector-freighter Alex Toponce, who took $20,000 out of his claim, and heard that Alder Gulch averaged $1,000 to the running foot, or about $5,000,000 to the mile.

They never did locate the quartz ledges that the Alder Gulch gold was supposed to come from. The channel in which the gold occurred was about thirty feet wide and did not follow the bed of the creek at all. There was very little pay dirt except in this . . . old creek bed. Sometimes the channel would be exposed but mostly it was covered up with dirt that had to be stripped off and this overburden was so deep in some places that we would sink a shaft and then drift on the channel and hoist the pay gravel with a windlass.

At least one engineer has conjectured that more gold remains in that area than has been taken out. That may also be true of other areas; sinking shafts and running tunnels nowadays is a very expensive gamble because of material and labor costs.

Last Chance Gulch and Alder Gulch were the largest bonanza areas in Montana but not the most spectacular in their richness. That was reserved to Confederate Gulch, about thirty-five miles southeast of Helena; a sensational strike was made there in 1864, and Diamond City, the camp, became known as the mining capital of western Montana. A single pan of gravel in these diggings sometimes held a thousand dollars of gold; and one shipment weighing two tons was worth almost a million.

There were other big gold rushes, including southern Arizona, the Black Hills, and, at the end of the century, Tonopah; but these, and all the others, are pretty colorless and feeble when compared to the hysterias and the melodrama of the Klondike. Quiett calls it the most "bizarre" of them all; it certainly was the most hazardous, frenzied, and sensational. The day a ship entered Seattle's harbor with a ton and a half of gold the excitement began, and it steadily mounted until it became almost uncontrolled hysteria, fed by an "astute and widespread publicity campaign. . . . News despatches, special articles, telegrams, letters, diplomatic messages to foreign governments," until Seattle was "a madhouse for crazy gold-seekers. Every train brought its load. . . . Hotels overflowed. Livery barns sold places to sleep in the hay." It has been estimated that soon after the news of gold in Alaska reached the States, between two hundred thousand and a quarter of a million people headed for the north and that fifty thousand actually reached the interior, though the expense of getting there ranged from $500 to $10,000, depending on the quality and the amount of the baggage taken. The terrible passes, the White and the Chilkoot, would separate the weak from the strong, and in the late fall of 1897 4,000 dead horses along the White Pass trail would turn back those who hadn't the stomach for practically anything.

On the ships going north the gold seekers took an outlandish assortment of goods, because the merchants in Seattle were getting rid of their unsaleable merchandise to greenhorns. One "tenderfoot recalls how he purchased from a fly-by-night merchant a heavy and expensive case of desiccated eggs which, when opened in Alaska, turned out to be nothing but cornmeal." The "eager Klondikers were loaded with a variety of luggage, including what they fondly believed to be necessities—shotguns, six-shooters, fishing-

Virginia City, Nevada, then . . .
. . . and now

Idaho City then . . .
. . . and now

Silver City then . . .
. . . and now

Virginia City, Montana, then . . .
. . . and now

Leadville then . . .
. . . and now

Cripple Creek then . . .
. . . and now

Tombstone then . . .
. . . and now

rods," and a whole assortment of worthless gimmicks that were guaranteed to make easy the discovery of the richest placers. One Englishman took his valet. Another, "going for a jolly good time, you know," carried a lawn-tennis set, a bull pup, and two Irish setters. A third boasted that he had in his luggage thirty-two pairs of moccasins and a case of pipes. O'Connor tells of proposals made to Mike Mahoney when he came down to Seattle. Women eager to get up there, even if they had to crawl, offered him their virtue, or what they had left of it, and everything else they had if he'd take them back with him. A blonde told him she had saved a thousand dollars and would cook and wash for him, run with his dogs in his dog teams, and sleep with him. She would even go so far as to marry him, if he insisted on it, but thought such an extreme unnecessary. O'Connor does not explain why, if she had a thousand dollars, she needed an Irishman's help to get up there.

When the gold seekers poured across the Sierras

to the Washoe some of them sang, "Farewell, ole Californee, I am going far away where gold is found in bigger hunks and brighter hunks they say!" When they rushed from there to British Columbia they were singing, "Oh, I'm going to Caledonia, that's the place for me; and I'm going to Fraser River, with the goldpan on my knee!" Peter Sapp, as he pushed north, now almost seventy years old and after fifty years in the diggings still as poor as a church mouse, was singing in the ear of a blowsy female whose actual name was Jacqueline Dopp but who was known in the trade as High Bosom Sal, "Of all the diggers I have known there's none now left but me, and the only thing I'm fitted for is president to be! The people cry as I pass by, There goes a traveling sign, that's old Pete Sapp a bummer of the days of forty-nine!"

Quiett is not alone in saying,

Never was there such a rush as this. Thousands upon thousands who had never known anything rougher than clerking behind a counter or keep-

Skagway, Alaska

Unloading yard at Dyea

ing a set of books were thrown into this land of hardship, and more kept coming, lured by the stories of easy wealth sent out by the railroads and city publicity bureaus. It was a panic of crazy gold-seekers. "I was in Salmon River when 10,000 went in but it was nothing like this," an old-timer would say. "I was in Leadville too, but it was nothing like this. Here people have lost their senses and their heads. . . ." It was said that not 10 per cent of those who started out the first year got through.

Take, for instance, the job of making small boats to cross bodies of water on the trail: we who alone have logged Douglas fir and operated a small sawmill shudder at the strength and endurance demanded by this rush and the tasks set before it. Two men cut trees off seven feet from the ground and laid two beams across each of the two stumps; and across the beams a log fifteen or eighteen inches in diameter was laid; and under this a man stood on the ground and another man stood on the beams above, and with a whip-saw (a huge

unwieldy rip saw) about seven feet long in the blade the two men, pushing and pulling, sawed boards—and what boards they were! And after days or weeks of this back-breaking work they made a boat of these boards and calked it with rags—and what a boat it must have been! With a clear view of that man standing above, and the man below, pulling and shoving a dull saw through a length of twelve or fifteen feet of tough green log we can easily believe the story that mining engineer Rickard tells—of how a

young fellow packed the 1800 pounds constituting his outfit from Dyea over the Chilkoot Pass to Lake Lindemann, where he built a boat. Going down the rapids to Lake Bennett, he ran against a rock and lost everything. Thereupon he walked back to Skagway and procured another outfit, which he carried, as before, over the pass. He then built a boat again, and descended the rapids, only to strike the same rock, . . . Going ashore, he blew out his brains.

A store at Sheep Camp on the trail

Sourdoughs, the Klondikers were called: "In every old prospector's cabin, on a shelf behind the stovepipe, there was a bowl of sour dough saved from the last baking so that it might serve as yeast, and whenever a lump of sour dough was taken out of the bowl, another lump of fresh dough was put in so that it too could sour and be used as leaven." One of them said, "Any man who spent a summer shoveling gravel and sticky red clay with mosquitoes biting 24 hours a day and subsisting on beans, dried fruit, and sour dough bread knows all there is to know about hell." He might have mentioned the weather:

In the new cabins, before the wood was dried out, the log walls would be damp when the Klondike stove was glowing, and covered with a film of ice when the fire died down. At 60 degrees below zero the stove would roar like a forge, and the wood put into it seemed to melt like ice, but it produced little heat. Boiled rice might be burned at the bottom of the pot and raw at the top. . . . Frozen feet were common, and many a miner saved one from amputation only by immersing it in kerosene for four or five hours.

In 1897-98 the temperature went to 54 below at Dawson. Experienced miners judged it by the facts that the mercury froze at 40 below; kerosene at 35 to 55, depending on its quality; St. Jacob's oil at 75, and the best Hudson's Bay rum at 80. Twenty-five below was said to be invigorating, 40 was chilly, and 50 was a bit cold. Since in such weather a person's face felt numb it was difficult for him to tell when it was frostbitten; so they watched one another's faces for the appearance of a patch as white as snow. It was commonplace for frost to collect on lashes, brows, and whiskers, and mustaches sometimes had a row of icicles, like an eave. Falling snow was so thin as to be almost invisible, and gave to mountainsides the appearance of being carved from flawless marble. In the summer the sun shone twenty hours a day, and from the mountains above Dawson could be seen for twenty-four hours.

Out from Dyea, the Chilkoot was called Poor Man's pass because to Dyea came the poorer gold seekers who had meager equipment, could not afford horses or dogs, and had no money to hire Indian packers. Dyea, a splotch of cheap shacks, was on the beach next to Dyea River, and had so little money in it that the gamblers and con men there were strictly second rate. The trail out of Dyea for the first few miles was across meadow

The Scales, where packs
were weighed by porters.

and through a woodland of willow, birch, cotton-wood, and spruce. As the trail became more difficult, it was, like the Oregon Trail, littered with stuff thrown away, for most of the people hadn't even a child's notion of what to take with them, and so abandoned "trunks of every description,

at the base of the mountains, the last spot where wood could be gathered. Usually several hundred persons were camped there, looking up at Chilkoot, which they had to climb; the path upward was in wintertime a succession of ice shelves, and for three or four months in the summer a night-

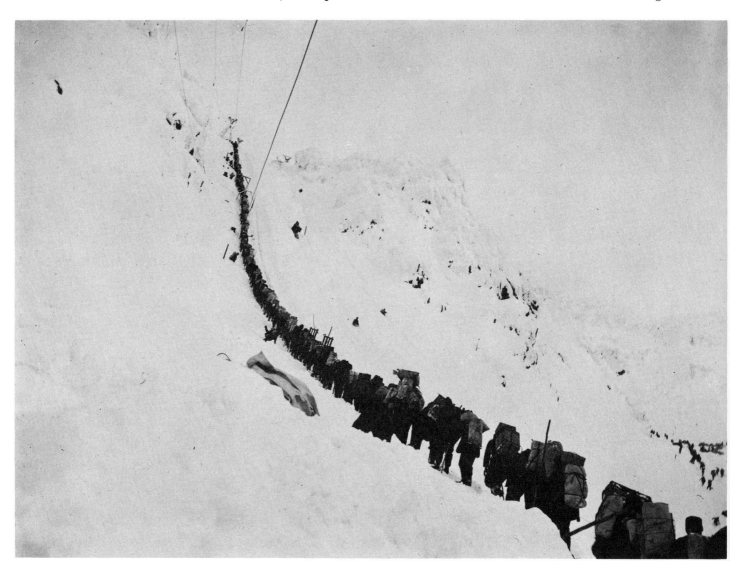

Climbing the Chilkoot Pass

many of them filled with jewelry and trinkets and framed pictures." Also thrown away were dozens of pairs of heavy rubber boots. Five miles out of Dyea was Finnegan's Point, a collection of tents, the inevitable saloon, a blacksmith shop, an eating place. At this point the trail entered a canyon for two miles, at the far end of which was Pleasant Camp; it then climbed to Sheep Camp,

mare of shale and granite almost as slick as grease. The Scales was at the bottom, where the packers weighed their packs. The climb up was a steep 30 percent; in the worst spots a man had to go on hands and knees, at least at certain times of the year, and a horse could ascend only if pulled forward by a rope, to keep it from tumbling or slipping backwards. Across the Chilkoot Glacier horses were blindfolded and dogs were carried. At the top was a fairly level area of snow and ice, where vast piles of luggage were temporarily stored; and beyond it was another glacier, down which packs were scooted on canvas. At the bottom was Crater Lake, across which it cost a cent a on a 35-degree incline, with only two resting places in the next four miles, the first under a ledge known as Stone House, and The Scales, where the rate for human packers jumped to a dollar a pound. Beasts with packs could not climb beyond that point. "Thousands of tons of outfits, half-

Another view

pound to ferry goods; and after four miles it was Long Lake, and another cent a pound; and then a portage of several hundred yards to Deep Lake, which was a mile across. Packs were then carried overland to Lake Lindemann, and from there a trail led to White Pass. From Dyea to Lindemann was about twenty-seven miles.

Berton says the trail from Sheep Camp climbed hidden by the ceaselessly falling snow, were piled here, waiting for their owners to gather stamina for the supreme effort of the last climb. Most of the men in that line were greenhorns, dressed in wool and fur and sweating and freezing by turns. Unable to take their sweat-soaked garments off, much less bathe, stung all day by the driving winds," and full of cold beans and soggy flapjacks, they were often doubled over by belly cramps and paralyzed by dysentery; but "filthy, stinking, red-eyed, and bone-weary" they still moved slowly up and up, in the Chilkoot lock-step. Most of the trail to the summit was so narrow that there was no getting out, once in; the men climbed almost back to front, the slowest setting the pace. Accident or severe illness or sudden violent storm might force the entire line to stop. It took the average of these men not one journey or two or three, but, says Berton, an average of three months to get their baggage over the long trail. The stampede now was a slow crawl and for many of them a complete collapse. If a man for dysentery or other reasons had to leave the single file he might not be able to squeeze back in all day long. It took them about six hours to climb a thousand feet with only a fifty-pound pack. What a scene it was! What a test of men! Berton tells of an Indian packer who reached the summit with a 350-pound barrel on his back, an incredible tale; and of a Swede who "crawled up on his hands and knees with three huge six-by-

four timbers strapped to him." An Iowa farm boy on a wager carried a 125-pound plow up the slope, though heaven alone knows what he thought he would do with it. A Swede named Anderson and a Siwash Indian packer called Jumbo, driven forward by a wager, climbed from The Scales to the summit with an unbelievable three hundred pounds. Berton grew up in that land and these are his stories.

When a day of toil was done most of the men were too tired to cook their flapjacks and beans thoroughly, often too tired even to try to make a fire. Tempers were short. Sickness was common. Says Quiett,

What a terrible trail it was, and how those who went over it labored and suffered! There were fallen trees and giant boulders under the snow for men to stumble over. Deep crevices endangered the lives of those who left the beaten track. When the snow had melted, men had to crawl on their hands and knees over the rocks. Roaring streams of water rushed along 500 feet straight below, and if ever a pack was dropped over the side of the trail, it could not be recovered. . . .

Summit of Chilkoot Pass

Summit of White Pass

Once at the top of the Chilkoot Pass, the Chee-chakos [Indian name for the tenderfeet] had only begun their perilous journey.

In certain times in winter a man might crash through and sink to his shoulders in soft slush; in summer he might drop through what looked like safe ice and land on rocks. Avalanches were a constant threat; April 27, 1898, at The Scales, seventy sleeping people (some say seventy-two) were buried and all but ten perished. And among the hazards were the gamblers. During periods of bitter cold on the last steep ascent about fifteen hundred steps were cut into the ice, with guide ropes and a toll charge. Every little way there was a much wider step, and here gamblers some-times rested and tried to lure the unwary into shell games.

The ascent, says Rickard,

of heavily laden men along a trail through the snow would have been difficult and tiresome at any time, but when it was made by a frenzied mob of wayfarers treading on each other's heels it be-came so difficult for some individuals as to be tragic. If the unholy pilgrimage to Dawson was marked by the horrors of privation and death, it was due largely to the inexperience of most of the pilgrims. Among them were old men and

immature boys; also women, both old and young. . . . Most of the men, and women, that took the trail were unfamiliar with snow and with moun-tain-climbing. People from the cities, unused to open-air life, unaccustomed to carrying loads, wholly ignorant of how to take care of themselves, in a frenzy to reach Eldorado.

And what did they have in their luggage? Fresh eggs, Berton says; whisky (of course), and since this had to be smuggled past Canadian customs officials the merchants in San Francisco contrived ingenious containers, such as a kerosene tin, out of whose spigot came coal oil, sure enough, but in whose belly was whisky. Berton says kegs of liquor were hidden in bales of hay, but these the officials found by probing with a steel needle on an iron rod. "Whisky and silk, . . . pianos, live chickens and stuffed turkeys, timber and glassware, . . . all went over on men's backs."

At the top of the pass the first winter there were acres of piled goods, snowed over and buried, with dozens of poles sticking up to mark each separate pile. Seventy feet of snow fell on the summit that winter and most of the goods were still there until the snow melted. When the skies cleared a man on the summit could look back down toward Dyea, or ahead and down, and see the unbroken line of men like a dark ribbon or rope up over the summit; but during most of the winter season the skies were dark as snow fell, and visibility was cut to a few feet. It is said that as much as six feet of snow fell in a day and a night, and once the storm never opened up for two months. Nevertheless the Canadian police stuck to their posts. They checked 22,000 men "across the pass that winter"—and a few women—"stocky soubrettes heading for the dance halls of Dawson . . . Indian girls who carried seventy-five pounds on their backs"—a German woman almost seventy; a middle-aged woman, apparently alone, who pulled a sled with a stove on it, so that she could have hot meals. Arizona Charlie Meadows was there with a portable bar; the price of a drink of whisky rose with the trail—two-bits at Canyon, four-bits at Sheep Camp, and six-bits at The Scales. "There were seldom fewer than fifteen thousand people in Sheep Camp," with tents so close together it was almost impossible to walk among them. In a few makeshift sheds "a man could buy a woman for five dollars or a meal of bacon, beans, and tea for two-fifty." There were fifteen huts, all called hotels, the best of them known as the Palmer House, named for a Wisconsin farmer who had paused there with his wife and seven children and gone into business.

"He and his family fed five hundred people a day and slept forty of them each night, jamming them so tightly together on the plank floor that it was impossible to walk through the building after nine in the evening."

If Indian girls carried seventy-five pounds on their backs up the trail, their men did twice as well. A Chilkoot Indian packer had a wide band of cloth from the pack around his shoulders and a second band across his forehead; and he carried from a hundred to a hundred and fifty pounds, while his wife trudging behind him carried seventy-five, and his children, if he had any, carried from twenty-five to sixty each. At the beginning they charged five cents a pound but it eventually climbed to forty cents. The "early tenderfeet had their first encounter with the Alaska Indians when they tried to engage them to help carry their supplies over the trail." The packers were a capricious and unreliable lot, with "no understanding regarding the sanctity of a contract . . . they were dirty and most unattractive; some had their faces painted black and rubbed with balsam and grease, and, as William B. Haskell says, their language was 'a confusion of gutturals with a plentitude of saliva—a moist language with a gurgle that approaches a gargle.' "

The treatment of horses is the most shocking part of this gold rush. Driven by greed and des-

A miner's shack

peration the men treated their helpless dumb brutes with a ruthlessness that brought them down to the level of their animals and far below it. "Pack animals," says Berton, "were so scarce that even the poor ones sold for six or seven hundred dollars. And although the cost of their feed ran as high as one hundred and fifty dollars a ton, each animal could earn forty dollars daily in packing fees before he collapsed." On the trails

projecting ledges, would slip or be knocked off, and would fall five hundred or a thousand feet into the gorge below.

"The desert Sahara," wrote Hellprin in the London *Times,* after he had climbed the Chilkoot, "with its line of skeletons, can boast no such exhibition of carcasses. Long before Bennet [sic] was reached I had taken count of more than a thousand unfortunates whose bodies now made part

Dead Horse Trail

where horses were used the melting snows formed hundreds of mudholes, into which a horse or an ox would sink almost out of sight, or even now and then out of sight. Such beasts were abandoned; they were not even shot but were left to die. In some cases they were actually there in the holes, alive, as men and other beasts walked over them, using them for corduroy. One man was so enraged that he built a fire under his ox and burned it alive. Some ascents, such as Devil's Hill, Porcupine Hill, and Summit Hill were so dangerous that unwary horses not carefully choosing their footing, or who struck their packs against

of the trail. . . . The poor beasts succumbed not so much to the hardships of the trail as to the lack of care and the inhuman treatment which they received at the hands of their owners." "Every man we meet," Adney writes,

tells of the trials of the trail. . . . I saw one halfway up a hill asleep on his pack, with his closed eyes towards the sky and the rain pattering on his face, which was as pale as death. . . . A little way on three horses lie dead, two of them half buried in the black quagmire, and the horses step over their bodies, without a look, and painfully struggle on. . . . The cruelty to horses is past belief; yet it is nothing to the Skagway trail, we hear.

There are three thousand horses on the Skagway trail—more to kill, that's about all the difference. Sheep camp is filling up with broken-down brutes. . . . Their backs are raw from wet and wrinkled blankets, their legs cut and bruised on the rock, and they are as thin as snakes and starving to death.

Greever says that "Probably three thousand horses and mules were killed on the White Pass route." We have found no estimate of the total killed on all the trails. Captain Goddard says that "I could walk on dead horses for a solid mile, and I took pictures of the awful condition of the dead on this trail that were so terrible that I have never made a print from the plates." That is too bad, for the inhumanity of man to man and of man to beast should be fully documented.

Dogs were another matter, probably because they fawn on their masters. "They were more than dogs; they were pals, and the unwritten law of the Yukon was, 'Never refuse a drink. Never kick a dog.'" When their feet became sore from traveling over the ice their masters made little moccasins for them, of canvas or hide; and his cry to them when he wanted them to move on was "Mush!" from the French *marchez*. This became the common term shouted at anyone in the Klondike who was urged to move; and "Mush on" was a term of encouragement to those who were dispirited and about to give up. Many writers have mentioned the silence of the northern lands—one Klondiker said it took him three years after he returned home to get accustomed to city noises.

Greever gives a story told by R. M. Crawford about dogs and silence. In Dawson and Skagway there was never any wild shooting up and down the streets, as in cow towns and mining camps in the States, by men who thought to show off their heroism by frightening defenseless people. At midnight July 3, 1897, some men fired revolvers to welcome in the Fourth. Then other men began to fire their guns, and soon all over Dawson fireworks were exploding. The dogs, "who had lived in the silent North all their lives and had never heard any loud noises, were so frightened that they broke their leashes and never stopped running until they jumped in the river. Hundreds swam to an island five miles away, which was afterwards known as Dog Island."

Did the 10 percent prosper who made it over the passes? A few of them did. A few became rich, including the Scot Alex MacDonald, who

The legend says: George Dean, who died seven hours before rescue party arrived. John F. Hueston was on the point of dying. Thiery still had some strength and no doubt would have lasted a week longer. The portion cut out of Dean's thigh was in the pot cooking when rescuers arrived.

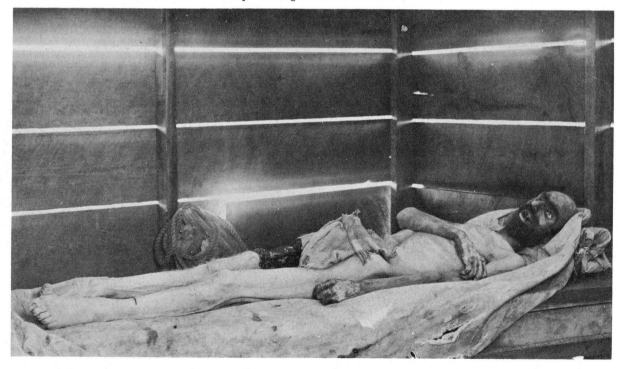

would be known as King of the Klondike, yet who was working for ordinary wages when the news first reached him. The common miner in some areas found his labor even more exhausting than climbing the Pass, for the overlays were sometimes deep, and frozen solid fifteen or twenty feet down; and after they stripped it off and got down fifteen feet they found themselves working in black muck. One advantage they had which those in diggings in the States did not have: they could tunnel back under the ten or fifteen feet of frozen overlay, with no risk of being buried by a cave-in.

Most of the Peter Sapps and Littlebury Shoots in the diggings made little more than enough to support themselves. Some, says Greever, saw so much gold dust that it no longer impressed them —at least not the way a broom did, which they could buy for $16 to sweep their cabin, or fresh eggs at $1.50 each, or fresh milk for a dollar a pint. Some enterprising men who never lifted a shovel were more prosperous than those who dug themselves, as John Matson did, into old age and a lonely death miles from the nearest prospector. One man brought a cow up from Seattle, sold her first milking for $40, and her milk thereafter at a dollar a pint on bars in the saloons. Judge Maguire bought a newspaper fresh from Seattle and charged $2.50 a person to hear him read the account of the sinking of the battleship Maine. He rented a hall and gave readings each evening at seven, nine and eleven. That was a lot easier than taking out an ounce of gold while toiling ten or twelve hours deep in a black hole in zero weather.

Dawson by the light of the Aurora Borealis

PART II

Life in the Camps

The Camps and Their People

There were more mining camps than most people seem to realize. According to the Geological Survey's *Mineral Resources of the United States, 1907,* there were, in 1866, over 500 organized mining districts in California, over 200 in Nevada, and more than 100 each in Idaho, Oregon, and Arizona. There were about 50 each in Montana, New Mexico, and Colorado—or, as Shinn says, "a total of 1100 camps in the Far West." They weren't all a "howling wonder of the western world," which is what a writer has called Tombstone, but they were pretty lusty places.

After returning East, says Harlow, a stage driver was interviewed. He was asked if Deadwood was as bad as it was said to be. "Worse," he said. "I don't know what the papers say about it, but I know it's worse'n any language kin tell. It's the oneryest place this side of hell. There's no law and no Sunday. Every man's his own court, and his revolver is his lawyer, judge, and jury, and executioner—'specially executioner. The gamblin', drinkin' and fightin' goes on all the time, day and night. You wouldn't know when Sunday comes around if you didn't put it down in a book."

As for robbing of the stages, that was a

reg'lar business, and is done systematic. It's done this way. The clerks in the banks and offices and stores in Deadwood and Custer can easy find out when there's money to go in the stage, and they notify the thieves. In with 'em, you see—have reg'lar partnerships. Astonishes you, does it? Well, it'll surprise you still more when I say I've even known the agents of the stage company to help rob the box. . . . The thieves are always posted, and never make a mistake and stop a coach without money in the box.

A bronze tablet on the main street of Columbia, California, tells the curious visitor that gold was discovered there in March, 1850, one thousand feet south of that point; and that a month later

the population was eight thousand. It is said that Columbia had everything that a bonanza camp should have—"thirty saloons, one hundred and forty-three gambling layouts, fifty-three stores, four hotels, four banks, two theaters, and three express offices." With a good boot-hill graveyard that's about right. One of the first candidates for "boot hill" was a barroom bum who, in a duel with a gambler, was shot through the head. Though "the brains exuded and considerable proportions of it were spattered on the ground," to the "amazement of all, the man recovered, and in three weeks was walking the streets. There were some tough hombres in those hills."

In a camp, as Mark Twain tells it (he probably had Virginia City in mind),

There were military companies, fire companies, brass bands, banks, hotels, theaters, hurdy-gurdy houses, wide open gambling palaces, political pow-wows, civic processions, street fights, murders, inquests, riots, a whisky mill every fifteen steps, a Board of Aldermen, a Mayor, a City Surveyor, a City Engineer, a Chief of the Fire Department with First, Second and Third Assistants, a Chief of Police, City Marshal, and a large police force, . . . a dozen breweries and half a dozen jails and station houses in full operation, and some talk of building a church.

A writer has pointed out that it was a habit with some newspapers to say that this camp or that was the most hell-roaring bonanza of them all. Of Nevada City on Alder Gulch, a Montana paper said, for instance, that its "turbulent and amazing life, when thousands of seething, struggling and money-mad human beings of every nationality and country on earth, swarmed its single street by day and night, was probably without parallel on the continent." Newspapers all over the West would have challenged that claim. Dimsdale, who knew Nevada City, Bannack, and Virginia City as well as he knew his own newspaper, put it this way:

Let the reader suppose that the police of New York were withdrawn for twelve months, and then let them picture the wild saturnalia which would take the place of the order that reigns there now. If, then, it is so hard to restrain the dangerous classes of old and settled communities, what must be the difficulty of the task, when, tenfold in number, fearless in character, generally well armed, and supplied with money to an extent unknown among their equals in the East, such men find themselves removed from the restraints of civilized society . . .?

Of Lusty Leadville Caroline Bancroft says, "The town's character was typical of any early-day mining camp—ramshackle, rough, dirty, boisterous and devil-may-care, but remarkably honest. There was almost no law or order except that of mutual consent arrived at by the founders of the mining district." In 1879 a newspaper said Leadville had "19 hotels, 41 lodging houses, 82 drinking saloons, 38 restaurants, 13 wholesale liquor houses . . . 12 blacksmith shops, 6 livery stables, 3 undertakers, 21 gambling houses, 4 theatres, 4 dance halls, and 35 houses of prostitution."

Some who saw the camps saw in them nothing but evil. Helper, in 1855, said of the California camps that there was absolutely no good in them —the country was a cultural desert, the manners of the people were beyond description, there were murders around the clock, the sick and dying lay where they fell, uncared-for, and losses by fire, flood, and shipping were fantastic. Shinn, who knew the life in the California camps and presents all sides of it, felt that Parsons in his life of Marshall did not exaggerate when he wrote:

It was a mad, furious race for wealth, in which men lost their identity almost, and toiled and wrestled, and lived a fierce, riotous, wearing, fearfully excited life; forgetting home and kindred; abandoning old, steady habits; acquiring all the restlessness, craving for stimulant, unscrupulousness, hardihood, impulsive generosity, and lavish ways, which have puzzled the students of human nature who have undertaken to portray or to analyze that extraordinary period.

A true account of it all, Parsons said, would be "so wild, so incredible, so feverish and abnormal" as to resemble a Walpurgis Night. All mixed together you had

shrewd, energetic New England business-men . . . rollicking sailors . . . Australian convicts and cutthroats . . . Mexican and frontier desperadoes . . . hardy backwoodsmen . . . professional gamblers, whiskey-dealers, general swindlers . . . brokendown merchants, disappointed lovers, black sheep . . . professional miners from all parts of the world . . . stir up the mixture, strongly season with gold-fever, bad liquors, faro, monte, rougeet-noir, quarrels, oaths, pistols, knives, dancing, and digging, and you have something approximating to California society in early days.

Shinn approves what Parsons says, and throws in some observations of his own:

The typical camp of the gold prime of '49 was flush, reckless, flourishing, and vigorous. Saloons and gambling-houses abounded; buildings and whole streets grew up like mushrooms, almost in a night. Every man carried a buck-skin bag of golddust, and it was received as currency at a dollar a pinch. Everyone went armed, and felt fully able to protect himself. A stormy life ebbed and flowed through the town. In the camp, gathered as one household, under no law but that of their own making, were men from North, South, East, and West, and from nearly every country of Europe, Asia, South America.

We think there may be a little exaggeration in some of that—not all the men were armed, not all of them felt able to protect themselves, not all of them carried a bag of gold dust; but we see no exaggeration in what he says next:

But the vital waste and destruction of the struggle overwhelm the observer with pity. The mines were no place for weaklings; even strong, healthy, and earnest men often sank beneath the wearing excitement of that fervid life. But these rough and busy men, whose lives were so often only the saddest of tragedies, established and enforced a code of ethics . . . and in the end created a system of jurisprudence that has won the approval and endorsement of the highest courts of the land. Sturdy, keen-witted, courageous, independent, they left the impress of sterling natures upon all their primitive institutions.

It appears that many of the miners developed a genuine love affair with their camp, and for them Goodman's verses on Virginia City (Nevada) are the classic of nostalgia:

In youth, when I did love, did love
   (To quote the sexton's homely ditty),
I live six thousand feet above
Sea level, in Virginia City;

MINER SICK.

MINER'S DREAM.

MINER'S SLUMBERS.

FRIENDS IN COUNTRY.

MINER COOKING.

LETTERS FROM HOME.

WASHING DAY.

MINER'S CLAIM.

MINER'S CABIN.

MINER'S EVENING.

FRIENDS IN CITY.

SATURDAY NIGHT.

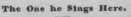

THE HONEST MINER'S SONGS.

The One he Sung at Home.

TUNE—*Susannah.*

Like Argos of the ancient times,
  I'll leave this modern Greece;
I'm bound to California Mines,
  To find the golden fleece.
For who would work from morn till night
  And live on hog and corn,
When one can pick up there at sight
  Enough to buy a farm?
    CHORUS.— Oh California! that's the
      land for me,
  I'm going to California the gold dust for
    to see.

There from the snowy mountains side
  Comes down the golden sand,
And spreads a carpet far and wide
  O'er all the shining land;
The rivers run on golden beds,
  Oe'r rocks of golden ore,
  The valleys six feet deep are said
To hold a plenty more.
    Oh California! &c.

I'll take my wash bowl in my hand,
And thither wind my way,
  To wash the gold from out the sand
In California.
  And when I get my pocket full
In that bright land of gold,
  I'll have a rich and happy time:
Live merry till I'm old.
    Oh California! etc.

The One he Sings Here.

TUNE—*Irish Emigrant's Lament.*

I'm sitting on a big quartz rock,
  Where gold is said to grow;
I'm thinking of the merry flock,
  That I left long ago:
My fare is hard, so is my bed,
  My CLAIM is giving out,
I've worked until I'm almost dead,
  And soon I shall "peg" out.

I'm thinking of the better days,
  Before I left my home;
Before my brain with gold was crazed,
  And I began to roam.
Those were the days, no more are seen,
  When all the girls loved me;
When I did dress in linen clean,
  They washed and cooked for me.

But awful change is this to tell,
  I wash and cook myself;
I never more shall cut a swell,
  But here must dig for pelf.
I ne'er shall lie in clean white sheets,
  But in my blankets roll;
An oh! the girls I thought so sweet,
  They think me but a fool.

The site was bleak, the houses small,
The narrow streets unpaved and slanting,
But now it seems to me of all
The spots on earth the most enchanting.*

Some of the men, like Peter Sapp, who had the souls of minor poets, loved certain things in the camps—and they were in many of them. They made up a kind of litany of names—Hoodoo Bar, Cut Throat, Randy Doddler, Nigger Slide, Fiddle Creek, Jackass Hill, Shirttail Diggins, Bloody Run, Bladderville, Gouge Eye, Humbug Creek, Red Dog, Tenderfoot Gulch, Lost Horse Gulch, Whisky Gulch, Gulch of Gold, Mad Mule Gulch.
. . .

It must have been in Golden Valley or on one of the Gold Hills, they liked to think, that a drunk staggered from a saloon in pitch-dark night and went weaving homeward. In the dark he stumbled and fell, and his hands, feeling around, told him that he was on the brink of a mine shaft

---

* Walter Van Tilburg Clark, who has the eye of a poet, has recently purchased in Virginia City a house almost a hundred years old, only a few steps from the camp's famous church.

or a precipice, and that only a small outthrust of rock, that seemed ready to fall if he dared move, was between him and a plunge to death. All night he clung there, trembling, praying, thinking of the folks back home, and offering his terrified soul to God; all night his paralyzed hands clutched the stone and not once did he move; and when daylight came and he was able to see, there was below him an awful chasm two feet deep and about a foot across——

It would have been a litany of golden things if Peter and the others could have known them all—Goldfield, Goldenville, Golden Valley, Golden Star, Golden Slipper, Golden Aster, Golden Mile, Golden Messenger, Golden Curry, Gold Stream, Gold Road, Gold Hill—oh, at least a half-dozen Gold Hills—and Jesus Maria, Loafer Hill, Slapjack Bar, Bogus Thunder, Ground-hog Glory, Golden Chariot and Hell's Delight. Of all the famous mines Peter and his friends would know only the names—Potosi, Oro Fino, Poorman, Idaho-Maryland, the Argonaut, the Carson Hill, the North Star, the Morning Star, the Consolidated—the Ophir and Sun and Moon and Fourth of July, the Caribou and the Golden Chariot. . . . But while the names danced in his mind Peter lay on a crude bunk back in a cave, on a mattress of juniper boughs, the four pine legs of the bunk standing in four cans of water, to keep the insects from swarming over him during the night, and out of his mouth, as he lay on his back and snored.

Since most of the camps were built in gulches and on steep slopes the buildings and shacks were in tier on tier above or below one another, so that a man couldn't stand on his own doorstep and "spit tobacco juice without putting out the fire in his neighbor's chimney."

Borthwick arrived in Hangtown to find the one long street

in many places knee-deep in mud, . . . plentifully strewed with old boots, hats, and shirts, old sardine-boxes, empty tins of preserved oysters, empty bottles, worn-out pots and kettles, old ham-bones, broken picks and shovels, and other rubbish too various to particularize. Here and there, in the middle of the street, was a square hole about six feet deep, in which one miner was digging, while another was bailing the water out with a bucket, and a third, sitting alongside the heap of dirt which had been dug up, was washing it in a rocker.

In a Colorado camp everything

was strung along through the gulch. . . . There were a great many tents in the road and on the side of the ridge, and the wagons were backed up, the people living right in the wagons. Some of them were used as hotels; they had their grub under the wagons, piled the dishes there, and the man of the house and his wife would sleep in the wagons nights. They would get some rough boards and make tables where the boarders took their meals, and those who did not want to board did their own cooking. The gamblers would have tables strung along the wayside to take in the . . . unwary miner.

Traffic in the camps had no right-of-way but took either side or the middle of the street, as it pleased. There were no fixed places for parking. The streets in most of them were hideously cluttered, dusty, often filthy; and as likely as not livery stables opened on the main street and manure was pitched into it, to accumulate with that of the thousands of mules, horses, and oxen that freighted up and down it almost every day. It was not unusual for dead beasts to lie in the streets until they stank, and in some of them hogs ran loose to root where they pleased. The odors were sometimes more than the fastidious visitor could stand—a blend of livery stable, dead beast, chicken feathers, unclean dogs and cats, and unwashed humanity, for bathing was not even a weekly habit with most of the miners. Usually the camp's newspaper kept its readers informed of the major troubles in the streets, such as runaway-teams, with terrified passengers spilling "out behind." The Montana *Post,* for instance, gave news of one wild coach that went off the road and plunged into a deep gulch; while in "our own town, a span of horses, with half a wagon behind them came dashing through Jackson street," tearing down hitching posts, a store sign, and smashing a crate of glass.

The larger towns, such as San Francisco and Denver, were no exceptions. Richardson found Denver

society . . . a strange medley. There were Americans from every quarter of the Union, Mexicans, Indians, halfbreeds, trappers, speculators, gamblers, desperadoes, broken-down politicians and honest men. Almost every day was enlivened by its little shooting match. While the great gambling saloon was crowded with people, drunken ruffians sometimes fired five or six shots from their revolvers, frightening everybody pell-mell out of the room but seldom wounding any one. One day I heard the barkeep politely ask a man lying upon a bench to remove. The recumbent replied to the request with his revolver. Indeed firing at this bar-tender was a common amusement among the guests. At first he bore it laughingly, but one day a shot grazed his ear, whereupon . . . he buckled on two revolvers and swore he would kill the next man who took aim at him. He was not troubled afterward.

If it was that simple one wonders why he took so much of it.

A number of persons have told us, in the preceding pages, that there were *all kinds* of people in the camps. We are going to take a close look at them, and sort them out, as it were, into recognizable categories or types.

M. H. Hayne of the Mounted Police in the Klondike thought "There is certainly nothing heroic about the ordinary miner; he risks his life for no scientific enterprise; he faces danger to conquer no foe; he desires neither notoriety nor glory; he merely seeks to fill his own pockets." He was trying, of course, to fill more than his pockets.

There is a whale of a difference between the loss of individuality in conformity, as we see it today (*Why Conform?* asks a recent title), and life in the camps. As Borthwick saw it, "A man's actions and conduct were totally unrestrained by the ordinary conventionalities of civilized life, and, so long as he did not interfere with the rights of others, he could follow his own course, for good or for evil, with utmost freedom." It was the unrestrained course toward evil that led to the vigilantes.

For Dimsdale, a newspaper editor,

The one great blessing is perfect freedom. Untrammeled by the artificial restraints of more highly organized society, character develops itself so fully and so truly that a man who has a friend knows it and there is a warmth and depth in the attachment which unites the dwellers in the wilderness that is worth years of the insipid and uncertain regard of so-called polite circles, which, too often, passes by the name of friendship. . . . Those who have slept at the same watch-fire and traversed together many a weary league, sharing hardships and privations, are drawn together by ties which civilization wots not of.

Two miners before their fire, in the Sierras

Many visitors were impressed by the unaffected hospitality. Said the Reverend Walter Colton:

I have never been in a community that rivals Monterey in its spirit of hospitality and generous regard. Such is the welcome to the privileges of the private hearth, that a public hotel has never been able to maintain itself. You are not expected to wait for a particular invitation, but to come without the slightest ceremony, make yourself entirely home, and tarry as long as it suits your inclination, be it for a day or for a month. You create no flutter in the family, awaken no apologies, and are greeted every morning with the same bright smile. It is not a smile which flits over the countenance, and passes away like a flake of moonlight over a marble tablet. It is the steady sunshine of the soul within.

That is probably a fair definition of what is meant by the *spirit* of the old West.

Nine months later he wrote in his journal:

Their hospitality knows no bounds; they are always glad to see you, come when you may; take a pleasure in entertaining you while you remain; and only regret that your business calls you away. . . . If I must be cast in sickness or destitution on the care of the stranger, let it be in California; but let it before American avarice has hardened the heart and made a god of gold.

And still later he was writing:

There is no need of an Orphan Asylum in California. The amiable and benevolent spirit of the people hovers like a shield over the helpless. The question is not, who shall be burdened with the care of an orphan, but who shall have the privilege of rearing it. . . . A plain, industrious man, of rather limited means, applied to me to-day for the care of six orphan children. I asked him how many he had of his own; he said fourteen as yet. "Well, my friend," I observed, "are not fourteen enough for one table, and especially with the prospect of more?" "Ah," said the Californian, "the hen that has twenty chickens scratches no harder than the hen that has one."

It may be true that Colton was a bit too impressionable, or that he was an unusually thoughtful and thankful guest, or that the people of Monterey were far above average. But Bancroft saw the people there and elsewhere and thought them

Simple, honest, earnest, they affected nothing, and in the direction of self-government, attempted nothing which they failed to accomplish. Here was a people who might give Solon or Justinian a

lesson in the method of executing justice . . . the California miners . . . have shown to the world more than any other people who ever lived how civilized men may live without law at all.

Like Colton, he overstates it.

Taylor reports that among

the number of miners scattered through the different gulches, I met daily with men of education and intelligence, from all parts of the United States. It was never safe to presume on a person's character, from his dress or appearance. A rough, dirty, sunburnt fellow, with unshorn beard, quarrying away for life at the bottom of some rocky hole, might be a graduate of one of the first colleges in the country, and a man of genuine refinement and taste. I found plenty of men who were not outwardly distinguishable from the inveterate trapper or mountaineer, but who, a year before, had been patientless physicians, briefless lawyers and half-starved editors.

It has been said that Bayard Taylor had a tendency like Bret Harte to glamorize the miner. That is not true of Trimble, who says in his highly praised book that

for the observer wishing to be impartial, a great deal depends upon one's point of view. If ·he undertakes to apply to mining communities the conventional standards of conduct which ruled in the sixties in quiet villages of the East, he will find sufficient transgressions to shock him. . . . the impartial student, without in the least denying or seeking to palliate what was ugly, will not overlook essential traits of manhood, but will remember that most of the mining populace were young men, far from the restraints of home; that they had come, many of them, from the less exhilarating atmosphere of lower altitudes, . . .

The average of altitudes was hardly lower in the California camps. Richardson, like Colton and some others, found that the "people are warm and demonstrative. One of them going back to the East is surprised at the general coldness and formality. He fancies that his old friends have never thawed out from the freezing their fathers got on Plymouth rock." He thought that the rugged frontier life took the nonsense out of people. "It teaches practical sagacity, rare judgment of men, quick detection of shams, ready weighing of a stranger's capacity, and generous trust in the trustworthy." He learned that in miners' talk a bilk was an impostor; to slop over was to make foolish errors.

A typical barber shop

But for the editor of the San Francisco *Alta* the miners were simply men who hoped to get rich.

Bright visions of big lumps of gold and large quantities of them, to be gathered without any severe labor, haunt them night and day. . . . hope to find a land where the inevitable law of God that "man shall earn his bread by the sweat of his brow" has been repealed, or at least for a time suspended. They come here with this hope, and it takes but a few short weeks to dispel it. . . . Temptations are about them on every hand. They drink and they gamble. They associate with men who, in their eastern homes, would be shunned by them as the worst of their kind. . . . they sink lower and lower, until they become thieves, robbers and desperadoes.

So now we have two views of it, and possibly both are a bit extreme. "I confess, without shame," Mark Twain wrote,

that I expected to find masses of silver lying all about the ground. I expected to see it glittering in the sun on the mountain summits. I said nothing about this, for some instinct told me that I might possibly have an exaggerated idea about it.

. . . Yet I was as perfectly satisfied in my own mind as I could be of anything, that I was going to gather up, in a day or two, or at furthest a week or two, silver enough to make me satisfactorily wealthy—and so my fancy was already busy with plans for spending this money.*

What did the miners look like? Nearly all of them wore beards. Henry Page wrote his wife that not more than one man in a hundred had a shaven face, though some of them cut a part of the hair away from their mouths so that they could more easily feed themselves. Some early drawings show them in the evening, sitting around a miner's lamp, made by filling a bowl with sand and pouring in melted bacon or other grease, with the wick standing in the sand. "A razor," Page wrote, "has not been used by any of us since we left St. Joseph, and with the help of shears to keep the road open to our mouths, and a little trimming to keep our faces a little ship-shape, we are at the top of the fashion. . . . The truth is, something looks wrong if I see a white shirt or a shaven face——"

———

* Mark Twain was, of course, on the Comstock, where the chief metal was silver.

In heavy boots, flannel shirt, and trousers saturated with the muck of bar or hillside—never shaved, seldom washed, their tangled hair matted and long, they were tough lookers, all right; but Capron, in his history of California, agrees with Taylor that a lot of them came from the "learned professions. Whoever attends one of their miners' meetings will be convinced that neither the fools nor the drones go to the mines." For it was in the mines that the curious visitor could see "the grave divine, the skilful physician, the shrewd lawyer, the professor, the philosopher, the gentleman of leisure, and the student, as well as the farmer, mechanic, and common laborer." Just the same, Hugo Reid, as quoted by Dakin in *A Scotch Paisano,* said, "Don't go to the mines on any account. They are loaded to the muzzle with vagabonds from every quarter of the globe, scoundrels from nowhere, rascals from Oregon, pickpockets from New York, accomplished gentlemen from Europe, interlopers from Lima and Chile, Mexican thieves, gamblers of no particular spot, and assassins manufactured in Hell. . . ." So he wrote a friend.

After one has searched a lot of books written by persons who were there one concludes that just about every type was to be found in the mines, except the delicate aesthete, the hothouse poet, and the loafer with no skill as either robber or gambler—from the one who had had his ears cropped as a youth and wore his hair trained down the sides of his head to try to hide the mutilation, while scowling sullenly and fiercely

round him, and the Kentuckian named Bedbug, who petitioned a legislature for a change of name, saying that if he were to marry he could never endure the sight of little Bedbugs crawling around, to the mysterious melancholy prospector who had never been heard to speak, and the man with the cow who followed him as a dog might, and who spent half his time trying to keep other men from milking her. We shall now look at a few of the more obvious types.

There was the madam* of a whorehouse, middle-aged and fat, with dewlaps under her jaws, who took to her huge bosom, for consolation and profit, a slender, dapper pimp, and gave him a room in her house, where he could sleep till noon every day. When the respectable people of the camp refused to accept him as a businessman, or even to speak to him, she was terribly hurt. It seems probable that she actually could not understand why he was rejected socially. And there was Gold Pan Bessie who mixed her profession of light-loving and charity and made quite a character of herself.

Borthwick singled out as a type the Missourian,

who wore the same clothes in which he had crossed the plains, [and who seemed to be] keeping them to wear on his journey home again. . . . As for their persons, they were mostly long, gaunt, narrow-chested, round-shouldered men, with long, straight, light-colored, dried-up-looking hair, small thin sallow faces, with rather scanty beard and moustache, and small grey sunken eyes, which seemed to be keenly perceptive of everything around them.

_____

* For madams see "Girls of the Line."

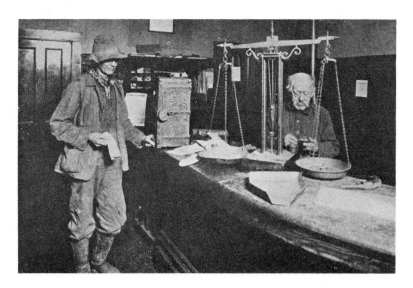

Miner selling his gold dust

In all camps there was the showoff, the person of shallow consciousness eager for attention and the limelight, like one named Brady, of whom the Boise *News* said that he had got out a "poster announcing that he will begin at 10 o'clock on Wednesday next at the Miner's Exchange, to walk '100 consecutive hours without sleep or rest.' Anybody can see him walk for fifty cents." Two weeks later the newspaper reported that he had walked the hundred hours and finished in fine form.

There was in every camp the politician, the man itching for public position so that he could have power over people and promote his fortunes. Many of those elected to public office were misfits or scoundrels; and now and then one seems to have been a lunatic. On March 10, 1866, the *Avalanche* set in black type the words, "WANTED IN IDAHO A governor that is a man, and not an old imbecile as we have now. . . . He is in Boise City raving at those who choose to differ with him. . . . We don't want any importations, but men of I.T. . . .* We want *something* or a simonpure vacuum or nonentity, not a living nothing."

There was the professional killer. In the Southwest, as Rickard points out, the men in the sixties were all outlaws, except a few ranchers. They were, says Pumpelly in his *Reminiscences,* "refugees from the vengeance of the San Francisco Vigilance Committee . . . and there were escaped convicts from Australia." Murder, he says, "was the order of the day; it was committed by Americans upon Americans, Mexicans, and Indians; by Mexicans upon Americans; and the hand of the Apache was, not without reason, against both of the intruding races." In the camps, Woods says, the miner who intended to get drunk and pick a fight put on a red flannel shirt, and was usually called Red. Possibly it was a couple of Reds reported by the *Statesman*: "Quite a little fight took place on Main street this afternoon between Tom and Jessy. Tom knocked Jessy down, gave him two beautiful black eyes, and Jessy bit Tom's finger nearly off."

Every bonanza camp had an immense horde of parasites whose only purpose in the area was to prey on the unwary and the weak. They included professional gamblers, bunco steerers, decoy ducks, spielers, cappers, shills, thimbleriggers, sneak thieves, forgers, brawlers, blacklegs, swind-

lers, arsonists, cutthroats, bandits, and pimps, who were almost the least offensive of the lot. It is said that in some areas a pimp was called Mac or Mack. "Hey, Mac," must have been an insult equivalent to a slap across the cheek.

Bret Harte tried to glorify the gambler. It is true that many of them were smooth well-mannered sharks with the charm of the accomplished psychopath. Since it was their job to lure the unwary to their gaming tables they cultivated an impression of sincerity and friendliness. There were so many kinds of deadbeats plotting to separate the simple miner from his gold dust that the miners sought ways to secure their treasure. And the buyers of dust were also the prey of the crim-

A dude arrives. Leslie's Newspaper, 1879

inals. One of these, Harlow says, "who carried a small trunk for his dust is said to have kept three live rattlesnakes in it as a pleasant surprise for any robber who might open it." We doubt that the hardier robbers paid much attention to rattlesnakes.

---

* Idaho Territory.

There were also dealers in bogus dust. Most of the bogus, McConnell says, was manufactured in San Francisco "by simply cutting bars of lead into small particles which resembled gold dust in size; these were then galvanized with gold by means of a special process, and so perfect was the deception that nitric acid, as usually applied, would not reveal the fraud." But bogus dust and counterfeiting were a minor evil compared to the ruin of thousands of people by the promoters of worthless stock. This was especially vicious on the Comstock and led, of course, to the financial collapse on the Pacific coast, as well as to enormous fortunes for the manipulators, including the four Irishmen, Mackay, Flood, Fair and O'Brien.

But in contrast to the evil life of the parasites were the thousands of fairly honest and hard-working people, who with thrift and long hours tried to accumulate a modest substance. The story Elisha Brooks tells in *A Pioneer Mother of California* is no more than a typical frontier story, as we know who were born and reared there. His family had a few cows in the area of Bidwell's Bar, and because the father was an invalid the twins Elisha and Elijah did the work. They left their beds at 3:00 A.M. in the summer, at 4:00 in the winter, to go into the dark and milk the cows; and then with cans of milk behind their saddles they delivered to people up and down the river, getting back home around nine in the morning. That was about the hour when some of the pimps and shills and sneak thieves were going to bed.

In every camp there were eccentric persons who today would be known as "characters" and who in those times took the place, in a small way at least, of television. We shall now look at a few of the outstanding ones, beginning with Borthwick's account of The Flying Dutchman.

Borthwick was sketching one day when

I became aware that a most ragamuffinish individual was looking over my shoulder. He was certainly, without exception, the most tattered and torn man I ever saw in my life; even his hair and beard gave the idea of rags, which was fully realized by his costume. He was a complete caricature of an old miner, and quite a picture by himself, seen from any point of view.
The rim of his old brown hat seemed ready to drop down on his shoulders at a moment's notice, and the sides, having dissolved all connection with the crown, presented at the top a jagged circumference, festooned here and there with locks of light brown hair, while, to keep the whole fabric from falling to pieces of its own weight, it was bound round with a piece of string in lieu of a hat-band. His hair hung all over his shoulders in large straight flat locks, just as if a handkerchief had been nailed to the top of his head and then torn into shreds, and a long beard of the same pattern fringed a face as brown as a mahogany table. His shirt had once been red flannel—of course it was flannel yet, what remained of it—but it was in a most dilapidated condition. Halfway down to his elbows hung some shreds, which led to the belief that at one time he had possessed a pair of sleeves; . . . There was enough of his shirt left almost to meet a pair of—not trousers. . . . He must have had trousers once on a time, but I suppose he had worn them out; and I could not help thinking what extraordinary things they must have been on the morning when he came to the conclusion that they were not good enough to wear. . . . His boots . . . were more holes than leather. . . .

He was a man between thirty and forty, and, notwithstanding his rags, there was nothing in his appearance at all dirty or repulsive; on the contrary, he had a very handsome, prepossessing face, with an air about him which at once gave the idea that he had been used to polite society.

As indeed he had been; for The Flying Dutchman was in fact a doctor of medicine and spoke a number of languages. His mining crew were two Americans, two Frenchmen, two Italians, and two Mexicans. One day Borthwick watched them putting in a wing dam:

The Americans, the Frenchmen, the Italians, and the Mexicans, were all pulling in different directions at an immense unwieldy log, and bestowing on each other most frightful oaths, though happily in unknown tongues; while the directing genius, the Flying Dutchman, was rushing about among them. . . . He spoke all the modern languages at once, occasionally talking Spanish to a Frenchman, and English to the Italians, then cursing his own stupidity in German, and blowing them all up collectively in a promiscuous jumble of national oaths.

It is Delano in his account of his life in the diggings who tells us about Peter the Hunter. Peter had a French father and a Sioux mother; about fifty-eight years old in 1850-52. "His tall, gaunt, attenuated frame, his restless black eye, his thin, sunken cheeks, attested a life of hardship and exposure, . . ." He was a fine shot and followed the chase for a living. He was married to a Catholic squaw, by whom he had three daughters and a son, and the son was absolutely the apple of his eye. With utter devotion he was teaching him how to shoot and track and hunt and trap, but one day the lad died of an unknown disease. This almost broke Peter's heart. The best he could do then was to teach his daughters, all of them wild lovely things; and they became such expert riders and shots, and were so graceful and fearless, that many men coveted them. Among these men was a small group of Frenchmen who pressed their attentions on the girls and tried to intimidate them, until Peter became afraid of them. Delano told him to pitch his tent close by his, and he and his friends would help him drive the Frenchmen off.

It was in the spring of 1850 that Peter was out riding with his daughters—how the four of them loved to race their fleet ponies over the fragrant hills, gathering flowers or fruits in season, or grouse and wild turkey! High in the hills, when Peter was crawling forward to an eminence, to spy out wild game, he came face to face with the dreaded grizzly. The monster seized him at once, and having only his knife on him he plunged it again and again into the beast, trying to reach its heart. The bear tore Peter's scalp from his head, and with its long claws raked deep furrows down over him; and in the struggle they both fell and rolled some distance down the hill. His girls had come up, and after watching the bloody struggle for a few moments the oldest one stood paralyzed, but the second one leapt to the fight. With the nimbleness of a wolf she leapt round and round the two in the death struggle, until she got a chance to fire at the beast's head and then into one of its ears. The girls then took their father, red all over with his and the bear's blood and half dead, to their tent, and the three of them nursed him for weeks. When Delano last saw them Peter was still refusing offers of a "home" for the girls from white men, for he knew what kind of Christian home they gave to women on whom they looked as squaws. One morning Delano looked over to the spot where Peter's tent had stood. It was gone. He and his daughters were gone, and he never saw them or heard of them again.

A third queer one was named Austin Savage. Savage had been in the California camps but had followed the rush to Nevada and from there to Idaho, where from the first he was hated by many because of his hatred of the Rebels. He became

a recluse and filled his shack with animal pets. "I have never known a man in private life who possessed more power with animals. His horses, cats, and dogs were trained to circus pitch!" Savage was not only frugal; he was "out-and-out stingy. One of his sisters was 'sealed' to Brother Angell, to whom is given the credit for designing the Mormon Tabernacle, the huge inverted soup tureen at Salt Lake." Savage became the internal revenue officer in the Boise area and occupied a building that was a feature of the town—a one-story shack lined with tar paper, 7 feet wide and 12 feet long, with a door and two windows. It stood on a large wagon whose wheels were kept "well greased. . . . Mr. Savage received nine hundred dollars each year as rental from the Government on the assurance that the place was fireproof." When an inspector stopped on his round to have a look at it he was appalled: there were other frame shacks all around it. Savage told him to compose himself and have a cigar.

"The inspector sat down and lighted a cigar. Mr. Savage closed the door and then tugged on a rope which hung overhead. 'I want to say,' continued Mr. Savage, 'that this is the only fireproof building in the territory. In fact, it is the only hygienic structure in all Idaho. You will agree with me in five minutes.' . . . All of a sudden the building began to rock from side to side and then came a wild gallop. The inspector grabbed Mr. Savage, but Savage held him down in his chair. For two minutes everything was riot and confusion. Then came a jolt, and the shanty lay in calm. Mr. Savage laughed in the inspector's white face, opened the door, and said, 'Step outside, sir!' The inspector fairly leaped out. To his surprise, he found himself at the public square, facing a statue of George Washington carved in wood."

Two horses were hitched to the wagon on which stood the building that rented for nine hundred dollars annually, or about five or six thousand in today's dollar. We can imagine what a flabbergasted man the inspector was, for in two minutes of wild ride he had jolted halfway across town. Savage now blandly explained that just around the corner from his customary place was Agnew's livery stable, and that in case of fire all he had to do was yank the rope, and in less than a minute a team was hitched to his wagon and he was rolling away to safety. "Good God!" snorted the inspector. "Couldn't you have said so without giving me heart trouble? I wouldn't take that ride again for a million!" That is what Horace Greeley is reported to have said, after Hank Monk brought him across the Sierras.*

Of the persons who discovered the great mining areas in the West, one of the most eccentric in his manners and appearance was Henry Thomas Paige Comstock, the tall long-legged cadaverous fellow with the loud voice and the load of bluff, who is more commonly known as Old Pancake. He might have become a millionaire if he had had any sense of management; but he managed his mining claims as ineptly and stupidly as he managed his women. This heartrending account of his marriage is based on Dan DeQuille, whose honest name was William Wright, in his book *The Big Bonanza*. Like Borthwick, Delano, and Donaldson, he was himself a member of the mining camp and knew Old Pancake well.

There came to the Nevada diggings a mousy Mormon named Carter, with a wife; and at once Comstock's lonely soul "went out toward her as she sat there in the end of the little canvas-covered wagon, mournfully gazing out from the depths of her calico sunbonnet." Either Old Pancake knew more about the ways of a maid than most men know or she was awfully sick of her husband, for in no time at all he was sitting on the wagon-tongue and she was combing the filth out of his hair. It could be, of course, that he had told her in his booming way that he had a mine worth millions.

The next day he headed over the hills with her to Carson and found a preacher to marry them, for Pancake's opinion of the Mormon marriage ceremony was so low that he ignored it; and he was getting squared around for a honeymoon when the enraged husband appeared on the scene. Possibly because he was only a "little sore-eyed fellow" he was willing to make a deal: he sold all his rights in his wife for a horse, a revolver, and sixty dollars. He was making off with his bargain —and it was a bargain, as the reader will see— when Pancake called him back and said he wanted a bill of sale. He'd had a little experience with sharpers and he didn't want to pay for her more than once.

Before he had time to savor the honeymoon Pancake was summoned to San Francisco on busi-

ness, and leaving his bride at Carson he set forth; but at Sacramento he heard that his bride had taken a powerful fancy to a younger and handsomer man. Picking up one or two old cronies Pancake turned back and in Placerville waited for his wife and the wife-stealer to come along. They came, and they had walked all the way across the Sierras. Instead of shooting the thief, as he had good reason to do, Pancake sat down and tried to reason with him. The woman, shifting her interest again, said she was ready to settle down with Pancake and be as good a wife as the Bible ever told about. There are some pretty good ones in that book.

After a drink or two Pancake felt so elated that he thought to bring his wife forth to public gaze and tell his friends what a model of virtue she was. But when he went to the room where he had left her she was gone again. It was plain that she had climbed through a back window, and even such a moonstruck halfwit as Old Pancake could figure it out that she hadn't left Placerville alone. So with his cronies he mounted and took off in what he hoped was pursuit, proclaiming along the way a hundred-dollar reward for the capture and return of the runaways. To be sure of getting his bride back he hired horses and sent other men after them. One so employed was a miner dead broke and he meant business. The next day this man "walked the runaways into Placerville in front of his six-shooter," and Pancake didn't bat an eye when he paid the reward. He then locked his wife in a room and stretched his long legs for another talk with her. The wife-stealer was also locked up while the men of the camp tried to decide what to do with him. When night came they told him they were going to hang him, but if he could manage to escape he should do it at once, and move with the speed of the coyote. Having a perfectly clear idea of what they meant he was soon gone and was never again seen in that area. By riding herd on her Pancake kept his wife in sight a few months, but the next spring, when "the little wild flowers were beginning to peep out about the rocks," she ran away with another miner, and after coming at last to the kind of life she was born to fill she settled down in a lager-beer cellar in Sacramento.

The runt of a Mormon husband also came to a

* See "Transportation."

sordid end. He found another wife somewhere but when she learned that he had already sold one wife she sued and won a divorce and married another man. Carter, while with her, had staked some mining claims in her name, as was the custom with men who had women, and now he and Winnie, the second husband, came to open war. In a gun duel Carter shot three fingers off Winnie's left hand, and Winnie shot Carter dead. In another duel soon afterward Winnie had some fingers shot off his right hand.

The women like the men were of all kinds except the frail and clinging. Those in California were thought to be "wanting in figure; they too much resemble the mold of their clipper ships, very straight and flat lines. They dance polkas and quadrilles, the ladies calling across the set to those opposite; loud laughter when anything witty was said; and romping rather than dancing. It reminded me of a servant's new-year's merry-making in England." They were a hardy lot: Root in his book on the overland stages says the women who kept overnight stands for coaches would ride fifty miles to and from a dance and think nothing of it. In a Marysville newspaper a woman advertised for a husband, and after saying that she could cook, spin, weave, and feed the pigs, she gave her terms: "Her age is none of your business. She is neither handsome nor a fright, yet an *old* man need *not* apply, nor any who have not a little more education than she has, and a great deal more gold, for there must be $20,000 settled on her before she will bind herself to perform all the above. Address to Dorothy Scraggs, with real name." She was obviously overpriced but possibly she found a mild unaggressive little fellow like Peter Sapp and made life miserable for him.

Elsewhere we present such camp types as the harlot, the adventuress, and the angel of mercy; here we point out that eccentrics seem to have been almost as common in the female as in the male sex. In the chapter, "Transportation," the reader will meet Charlie Parkhurst, one of the more famous stagecoach drivers who during all her years in the West concealed the fact that she was a woman. How many women passed themselves off as men nobody will ever know. In Idaho was the rather pathetic case of Joe Monahan, a small person with a rather feminine voice. Joe's sex seems never to have been suspected during all

the years she lived alone on Succor Creek, near Silver City, not in a house or even a shack but in a dugout in a hillside, a bunk bed filled with hay for a mattress, a dirt floor, dirt roof, a few utensils, and bedding of castoff stuff, like her garments. She raised a few pigs and chickens and cattle. In December of 1903 Joe undertook to drive her livestock to winter pasture on the Boise River, fell ill and died. Only then was her sex discovered. One who knew her said Joe was a "small, beardless

Joe Monahan. From a Buffalo paper, January, 1904

man, with hands, feet, stature, voice of a woman." On her death, letters were written to Buffalo to find her relatives, for it was known that from time to time Joe had sent a letter to that city. In a Buffalo newspaper appeared the photograph shown here. A story sprang up after her death that Joe was the notorious Kansas murderess Kate Bender, but those who had known the recluse many years didn't take the rumor seriously.

Another remarkable woman in the Idaho camps was Peg-Leg Annie, whose story is now more legend than fact. She is said to have been the owner of a restaurant and lodging house in a mining camp, when she set out with a woman friend to cross Bald Mountain in midwinter. When after three days and nights there was no sign of them a searching party was sent. Annie is said to have been crawling around on frozen feet and hands, out of her mind from cold and hunger. Her friend was dead. The legend says that Annie was filled with whisky and a man named Tug Wilson, using a jackknife and common meat saw, amputated her legs between ankles and knees. The fact is that a surgeon was sent

Diamond Tooth Lil

for to Mountain Home, about eighty miles distant. Annie became a famous character in the Atlanta-Rocky Bar area. One who visited her not long before her death says that the woman had suspended from the ceiling, above her bed, a contraption with two leather loops, through which

she could thrust her legs, to help her move and maneuver. A woman who visited her says she had a live chicken in bed and that the odors in the room were sickening. Whether fact or legend we don't know but the story goes on to say that for many years her man was an Italian, who had a saloon next to her restaurant, and that when he said he was off to Italy to see his people, Annie sent with him her savings, for deposit in a San Francisco bank. Though she never saw man or money again to the end of her life, she nourished the comforting fantasy that on his way out he had been robbed and killed.

Diamond Tooth Lil (Mrs. Evelyn Hildegard seems to have been her name, but in various camps she lived under different names) was another famous mining-camp character in her last years; in the picturesque illiteracy of her speech and writing she could stand as godmother to a thousand of them. For years she was a pal of some sort to Diamond Field Jack, an adventurer with a past as mysterious as her own. In a letter written in her last years to one who asked for information about her she said, in part:

In regards to Jack he s been bothering me early mornings, around 6 A.M. . . . He is all in cant hardly breed. So I havend seen him since, he has no address. As he moves so much. He owes me a few dollars, wich I dont care, and he is so mean, uses bad language and he wont give me no decent answer he is gone there is no way I can find out were he lives, as he talks foolish. You know he must be real old. When I met Jack in Goldfield I had a dance hall. Called the Nevada Club across from A. Jacks Cafe. . . . Gambler name Heins killed a Rushen Count in this Cafe in 1906. I was only in Goldfield 9 months. Gains-Nelson fought there. . . . I dance my way from Youngstown, Ohio to San Francisco when I was 13 years old. . . . I worked Barbary Coast I worked for Johny Crowly called the Black Cat. After the fire of Cours I traveld. . . . I have run a business since I was 13. . . . I mean now and then I worked nights in Caberais run my business days and nights out. . . . I had chances for Alaska but my ex husband went to Nome, Alaska. . . . I use to belong to the Elks Club and the club use to protect me from him, as I was divorced from him, but he followed me around for 7 years. He shot me in El Paso, Texas. . . . I need is my health. Money is no good today. Ive made a 100,000 in Boise, and Ive made the same out of Boise. I lived well I had a real life injoyed life. Ive never missed nothing and am still alive and always injoy life. . . .*

_____

* The letter appears in MacLane; excerpted here by permission of the Pandick Press, Inc., 22 Thames Street, New York.

Some writers have put the prospector among the frontier's eccentrics. Whether he belongs there or not it certainly is true that we owe him a salute and a better understanding of the kind of man he was, because without him there would have been no gold rushes, and no hundreds of books written because of his discoveries, including this one. Like the free trapper of a little earlier time he was a man of amazing self-reliance, resourcefulness, and courage, and we have tried to present here a fuller photographic image of him than has yet appeared in a book. In his *Romance of Mining* Rickard observed that "the kind of man who made a good prospector is becoming uncommon, because he was essentially the product of the frontier . . . he was physically strong, self-reliant, adventurous. . . . he hated to work for wages, because he loved independence; he wanted to be his own master. . . . for him the old days are gone; in their stead is a world of red tape and petty irritations"—of the "great society," Rickard might have said if he had written today.

The first mention of the prospector for precious metals was in expeditions sent out by the pharaohs; that "he had an important position is shown by the fact that only two such men were employed on each of four expeditions in which from 300 to 450 men participated." The ancient prospector, or fossicker, as he is called in Australia, meaning one who rummages among rocks, didn't have the knowledge of the prospector today, who eats cheesecake instead of sourdough, shaves every morning, and knows that water containing a good bit of salt reacts with silver sulphate to form the almost insoluble chloride of silver, a waxen stuff known to miners as horn silver. Just the same, the prospector in Arizona was known as a chlorider.

As Rickard, geologist and mining engineer, and a bit of a prospector himself, tells the story, the "seeker after ore looks for 'float,' the bits of mineral or vein-matter that have the been broken, by frost or other agencies of erosion, from an outcrop, which is the part of a vein or lode that projects above the surface because it is harder than the enclosing rock. Such hardness is due commonly to the presence of silica, or quartz, a cousin of flint. The prospector may follow traces of relatively insoluble metals, such as gold, . . . either in the wash of the creek or in the soil on the hillside." The early prospectors knew that much.

They scanned the whole world around them for signs of mineral, in the diggings and home-building of his animal friends, such as the ants and all the burrowing beasts. Their mounds have led to the "finding of much ore." The ancients told fan-

Jean Heazle, last resident of DeLamar, Idaho

tastic tales about gold-digging ants, not as large as dogs but as large as foxes, which, we must assume, excavated gold nuggets smaller than melons but larger than oranges. "The American miner in the arid region of the Southwest will pan the earth of the ant-heaps. . . ." One would think that all the millions of ant mounds had been panned, but is told that they have not been.

In the ancient world and the Middle Ages the dowsers wandered "hither and thither at random through the mountain regions." All that, as Rickard says, was hocus-pocus, but we still have the dowsers with us, and probably shall have, as long as astrologers are called in to advise Presidents. Paracelsus told the world as early as 1541 that the divining rod had deceived and squandered the time of many miners. It is more sensible to store the rods away with the ancient cures of rheumatism and look "for outcrops of quartz, because quartz is usually, but by no means always, the matrix of the precious metals. If the quartz be iron-stained, that is a good sign, because minerals containing iron, notably pyrite and chalcopyrite, in common sulphides of iron and copper, are usually associated with gold and silver. . . . Among the signs, . . . by which the prospector is guided is the colour of the rocks. Red means the oxidation of iron. . . . A green tint may be due to the oxidation of copper . . . but the experienced prospector knows that a very little copper may suffice to colour a whole hill-side." Indeed, as both iron and copper do, in countless places all over the West.

When it is borne in mind how many prospectors there were, searching river- and creek beds, ravines and canyons, and even the highest mountains, and how few found diggings of substantial value, it seems likely that a lot of them didn't know what to look for. We once heard an old-timer who had done some prospecting say that he could spot gold in any kind of rock. We took him to J. A. Harrington, prospector, Klondiker, ex-U.S. marshal, who had a fine collection of ores, a few of which were very rich in gold. These were spread upon a table and the man who knew gold was told to choose the half-dozen that he thought richest in gold. He stared at them, hefted them, turned them over and over, and took more than two hours to make up his mind, and at last chose a half-dozen that had almost no gold in them.

That silenced him for weeks, for the prospectors as a tribe were reticent, antisocial, lonely, suspicious; they picked up a grubstake and tools and vanished, with a burro or two, though often with no animal at all, for weeks or months, again and again, through a long lifetime, until at last they were too old and broken to try any more. Some of them in the Klondike rush died trying to climb the Chilkoot and White passes. Now and then there was a different sort, such as Dr. Atkinson, a "most original character," according to Granville Stuart. With a companion or two and a pair of glasses in his baggage he would climb to mountaintops and gaze round him, and say that all of it looked mighty promising. But Stuart "never knew of his digging a hole or panning a pan of dirt." Closer to the type was a tough griz-

zled fellow named McClellan, who had a Blackfeet squaw. She had so many relatives coming in day and night for anything they could eat, beg, or steal that Mac said he guessed he would die poor, but shouldn't complain, for what it cost him was a trifle when measured against all the comforts she gave him. He is reported to have said, "I've got as good a thing as I want." Men interpreted that to mean a gold mine. Greever tells us that one day Mac was in camp buying supplies, in thirty-five below zero weather, and that when he set off for his shack "about twelve hundred men" followed him.

Jim Baker was more mountain man and scout than prospector but he did spend a lot of time looking for gold; we include him here because for us, better than any other, he had the prospector's face. A group of emigrants on their way to California gave him some rich float; as well as he

A prospector in Arizona killed by Apaches

could tell from what they said they had found it on a mountain south of Hahn's Peak in northern Colorado. Baker went looking for it and for a lot of "lost mines," but, like Jim Bridger before him, he brushed the facts aside when he wanted to tell a good story. That was true of prospectors in general, and those who hunt for lost mines would do well to remember it. Ed Schieffelin, discoverer of the Tombstone area, was more the typical prospector than such discoverers as Pierce and Marshall. A newsman who saw him in 1876 thought him about the queerest-looking human specimen he had ever seen—a gaunt six feet two, about thirty but looking forty or older: his "black hair hung down below his shoulders, and his full beard, a tangle of knots, was almost as long. His clothes, including his hat, had been so patched with pieces of rabbit skin that he resembled a scrofulous fur-bearing animal."

Possibly it was one much like Ed who left on sandstone a message that was found among scrub oak in the Black Hills, in the spring of 1877, by a man digging for ore. The piece of stone was about twelve inches square, with words scratched on each side.

Came to these hills in 1833 seven of us
De Lacompt Ezra Kind G. W. Wood T Brown
R. Kent Wm. King Indian Crow all ded but
me Ezra Kind—Killed by Inds beyond the
high hills got our gold June 1834.

Got all of the gold we could carry out
ponys all got by the Indians I have lost
my gun and nothing to eat and Indians
hunting me

It is speculated that the Indians found him, for if he had come out, a Black Hills gold rush might have preceded the California rush by fifteen years.

A quite a few of the prospectors never came out; a quite a few went insane. "Within this last week," Greeley wrote when in Denver, "we have tidings of one young gold-seeker committing suicide, in a fit of insanity, . . . two more found in a ravine, long dead and partially devoured by wolves; while five others, with their horse and dog, were overtaken, some days since . . . by one of those terrible fires . . . which sweep the hilltops. . . . "

Quiett tried to suggest the myth of the "mountain of pure gold discovered by a prospector crazed by the heat, or frozen by the cold, or full

of virulent coffin-varnish. Always he wandered in circles or ellipses or in jig-saw puzzle outlines until he staggered into civilization and, holding up his pouch of gold, fell exhausted in the Red Eye Saloon, the original face on the bar-room floor. When he came to, he could only mutter incoherently; . . . finally died babbling a few last words

Jim Baker, a typical early western face

about vast golden treasure and leaving an incomprehensible map scrawled on the back of an old envelope. And that was the signal for half a century of optimistic parties reeling over the desert in search of the lost Gum Drop Mine." According to Quiett, so many "inexperienced prospectors have been lost in the desert, so many have died of thirst, that the Nevada legislature long ago passed a law requiring any train to stop if signaled by a man on the desert. . . . That law has saved many lives."

The prospector most likely to strike it rich was

the one who spied on the others. James Graham Fair, one of the four Comstock bonanza kings, known as Slippery Jim, was that kind. Lewis says that in his early years when out hunting for gold he saw Indians trading a few pieces of rich ore for a jug of whisky. Slippery Jim, who knew that a mine already found was worth far more to a lazy man than one still unfound somewhere in ten thousand gulches and gullies, craftily hid out and watched and waited, and trailed the Indians to their secret spot. Lewis says that Fair and a partner, equally slippery, took $180,000 from that spot in a few weeks.

There aren't many prospectors going out any more. Aware of that fact, the Denver Chamber of Commerce solicits funds to grubstake a few of them, and at the time Rickard wrote about it, forty-three locations had been found that were promising. Of precious metals a great deal more remains in the earth, the experts think, than has been taken out. Today's prospector has far better equipment than his predecessor had; this includes the magnetometer which Edison invented and which has many times been used successfully —as when it indicated ore under eighty feet of glacial clay and quicksand which for eighteen years defeated all efforts to sink a shaft through them. The body of ore revealed was the famous Falconbridge "at a point where it was a hundred feet wide."

Typical prospectors

Typical prospectors

Using his dog as a pack animal

CHAPTER 7

Transportation

As mines were opened here and there in the Far West various persons came up with plans to get people across the vast distances. A firm in St. Louis considered a wagon stage that would carry 120 persons at $200 per head; but a man named Birch knew that would be too slow and too shattering to the human nervous system. He figured that if a man named Todd "could deliver mail to Stockton at a rate of four dollars per letter, carry gold-dust back in butter kegs at other such rates of profit . . . and divide sometimes as much as a thousand dollars a day with his partner, surely the prospects were bright enough in the business of transportation." There was agitation in the East for a railway across the continent to bind East and West with "iron bands" but a railway would take twenty or thirty years to build. Birch and some others saw that it was a time for wheels.

When Horace Greeley came out there were enough wheels to impress him: "Such acres of wagons! Such pyramids of extra axletrees! Such herds of oxen! Such regiments of drivers and other employees! No one who does not see can realize how vast a business this is, . . ." The Southern Overland had more than 100 Concord coaches, 1,000 horses, 500 mules, 150 drivers. "The deadly deserts, through which nearly half its route lay, the sandstorms, the mirages, the hell of thirst, the dangerous Indian tribes . . . made it a colossal undertaking." In six weeks, in 1865, 6,000 freight wagons passed Fort Kearney. Frank A. Root counted in one day 888 westbound wagons, drawn by 10,650 oxen, horses and mules, between Kearney and Julesburg. In 1865 over 21 million pounds of freight were shipped from Atchison alone, requiring 4,917 wagons, 8,164 mules, 27,685 oxen, and 1,256 men to handle. The Conestoga wagons cost $800 to $1,500 each; the harnesses from three to six hundred for a ten-mule team, and the mules from $500 to $1,000 a pair. All freight was as-

sessed by the pound: between Atchison and Denver, a distance of 620 miles, flour was 9¢, sugar 13½, whisky 18, glass 19½, trunks 25, and furniture 31.

Ben Holladay became, of course, the biggest man of them all. After Bowles journeyed west with Schuyler Colfax, Speaker of the House, he published a book to tell what he had seen:

His whole extent of staging and mail contracts—not counting, of course, that under Wells, Fargo & Co., from Salt Lake west—is two thousand seven hundred and sixty miles, to conduct which he owns some six thousand horses and mules and about two hundred and sixty coaches. All along the routes he has built stations at distances of ten to fifteen miles; he has to draw all his corn from the Missouri River; much of his hay has also to be transported hundreds of miles; fuel for his stations come frequently fifty and one hundred miles; the Indians last year destroyed or stole full half a million dollars' worth of his property. . . . Mr. Holladay visits his overland line about twice a year, and when he does, passes over it with a rapidity and a disregard of expense and rules, characteristic of his irrepressible nature.

He was in a hospital in San Francisco one day when he heard that the Congress did not intend to renew his mail subsidies. Exploding oaths as he climbed out of bed, and shoving aside protesting doctors and friends, he summoned a stage with six of his best horses and set off on a journey to the East at a pace so furious and unrelenting that it made a record and became an international sensation.

Banning says of him, "With coaches by the hundreds, horses and mules by the thousands, it was believed that his staging concern alone was the nation's, if not the world's, greatest enterprise ever owned and controlled by one man." It made him millions, but by the spring of 1864 the Indians had practically declared a war of extermination;

Overland transportation before the stagecoach

and the story of their massacre of passengers and station-keepers, not yet fully written, would make one of the bloodiest chapters in American history. For weeks at a time stages on some of the Holladay routes did not dare run at all, and when the governor of Colorado asked for troops to be sent in his request was ignored. The Indians killed so many of his passengers and destroyed so many of his stations and stages, and the American press heaped upon his head such virulent criticism, that at last he was glad to get out of the business.

"He was perhaps the most unpopular stage proprietor that ever lived."

James E. Birch was the ablest of the first men who put wheels under an expanding nation; before the age of thirty he was wealthy, had built an elaborate mansion in New England, and died soon afterward with four hundred others when the steamer *Central America* sank September 12, 1857. Holladay brought the whole thing to its climax—and what a sight the main staging areas must have been!—as they stood four abreast, filling, as in Sacramento, a whole street for a distance of nearly a hundred yards, while passengers and baggage were being loaded into them, and the horses, the finest money could buy,* pawed and snorted, as grooms tried to quiet them.

If Holladay was a marvel among stagers, the Concord coach was outstanding among vehicles. It was a far cry from the California oxcart, which, Colton tells us, was

quite unique and primitive. The wheels are cut transversely from the butt-end of a tree, and have holes through the center for a huge wood axle. The tongue is a long, heavy beam, and the yoke resting on the necks of the oxen, is lashed to their horns, close down to the root. . . . On gala days it is swept out, and covered with mats; a deep body is put on, which is arched with hoop-poles, and over these a pair of sheets are extended for covering. Into this the ladies are tumbled, when three or four yoke of oxen, with as many Indian drivers, and ten times as many dogs, start ahead.

Compared to that the Concord was in the jet age of wheels; and those who made it, as Banning says, combined

the Arab's knowledge of the horse to the mechanical perfection of the wheel as developed by the craftsmen of Concord. . . . The conveyance itself, as one cannot be reminded too often, was not the worn, torn, dead thing of our museums, much less the grotesque rattletrap of our fiction, drama, and other oft-profaning arts, but a resplendent and proud thing—a heavy but apparently light thing of beauty and dignity and life. It was as tidy and graceful as a lady. . . . With trim decking and panels of the clearest poplar, and with stout frame of well-seasoned ash, this body, in all of its tri-dimensional curves, fairly held itself together by sheer virtue of scientific design and master joinery, for very little iron was used; and this, where needed, was iron only of the best Norway stock.

The greatest single marvel in this coach was the thoroughbraces—

———

* Some cost as much as $2,000, which would be about $10,000 now.

two lengths of manifold leather straps of the thickest steer hide—suspended this heavy structure, allowed it to rock fore and aft, and, incidentally, to perform a vital duty beyond the province of any steel springs. It was a function of such importance to the Western staging world that we may hazard the contention that an empire, as well as the body of its coach, once rocked and perhaps depended upon thoroughbraces. Without them, at least, there could have been no staging to meet a crying need. Without them over the rugged wilderness, any vehicle carrying the loads that had to be carried, maintaining such speeds as the edicts of staging demanded, would have been efficient only as a killer of horses. For thoroughbraces, while they served the purpose of springs to a very adequate extent, had the prime function of acting as shock-absorbers for the benefit of the team. By them violent jerks upon the traces, due to any obstrucation in the road, were automatically assuaged and generally eliminated. It was the force of inertia—the forward lunge and the upward lurch of the rocking body—that freed the wheels promptly from impediment, and thus

averted each shock before it came upon the animals. . . . Each spoke of the Concord wheel, hand-hewn from the clearest ash, was the result of a series of painstaking selections, carefully weighed and balanced in the hand, and fitted to rim and hub . . . as snugly as any surface joint in the best of joinery. So well seasoned was the whole that it was practically insensible to the climatic changes in any part of the world.

The first Concord rounded the Horn and appeared in San Francisco, "drawn by six splendid bays," on June 24, 1850. Thought by many to be the "only perfect passenger vehicle in the world, it was an epoch-making Yankee creation whose excellence, in the light of its purpose, had never been, and was never to be, duplicated by other manufacturers. Abbot, Downing & Company alone knew how to make them." Having been making them, or their forerunners, since 1813, by 1858 their plant covered four acres and called

Mount Shasta in the background

Arrival in a California camp

itself the largest factory in the United States. A month after the first coach arrived a newspaper was telling its readers:

The stages are got up in elegant style, and are each arranged to convey eight passengers. The bodies are beautifully painted, and made watertight, with a view of using them as boats in ferrying streams. The team consists of six mules to each coach. The mail is guarded by eight men, armed as follows: Each man has at his side, fastened in the stage, one of Colt's revolving rifles, in a holster below one of Colt's long revolvers, and in his belt a small Colt's revolver, besides a hunting knife; so that these eight men are ready, in case of attack, to discharge 136 shots without having to reload.

Edward Hungerford tells us that it took eight steer hides for one set of thoroughbraces and boots, and that the "Bodies were vermilion, running gear a bright, rich yellow with black trim. And the door panels were the work of qualified artists . . . New England landscapes. The useful life

span of a Concord was nearly that of a man." A coach weighed only about a ton, yet was built to carry fifteen passengers* and their luggage at the top speed of the fastest horses over rough mountain roads.

Borthwick left us a picture of the stages in Sacramento, ready to go:

The teams were all headed the same way, and with their stages, four or five abreast, occupied the whole of the wide street for a distance of sixty or seventy yards. The horses were restive, and pawing, and snorting, and kicking; and passengers were trying to navigate to their proper stages through the labyrinth of wheels and horses, and frequently climbing over half-a-dozen wagons to shorten their journey. Grooms were standing at their leaders' heads, trying to keep them quiet, and the drivers were sitting on their boxes, or seats rather, for they scorn a high seat, and were swearing at each other in a very shocking manner, as wheels got locked, and wagons were backed into the teams behind them, to the discomfiture of the passengers on the back seats, who found horses'

_____

* Apparently that many were sometimes crowded in.

heads knocking the pipes out of their mouths. In the intervals of their little private battles, the drivers were shouting to the crowds of passengers who loitered about the front of the hotel. . . . On each wagon was painted the name of the place to which it ran; the drivers were also bellowing it out to the crowd. . . .

Seeing a man who looks lost the driver to Nevada City

drags him into the crowd of stages, and almost has him bundled into that for Nevada City before the poor devil can make it understood that it is Caloma he wants to go to. . . . There certainly was no danger of any one being left behind; on the contrary, the probability was, that any weak-minded man who happened to be passing by, would be shipped off to parts unknown before he could collect his ideas. . . .

At last the solid mass of four-horse coaches began to dissolve. The drivers gathered up their reins . . . cracked their whips, and swore at their horses; the grooms cleared out the best way they could; the passengers shouted and hurrahed; the teams in front set off at a gallop; the rest . . . all in a body, for about half a mile [then] spread out in all directions to every point . . . I found myself one of a small isolated community, with which four splendid horses were galloping over the plains like mad. No hedges, no ditches, no houses, no roads in fact. . . . in some places very narrow . . . covered with stumps and large rocks . . . full of deep ruts and hollows, and roots of trees spread all over it . . . all sense of danger was lost in admiration of the coolness and dexterity of the driver as he circumvented every obstacle. . . . He went through extraordinary bodily contortions, which would have shocked an English coachman out of his propriety; but, at the same time, he performed such feats as no one would have dared to attempt who had never been used to anything worse than an English road. With his right foot he managed the brake, and, clawing at the reins with both hands, he swayed his body from side to side to preserve his equilibrium, as now on the right pair of wheels, now on the left, he cut the "outside edge" round a stump or a rock; and when coming to a spot where he was going to execute a difficult maneuver on a piece of road which slanted violently down to one side, he trimmed the wagon as one would a small boat in a squall, and made us all crawl up to the weather side to prevent a capsize.

British artist Borthwick was on his way to Placerville; he says they covered the sixty miles in about eight hours.

For Mark Twain it was like this:

And when at last he grasped the reins and gave the word, the men sprung suddenly away from the mules' heads and the coach shot from the station as if it had been issued from a cannon. How the frantic animals did scamper! It was a fierce and furious gallop—and the gait never altered for a moment till we reeled off ten or twelve miles and swept up to the next collection of little station huts and stables.

So we flew along all day. . . . fifty-six hours out from St. Joe—THREE HUNDRED MILES!

Exactly what was the ride in a Concord like? There were all kinds of passengers and most of them left descriptions of their ride. Mrs. Vance Kimball, who had a station out from Salt Lake City, says the thing in passengers that most annoyed and astonished her was their ignorance in taking care of themselves. Drivers kept themselves clean when that was possible. "But the passengers! I hated to see them come into my house!" She thus spoke to Mr. Banning of *Six Horses* when she was a hundred years old. She then "shuddered and mentioned the unmentionables." Knowing that the wild ride drove some passengers into hysterics, even into lunacy, we can easily imagine what a sight they were when at last they tumbled out.

According to Richardson, "Empty, it jolts and pitches like a ship in a raging sea; filled with passengers and balanced by a proper distribution of baggage in the 'boot' behind, and under the driver's feet before, its motion is easy and elastic. Excelling every other in durability and strength, this hack is used all over our continent and throughout South America."

We have quoted Sir Henry Huntley on food and hotels. He thought the food a mess of nauseous horrors and this is what he thought of the coach:

Of all public conveyances I ever encountered, the stages here take the lead, in discomfort and vulgarity. The carriage is either a lumbering old-fashioned thing, or . . . it is a waggon on springs; as many persons as can jam themselves in, do so; they have no regard for those already in possession of seats; but, as there are no divisions, the new passenger coolly gets into the vehicle, and, placing himself between two others, sits down, and relies upon his own weight making the other two sufficiently uncomfortable, to aid him in establishing himself between them.

It is too bad that nobody described the exchange of glances and stares.

"I must say," Barnes exclaims, "there are places where crinoline is out of order and babies become a downright nuisance! . . . found myself *vis-a-vis* with a grass widow and four children under eight years of age. . . . Each had a specific want at least once in twenty minutes. . . . covered with molasses and the stage with crumbs. . . ."

For Mrs. D. B. Bates, as quoted by Manning, riding in a stage was "perfectly awful." She tells of a woman who staggered into a hotel just before midnight "looking more dead than alive. She was leading a little girl of about seven years of age, who was in the same plight as her mother. They were both covered with bruises, scratches and blood, with their garments soiled and torn." Their stage had overturned; in it had been nine Chinese, who, with their hair reaching to their heels, had been quite a sight "rolling and tumbling about in the most ungraceful manner." Another woman said she knew at last what hell was like; she had had twenty-four days of it. At least one man went insane from lack of sleep, and was bound fast in one of the boots and put off at the next station.*

Donaldson says the suffering endured on a long journey can

never be adequately described. I have ridden continually for days and nights sitting upright with my knees jammed against the knees of another unfortunate and both of us suffering the agonies of the damned. Try to realize what it meant when you are told of the overland stage journey from St. Joseph, Missouri, to Sacramento, California. Twenty-four days and nights of ceaseless travel and the passengers unable to recline; nothing to do but maintain a sitting position. If a passenger was ill or tired and chose to lay over at a station,

_____

* Readers interested in a description of a long coach journey can find Dr. J. C. Tucker's in *Six Horses*.

Indians and bandits were the stagecoach's worst enemies

Stage station between Boise and Idaho City

the chances were that the following coach would be crowded, and he might remain for days awaiting a coach with a vacant seat.

In areas of deep snow, travel by coach in wintertime was even more dangerous.

The beaten road, or grade, for sleighs or wagons over the snow as hard as a floor, but one inch to the wrong side and horses and man would disappear in the soft snow. I have frequently seen a horse drop off the grade, and it was a circus to get the beast back to the surface. Every male passenger would get out, tug, pull, swear, and yell at the poor brute, . . .

Sometimes a passenger put his impressions in verse:

Creeping through the valley, crawling o'er the
    hill;
Splashing through the branches, rumbling by
    the mill;
Putting nervous gentlemen in a towering rage.
What is so provoking as riding in a stage?

Spinsters fair and forty, maids in youthful
    charms,
Suddenly are cast into their neighbors' arms;
Children shoot like squirrels darting through
    a cage—
Isn't it delightful, riding in a stage?

Feet are interlacing, heads severely bumped,
Friend and foe together get their noses
    thumped;
Dresses act as carpets—listen to the sage:
"Life is but a journey taken in a stage."

In Idaho, during the years 1862-1875, there were mud wagons, Donaldson says, in addition to Concords. A Concord held sixteen to eighteen persons, inside and out. He says that at a main station passengers were started off in a handsome Concord but before long were hustled into mud wagons, a low strongly built vehicle that swung, like the Concord, on thoroughbraces. Riding in a mud wagon was much rougher than in a Concord. After 1869 all vehicles for passengers were equipped with the Sarvin wheel, from which spokes could be taken or put in without removing the steel tires. Other vehicles for short roads and bad hauls were the jerky, the buckboard, and the dead-axle wagon.

Even if a ride was a journey through hell, a stage line

during those early days was splendidly equipped, having over 100 of the very best Concord stage-coaches, 1,000 horses, 500 mules, and 750 men, of whom about 150 were employed as drivers. The fare across the continent by stage was $100 in gold. On each of the first coaches that departed from either end of the line was a correspondent of one of the great New York dailies. . . . The coach going east from San Francisco made the trip through to St. Louis in a half-hour less than twenty-four days. The first stage west made the distance in an hour or two less time.

The relay stations were twelve to fifteen miles apart but the home stations, with meals and lodging, were fifty to sixty miles apart. "The stage stations were one-story log houses with dirt or mud roofs, the men and horses sleeping under one shelter." The home stations were also of logs, and usually were owned and managed by a man and wife.

They were rich in little save dirt. The meals were uniformly bad and one dollar each. Bacon* and "white-lead" bread were the staple articles of food. . . . Pie was another staple article, and such pie! It consisted of a sole-leather, lard-soaked lower crust, half-baked, with a thin veneer of dried apples daubed with brown sugar. A large pot of mustard containing an iron spoon which had partially succumbed to the attack of the vinegar always decorated the center of the table. Ten times each meal you knew it was a standby because ten times arose the call, "Pass the mustard, Jedge. . . ." The table was of rough pine boards and the benches or chairs were equally rough. The table furniture was of ironstone ware and tin, with iron spoons and heavy knives. . . . The bread was the result of "self-raisin'" flour, nearly white lead in consistency. The coffee and tea were peculiar to the country. I never tasted anything quite so bad in any other part of the world.

But, Donaldson continues, those who managed the stations

were good, decent people, charitable and attentive. [Their profits were but meager] compensation for the hermit existence forced upon them.

Bancroft thought

There never was any stage service in the world more complete than that between Placerville and Virginia City.† A sprinkled road, over which dashed six fine sleek horses before an elegant Concord coach, the lines in the hands of an expert driver, whose light hat, linen duster, and lemon-colored gloves betokened a good salary and an exacting company, and who timed his grooms and his passengers by a heavy gold chronometer watch, held carelessly, if conspicuously, on the tips of his fingers. . . .

"At each station," according to Root and Connelley, "there was annually consumed from forty to eighty tons of hay. . . . At the various stations it required about 20,000 tons annually to supply them, costing an average of about twenty-five dollars per ton, or, say, $500,000 per annum. In grain, each animal was apportioned an average of twelve quarts of corn daily, which then cost from two to ten cents a pound."

In areas of deep snow "square wooden slabs," Harlow says, were attached to the feet of horses to keep them from sinking to their bellies, but because wet snow clung to the wood, steel plates came into general use, sometimes with a coating of rubber. "The plates were nine inches square, and fastened to the hoofs with screws and straps. The horses cut their legs with the edges of the plates at first, but they soon learned to spread their feet apart in walking."

Though there were accidents of many kinds they were surprisingly few, considering the roughness of the roads, the steepness of many descents, and the speed at which the beasts were driven. Sometimes a coach was blown over or off the road "by the fierce wind-blasts of the Sierra winters." There were runaways down dangerous grades when brakes failed or beasts were frightened. Drury tells of the time when six horses took fright and ran, throwing the driver off the seat. Inside the coach a man named Boyle looked out, and quickly grasping the situation and knowing that if the beasts continued to run every passenger would probably be killed, he shot one of the lead horses, which then piled up the other five. "Boyle's action saved the lives of all aboard."

Sometimes the journeys ended in tragedy, "dark, bloody and pathetic as ever found expression in

---

* Donaldson does not mean bacon as we know it, but fat unsmoked hog-belly, much of it with no lean meat in it.
† This is U.S. 50 today across the Sierras; Hank Monk was one of the drivers over it, whom see, below.

Stage stop at Iron Springs, Arizona

tale or story. To many a station did the old coaches come down the trail like the wind, sore beset by bloodthirsty savages. . . . Sometimes the driver was dead and the passengers were maimed. More than once the coach was left surrounded by dead and scalped travelers, . . ." The authors quote David Street, who said that Holladay's

drivers and stocktenders were the best. No storms, no dangers could daunt them. . . . Many heroic deeds by agents, drivers and passengers could be recounted—of facing blizzards, plowing through snow-banks, the dangerous snow- and landslides; swimming coach and team across swollen streams; . . . facing Indians on the war-path; drivers and messengers shot from the box by Indians and road-agents. I have known coaches to come in to the station with the driver dead in the front boot, the mail soaked with his blood. I recall instances where employes traveling with coaches attacked by Indians have kept up a fight for a whole day and part of a night.

Accidents occurred, Harlow says, no matter what the skill of the drivers—caused by landslides, stumbling horses, or droves of wild Mexican cattle.

One day a large grizzly bear popped out into the road and scared a team; they reared, then partly ran around the coach, and broke the pole, while the passengers leaped off and ran in all directions. A stage at the top of Johnson's pass toppled over the rim of a canyon and lodged in the boughs of a tough-fibered Sierra pine; the passengers dropped from limb to limb to the earth. It was a

thousand feet further to the bottom of the canyon, where they would have gone had it not been for that tree. . . . When a Pioneer coach from Virginia City to Placerville was upset and rolled down a twelve-foot bank in 1863, a ton of silver bars lying on the floor was hurled to and fro in the vehicle and inflicted serious injuries on the four passengers, . . .

Whether or not there were accidents depended largely on the drivers. As a group they have been eulogized and they have been damned. Some of them must have rubbed Mark Twain the wrong way: when passengers spoke to a driver, he says,

they received his insolent silence meekly, and as being the natural and proper conduct of so great a man; when he opened his lips they all hung on his words with admiration (he never honored a particular individual with a remark, but addressed it with a broad generality to the horses, the stables, the surrounding country *and* the human underlings) ; when he discharged a facetious insulting personality at a hostler, that hostler was happy for the day; when he uttered his one jest—old as the hills, coarse, profane, witless, and inflicted on the same audience in the same language, every time his coach drove up there—the varlets roared, and slapped their thighs, and swore it was the best thing they'd ever heard in all their lives. And how they would fly around when he wanted a basin of water, . . . or a light for his pipe!—but they would instantly insult a passenger if he so far forgot himself as to crave a favor at their hands.

Artemus Ward also took a dim view of them:

The driver with whom I sat outside informed me as we slowly rolled down the fearful mountain road which looks down on either side into an appalling ravine that he has met accidents in his time and cost the California Stage Company a great deal of money, "because," said he "juries is agin us on principle, and every man who sues us is sure to recover. But it will never be so agin, not with *me*, you bet!" "How is that?" I said. "Why, you see," he replied, "that corpses never sue for damages, but maimed people do. And the next time I have an overturn, I shall go around and keerfully examine the passengers. Them as is dead, I shall let alone; but them as is mutilated, I shall finish with the king bolt. Dead folks don't sue. They ain't on it." Thus, with anecdote did this driver cheer me up.

The drivers were a remarkable tribe of men, and felt, it may be, that they were privileged. Richardson found that "The driver is invariably a character; always intelligent, often entertaining and witty, never any respecter of persons. There is a story of one, with a clergyman upon the box beside him, who swore long and loud at his balking horses. 'My friend,' expostulated the preacher, 'don't swear so. Remember Job; he was severely tried, but never lost his patience.' 'Job, you say? What line did *he* drive for?' "

The driver of those days, says Jackson, "drove like Jehu (the record proves it), scorned the yawning abysses so close to his rolling wheels, swore like a trooper, and brought the stage in on time. Now and then he drank too much. Swaggering was his specialty. Sometimes he died in line of duty, always he risked death a dozen times a week. In short, he was a very devil of a fellow."

Demas Barnes, a greenhorn from the East, was for some reason allowed to take the lines:

The horses were fiery, lively fellows, the leaders real live mustangs. Presently they took it into their heads that it was time to go, and on they started. Before I could gather the reins they were in a keen run down the hill; I pulling, yelling, and bobbing about on the high stage box like a rush in a wind storm. The harness of the wheel-horses had no breeching, and I did not think to place my foot on the brake, which is the only way to hold back in this country. So down they went, pell-mell, until I threw the leaders, and the other four horses, entangled in the harness, were all piled together and the stage over them. I had succeeded in breaking the pole, the harness, and almost everything breakable, and furthermore in convincing myself and fellow-passengers that driv-

ing six wild horses down a mountainside was not exactly in my line.

The wonder is that there were so few serious accidents. Banning says that in California, 1855-60, the average was no more than one a month, with about twenty-five passengers injured and one killed. That is no worse than persons are doing nowadays in automobiles. The skill of the drivers and horses and the strength of the coach are indeed a marvel for anyone who has seen some of the roads that they traversed at top speed. As Banning points out, some of the turns were so sharp that six horses and the coach might at the same moment fill a piece of roadway in the form of S. When his book was published in 1928 one of the more famous roads had "been carefully preserved in all of its essentials." Adams tells of a driver in Silver City, known as Little Mac, who was so skilful with a string of fourteen horses (seven teams in tandem) that he "could make his swing team jump the stay chains to negotiate the sharp turns." Only those who have watched experts handle freighting teams can realize what a feat that was.

For our appalled Sir Henry the road was "often so much on the incline, that the driver calls out, 'Sit up to windward,' and every one scrambles to the high side of the vehicle; I have seen a wheel actually off the ground, and brought to it again by 'sitting to windward;' the driver lighting his cigar at the time." Maybe some of the drivers were a bit of a showoff, but as Glasscock and others have pointed out, "the guiding of six-horse stages, frequently at a run, around the curves of those mountain grades required trained hands and eyes, strong arms, and quick judgment. . . . The three-day schedule over the one hundred and sixty-two miles from Sacramento to Virginia City in the first great year of travel was cut to eighteen hours"—an incredible feat, for half that distance was across the high Sierras. "On one occasion, to serve three wealthy mining operators in an emergency, the run was made in twelve hours and twenty minutes, an average of about thirteen miles an hour."

And the drivers had to be bold men, for their enemies were Indians and road bandits, all along the way. Drury knew a stage owner whose coaches "were stopped so many times up in Montana that Wells-Fargo threatened to take their business away from him and start an opposition line. This made him so angry that he called his drivers together and told them that they were a pack of cowards, and said he'd like to see anybody stop *him*. That was before shotgun messengers were common and every driver was supposed to look out for himself." So the enraged owner drove "out of Helena in an old-fashioned mud-wagon, with a shotgun resting across his lap and two passengers in behind. About seven hours later one of the passengers drove the wagon back to town. The money-box was gone and Preston was lying in the bottom of the wagon with both of his hands literally shot off at the wrist. He refused to throw up his hands when told to. The last time I saw Preston he had two brand-new wooden hands, with black kid gloves on them."

Most famous of all the drivers was, of course, Hank Monk, who brought the first coach mail by the Central Overland to its western terminus. He became famous in part because of his superb driving, his way of talking to the horses, and his twenty years across the Sierras without an injured passenger; but chiefly because Horace Greeley had an itch to be president and decided to have a look at the West. "He wore a rusty and well-worn white coat, and carried a still rustier and still more worn umbrella—blue cotton, with a bulge in the middle. His roll of blankets was shoved under a seat; his precious trunk, filled with manuscripts, was strapped in the after-boot." Sometimes on his long journey he slept in a tent, sometimes in a wagon, sometimes under a wagon. Arriving in the wild town of Denver, well-shaken, wrenched and

bruised, he put up in a sumptuous hotel with a canvas roof, a lean-to shack for a kitchen, and a bunk that had neither mattress nor pillow. He was pitched from the coach into a river, and still again, in an "avalanche of mail bags"; and the after-boot burst wide open and spewed forth the precious trunk and manuscripts into the swift current of the Sweetwater. A little farther, on the road to manifest destiny, the coach met a wagon on a sharp curve, and if the driver had not been able to control the teams "we might have been pitched headlong down a precipice of a thousand feet." It was all fine education for the man who had told young men to go west. And, of course, before he was finished with his ordeal he had to meet Hank Monk.

Whether the Greeley-Monk story was authentic or was started by Mark Twain or some other wag we don't know. It must have been told more than ten thousand times. Richardson, out west for a look around, told it in his book and followed it with a second tale—of the time Monk drove at such a furious pace over his dangerous mountain route that he scared the living daylights out of a "judicial personage" who cried in anger, "I will have you discharged before the week is out! Do you know who I am, sir?"

"Oh, yes," Monk is said to have replied. "But I'll take this coach into Carson City on time if it kills every one-horse judge in the State of California."

I can tell you a most laughable thing indeed, if you would like to listen to it. Horace Greeley went over this road once. When he was leaving Carson

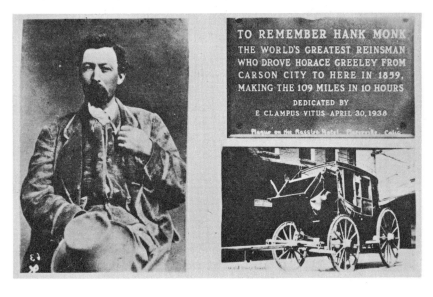

TO REMEMBER HANK MONK
THE WORLD'S GREATEST REINSMAN
WHO DROVE HORACE GREELEY FROM
CARSON CITY TO HERE IN 1859,
MAKING THE 109 MILES IN 10 HOURS
DEDICATED BY
E CLAMPUS VITUS APRIL 30, 1938

City he told the driver, Hank Monk, that he had an engagement to lecture in Placerville, and was very anxious to go through quick. Hank Monk cracked his whip and started off at an awful pace. The coach bounced up and down in such a terrific way that it jolted the buttons all off of Horace's coat, and finally shot his head clean through the roof of the stage, and then he yelled at Hank Monk and begged him to go easier—said he warn't in as much of a hurry as he was awhile ago. But Hank Monk said, "Keep your seat, Horace, and I'll get you there on time"—and you bet you he did, too, what was left of him!

Mark Twain tells the story thus and repeats it *three* times; and says that in six years of crossing the Sierras he had heard it either 481 or 482 times. Everybody told it, including the Chinese and the Indians; and he was sure that it was in the Talmud and in "the inquisition in Rome," and in nine different languages. "And what makes that worn anecdote the more aggravating, is, that the adventure it celebrates *never occurred.*"

Cross in *The Early Inns of California* says that it was not in a Concord but in the Salt Lake Mail wagon that Monk took Greeley across the Sierras, and not to Placerville but to Eight Mile House. According to the San Jose *Pioneer* Monk wrote Greeley to remind him of their ride together, and Greeley replied: "I would rather see you ten thousand fathoms in hell than give you even a crust of bread. For you are the only man who ever had it in his power to put me in a ridiculous light before the American people, and you villainously exercised that power." A sense of humor was not one of Greeley's outstanding traits. Anyway, it was Mark Twain in *Roughing It,* and not Hank Monk, who made the ride notorious and had half the nation laughing at the editor of the *Tribune.*

Henry James Monk, says Drury, "was a chum of mine at Carson City. I can see him now, in his travel-worn suit of corduroy, beneath his battered old Stetson. . . . I remember how heartbroken the veteran was when he had the first stage upset of his long career." His coach overturned, spilling a dozen passengers, of whom eight were on top; and though none was seriously injured, "Hank was never the same jovial fellow again, and he died not long afterward." He would have become known as a driver if there had been no Mark Twain to make him famous, for his Sierra route

was one of the toughest in the West and he was undoubtedly one of the great drivers. It is said that he never failed to take mail and passengers through on time. Such a feat can be measured against the fact that the paved highways over the Sierras have time and again been closed to traffic, in spite of enormous snowplows and other road-clearing equipment. In fact, enormous buses have literally been buried by snowstorms and stranded; and if Hank never injured a passenger that is more than some drivers of buses can say.

Beebe found somewhere the statement that before alcohol became Monk's master he could turn his six horses and a coach on a dead run in a street, with every one of the six reins seeming to be loose. Haskins found another tale, that when descending a steep grade with his coach full the brake bar became useless. If Hank had not put his teams into a dead run down the mountain the coach would have run against and thrown the wheelers and the whole thing would have been brought up in a pile. It is said that when some of the passengers tried to leap out Hank shouted at them to sit still, and that one of them said, "We are making pretty fast time but Hank has hold of the lines and it's all right."[*]

There were other great drivers—Grinnel, Blair, Curly Dan, David Taylor, Crowder, Bill Norris, Hoffman, Putnam, Sickles—and possibly Charlie Parkhurst. Bowles rode under one of them one morning, and Speaker Colfax of the House said it took more talent "to drive a stage down the Sierras as we were driven than to be a member of Congress." It was not on a Sierras road where

For bold daring and brilliant execution, our driver this morning must take the palm of the world. . . . For several miles the road lay along "the hog's back," the crest of a mountain that ran away from the point or edge, like the sides of a roof, several thousand feet to the ravines below; so narrow that, pressed down and widened as much as possible, it was rarely over ten or twelve feet wide, and in one place but seven feet; and winding about as the crest of the hill ran;—and yet we went over this narrow causeway on the full gallop. . . . the road ran down sixteen hundred feet in the two miles to the hotel and it had thirty-five sharp turns in its course. . . . Mr. Foss has driven it, after this style, for many years, and never had an accident.

No doubt there were more drivers than one

---

[*] Readers interested in a description of the way Monk talked to his horses can see the *Land of Sunshine* magazine, XI, Dec., 1899.

named Charlie. Browne crossed the Sierras with one named Charlie, and asked, as they "bowled along the edge of a fearful precipice," "Do many people get killed on this route?" The driver is reported to have said, " 'Nary a kill that I know of. Last summer a few stages went over the grade but nobody was hurt bad—only a few legs'n arms broken. Them was opposition stages. . . . Drivers only break their legs a little on this route. . . . our company won't keep drivers as a genr'l thing that gets drunk and mashes up stages. Get aeoup, Jake! Git alang, Mack! 'Twont pay; 'tain't a good investment.' " In the dark how could he tell where he was? " 'I can tell where the road is by the sound of the wheels. When they rattle I'm on hard ground; when they don't rattle I gen'r'lly look over the side to see where's she's agoing.' "

It has been speculated that Browne was riding with Charlie Parkhurst, but so far as we know this Charlie never had a route across the Sierras. Farish, who rode with her, says she was "as good a driver as could be found. Disguised in male attire, she drove a stage for twenty years, from Knight's Ferry to Stockton, without anyone suspecting her sex. About 1870 I made the trip with her. She was quite tall, broad shouldered and compactly built. Her face was much bronzed from exposure and she spoke in rather a masculine falsetto voice."

Most of the drivers, says Banning, were enigmas. That may overstate it for most of them but not for this Charlie, whose life as a driver began in New England when a man named Balch took her as a child out of a poorhouse, promising to make a man of her. He taught her to drive a team, then a four, then a six, and when still in her teens she is said to have been an expert with six heavy leather reins between her fingers. Birch and Stevens had known her back there, and after establishing their stage lines in California they sent for her. When, says Banning, "runners were crying out their routes and all men swearing as they would never have sworn save to accompany this grand male chorus, Charlie was there, somewhere among them." To what extent she has become a legend it is hard to say; we have searched far and wide trying to find a photograph of her, and to learn more about her than Farish left in the record.

He says not only that her skill as a driver was equal to that of any of the men but that she had nerves of steel and was braver than most of them. In support of such statements he says that she was driving across the Tuolumne River in high water when she sensed that the bridge was being carried away under her, and so "sharply plied the lash to her faithful animals, who sped forward with a bound only reaching the other shore as the bridge was swept away, . . ." He is also the source of the story that when stopped by road bandits and told to throw down the express box she threw it down, but said, "The next time you come to rob me I'll be fixed for you." Two weeks later when armed men tried to take the box "she shot the leader, applied the lash to her horses and escaped without harm. The body of the highwayman was found by a searching party in an old tunnel near the scene of the shooting."

The human male is likely to stand tall and strong and glowing in his sense of chivalry when he hears of a woman who passed as a man and in feats of daring was a man with the best of them. Most writers who mention her have accepted Farish as gospel. In her old age she suffered from rheumatism, and buying a ranch in Salinas Valley lived alone and died there. "In preparing the body for burial, to the utter surprise of every one it was discovered, for the first time, that she was a woman." Jackson says that the story of her became a newspaper sensation all the way from California to New England. Beebe and Clegg, who are not men to be frightened by legends, say that a doctor after her death certified that she had been a mother, and that according to Browne she did no more than to chew an extra large quid of plug tobacco when she drove on dangerous road. But such tales as that, and the one that after payday she was always broke because she drank so much whisky and squandered so much throwing dice for two-bit cigars, have the strong odor of legend.

Of all the stage drivers apparently only one proved faithless to his trust. Known in the records as Frank Williams, he is the subhuman creature who deliberately drove his loaded stage into ambush, where it fell into the hands of waiting road bandits. The theft of sixty thousand in gold and the murder of five passengers shocked the whole country. As to the privileged position of these drivers, which so outraged Mark Twain, there can be no doubt: Bowles, an enlightened observer, says that at the way stations the driver was king: he "has a 'dreadful winning way' with him, both for horses and women. The philosophy of it I do not understand, but the fact is universal and stubborn; he is the successful diplomat of the road; no meal can begin until he is in place; and there are no vacant seats where he drives—even the cold night air would not send our girls off the box, and inside, during the long ride." One wonders how many of the infatuated lovelies rode at the side of Charlie Parkhurst.

The stage driver may have been king of the distances, but it was the pony express that captured the heart of the country. Visscher, whose book was published in 1908, says a standing joke in California just before the express came into being was that the term of a member of Con-

Pony express riders

gress from that state might expire before he could reach the capital. The pony express had such a brief life (about a year and a half) and made so few round-trip journeys across the continent (about 150) that Mark Twain said, a bit wistfully, "So sudden is it all, and so like a flash of unreal fancy, that but for a flake of white foam left quivering and perishing on a mail-sack, we might have doubted whether we had seen any horse and man at all." But as a historian has pointed out, it so completely took the national fancy that "scribes, in regular relays, have been riding the tolerant animal for nearly seventy years"—indeed, the pony has become an "immortal Pegasus" and is still racing from station to station in motion pictures and books. And it certainly is

true that during the year and a half the pony feet were running they awakened the nation to the need of a telegraph line and a railway into the Far West.

Though the ponies covered nearly two thousand miles in ten days and outran the famous Butterfield coaches on the southern route, they carried only about a hundred letters on an average run, whereas a coach carried thousands. The rates were high—from one dollar to five dollars per half ounce, and this didn't pay more than a part of the cost; it is said that the total revenue was about a hundred thousand, and the costs a half a million. One company gathered four hundred picked horses, chosen for speed and endurance, and from many applicants chose eighty young men, all of them small, lean, and wiry, and some of them hardly more than boys. This is what Mark Twain saw when he looked at them:

The pony-rider was usually a little bit of a man, brimful of spirit and endurance. No matter . . . whether it was winter or summer, raining, snowing, hailing, or sleeting, or whether his "beat" was a level straight road or a crazy trail over mountain crags and precipices . . . he must be always ready to leap into the saddle and be off like the wind! . . . He rode fifty miles without stopping, . . . He rode a splendid horse that was born for a racer and fed and lodged like a gentleman; kept him at his utmost speed for ten miles, and then, as he came crashing up to the station where stood two men holding fast a fresh, impatient steed, the transfer of rider and mail-bag was made in the twinkling of an eye. . . . The rider . . . carried no arms—he carried nothing that was not absolutely necessary. . . . There were about eighty pony-riders in the saddle all the time, day and night, stretching . . . from Missouri to California, forty flying eastward, and forty toward the west. . . .

"Here he comes!" . . . Away across the endless dead level of the prairie a black speck appears against the sky, and it is plain that it moves. Well, I should think so! In a second or two it becomes a horse and rider, rising and falling, rising and falling—sweeping toward us nearer and nearer, . . . and the flutter of the hoofs comes faintly to the ear—another instant a whoop and a hurrah from our upper deck [stagecoach], a wave of the rider's hand, but no reply, and man and horse burst past our excited faces, and go swinging away like a belated fragment of a storm!

When the first of them dashed into San Francisco the hook and ladder and engine companies were out to welcome the hero; and the newspapers

The saddle. Some glamorized photos in books are not of authentic saddles.

reported that the multitude cheered "till their throats were sore; the band played as if they would crack their cheeks, . . ." It took seventy-five ponies to make the run from the Missouri in ten and a half days. The riders, Harlow says,

dressed according to their own taste, but the usual costume included a buckskin hunting shirt, cloth trousers, boots or shoes and cap or felt hat. Some had a full buckskin suit with the hair on for cold or rainy weather. . . . The letters were carried in a *mochila* or pair of saddle bags which had a hole in the front center, fitting over the saddle horn. This had four pockets or *cantinas,* one in each corner, so that one on each side was in front of and one behind the leg of the rider. Three of these pockets were locked and opened at the military posts and Salt Lake City, and under no circumstances anywhere else. The fourth contained the way mail, and each station keeper had a key for it.

Visscher says a bundle of letters might not be any larger "than an ordinary writing tablet" and that they were written "on the thinnest tissue paper. . . . no silly love missives among them . . . business letters only." He says that only one mail was lost: the rider was shot off his horse by Indians and the pony escaped with the pouch.

As for the records they are close to the fantastic for anyone who has ridden fifty miles in a day. The fastest run is said to have been in March, 1861, which carried Lincoln's inaugural address from St. Louis to Sacramento, 1,980 miles, in seven days and seventeen hours. Jack Powers (Los Angeles Jack) who, according to Horace Bell, was "like a lion walking among rats," laid a wager, which Jackson accepts as "well-authenticated," that he could ride 150 miles in eight hours. Using twenty-five horses and taking two rest periods of seven minutes each, he covered the distance in a shade over six hours and forty-three minutes. Anyone who thinks a jet plane is a miracle compared to that ought to try it.

At the time Visscher wrote, nobody knew who held the record for endurance or length of ride. Visscher is authority for the statement that F. X. Aubrey, for a wager of one thousand dollars, undertook to ride from Santa Fe to Independence in six days:

he sent half a dozen of the swiftest horses ahead to be stationed at the different points for use in his ride. He left Santa Fe in a sweeping gallop, and that was the pace kept up during every hour of the time until he fell fainting from his foam-covered horse in the square at Independence. . . . He would have killed every horse in the line rather than to have failed. . . . It took him just five days and nineteen hours to perform the feat, and it cost the lives of several of his best horses. After being carried into a room at the old hotel at Independence, Aubrey lay for forty-eight hours in a dead stupor.

Jim Moore was another who made a remarkable ride. Harlow says he

left Midway station one June night, headed westward with Government dispatches in his bag marked "urgent." At the end of his run, old Julesburg, 140 miles distant, he found that the rider scheduled to go eastward a few minutes later had been slain. . . . With only ten minutes' rest, Moore started back and completed his 280-mile round trip to Midway in fourteen hours and forty-six minutes, or at an average of eighteen

Diamond Springs Pony Express Station

miles per hour. Jack Keetly in the same manner, though not quite so speedily, doubled back over two men's stages, doing 340 miles in thirty-one hours, and riding the last five miles asleep in his saddle.

Sometimes a rider wrote his own story. This is W. S. Lowden:

Here my race commenced. I sprang into the saddle with the bags, which weighed 54 pounds. I changed horses nineteen times between Tehama and Shasta, touching the ground but once. This was at the Prairie House, where Tom Flynn, the man in charge of my horse, was actively engaged in a fight with the keeper of Wells, Fargo's horse and had let mine get loose. I rode my tired horse a little past where the fight was going on, sprang to the ground, caught the fresh horse by the tail as he was running away from me and went into the saddle over his rump at a single bound. . . . All the other changes I made while the horses were running, the keeper leading the horse I was to ride and riding his extra horse. I would make myself known with a whistle about half a mile before reaching the change, which gave ample time to tighten the cinch and start the fresh horse on the road. . . . I reached Shasta, sixty miles, in two hours and thirty-seven minutes. This was to end my ride [but the next rider was not able to go, and so Lowden took his place]. I had nine changes of horses between Shasta and Weaverville, and reached the latter place in five hours and thirteen minutes from the time I left Tehama. From Shasta to Weaverville was run after dark, with a light snow falling, but when I reached the mountains and had my favorite horses to ride—Wild Cat, Comanche, Greyhound, Pompey and Jack—snow did not make much difference in speed. . . .

The most famous of the pony express riders is, of course, Bob Haslam, who achieved, says Harlow,

what seems to us the most heroic feat of all in his great ride that spring when the Indians were ravaging Nevada. His course lay between Friday's Station and Buckland's, later known as Fort Churchill. When he reached Reed's Station going east that day, he found no horses, they having been seized by the volunteers to use with the troop fighting the militant reds. Bob fed his horse and rode on to Buckland's, fifteen miles. This was the end of his seventy-five mile course, but he found the next rider, Johnny Richardson, so alarmed at the Indian danger that he positively refused to go out. Superintendent Marley was at the station and offered Bob $50 to ride Richardson's route. Within ten minutes he galloped away on a lonely ride of thirty-five miles through alkali dust in the sink of the Carson. A quick change and a spin of thirty-seven miles more to Cold Springs; and finally, a thirty-mile stretch to Smith's Creek, where Jay Kelly took the bags from him and sped eastward. He had ridden 185 miles without stopping save to change mounts.

On another ride he is said to have covered 384 miles with only a rest of nine hours.

Compared to the "whips" of the stage lines and the riders of the express, those who handled the freight teams may seem very dull indeed, but packing and freighting in the American West is a neglected and very colorful and dramatic chapter of history. There was, for instance, the important and dangerous task of transporting from the mines the enormous riches in gold and silver.

The position of the express messenger became fully as important as that of the driver; for there he sat at the side of the driver, with a short double-barrel shotgun loaded with buckshot across his knees. Road bandits became so numerous and robberies so frequent that miners and others journeying with their own dust sought ways to outwit the bandits. One ruse was to use mules instead of a stage: part of the filling of an aparejo was removed and a bag of gold dust put in each side of it. A mule might carry $3200 worth at $16 an ounce, and sometimes there were

cured; and when all is cast loose, each man removes his half of the cargo and places it on the ground. Another mule is then led up to the same spot, and unpacked in like manner; the cargo being all arranged along the ground in a row, and presenting a very miscellaneous assortment of sacks of flour, barrels of pork or brandy, bags of sugar, boxes of tobacco, and all sorts of groceries and other articles. When all the cargoes have been unpacked, they then take off the *aparejos,* or large Mexican pack-saddles, examining the back of each mule to see if it is galled. The pack-saddles are all set down in a row parallel with the cargo, the girth and saddle-cloth of each

Ruby Valley Pony Express Station

mule trains of forty or fifty beasts, with an armed guard. So far as is known, no treasure shipped in this manner was ever taken. Another trick was to pack beasts with dry hides under which gold dust was hidden. Borthwick saw a Mexican packtrain with

from thirty to fifty mules, and four or five Mexicans to attend them . . . just as they were about to unpack and make their camp. . . . Two men unpack a mule together. They first throw over his head a broad leathern belt, which hangs over his eyes to blind him and keep him quiet; then, one man standing on each side, they cast off the numerous hide ropes with which the cargo is se-

being neatly folded and laid on top of it. The place where the mules have been unpacked, between the saddles and the cargo, is covered with quantities of rawhide ropes and other lashings, which are all coiled up and stowed away in a heap by themselves.

Donaldson says that mule trains numbering as many as sixty-five and each with a cargo of three hundred pounds carried the ore over dangerous trails from the Idaho mines.

It took a lot of freighting by wagon and muleback to bring supplies to the hundreds of inland camps. It has been estimated that in 1866 in the

Alex Toponce, typical freighter

Virginia City, Montana, area there were 2,500 men, 3,000 teams, and 20,000 oxen and mules hauling supplies from Fort Benton. Alexander Toponce, a freighter to these camps, tells in his reminiscences how in Brigham City, Utah, he saw an enormous dressed hog that weighed over six hundred pounds. Though he was already well loaded with food and tools from Salt Lake City he bought the hog for six cents a pound and spreadeagled it on top of the wagon cover across the strong bows. He hauled it in freezing weather all the way to Virginia City and sold it for a dollar a pound.

Compared to modern highways, which begin to fall to pieces in no time at all, the roads across the Sierras and other mountains were marvels of engineering. To grasp the incredible enterprise and skill of it the traveler over U.S. 40 or 50 today has only to reflect on the fact that teams of ten to twenty mules were pulling wagon-trailer loads of 84,000* pounds, and the wagons extended almost mule nose to tailgate over the entire hundred miles from Virginia City to Placerville. If a wagon broke down and had to move out of the line it sometimes waited for hours before there was space enough for it to move its twenty mules and wagon and three or four trailers back into the line. This freightline, like others across the Sierras and elsewhere, was moving huge piles of

* A memo attached to a photograph sent us by the Southern Pacific Railroad says: "Before arrival of the Iron Horse, the mule was the greatest traction factor in the development of the great Southwestern country. Camels, oxen and horses were tried but only the mule could be adapted successfully to conditions of the country for use in heavy teaming. . . . Some of the freighting outfits were awe-inspiring affairs. The team might be anything up to 24 mules driven with uncanny skill by a single 'jerk line' in the hand of the 'freighter'. Some of the freighting wagons had wheels fully eight feet high and capacity for half a carload of goods. Usually four wagons, diminishing in size to the end vehicle, were pulled by a team of 16 mules."

Across the Sierras

A 20-mule team took some handling

machinery, lumber, and merchandise of all kinds to the camps. A Sacramento newspaper said that in 1863, 2,772 teams with a total of 14,652 animals were employed on the freight routes to the Comstock area.

Because of heavy losses to the stage and ore-train robbers, some of the miners came up with the idea of milling the silver into 750-pound balls, which would be too heavy for bandits to make off with. As late as 1964 the press carried the story of a 629-pound ball of Nevada silver that was trucked all the way from Carson City through eleven state capitals to Philadelphia, with multitudes viewing it all the way across the continent.

Silver City, Idaho, main street

The burro, so indispensable to the prospector, has been one of the most abused animals of all the animals that the *human* animal has abused. The punishment some of them took is almost beyond belief.

CHAPTER 8

Food, Clothes, and Lodging

Some of the visitors to the California diggings were horrified by what they saw, and none more than Sir Henry Huntley, an English prig, who probably looked through a monocle at the people, their lodgings, and their food. And above all at their food, for it was this that drew his most prodigious sneers: "After one has eaten of a number of messes, then comes in a really good roasted joint, which one has no appetite to partake of. After dinner to the general drawing-room —no tea, a melancholy want!" It must have been a high-class home that had a drawing room. Usually with Sir Henry it was "Tripe and pork! What a combination of nauseous horrors!" The meals, he thought, were "very often sorry affairs both in the material and in the society; every labourer, miner, teamster, or other person who happens to be at the house, sits down together; certainly, each of them has washed, at the general basin, his hands, and brushed his hair with the general brush hanging on the wall of the supper-room; in every other respect each is as he came from his work, dirty enough." There can be little doubt of that but one wonders what on earth Sir Henry was doing there.

"Oh! this American table-d'hôte! I would rather live in a tent under a tree, than be exposed to its disgusting combinations. . . . I dined to-day with Mr., commonly called 'Colonel' G--n. Salmon, mutton, pudding, and coffee, all crammed down one's throat together; then cigars and brandy, in a small room. Made an excuse to get away, and walked about in the air, till the execrable smell of tobacco had left my clothes." He thought that "the cookery is all grease and bad meat; liver and pork—the latter often from the groveller of the grave-yard, and always having the run of the slaughter-house—are here held in much esteem." But, blessed day, he at last dined with people in a house "so like that of English people—clean and comfortable. I change dirty living, innumerable rats running about my bed even, and six squalling, ill-bred children, for the cleanest ménage, two very good babies, an adjoining fowl yard, and a yelping dog."

Marryat, another English visitor, met Sir Henry in California, and they probably shuddered together at the horrors around them. For Marryat didn't like the food any better. He said that "could the traveller by any possibility be present at each and every hostel at the same moment, he would find a stereotyped bill of fare, consisting, with little variation, of a tough beefsteak, boiled potatoes, stewed beans, a nasty compound of dried apples, and a *jug of molasses.*"

The nauseous horrors at which Sir Henry stared were on the tables of the best hotels and in the homes of some of the wealthiest citizens. It seems likely that he would have fled over the mountains toward England if he could have seen Peter Sapp prepare his breakfast. The cooking utensils alone would have put him to flight. Though unlike certain Indian tribes who let the tongues of dogs wash their dishes for them, the habits of Peter and the average miner were just about as filthy; we speak as one who long ago saw the utensils in a miner's shack, in a trapper's shanty, and in the wagons of sheepherders. Peter's cooking tools were just what Helper saw, a frying pan and a pot, "neither of which, except in rare instances, is ever washed." The miner "boils beans by the potful; his flapjacks are of flour and water cooked in grease, if he has any, or without." Cooked in grease, or without, they were enough to double any miner up with bellyache. Helper saw that miners had a bottle of molasses, "unstopped, and containing an amount of flies and ants," which they seemed to be unaware of. For such as Peter, a jug of molasses, even with flies, was a luxury. The miner, said Helper, "is not very squeamish

137

about his diet." It was unspiced fare, except for salt when they had it. Borthwick says he was told that no coyote would eat the flesh of a dead Mexican, because it was so infused with the sharp bite of the pepperpod or chili colorado. He also heard the story that cannibals would not eat captured white sailors because of the taste of salt in their flesh.

Toponce, a freighter, gives us a good picture of a lumber camp:

There were only two men to do the cooking for two hundred men and over. One had three big potash kettles, holding forty or fifty gallons each, walled up in the furnace. His part of the business was to cook pork and beans and make tea. The other man took care of the big round ovens outside, and baked the bread. There were no dishes to wash around the shanty. The fellow who cooked the pork and beans, sliced the pork up on a board, and the other fellow sliced the bread. Every fellow had to furnish his own tin cup. He would walk up to these fellows' tables and get a slice of bread and a chunk of meat, with which to make a sandwich; then dip into the pot of beans, and if he wanted any tea he used the same cup to dip into the kettle of tea.

Some of the cooks were so dirty that one diner, we are told, feigned sickness, after having watched his landlady "fondling the dogs and cats, and shortly afterwards without washing, thrusting her hands into the flour and mixing up the pan of biscuit." Those who knew nothing about her habits said no cook in the area could make such biscuits! Well, many a frontier wife mixed dough without cleaning away the filth under her fingernails.

Salt pork, beans, coffee, and flour cooked in some form were the principal foods in the diggings. Most of the men labored on beans, salt pork, and coffee month after month. The coffee-pot was usually a fourth or a half full of grounds, for usually coffee was dumped in, followed by water, and nothing was dumped out until the pot was well filled with dregs. The pot was set on the open fire and heated and reheated, often many times during the day. Flour was mixed with water and a little salt, stirred, poured into a pan swimming with sow-belly fat, and fried; or flour and water were mixed and baked under hot ashes. If there was no baking soda, seidlitz

Mrs. Gramb's boarding-house, Silver Reef, Utah, in the 1870's.

Typical boardinghouse
kitchen.

Butcher shop, Park City,
Utah, 1905.

Typical eating places for the Peter Sapps.

powders were used to "raise the bread." On such a diet as this whole camps were sometimes laid low with scurvy.

When Borthwick arrived in Placerville he found that in the boardinghouses "miners sat down three times a day to an oilcloth-covered table, and in the course of about three minutes surfeited themselves on salt pork, greasy steaks, and pickles." A hotel had the "extra luxuries of a table-cloth and a superior quality of knives and forks." Welsh cooks made dampers. After mixing dough of flour and water and a pinch of salt a fire of hardwood was raked down, and the dough was laid on hot ashes and covered over with ashes. Practice told by feeling the crust when the dough was cooked through. It was a very heavy and tasteless bread but it retained a part of its moisture for a week or so, and with

beans "stuck to the ribs" of hungry men in the diggings.

Richardson, who crossed the plains in 1859, makes this astounding statement: "The experience of every mining region demonstrates that salt pork is the most nutritive and stimulating diet for miners, whose labor is the most exhausting in the whole world." According to him it was pork, pork everywhere; and he tells a story that possibly began with Mark Twain. A landlord set before a diner pork and mustard and nothing else. Observing that the diner was staring in turn at the two the landlord asked, "Sir, will you have some pork?" The man shuddered and said he couldn't stand the sight of pork. "Then," said the landlord, "help yourself to the mustard."

Delano says that sometimes the flesh of cow was

BILL OF FARE.

WARD HOUSE, RUSSEL & MYERS, Proprietors.

THURSDAY, DECEMBER 27, 1849.

Soup.

Ox Tail..$1 00

Fish.

Baked Trout, White and Anchovy Sauce.........................$1 50

Roast.

Beef...................... $1 00 | Mutton, do. $1 00
Lamb, stuffed............ 1 00 | Pork, Apple Sauce 1 25

Boiled.

Leg of Mutton, Caper Sauce.... $1 25 | Corned Beef and Cabbage...... $1 25
              Ham $1 00

Entrees.

Curried Sausages, a mie........ $1 00 | Tenderloin Lamb, Green Peas... $1 25
Beef, stewed with Onions...... 1 25 | Venison, Port Wine Sauce...... 1 50
         Stewed Kidney, Sauce de Champagne... $1 25

Extras.

Fresh California Eggs, each.....................................$1 00

Game.

Curlew, roast or boiled, to order..............................$3 00

Vegetables.

Sweet Potatoes, baked........ $ 50 | Irish Potatoes, mashed........ $ 50
Irish   do.   boiled .... 50 | Cabbage....................... 50
              Squash......................... $ 50

Pastry.

Bread Pudding............... $ 75 | Rum Omelette.............. $2 00
Mince Pie ....   75 | Jelly do. 2 00
Apple Pie...................   75 | Cheese...................   50
Brandy Peach .... 2 00 | Stewed Prunes   75

Wines.

Champagne................. $5 00 | Claret..................... $2 00
   do.   half bottles........ 2 00 | Champagne Cider............ 2 00
Pale Sherry................ 3 00 | Porter..................... 2 00
Old Madeira................ 4 00 | Ale........................ 2 00
Old Port, half bottles........ 1 75 | Brandy, per bottle.......... 2 00

☞ BREAKFAST—From half-past 7 to 11, A. M.
☞ DINNER—From half-past 1 to 6, P. M.
☞ TEA—From half-past 6 to 12.

Monson and Valentine, Book and Job Printers.

served, that had come two thousand miles on the overland trail.

There was not a particle of fat in the steak to make gravy, nor was there a slice of bacon to be had to fry it with, and the flesh was as dry and hard as a bone. But a nice broiled steak with plenty of gravy I would have—and I had it. The inventive genius of an emigrant is almost constantly called forth on the plains, and so in my case. I laid a nice cut on the coals, which, instead of broiling, only burnt, and carbonized like a piece of wood, and when one side was turned to cinder, I whopped it over to make charcoal of the other. To make butter gravy, I melted a stearin candle, which I poured over the delicious tid-bit, and, smacking my lips, sat down to my feast, the envy of several lookers-on. I sopped the first mouthful in the nice-looking gravy, and put it between my teeth, when the gravy cooled almost instantly, and the roof of my mouth and my teeth were coated all over with a covering like hard beeswax. . . .

A hungry man watching him began to drool, and so Delano shared his feast with him; but after the man rolled a large bite "about in his mouth a while" he spit it out with a great oath.

Some of the miners turned to verse to express the joys of feasting:

The beefsteak is of leather,
And the pies are made of tin,
The bread you could not cut it with a sword.
The butter wears side whiskers,
And the coffee's pretty thin,
In that little one-horse hash-house where I
    board.

Another poet became awfully tired of dried-apple pie:

I loathe! I abhor! detest! despise!
Abominable dried-apple pies;
I like good bread, I like good meat;
Or anything that's good to eat;
But of all poor grub beneath the skies
The poorest is dried-apple pies.
Give me a toothache or sore eyes
In preference to such kind of pies.

Hangtown Stew was popular in Hangtown after it became Placerville. It is said that after striking it rich a prospector swaggered into a plush beanery and asked what was the most expensive dish in the house. Told that oysters were, he ordered a pan of them, with eight or ten eggs mixed in, all of it garnished with strips of bacon.

No less savory for Kelly was a steak the way the Indians prepared it. A piece of flesh was skewered or tied up in the animal's own skin, and placed on very hot embers; "if carefully tended, before the hide is burned through the meat is thoroughly done, juicy and savoury beyond conception, being stewed in its own peculiar gravy."

Delano tells how the miners prepared raccoon: "First catch your coon and kill him, skin him, and take out the entrails; cut off his head, which throw away; then if you have water to spare wash the carcass clean. Parboil an hour to take out the strong musk, then roast it before the fire on a stick. While it is roasting, walk ten miles, fasting, to get an appetite, then tear it to pieces with your fingers and it will relish admirable, with a little salt and pepper, if you happen to have them. A tin cup of coffee without milk, taken with it, makes, under the circumstances a feast fit for the gods."

Buffalo, Greeley reported, was a

general favorite, though my own experience of it makes it rough, dry, wooden fiber, only to be eaten under great provocation. I infer that it is poorer in spring than at other seasons, and that I have not been fortunate in cooks. Bear, I was surprised to learn, is not generally liked by mountaineers. . . . The black-tailed deer . . . is a general favorite; so is the mountain hen or grouse; so is the antelope, of course; the Elk and mountain sheep decidedly less so. None of our party liked horse nor knew any way of cooking it that would make it really palatable. . . . Our conductor had eaten broiled wolf, under compulsion, but would not recommend it; but he certified that a slice of cold boiled dog—*well* boiled, so as to free it from rankness, and then suffered to cool thoroughly—is tender, sweet, and delicate as lamb.

Bryant said that "No delicacy which I have ever tasted of the flesh kind can surpass this when properly prepared"—that is, the tenderloin of fat young buffalo cow. The liver, tongue, hump ribs, and the part which plainsmen called the marrow gut, were also favorites. When eating tasteless food that he could barely chew Peter, for comfort, would force his thoughts to the Digger Indians, who had brought to the camp, on his overland journey, crushed serviceberries mixed with pulverized grasshoppers, or red ants, heated on hot stones and crushed to a powder. In California he soon became tired of a reddish berry called manzanita, or a tiny apple, which grew on a curious small tree that shed its bark once a year. And like Littlebury Shoot and his other companions who did well to find enough gold for sow belly and beans he became accustomed to a mouth lined with cold tallow, with which he sometimes cooked his almost indigestible flapjacks. Most of the miners needed fruits and vegetables as they needed nothing else on earth. Delano found that the wild onion was abundant: "They were about the size of a hickorynut, and covered with a close net-work, which is stripped off like a husk, leaving the onion clear and bright, and equal in flavor to any I ever ate." But many of the men rejected such wild fruits and vegetables as could be found, believing that they had no strength in them for the arduous digging.

Bowles and Richardson and others have left records of the fantastic Chinese banquets:

There were three hundred and twenty-five dishes. . . . Gov. Bross religiously tasted every one. Here are a few: bamboo soup, bird's-nest soup, stewed seaweed, stewed mushrooms, fried fungus, banana fritters, shark fins, shark sinews, reindeer sinews, dried Chinese oysters, pigeons, ducks, chickens, scorpions' eggs, watermelon seeds, fish in scores of varieties, many kinds of cake, and fruits ad infinitum. . . . Neither butter nor milk is used in cooking. The Celestials drank champagne and claret as if to the manner born. At every sip, each guest bowed seriatim to every other person at the table.

Good food was to be had for those who could afford it but it was a wide world between the Peter Sapps and the bonanza rich. On the Comstock the International Saloon's sign said that "Jim Gray, the Handsomest Man Living, runs this institution by day"; it forced the Black Crook Saloon to lively competition, which included six pretty girls "Direct from New York by steamer" as well as "the finest wines, liquors and cigars" and an Oyster Bar "where will be found every luxury obtainable in this market." It was no unusual thing, says Taylor, to see a group of miners who had made a lucky strike, and who "never before had a thought of luxury beyond a good beefsteak and a glass of whiskey, drinking their champagne at ten dollars a bottle, and eating their tongue and sardines, or warming in the smoky camp-kettle their tin canisters of turtle-soup and lobster-salad." But sometimes, says Rickard, the miner who found rich diggings

couldn't eat expensive food, after years on pork and beans. He tells of lunching with that successful miner and millionaire, Thomas F. Walsh, "at the Elysee Hotel, on the Champs Elysées, in Paris. It was no Elysian Fields to him. While I enjoyed an *omelette aux fines herbes* and some *fraises de bois,* he was restricted to a diet of tea, toast, and tomatoes."

Beebe, an authority in these matters, says that in Goldfield the Palm Grill had practically everything on the menu that could be ordered in the Palace in San Francisco or at Rector's in New York, including desert quail and twenty champagnes on the wine card, not to mention sand dabs, antelope steak, and Cape oysters from Buzzards Bay.

Wolle dug up a newspaper story on a grand dinner at the Teller House in Central City—"The most elaborate bill of fare we venture to say that has been provided by any hotel west of the Mississippi—fishes, veal, beef, mutton, pork, oysters in every style, turkey, mallard duck, mountain grouse, prairie chicken, wild turkey, antelope, venison, buffalo, Rocky Mountain black bear, with entrees, vegetables, relishes, side dishes, pastries, puddings, ice-cream, jellies—" a dinner "incomparably greater than has ever before been witnessed in the west," at five dollars per head. When the famous Teller House was cleaned up in 1932 it was found that under ten layers of wallpaper there were "classic murals."

Whether that was the grandest dinner west of the Mississippi up to that time no man can say, for there were many elaborate dinners in various mining camps for the bonanza princes and their guests. Before that dinner there was in Russell Gulch a banquet that included "Bean soup; Brook trout, a la catch them first; Antelope, larded pioneer style; Biscuit, handmade, full weight, a la yellow; Beans, mountain style warranted boiled forty-eight hours a la soda; Dried apples; Coffee in tin cups, to be washed clean for the occasion." That, of course, was only a feast for the riff-raff in the diggings, who helped for the occasion to make a flag out of blue jeans, a "biled" shirt, and red flannel drawers.

The food in Virginia City, Nevada, in the expensive places is said to have been as fine as any in New York. Possibly there were no finer wine cellars in the country, but this camp had a whole stable of multimillionaires. Later, in Tombstone, there were two spots for the gourmets, the Can Can and the Maison Doree. Quail on toast, one writer reported after sampling it, was about the tastiest morsel that side of Paris.

For drink with their meals, Peter Sapp and his friends had the kind of coffee we have described, or water, though water in some camps was poisonous. In Nevada's Virginia City and Gold Hill the water was so heavy with alkali that alkali sickness became a stock joke in the camps. There were millions of gallons of pure cold water in the Sierras only twenty-five miles west as the crow flew but a valley 1,700 feet deep lay between the water and the camps high on the Washoe. Bringing the water to the camps was a feat in engineering so spectacular as to be impressive even in the jet age. An order, for instance, was given to a San Francisco foundry for iron pipe that would make so many turns and dips and meanderings down the mountains, across the valley and up the mountains that thousands of different moulds had to be made for casting the pipe. And all that iron pipe was hauled over the Sierras on wagons. Little wonder that cannons exploded, brass bands blared, and fireworks filled the sky when at last cold water without a trace of alkali gushed into the thirsty city! For the people had been drinking water that contained not only the lime of the Nevada desert but arsenic, graphite, copperas, potable gold, and liquified silver. That, anyway, is what Professor Greever says—and that "hundreds died from it." In the camps in arid regions, as Editor Drury points out, bathing was not overdone. Candelaria had to haul its water eight miles but it is a libel, Drury says, tongue in cheek, to argue, as some have, that barkeeps threw into a cocktail glass a piece of glass instead of ice.

Mark Twain became acquainted with a famous drink when a waiter "poured for us a beverage which he called 'Slumgullion,' and it is hard to think he was not inspired when he named it. It really pretended to be tea, but there was too much dish-rag, and sand, and old bacon-rind in it to deceive the intelligent traveler."

Haskins, who was a Forty-niner, gives a typical bill of fare at a better restaurant, where choice of cabbage, sweet and Irish potatoes, and squash were offered, and choice of eight desserts, all at what appear to be reasonable prices. At the Eldorado in Hangtown a menu offered beef, wild;

beef up along, a la mode, and tame, with one potato, "fair size"; baked beans, plain, at 75 cents, greased for $1.00; hash, low grade, for 75 cents, and whole jackass rabbit for $1.50; and grizzly roast at $1.00 and fried at 75 cents. Under pastry was rice pudding, plain, 75 cents, with molasses, $1.00, with brandy peaches, $2.00. A "square meal" with dessert was offered at $3.00.

One is startled to learn that famine struck some of the camps, for advertisements frequently appeared in some of the camp newspapers that told of huge shipments of food. Thus in the *Avalanche,* August 3, 1867, a merchant said he had just received 100,000 pounds of California flour; 50,000 of bacon; 20,000 of lard; 20,000 of beans; 50,000 of Humboldt salt; 5,000 of peaches; 2,000 of cheese; 200 cases of "can fruit"; 100 kegs syrup; 150 barrels sugar; 20,000 chemicals; 5,000 of green and 3,000 of ground coffee; 200 barrels of wines and liquors, 50 baskets of champagne, 50,000 cigars, 5,000 pounds of tobacco, and other items.

But famine, or at least extreme hunger, was almost a commonplace of camp life. In Grass Valley, in the winter of 1852-53, there was a "Hungry Convention." Food was in such short supply that a meeting was called, but it immediately, Shinn says, "degenerated into a wild burlesque. . . . A duly elected committee reported, declaring war upon San Francisco, and their resolve to have supplies thence 'peaceably if we can, forcibly if we must.'" The cause that time was unscrupulous hoarders and speculators. Sometimes the cause was deep snow.

In the severe winter of 1864-65 there was such a shortage of food in Virginia City, Montana, and Idaho City, Idaho, that flour climbed to $150 a hundred. In Virginia City five hundred armed men searched the camp for flour and seized what they could find, promising to pay for it at $27 a hundred, which was its price before it skyrocketed. They then distributed the flour in twelve-pound lots to those who could show need, and paid for it. In Idaho City hungry mobs rioted and looted.

Sometimes exorbitant prices were caused by stock manipulators or by dishonest bankers. After the California rush of '49 and '50 many businesses and banks collapsed. There had been "wholesale fraud among the great banking institutions, which dissipated or stole the people's money." Honest Harry Meiggs, a man of spotless reputation and a host of friends, had milked rich and poor alike in an enormous swindle, and with a million or two safely hidden took off for Peru. The cheats and swindlers on the Comstock brought collapse to the whole Pacific coast, and thousands to ruin.

The fact should be kept in mind not only that the dollar's purchasing power in those years was five or six times what it is today but that the smaller coins were spurned, even though a dime, say, was worth a fifty-cent piece today. Some writers have said that nobody bothered with a coin smaller than a half-dollar, but Bowles is probably nearer the truth, for he was there: "There is no coin in use less than a dime; one of these answers as 'a bit'; two of them will pass for two bits, . . . but a man who often offers two dimes for a quarter of a dollar is voted a 'bummer' . . . The currency of these States is gold and silver. Paper money has been kept out by the force of a very obstinate public opinion and the instrumentality of State legislation. Our national currency of greenbacks are seen here simply as merchandise; you buy and sell them at the brokers, for about seventy-five cents in coin to the dollar."*

As for clothing, the well-dressed dandy of '49 probably wore buckskin pants, fringed, of course, a red silk sash, a fancy buckskin shirt, trimmed with bell buttons, and a broad-brimmed felt hat. From his waist on one side hung a revolver, from the other side a knife; and his leather boots had a fresh shine. But the men in the diggings were not dandies. Most of them, Morrell thinks, "developed a dress and appearance of their own—red or blue open-necked woollen shirts, belts garnished with knives and pistols, trousers tucked into high boots, dingy slouched hats covering bronzed, bushy, bearded heads—and sought to differentiate themselves from the traders and parasites in the towns."

In this, as in nearly all matters, it is best to go to those who saw them. Says Colton:

A Californian is most at home in his saddle; there he has some claims to originality, if not in character then in costume. His hat, with its conical crown and broad rim, throws back the sun's rays from its dark, glazed surface. It is fastened on by a band which passes under his chin, and rests on a red handkerchief, which turbans his head, from beneath which his black locks flow out upon the wind.

The collar of his linen rolls over that of his blue spencer, which is open under the chin, is fitted closely to his waist, and often ornamented with double rows of buttons and silk braid. His trowsers [sic] which are fastened around his loins by a red sash, are open to the knee, to which his buckskin leggins ascend over his white cotton drawers. His buckskin shoes are armed with heavy spurs, which have a shaft some ten inches long, at the end of which is a roller, which bristles out into six points, three inches long, against which steel plates rattle with a quick, sharp sound.

But that is the native dandy, whose dress is far from that of the man in the diggings. Borthwick also saw the people—the Chinese

in all the splendor of sky-blue or purple figured silk jackets, and tight yellow satin continuations, black satin shoes with thick white soles, and white gaiters. . . . the lower orders . . . generally dressed in immensely wide blue calico jackets and bags . . . on their heads . . . wickerwork extinguishers, which would have made very good family clothes-baskets. The Mexicans . . . wore their natural costume—the bright-colored serape thrown gracefully over the left shoulder, with rows of silver buttons down the outside of their trousers, which were generally left open, so as to show the loose white drawers underneath, and the silver-handled bowie-knife in the stamped leather leggins. [Down-east Yankees preferred] their eternal black dress-coat, black pantaloons, and black satin waist-coat. . . . Those who did not stick to their former style of dress, indulged in all the extravagance of California costume, which was of every variety that caprice could suggest. . . . The prevailing fashion among the rag-tag and bobtail was a red or blue flannel shirt, wide-awake hats of every conceivable shape and color, and trousers stuffed into a big pair of boots. Pistols and knives were usually worn in the belt at the back, and to be without either was the exception to the rule. . . .

Let's take a man fresh from the diggings in Colorado and see what he wears: in February, 1861, the Territory was established, and the member elected to its legislature from California Gulch arrived and

laid down his blanket at the door of the house of representatives. He had been mining and possessed no other clothes than those he wore at his sluice-box in the gulch. His dress was a blue flannel shirt, trousers patched with buckskin, an old boot on one foot and a brogan on the other, an old slouch hat that he had slept in, the brim partly gone. His face was blackened by the smoke

_____
* This attitude toward hard and soft money has persisted in the West to this day, especially in the silver states.

of the campfire and furrowed by perspiration, his eyes hollow with fatigue and hunger, feet blistered by walking, hair tangled and beard yellow with dust. The next day we elected this same George Crocker speaker of the house. [The members of the house] batched in their little assembly room, took turns at cooking at the fireplace . . . carried water in pails from the creek, and slept on the floor.

It would seem to be a few light years between that spirit and the status-seeking spirit of the na-

An elegant costume of the time

tion today. In the camps it was ordinarily impossible to distinguish between the millionaire and the Peter Sapps, in their manner of dress. It has been said that if John Mackay and Jim Fair had stood among a hundred miners in the Crystal Palace saloon in Virginia City nobody could have identified these two multimillionaires by their clothes.

In regard to the women the matter seems to have been different. Borthwick says that the "numbers of female toilettes, of the most extravagantly rich and gorgeous materials, swept the muddy streets, and added not a little to the incongruous variety of the scene." For Bowles the

ladies generally dress in good taste. Paris is really as near as San Francisco as New York, . . . But the styles are not so subdued as in our eastern cities; a higher or rather louder tone prevails; rich, full colors, and sharp contrasts; the startling effects that the Parisian demi-monde seeks—these are seen dominating here. . . . Perhaps in no other American city would the ladies invoice so high per head as in San Francisco, when they go out to the opera, or to party, or ball. Their point lace is deeper, their moire antique stiffer, their skirts a trifle longer, their corsage an inch lower, their diamonds more brilliant. . . .

The Boise *News* observed that

We frequently, of late meet ladies in the streets who, from outward appearances, have abandoned the use of hoops as an article of wearing apparel. Although at first blush they presented rather a forlorn and shrunken appearance, yet as we become accustomed to see them more frequently, they really look more comfortable—taking into consideration the wintery state of the weather— than those who still indulge in hoops, which are purely a Summer luxury, and should be discarded in Winter, or else we are an old bachelor and don't know anything about it.

In the *Montanan* the editor was more disgusted:

Chemile, "shemlin," "chimi-drawers," "getin," "chimjupe," and "chemiloon" are among the proposed names for the new what-d'ye-call-ems introduced by the female dress reformers. We don't want to be captious about trifles, but the use of such abominable language ought to constitute sufficient grounds for a divorce.

In Idaho the *World* gave its recipe for making a fashionable woman:

Take ninety pounds of flesh and bone, mainly bones, wash clean and bore holes in the ears, bend

the neck to conform with the Grecian bend, the Boston dip, the kangaroo droop, the Saratoga slope or the bullfrog beak, as the taste inclines. Then add three yards of linen, one hundred yards of ruffles and seventy-five yards of edging, eighteen yards of dimity, one pair silk-cotton hose with patent hip attachments, one pair of false calves, six yards of flannel, embroidered, one pair Balmoral boots with heels three inches high, four pounds whalebone in strips, seventeen hundred and sixty yards of steel wire, three-quarters of a mile of tape, ten pounds of raw cotton or two wire hemispheres, one wire basket that would hold a bushel, four copies of the *World,* one hundred and fifty yards of silk or other dress goods, five hundred yards of lace, fourteen hundred yards of fringe and other trimmings, twelve gross of buttons, one box pearl face powder, one saucer of carmine and an old hare's foot, one bushel of false hair frizzed and fretted *á la maniaque,* one bundle of Japanese switches with rats, mice and other varmints, one peck of hairpins, one lace handkerchief, nine inches square, with patent holder. Perfume with attar of roses or "Blessed Baby" or "West End." Stuff the head with fashionable novels, ball tickets, playbills, weddingcards, some scandal, a lot of wasted time and a very little sage. Add a half-grain of common sense, three scruples of religion and a modicum of modesty. Season with vanity, affectation and folly. Garnish with earrings, finger rings, breastpins, chains, bracelets, feathers, and flowers to suit the taste. . . .

In Denver, ladies were strolling along the streets

arrayed in the newest, costliest silks from Stewart's made up in strict conformity with the latest fashions (crinoline and expanding hoop skirts). The daintiest bonnets are gracefully appended to the back of their dear little heads, and butterfly parasols have shielded them from the ardent rays of "old Sol" for months past. Keevil hats, Heenan neckties, patent leather gaiters and the complete get-up that goes to make up a dandy outfit no longer look strange to admiring savages.

There was "society" in such places as San Francisco, Denver, and on the Comstock. Many of the mining camps had no women in them at all; in others there were a few wives, an adventuress now and then, and the soiled doves, as we shall see on a later page.

The lodgings in the camps were about on a par with the food. Helper tells of the "miserable, filthy little hotels" in Sacramento; one advertised "Tiptop Accommodations for Men and Beasts"; another, "Eating is Done Here"; a third, "Good Fare and Plenty of It"; and still others told the weary traveler, "Replenish the Stomach in Our House"; "Rest for the Weary and Storage for Trunks"; "Come in the Inn, and Take a Bite."

Because building materials were scarce in many places the structures were often joined in row on row, with only one wall between every two; their colors as likely to be red, green, or orange as any other, or a combination of these; and their false fronts architecturally indifferent to everything around them, and of all shapes and sizes. These gingerbread fronts, the subject of much astonishment and outrage in many books, were really only billboards for signs; and if such a building were a saloon, the words "Saloon," "Whiskey," and "Bar" were all over its front. Under the signs were long wooden benches or stout chairs in rows where the patrons could sun themselves and squirt their tobacco juice. Because so many camps were in gulches and ravines the buildings were on hillsides, and as a consequence many of the privies were two-story structures.

A minister has left a description of the luxurious quarters to which he took his wife in Virginia City, Montana: "Our parsonage at Virginia City was built soon after our arrival there of en-

tirely green logs from the woods, and chinked with the mud nearest at hand. It was 13x16 feet, with dirt roof, one window filled with glass and the other two covered with muslin. A carpet was extemporized of cow skins, placed hair side up, and stretched tightly while green, and nailed to the floor." Well, if he had a plank floor he was a lucky man.

Visitors of fastidious tastes suffered dreadful shocks before more elegant structures were set up. A room given a couple in Montana for $125 a week in greenbacks (worth only $25 in gold!) must have left them limp and undone, for it was about the size of a clothes closet, with a crude bedstead made of planks, with small poles across for slats; a mattress stuffed with meadow grass or tule, and a pillow like it, that gave off many dry anxious grass sounds when they laid a cheek against it; an old blanket or two, a chunk of tree trunk to sit on when taking shoes off; and a privy out back which it were folly to describe, for no-

A privy still standing, Silver City, Idaho

Privies are between the two rows of houses

Two-story privy on a hillside

Miners' hillside dugouts

body could accept it as real who hasn't seen one. If it was an elegant hotel there might be a fragment of mirror, a towel that hadn't been used since its last cold wash in the creek, a piece of soap made of tallow and claybank, and an old gold pan as a basin to wash in. When the minister, above, found only one sheet on his bed and asked where the other sheet was he was told that it was in the dining room serving as a tablecloth.

Bishop D. S. Tuttle in his *Reminiscences* tells of having stayed in the hotel, given above, with the grass pillow, but his sojourn was a year later and he seems not to have known that the accommodations had been hugely improved. "We stop at the 'Planter's House,' the Fifth Avenue Hotel of the place. Our bedroom is about twelve feet square, and in it are one double, and two single beds. Mr. G. and I occupy the double. As there is need, the landlord sends whom he will into the cots. No lock is on the door; no wash bowl or pitcher in the room. Every morning we go down to the office to wash, wiping our faces on the office towel." Poor dear bishop, he should have seen the accommodations that cost only half as much! He does not mention bedbugs, lice, crickets, and other vermin, that were as common in most camps as grass pillows, bedsprings that were only pine poles, and filled-up privies. One Montana editor gravely told his readers that it had become a habit with some of the men to "take chunks of raw beef to bed with them to give the bugs a chance to feed."

In nearly all the camps the earliest lodgings were not of lumber or stone but of canvas, often ragged and shredded, or dugouts in hillsides, or slab or brush lean-tos; and for those more energetic or with the means to hire labor, crude log huts with dirt roofs and dirt floors, of which we give a generous representation here. Log huts with a ridge log and one or two other roof beams across which poles were laid, with six inches of earth thrown upon the poles, were, in areas of heavy rain, an unmitigated curse. Such a roof, after becoming thoroughly dried out in summer heat, would, under a sudden deluge, dissolve and disintegrate and pour in streams of mud between the poles and upon the earth floor below. In Virginia City, Montana, the earth roof of a saloon collapsed and a gambler sitting under it at a poker game was killed. Far better than such log huts roofed with earth was a dugout properly built in a hillside.

Peter Sapp's was not properly built but it served for a while. He tunneled back about ten feet into a steep hill face and then drifted right and left three or four feet, so that he had a dark earth chamber about eight by ten feet. In one corner he piled pine and juniper boughs and on these he spread the rags of his bedding. He found two or three chunks of old tree bole to sit on or to use as tables; and in a corner by the entrance he set his coffeepot, a black and battered thing always half-filled with dregs, and a battered frying pan, together with an old knife, a tin cup, and a tin plate. He built his fire at the entrance, to do his cooking, and when the wind was against

An Alaska cabin

Countless thousands of such cabins were built in the West

Typical miners' shacks

Typical miners

him his dugout was filled with pine wood smoke, which made a nice scent in his dreams all night long. In wintertime he went to the gambling halls for warmth and watched the men with richer diggings lose their gold dust; and about midnight he went to his cold black hole to crawl into his pile of cold bedding, without undressing, to shiver all night.

Those who came into sudden wealth advertised the fact with their houses, a few of which still stand, with their outlandish scrollwork and their heavy furniture built for the ages—their marble fireplaces, huge mirrors, crystal chandeliers, and monstrous beds and sideboards. The same kind of heavy furniture was in the saloons—the chandeliers, bars, mirrors, lamps, tables, with various murals and other incomparable works of art to beguile the drinking males, such as a nude woman in the clutch of a god, being abducted heavenward. Under the bar, Emrich says, was a rubber bulb which the barman could press to send a current of air against the painting in the area of her navel, to cause her to "palpitate ecstatically

and the customer to order another cooling round."

An old-timer who knew Boise as well as his own face says of John Henry Thomas Green's Missouri House that the chief furniture of the porch and office "consisted of tall split-seat hickory chairs and a vast number of deep wooden boxes filled with sawdust, in which the guests were wont to 'pflug' their tobacco juice (the proper word is 'ambeer'). The comfort of the guests demanded deposit boxes to the right, left, middle, and below." At its opulent peak Nevada's Virginia City commanded such attention that its plush International Hotel had a Presidential suite, and some New York and San Francisco newspapers afforded news bureaus in the silver camp. Wives of the rich men there, such as John Mackay's (whom see), summoned landaus with footmen when they felt an urge to call on a rich neighbor; and in the Washoe Club, for those who could afford it, the bill of fare offered everything from terrapin to tropical fruits. There were other famous hotels, a few of which still stand.

Boarders

Three famous hotels: the Idaho in Silver City; the Overland in Boise; the International in Virginia City, Montana Territory.

International Hotel
Virginia City, Nevada.

Gold Hill Hotel, Gold Hill, Nevada

CHAPTER 9

Camp Culture

Since in the camps there were so many saloons, gambling halls, brothels, and so much brutality and violence, it is sometimes assumed that there were no civilized refinements of any kind. On the contrary there was a more vigorous intellectual life than there is in some of the cities that the camps have become. It would be invidious to cite instances but a thorough examination of the early *Statesman,* published in Boise, reveals far more interest in books (for example) than that newspaper has today. Taking the camps as a whole one is astonished by the number of advertisements of books and bookshops in the mining camp newspapers. It was not at all uncommon for the editors to call the attention of their readers to the books available to them. The *Statesman,* for instance, said on March 19, 1868, "If there is any one thing for which those who have the moral and intellectual good of this community at heart ought to be thankful, it is the choice and high standard of literature supplied at the city

book store." In searching through the old newspaper files one is also impressed by the much higher quality of the newspaper fillers and editorials. Indeed, many of those mining-camp newspapers had an intellectual quality and vigor, a fearlessness in honest reporting, a devotion to justice and right, and a boldness in calling a spade a spade that have largely vanished from the newspapers of today. The ablest editors, if assembled as a group, would constitute one of the most remarkable bodies of men in American history.

We do not intend to argue that the camps were oases of culture. They were not. In most ways they were shockingly crude and rude and rampageous and a long way from the effeminate society of the nation today, but the essentials of which a civilization can be built were there. We are going to present such culture as the camps had in terms of newspaper fillers, language, children, animals, religion, Sunday, preachers and teachers, charity, and humor, for in their attitudes toward these things a people are pretty well reflected. Unlike some writers we shall have little to say about the theater: most of the men, if they had the price of admission, went to see *anything* that came along, no matter whether it was the bawdiest carnival fare or *Hamlet.* The theater, therefore, is not by any means as trustworthy an index of the hearts and souls of the people as their feeling toward animals and children and the destitute.

Reading Louis Dupuy was certainly not a common type in the camps but there were plenty of men who were much like him. He established the Hotel de Paris in Colorado, with exquisite French cooking and vintage wines, and if he didn't like the persons who came to his hotel to eat he didn't serve them. He had a fine library of

156

The public library in
Idaho City today.

several thousand volumes, and after a good dinner it was one of his joys to discuss philosophy, poetry, and other subjects with a few of his guests. That was not the kind of talk that Peter Sapp and his friends would have enjoyed, though in a vague way at least they were aware of poets and writers in the areas. The pursuit of gold seemed to awaken the literary genius not only in such as Clemens and Harte and Miller but also in such as jolly Jim Coffroth, Tuolume's favorite son and (he hoped) California's gift to the United States Senate. On the proper occasions he got up a madrigal or an ode or a high-

Four of the most famous newspapers. The *Enterprise* and *Epitaph* still occupy the original buildings; the *World* is in a structure more ramshackle than this renovated one; the renovated *Post* building is empty.

falutin' lyric, such as that immortal one for a blushing maid named Caroline. When Jim declaimed it to the blue skies above the gold fields even the frogs over in Calavaras were silent: Ye song clad Seraphs! Ye angelic train! That dwell amid the domiciles above, 'Mid stellate regions, that immortal reign, and wake the joyful saraband of love! It is said that the lines that brought the entire region to breathless attention were these: And thou bright compeer Truth! resplendent shrine! With tones auriferous wake the Heavenly Lyre! It is not known what Caroline thought of it, but a Judge Dorsey, after a moment of stunned silence, laughed his head off. He thought so highly of it for reasons that can be imagined that he laid five golden eagles on the desk of the Sonora *Herald* and told them to print the whole masterpiece.

The editors gathered from sources in libraries and other newspapers fine moral precepts, items about women, bits of humor of the kind the camps liked, or sometimes wrote fillers themselves. "If you want enemies, excel others; if you want friends, let others excel you. There is a diabolical trio existing in the natural man . . . pride, envy, and hate. Pride, that makes us fancy that we deserve all the goods others possess; Envy, that some should be admired, while we are overlooked; and Hate, because all that is bestowed on others diminishes the sum that we think due to ourselves." Three weeks later in the same newspaper: "Every man, if he would be candid, and sum up his own case as impartially as he would that of his neighbor, would probably come to this conclusion, that he knows enough of others to be certain that he himself has enemies, and enough of himself to be as certain that he deserves them."

Women came in for more than their share of ribbing: "Ye pining, lolling, screwed up, wasp waisted, putty-faced, consumption-mortgaged and novel devouring daughters of fashion and idleness, you are no more fit for matrimony than a pullet is to look after a family of fifteen chickens." Sometimes it was the man: "Jones was traveling with his wife and was so gallant in his behavior to his *cara sposa* that madame became uneasy and remonstrated against his attentions as too demonstrative for public observation. 'The duece,' said Jones, 'we're married, I suppose.' 'Yes,' said the lady, 'but judging by your deportment, folks will think that we are not married.' 'Well, what of it?' 'Why, not much to you,' said the careful dame, 'you are a man, but we women have our characters to take care of.' "

Most of the editors trained their heaviest guns on the pompous moralists. A poet by the name of Barker offered this gem to the Boise *News*:

One night, as old St. Peter slept,
    He left the door of Heaven ajar,
When through, a little angel crept,
    And came down with a falling star.

One summer as the blessed beams
    Of morn approached, my blushing bride,
Awakened from some pleasing dreams,
    And found that angel by her side.

God grant but this—I ask no more—
    That when he leaves this world of sin,
He'll wing his way to that bright shore,
    And find that door of Heaven again.

To which St. Peter-Editor replied:

Full eighteen hundred years or more
    I've kept my door securely tyed. . . .

I did not sleep, as you supposed,
    Nor left the door of Heaven ajar;
Nor has a "little angel" left,
    And gone down with a falling star.

Go, ask that "blushing bride" and see
    If she won't frankly own and say,
That when she found that angel babe,
    She found it by the good old way.

God grant but this—I ask no more—
    That should your numbers still enlarge,
That you will not do as before,
    And lay it to old Peter's charge.

But in the same issue the editor lifted this romantic gem from the *Atlantic Monthly*:

I watched her at her spinning;
And this was my beginning
Of wooing and of winning.

But when a maid opposes
And throws away your roses,
You say the case forecloses.

Yet sorry win one uses,
Who loves and thinks he loses
Because a maid refuses.

For by her once denying
She only means complying
Upon a second trying.

When first I said, in pleading,
"Behold, my love lies bleeding!"
She heard me half unheeding.

When afterward I told her,
And blamed her growing colder—
She dropped upon my shoulder.

Had I a doubt? That quelled it,
Her very look dispelled it,
I caught her hand and held it.

Along the lane I led her,
And while her cheeks grew redder,
I sued outright to wed her.

Good end from bad beginning!
My wooing came to winning—
And still I watch her spinning.

From the Cincinnati *Times* the *Avalanche's* irascible editor picked up this:

The man who wrote "Home, sweet home" never had a home. No, of course not. . . . The man who wrote "Old Arm Chair" never had an arm chair in all his life. . . . The author of "Take me Back to Switzerland" never was in Switzerland. . . . "Mother, I've come home to die" hadn't spoken to the old woman for years. . . . There is the author of "Old oaken bucket." There wasn't a bucket on the whole farm, water being drawn with a tin pail and a cistern pole. . . . "Hark, I hear the angels sing!" spent all his evenings in a concert beer saloon. Angels, indeed. . . .

Small fillers, picked at random, were of this sort: "A policeman in Buffalo has been fined $5 and costs for forcing a man in a horse car to give his seat to a lady. The Court said ladies had no more privileges than men, and, moreover, when treated civilly, were not profuse in acknowledgment." "Among the pensioners 'at the corner' there is an elderly lady who has always been contented to receive a five-cent piece. She astonished us yesterday, by returning it with some disdain, exclaiming, 'Why, yer honor, don't ye know everything is riz?' " "Rose Bingham, who threw her infant son out of a third story window in San Francisco, is but 12 years of age. She takes the matter but little to heart, and talks and laughs about it."

Alexander Dumas the younger was quoted:

Heaven, in its merciful providence, gave no beard to women, because it knew they could not hold their tongues long enough to be shaved. For the sake of women, men dishonor themselves, kill themselves; and in the midst of this universal carnage, the creature who brings it to pass has only one thought in her mind, which is to decide whether she shall dress herself so as to look like an umbrella or a dinner bell.

A prominent bachelor politician "on the Kennebec, remarked to a lady that soapstone was excellent to keep the feet warm in bed. 'Yes,' said the young lady, 'but some gentlemen have an improvement on that, which you know nothing about.' The bachelor turned pale, and maintained a wistful silence." An outraged husband whose wife had run away from him published this description of her: "My wife is about five feet high, has light brown hair, right eye out, the other is of a pale blue color, has a pimple on left temple, near eye, pug nose, has a scar across one arm, is stoop shouldered, has short, thick feet and easily makes the acquaintance of strangers." A man enamored of Miss Anna Bread inserted the following: "While belles their lovely graces spread And fops around them flutter, I'll be content with Anna Bread And won't have any *but her.*"

Language The language of the mining camps, as well as that of the cowboy, the trapper, and of other frontier types, can be found in various compilations. That presented here we have taken from books written by persons who were in the camps.

More than one of them was shocked by the profanity. Sir Henry, whom we have met before a plate of nauseous horrors, cried out,

How brutal the hotels are here! A bar room, behind the bar of which a man sells execrable spirits and wine to a congregation who do nothing all day but eat, drink, smoke, spit and blaspheme; it is no common swearing; the style is that of studied blasphemy. I defy the world to equal the dreadful language these California-Americans delight in, not when angry, more than when engaged in common conversation; it is insufferable; . . .

Poor Henry Huntley! He had no more notion of the psychology of profanity* than he had of a whore's morals.

Dame Shirley commented:

_____
* See our essay in *Thomas Wolfe and Other Essays*.

I think that I have never spoken to you of the mournful extent to which profanity prevails in California. You know that at home it is considered *vulgar* for a gentleman to swear; but I am told that here, it is absolutely the fashion, . . .

For the *Avalanche's* editor

The people of this land are certainly distinguished, to an extent unknown in other countries, except perhaps Great Britain, by profaneness. A stranger might infer from the tone of popular conversation, from the exclamations of excited individuals, from the clamors of anger and passion, that we acknowledge the Almighty for no other purpose than we might have a name to swear by, or a convenient expletive to fill up the chasms of discourse.

Dimsdale, the Virginia City (Montana) schoolmaster, observed that

One marked feature of social intercourse, and (after indulgence in strong drink) the most fruitful source of quarrel and bloodshed, is the all-pervading custom of using strong language on every occasion. Men will say more than they mean, and the unwritten code of the miners, based on a wrong view of what constitutes manhood, teaches them to resent by forse what should be answered by silent contempt.

Possibly the commonest, as well as the most insulting, expletive flung at man by man was the term "son of a bitch."

McClure in his *Three Thousand Miles through the Rocky Mountains,* published in 1869, gives some vivid vignettes of life in the bonanza camps. To be embarrassed, he tells us, was to be corraled; "git" was just about the most expressive term in Western speech. "It is the invariable word by which the hero of the whip and lines starts his teams"; and "You git!" was the "most emphatic notice that can be given any luckless chap, to leave the room or ranch, or to escape a revolver; and 'You bet' is the most positive manner of affirmation." Everything was an outfit, from a ship or packtrain to a pocketknife, a cat, horse, dog, or even a wife. A layout was "any proposed enterprise, from organizing a State to digging out a prairie dog." Anything that has been tried, "from running for Congress to bumming a drink, has been 'prospected' or 'panned out'; and 'he didn't git a color' expresses the saddest of failures."

"Git!" meant in modern slang to scram, to clear out; and "Git up and dust" was just a little more emphatic. "You bet" was a strong yes, equal to "That's so" back East. "None of it in mine" meant no to an offer; a "bilk" was a humbug; a "poor coot" was a man without luck or friends; "on it" meant in earnest, determined; "weaken" meant to fail in strength, courage, purpose; to "peter out" meant to give out, to fail, become exhausted; and "to corral" meant to control. Peter Sapp heard a man say, "It ain't up to snuff. The chap in thar he wouldda swung long ago without all the dead head speechifying and humbugging. Cuss me but I'm clean plum sick of this country, where they let cussed red devils and white wolves run over you without so much as slippin' the wind on a one of them when he's caught." He heard an Irishman say, "Oi tot he war a tinderfut jist arrove but whin Oi seen him trow mud wid his shuvel Oi knowed he war an auld hand. He cud boss a tousand shuvelers. Yez kin git foine preachers ivery day in the wake, but foine shuvelers is fur betwixt." Dod swat me, he knowed what was meant when a man said, "He's a ketch-colt." The favorite oath of Idaho's Governor Hailey was "By Goney." "May I chip in a little wind on this? By Goney, he fobbed my money and he better git, ya betcha. He cuts high shines and leaves me strapped, and though I ain't no man fer rues or after-claps, drot my soul you'n bet your bottom dollar that let's me out. A boomer, that's what." A boomer was a parasitic drifter, usually from the South.

"I hain't got no jawbone no more," meant that his credit was exhausted; that no longer could he buy grub or fodder. Delano says "God damn you" was the commonest expletive of the drivers and mule skinners. "Twig me, old pard," the miner said, offering his hand. "I slopped over. They's panned me down to bedrock and nary a color. I tell you I've come close to slippin' my wind." "I'm not heeled" meant "I'm not armed"; an "old rip" was a tarrier (terrier); a "burst" was a blowup with a pokeful of gold dust; and a "rough-looking case" was just about anything from a shaggy half-dead burro to the Flying Dutchman.

The *Avalanche's* editor thought that of all the things Artemus Ward had written the following was the "most appreciated":

There was many affectin' ties which made me hanker arter Betsey Jane. Her father's farm jined ourn; their cows and ourn squench't their thirst at the same stream; our old mares had stars on their forrerds; the measles broke out in both families at nearly the same period; our parients (Betsey's and mine) slept regularly, every Sunday, in the same meetin' house, and the nabers used to observe, "How thick the Wards and Peasleys air!" It was a sublime sight in the Spring of the year to see our several mothers (Betsey's and mine) with their gowns pinned up so they couldn't sile 'em, affectionately bilin' soap together and abusin' their nabers.

To the person who senses the genuine in vulgar speech Ward smells of the inkhorn and pans a pinched-out vein.

Children The same editor pointed out eighteen ways in which children were obnoxious to their elders: reading when elders were talking, or cutting their nails; staring at strangers, whispering in a meeting, leaving it before it closed. Reading aloud without being asked to; failing to give thanks for a present; calling attention to themselves; laughing at the mistakes of others, "joking others in company"; correcting their elders; commencing to eat as soon as they sit at table; answering questions put not to them but to others; and paying no attention to what elder persons say. To most of which many of today's youngsters would say, "Drop dead."

On the other hand Mark Twain tells of a female child two or three years old walking up a street in San Francisco, with a female servant, when a "huge miner, bearded, belted, spurred, and bristling with deadly weapons . . . barred the way, stopped the servant, and stood gazing, with a face all alive with gratification and astonishment. Then he said, reverently, 'Well, if it ain't a child!' He offered $150 if the servant would let him kiss the child." Mark Twain assures the reader that the "anecdote is *true*."

Colton was immensely impressed by fecundity, finding that of the Californians

remarkable, . . . It is no uncommon sight to find from fourteen to eighteen children at the same table, . . . There is a lady of some note in Monterey who is the mother of twenty-two living children. . . . There is a lady in the department below who has twenty-eight children, . . . What a family—what a wife—what a mother! I have more respect for the shadow of that woman than for the living presence of the mincing being who raises a whole village if she has one child, and

then puts it to death with sugar-plums. A woman with one child is like a hen with one chicken; there is an eternal scratch about nothing.

Apparently the reverend gentleman did all he could to promote the population crisis the world is in now.

It has been pointed out since his time, by Rebecca West and others, that there are no delinquent children, only delinquent parents. Many of the parents in the mining camps were delinquent. Various writers have pointed out that the children of the camps were "viciously precocious"—that some of them were proficient gamblers and drunkards before the age of eight or ten. Resisting all authority they fought like wildcats, and from the age of ten "sparked"—"sometimes even abducted—female children of their own years." That seems to lay it on a bit thick but it would be pretty naïve to imagine that the little darlings were clasping their hands in prayer at bedtime, or returning found articles to their owners.

"To give a faint idea of the precocity and waywardness of youth in this country" Helper tells of a lad nine years old who was already an expert gambler. Watching the boy play with men one evening he saw him, when charged with cheating, curse out a man "three times as large and four times as old" and shout at him, "God damn you, I'll shoot you!" All of the men became involved in a quarrel, some standing for the boy and some against him. When one man drew a knife another man shot him, the ball going through him and wounding a third man. The man who fired was then seized by enraged miners and hanged. Helper thought that the boy "may be taken as a fair sample of the rising generation in California."

Some of the boys before they were six years old were out in creek beds with an old pan trying to find colors. Others hunted for buried treasure—for Plummer's in the Virginia City and Bannack areas for instance, for it was believed with good reason that he had cached a lot of gold dust. In some camps it was a habit with the little rascals to slip after dark into the Chinese graveyard to steal the food that had been left on a fresh grave. And like the juvenile delinquents of today they loved to seize an especially long and tempting queue and try to yank its owner's head off, or to steal into the washhouse and swipe the laundry. Richardson says that in the handling and weighing of small amounts of gold dust about 25 percent was lost. "Every morning, little boys with shovels and pans gather up and wash out the sweepings from the stores, and sometimes realize five dollars apiece."

Then, as now, children sometimes killed themselves for reasons that seem trivial to adults. The Boise *News* reported that a lad named Naylor

committed suicide yesterday morning by shooting himself through the head with a rifle. He was only about 14 years of age, and there was no apparent cause for the dreadful act. On the evening previous the boy had been scolded or otherwise slightly punished for going off hunting and neglecting his work. This greatly incensed him. Shortly after he attracted the attention of his mother by dividing his money among his brothers and sisters. On being told to keep his money, he said he did not want it, and that he would never do another day's work for his father. He went to bed as usual, and next morning before breakfast, crawled out of the window, with the rifle and shot himself by placing the muzzle of the rifle against his head and pushing the trigger with his toe.

Indian lads also thought themselves shamefully treated. The same newspaper tells of Charlie, son of a Nez Perce chief, who at fourteen

wanted to take to wife a woman of the nation of bad repute; his father, thinking he was starting in rather early to raise and support a family, lectured him about it, and threatened to chastise him, and her both, if he did not put her away. So the young red Romeo, feeling his manhood insulted and outraged, shot himself through the "gizzard". . . . Such is life among the red-skins—only imitating the fantastic eccentricities of a highbred civilization.

Animals   Having no wife or children, some of the miners tried to soften their loneliness with pets—a dog or cat or burro, a cub bear, a coyote pup, a wounded bird, or a coon, skunk, or owl. Some of them, especially the prospectors, had a deeper affection for their burros than for anything else in the world. In the Colorado town of Fairplay, a burro named Prunes had a handsome monument erected to him, the inscription being made of marbles pressed into the slab before the concrete set. For the burro, one of the chief symbols of the American West, and

one of the most indispensable of all the things in its development, was cursed, beaten, overworked, starved, killed, but sometimes loved and sometimes painted, and once in a while given a line in what passed for poetry. In Cy Warman's verse with the oft-quoted lines,

> It's day all day in the daytime,
> And there is no night in Creede

"the meek and mild-eyed burro" fills one line.

But then as now possibly the greatest devotion was given to the dog. When on the way overland Kelly and his companions were faced with starvation he overheard the men saying that they would have to kill and eat Kelly's dog.

I gazed upon my dumb friend with a tearful eye and sickened heart; the more so as I fancied he looked wistfully in my face, standing in an attitude of dejection unusual to him, with drooping tail and hanging ears. I was unable to assent in words, but gave them silently to understand that I would interpose no obstacle; and no sooner had I done so than poor "Sligo" (so I called him), instead of coiling himself beside me as was his wont, slunk away to some distance, sitting in a mournful attitude, and watching our movements with a grave steadiness that perfectly unmanned me; impressing me with the steadfast conviction that his intuitive sagacity forewarned him of our cruel intentions.

It was clearly perceptible to all that his attachment and confidence were changed into fear and distrust, for no calling or coaxing would induce him to come nearer us; while if any approached him, he receded slowly. S---, who was the steadiest shot and had the best rifle, agreed to do the deed; and as he commenced loading, the poor brute betrayed increased uneasiness, moving and shifting restlessly as if about to run off, but finally sitting firmly still on a little mound, as if he had come to the determination of yielding himself up as a victim for the salvation of his master. The warm tears trickled freely down my cheeks, and I felt a disposition to go and embrace him when looking at him for the last time. As S--- raised his rifle to his shoulder, the poor animal at the same moment fairly confronted his executioner, throwing back his ears with a low piteous whine, and waiting his doom like a hero.

The coyote, cousin to the dog, and to the much larger and more savage wolf, had ways of playfulness and cunning that intrigued many of the people from the East who had never seen one. Of all the descriptions of it that we have read we think Mark Twain's is the most delightful:

The coyote is a long, slim, sick and sorry-looking skeleton, with a gray wolf-skin stretched over it, a tolerably bushy tail that forever sags down with the despairing expression of forsakenness and misery, a furtive and evil eye, and a long, sharp face, with slightly lifted lip and exposed teeth. He has a general slinking.expression all over. The coyote is a living, breathing allegory of Want. He is *always* hungry. He is always poor, out of luck and friendless. . . . When he sees you he lifts his lip and lets a flash of his teeth out, and then turns a little out of the course he was pursuing, depresses his head a bit, and strikes a long, soft-footed trot through the sage-brush, glancing over his shoulder at you, from time to time, till he is about out of easy pistol range, and then he stops and takes a deliberate survey of you. . . . But if you start a swift-footed dog after him, you will enjoy it ever so much—especially if it is a dog that has a good opinion of himself, and has been brought up to think he knows something about speed. The coyote will go swinging gently off on that deceitful trot of his, and every little while he will smile a fraudful smile over his shoulder that will fill that dog entirely full of encouragement and worldly ambition, and make him lay his head still lower to the ground, and stretch his neck further to the front, and pant more fiercely, and stick his tail out straighter behind, and move his furious legs with a yet wilder frenzy, and leave a broader and broader, and higher and denser cloud of desert sand smoking behind, and marking his long wake across the level plain! And all this time the dog is only a short twenty feet behind the coyote, and to save the soul of him he cannot understand why it is that he cannot get

perceptibly closer; and he begins to get aggravated, and it makes him madder and madder to see how gently the coyote glides along and never pants or sweats or ceases to smile; and he grows still more and more incensed to see how shamefully he has been taken in by an entire stranger, and what an ignoble swindle that long, calm, soft-footed trot is; and next he notices that he is getting fagged, and that the coyote actually has to slacken speed a little to keep from running away from him. . . .

Suddenly the coyote vanishes, and the dog returns to

[take] up a humble position under the hindmost wagon, and feels unspeakably mean, and looks ashamed, and hangs his tail at half-mast for a week. And for as much as a year after that, whenever there is a great hue and cry after a coyote, that dog will merely glance in that direction without emotion, and apparently observe to himself, "I believe I do not wish any of the pie."

Some of the camp people whose whole attention was not fixed on gold dust, gambling, and whisky became amateur naturalists; for in the animal world, whether in Arizona or Alaska, or in the broad realm between those two, there was a wealth of life unlike what they had known. It was a fascinating subject for newspaper editors besides Mark Twain. The *Statesman's* editor called to the attention of his readers that drugstore in Virginia City, as reported by the *Territorial Enterprise,* in which was to be seen

a very singular family, consisting of a large rattlesnake, a common ground squirrel and two diminutive mice. The snake rejoices in seven or eight rattles and appears to take great pride in its melodious tail. Although it flattens out its head and looks rather venomous, the snake does not attempt to strike at those who approach the cage. The squirrel seems perfectly fearless of the monster with which he is cooped up, but scampers over and around his snakeship, erecting himself upon his haunches and nibbling cheerfully at his food when within a quarter of an inch of the serpent's flattened head. The poor little mice appear utterly demoralized. They huddle into the corner of the cage, each trying to hide his head under the body of the other, and could plainly

be seen to tremble. . . . No racket could startle them from their corner; at every noise they only trembled more violently and exhibited greater anxiety. . . . Poor little devils, they felt that their days were numbered. We asked the foreign-looking man who owns these pets if the snake would eat the mice. Oh, certain! He not hungry now; when he feel hunger you shall see him eat them very quick.

Audubon's story of how a toad undressed itself was the kind of thing the editors printed from sources all over the nation, and apparently the kind of thing the subscribers liked to read:

He commenced by pressing his elbows hard against his side and rubbing downward. After a few smart rubs his hide began to burst open along the back. He kept on rubbing until he worked all his skin into folds on his sides and hips; then, grasping one hind leg with his hands, he hauled off one leg of his pants the same as anybody would; then stripped off the other hind leg in the same way. He then took his cast-off skin forward between his fore legs into his mouth, and swallowed it; then, by raising and lowering his head, swallowing as his head came down, he stripped off his skin underneath, until it came to his fore legs; then, grasping one of these with the opposite hand, by a single motion of the head, and while swallowing, he drew it from the neck and swallowed the whole.

But of all the birds and beasts of the West possibly none aroused deeper emotions in the camps—or in the Rocky Mountain camps anyway—than the formidable and almost-unconquerable grizzly bear. It was the only animal, Bowles says, "that our courage and our double-barreled shotgun were not equal to; and he is, indeed, next to the Indian, the terror of all hunters." A whole book could be compiled of the savage and deadly battles of man and grizzly, including those of Jed Smith, Hugh Glass, and Meriwether Lewis. There must have been hundreds of these hand-to-fur combats, often with the most horrible results to the men.

Kelly, who was on the scene early in the California rush, tells of an encounter. The bear was eating manzanita berries and minding his business when Kelly shot him: "The ball, as I subsequently found, glanced along the ribs, entering under the shoulder, and shattering some of the bones. I exulted as I saw him stagger and fall upon his side. The next glance, however, revealed him, to my dismay, on all fours, in di-

rect pursuit, but going lame." Kelly fled, dropping his rifle, and mounted his mule. He dug with his spurs but the beast refused to budge. But a moment later, seeing the bear, the mule took off at furious speed and brought Kelly "violently in contact with the arm of a tree, which unhorsed and stunned me exceedingly." Getting to his feet he saw that the monster was almost on him, and so shinned up a tree a few feet. The bear after a "fierce glance upwards from his blood-shot eyes" seized the trunk and tried to climb. It "succeeded in reaching the first branch . . . and was working convulsively to bring up the body, when, with a well-directed blow from my cutlass, I completely severed the tendons of the foot, and he instantly fell, with a dreadful souse and horrific growl, the blood spouting up as if impelled from a jet." The bear rose to its feet and limped round the tree and "kept tearing off the bark with his tusks." Kelly now drew his revolver and with another shot dropped the beast. It was not much of a fight compared to many of them.

Religion Though most of the camps had church buildings the spirit of formal religion was weak; the miners found it easy to hoot at pieties and parsons, when the newspapers set the example for them. The Boise *News,* for instance, just before Christmas in 1863, told gleefully of a parson's confession. The poor man had lost his wife and was left with a small child. One night when the child was crying and would not be comforted there entered the parson's room a lovely girl named Nancy, to hush the child's grief. When she leaned over the bed the parson kissed her. After making his shameful confession in church he "bowed himself upon his seat, like the mighty oak before the tornado," and the deacon, appropriately named Goodfellow, said he would forgive the parson but he wanted his brothers and sisters to know "that if I had no wife, and a pretty girl, like Nancy Stearns, should come into *my* room, and lean over my bed, and lean over me, I'd kiss her—sin or no sin."

The almost invariable habit of the newspapers was to give polite but empty approval of sermons by ministers and other soul-savers, but to find delight in exposing humbugs and impostors and to feel that most clergymen were frauds. The editor of the *Avalanche* told his readers that he had

Presbyterian church, Phoenix, 1879

learned from *Punch* that an Englishman had an invention called a patent pulpit, that could be

suspended over the preacher's head . . . in shape like a candle extinguisher. This was to descend so as, in exactly fifteen minutes, to completely extinguish the clergyman. . . . Eight minutes after the sermon had commenced a loud premature snap or click was to announce the commencement of the extinguisher's descent; and it is stated that upon several private trials no clergyman could be found who had the nerve to continue his sermon, after hearing this snap or click, over three minutes longer, such was the fear in his mind that his light would be hidden under the suspended bushel.

Two weeks later the same editor was telling of a child who, after her father went to war, learned praying from her grandfather, who was "short, and to the center." When at last her father, a deacon, came home she recognized him only as "man." The first morning he led the family in prayer, and prayed on and on, until his child rose from her knees and tapping him on the shoulder whispered in his ear, "Man, man, stop—you're froo!"

And five months later he was asking,

Does the Methodist Church take heavy risks against hell fire? This query occurred to us not long ago while listening to the remarks of our worthy friend who stands by the old M.E. He says and says sorrowfully, that as soon as an Eastern man is worth twenty-five or thirty thousand dollars, he, with his household, quits the Methodist Church; it isn't fashionable enough for him and his well-dressed wife and daughters; and they take first-class tickets in the Episcopal, or some other fashionable craft. We asked our friend the question, "If the reason was not, in part, that the more gorgeous vehicles on the heavenly line, having larger capital, take greater risks?" "No," said he, "the Methodists put some damned mean men through; it's only for the sake of being fashionable, and mostly the fault of the women."

About the same time the *Statesman* was complaining that of all places the wide world over, "Idaho City [forty miles north] has plenty of churches, but no preachers—several good houses of worship but no religious services. Boise City presents her case a little differently. We have good ministers and service every Sunday, but no churches." In some camps the ministers seem to have been intimidated by the criminals. Mrs. Mathews tells of the brutal treatment of a small child in Virginia City—of flogging it until it was covered with welts; of forcing it to stand until it screamed, while cold water was poured over its head. She lodged a complaint and a time was set for trial, but when she asked Bishop Whittaker to go with her to court he said he did not dare to, lest the child-beater burn his church down. Mrs. Mathews got no justice from the judge, who, she says, "committed suicide a few months after." The child, she tells us, disappeared.

But the spirit of the camps toward the churches and their hierarchies is best revealed, it seems to us, in a story told by Alex Toponce. At a time when he was working for a stage line the owner came to him and said to him and a driver named George, "You boys been dumping some of the U.S. mail out on the prairie; you'll have to be very careful this trip. You are going to have a fuzzy-shirt from Washington, D.C. as a passenger."

Silver City, Idaho

Columbia, California

Virginia City, Nevada

FAMOUS CAMP CHURCH BUILDINGS

Our Lady of Tears, Idaho City, in which mass was first celebrated in 1865. Being high on a hill it escaped the several fires that destroyed the camp. It is still in use.

In Virginia City, this was the first Episcopal church in Montana.

George said to put in an extra jug of whisky and he would take care of fuzzy-shirt. This man was a "general or colonel or something." George asked him if he and his aide preferred to ride in the front seat or the back seat: the two seats in the Concord faced one another. Fuzzy-shirt chose the rear seat. First-class mail was in sacks with a copper padlock; second-class with an iron padlock; and third-class with a "puckering string. About that time the Methodists, or some one like that, were sending an awful lot of Bibles and tracts through to the heathen in California" and all these were in the sacks with the puckering strings, about four hundred pounds to a sack. George piled these sacks on the front seat, with a rope around them, and when climbing a steep hill he furtively cut the rope and the great pile of heavy pouches fell in an avalanche upon fuzzy-shirt and his aide. "They yelled for help like sin, but we could not hear them because we kept yelling at the top of our voices at those six mules. . . . We

dug the general and his aide out . . . and gave them a jolt out of one of the jugs . . ." The general now chose the front seat. The bags were piled on the rear seat, again with a rope binding them, and when the stage was descending a steep hill George again cut the rope. "This time we could not hear them yelling because the brake was on and making an awful screeching. . . . The general was pretty nearly down and out, but he was able to sit up and take nourishment out of the jug. . . ." Looking at the pile of stuff that for a second time had almost crushed him, he asked: "Why don't you dump it? We're not going to have all those Bibles fall on us all the way to Santa Fe." So the general and his aide walked on ahead and out of sight, while George and Alex unpuckered the sacks and dumped the Bibles upon the prairie.

Sunday   The Nahl painting, reproduced here, of part of a camp on Sunday morning, is a

famous one that has been used many times; but as a California librarian said to us, Nahl was dominated by a wish to clean everything up and make it pretty. In a shanty we see a miner writing, in his journal or to the folks back home; to the right of him other miners are doing their Sunday wash; and on the left are some roisterers, still drinking as the sun comes up. "There is no store, shop or business place of any character closed that day," Bowles says. Sunday was "altogether the busiest of the week with shopkeepers, victualers, gamblers and whiskey dispensers." A few of the miners, after Sunday breakfast, dug out their "boiled shirt," slicked their hair back, put away their guns and other weapons, and slipped over to the churchhouse, if there was one, to hear what message the good Book had to give. But even so, it was probably the music they liked best.

"Last Sunday," observed the *Avalanche's* editor, "was an unusually lively day for Silver—boss packers, bull whackers, etc., acted as if they hadn't had anything to drink but Snake River for two months. Perpendicular humanity was a rare article 'in them parts' about 3 P.M." A week later the editor was complaining that "We have wasted considerable time listening to sermons." A tourist in that same year (1865) said that in mining towns on Sunday "the solemn echo of the tolling church bell was unheard." This was the day when miner and merchant settled what the one owed to the other; when the merchant presented a "villainous-looking paper, charged with figures; the other, a long greasy bag, charged with gold." His satanic majesty the Devil was "the ruling spirit and the spirit and the genius of the hour." Rotgut was "twenty-five cents at the bar; myriads of mammoth bottle-flies, pregnant with poison, sailing through the air; whisky the beverage, and monte the game; . . . a breathless corpse lay weltering in his blood; law is violated, and the victim is unavenged; the Sabbath closes with a clouded moon frowning on the sins of the day."

That picture is one-sided. The average miner, who didn't pan more than a few dollars a day, didn't have much to spend on whisky and gambling. Some of them were hoarding their earnings and sending them back home, and these usually spent Sunday washing and mending their clothes, repairing their tools, and baking bread and getting on another pot of beans. Some kept a diary, some strummed a banjo or sawed on an old fiddle,

some went fishing or hunting, and some toiled all day in their diggings, and especially if Saturday evening they had gambled away their week's earnings and were broke again. Some stayed drunk from early Saturday evening until Monday morning. "What tickled my fancy," says Kelly, "was to see the class of Christians I have been describing abstaining from work of a Sunday, which they invariably did, avowing their scruples of conscience about laboring on the Sabbath, while canvassing for a party to sit down to cheat, swear, and drink over a game of 'poker,' or listen to the filthy homilies of some hoary debauchee, who gloried the more in his audacious impiety because it was the Lord's-day."

Peter Sapp, who belonged with the more timid souls, usually wandered about the camp on Sunday, pausing to listen to a spell-binder, peering into gaming halls and saloons, and admiring and envying the bold predators who seemed to be wise in all the ways of the world. Perhaps he would be attracted by a loafer who had a whole bagful of tricks to keep him in beans and rum, a rough-looker with dirt in his hair and brows and beard, who pretended to be a prospector just in from the hills. Somewhere now and then he would get hold of a few pieces of rich gold quartz and putting on the act of a guileless and timid fellow he would sidle up to a prosperous-looking man, and showing the ore in his left palm would say, "I heerd ya know rock. What ya tink a-this?" The prosperous one, if he knew ore, would be thinking fast and big, and presently he would draw the man aside and stare at the rock under a better light. Casually, as if only slightly interested, he would ask the man where he had found it and the man would say yonder a far piece; and Peter, who had moved within hearing distance, would listen with both ears and wonder about men who seemed to be able year after year to raise a grub-stake with the same two or three pieces of ore.

Occasionally he would listen to a preacher. Peter would have agreed with the Reverend Walter Colton, the first Protestant minister in California, who thought that most preachers were pretty dull and humorless fellows. Monday, December 28, 1846, Colton made these observations:

This is the festival day of the Santos Innocentes, and is devoted by the lovers of fun to every kind of harmless imposition on the simplicity of others. The utmost ingenuity is exercised in borrowing, for every article lent has to be redeemed. Although aware of this, still, in a moment of forgetfulness, one succeeded in borrowing my spurs. . . . Two ladies performed feats that would have been difficult on any day. One borrowed money of a broker, and the other the rosary of a priest. It is rumored, but not credited, that a client has induced his lawyer to allow his case to be amicably adjusted; that a patient has actually persuaded his physician to permit the aid of nature in throwing off his disease; and that a customer has made a shopkeeper confess an imperfection in his wares. It is said, but doubted, that an old Spanish hidalgo, after being told that his son is engaged in marriage to a peasant girl, will probably sleep before he disinherits him. It is also said, though few believe it, that a wife, whose husband is going to sea, has consented that he shall take the family breeches with him. It is further stated, but on no good authority, that a political partizan has hesitated about voting for his candidate on account of his having been once sentenced to the penitentiary for sheep-stealing. Several other rumors are afloat, but they are not credited. One is that a disappointed lover has persuaded himself that his suit has been rejected without any parental interference; another is, that a young collegian has written a letter to his grandmother without quoting a word of Greek; another is, that a young clergyman has composed an entire sermon without anything about "Fixed fate, free will, foreknowledge absolute." Another is, that a man of giant intellect and profound erudition has selected as his life-partner a woman of sense; another, that a lady who has had an offer of marriage and rejected it, has kept it to herself; another, that an old bachelor has come to the conclusion that he is less captivating with the girls than he was when younger; . . .

The camps were sometimes overrun by salvationists eager to save souls. In one of them was a man ninety years old who swore that he had never smoked or drunk. On one occasion when two entranced women were admiring the colors of sobriety and health in his cheeks there was a terrific crash in the next room of the shanty, as of a ton of tin pans falling. The healthy old codger who had never drunk a drop told the women not to be alarmed. That was nothing, nothing at all. "It's jist that my ole man is drunk agin."

One of those who sought to reform the boozers was Sarah Pellet. When San Francisco had 537 places selling drink she invaded it, mounted a box, and turned loose such torrents of hellfire and damnation that the saloons were more completely emptied than by dogfight or gun duel.

Helper had seen "purer liquors, better segars, finer tobacco, truer guns and pistols, larger dirks and bowie knives, and prettier courtezans" in California than in any other land he had visited but he knew of "no country in which there is so much corruption, villainy, outlawry, intemperance, licentiousness, and every variety of crime, folly and meanness." It should have been fertile soil for the evangelist. Almost any day and in the same block or before the same hotel the evangelist called people to God, and the bunco man called them to his patent medicines; while up or down the street auctioneers were shouting the virtues of wornout mules or bogus jewels, a Mexican girl was grinding her hand organ and making eyes at the louts leering at her, and in the gutter a drunk flat on his back was trying to sing: Miss Ella she is twenty-nine, has taken two degrees, and tore her shirt-tail off behind so she can show her knees! . . . O a married woman gentle dove with nary tooth a-tall sets in a corner making love with some pimp at the ball! A dozen babies on the bed and all begin to squall; the mothers wish their brats were dead for crying at the ball!

A little distance from him, sitting up, his eyes out of focus, two guns at his waist and a knife about to fall out of its sheath, another drunk would try for a moment to hear what the preacher was saying and would then burst into song, to the tune of Caroline of Edinboro: You must buy two revolvers, a bowie knife and belt. Says you, old feller, jist stand off or I will have yore pelt. The greenhorn looks around about but not a soul can see; says he, there's not a man in town but what's afraid of me! . . .

Among those who tried to infuse a gentler spirit in the camps was Sarah Royce, mother of the philosopher, but though a few women were persuaded to come to her house to "have a sing" with her, and though she dug up a minister from somewhere to do a bit of preaching in her living room every second week, before long the hoots of the miners drove her back home. In Deadwood the Reverend Henry W. Smith supported himself with heavy labor, and in the evenings and on Sunday preached to anyone who would listen. When he decided to ride over to another camp to preach he was asked if he had a gun. He said he had a Bible, but a Bible didn't save Jedediah Smith and it didn't save him. A pretty legend says the Indians didn't see the Bible until after they had killed him, and then left him with the Holy Book in his hands.

Smith and Royce had about as much luck saving souls as any of them. Far more popular were the orators, who infested all the larger camps. Typical of them was an especially obnoxious blowhard named Copeland, who advertised in a camp that he would give an invocation to the American eagle. To do it in style he mounted a barrel turned bottomside up, and had the shades of Demosthenes and Daniel Webster pretty well abashed when the bottom suddenly fell out of the barrel. As the orator dropped into it his garments were snared by the nails that had held the bottom, and in his struggle to free himself he knocked the barrel over and inside it went rolling into the gutter. It had begun to rain. The gutter was flowing and it look as if the hapless orator might drown but he was dragged out, like a hog from a vat, thoroughly soaked and thoroughly silenced.

But for all their hooting at preachers the miners, and even the gamblers and whores, impressed visitors with their generous giving to the needy. M'Collum says,

The question is often asked, "How do people live who are sick, and out of money in California?" They get along better than the unfortunate and destitute anywhere else in the world. There is a fellow feeling there, a spirit of active, practical benevolence. Charity offerings are made upon a scale of munificence corresponding with everything else; $1,000 could be raised easier there, at a call of humanity, than $5 in any of our large villages; and these calls are frequent. I saw nor heard of no case of suffering there for want of money.

Dr. M'Collum's impression is supported by that of many other visitors. McIlhany tells of a young man who lost both legs and whose deepest wish was to go back home. McIlhany says he took him to a large gambling house, and had a singer there tell the story and pass a hat. "The result was that several hundred dollars in gold was in the hat. . . . The young man was so affected . . . that he shed tears." Mrs. Mathews wrote with a critical pen but she found that the people of California and Nevada "are very generous; they will divide their last potato with you, or give their last 'bit' to a charitable cause; and not many even stop to inquire whether it is for charity or not."

Humor The quality of the humor is certainly one measure of a people's culture. Constance Rourke has argued in her *American Humor* that because personal ties had been left behind "a vast sentimentalism overflowed along with hilarity"; that because of the hardships and heartbreak, the losses and failures, "the comic mood arose irresistibly. Some of the men loved practical jokes of a brutal kind—such as gently slipping into a sleeper's nostril something soaked with oil and touching a match to it. It is said that a dentist with a pair of pliers yanked a tooth out of a sleeping man, and, even more incredibly, that the editor of the Montana *Post* said, 'The joke is too good to lose, if the tooth was not.'" All that Rourke says there may be true, but it seems to be just as true of cowboy, logging camp, and frontier humor, and possibly of most American humor, no matter where.

Richardson observed that "The stage stations are named with sardonic humor. One is called Forest Grove, because there is not a single tree within fifteen miles; another, Cold Springs, because not a drop of water exists in the vicinity." In Oregon he came to a ranch known as the Six-bit, and of its owner says that he had a "frugal mind. . . . Just as an Indian was about to be hanged for murder, he mounted the scaffold and dunned the doomed man for six bits (seventy-five cents)." He says an "ex-governor and ex-senator was a passenger on the wrecked steamer Golden Rule. 'What did you save?' inquired a friend. He replied, 'I saved nothing but my character.' 'Then,' retorted the wag, 'you must have landed at San Francisco with less baggage than any other man who ever came to the Pacific coast.'" And he tells a story that seems to have, like so many, no known origin. A "burglar, . . . at midnight climbed up to a chamber-window, and cautiously opened it. The occupant, chancing to be awake, crept softly to the window, and just as the robber's face appeared, presented the smooth muzzles of two revolvers, with the injunction, 'You get!' 'You bet!' replied the house-breaker, dropping and running. There is no more pithy dialogue on record." Richardson published his book in 1869. Donaldson tells the story of James Reynolds, first publisher of the Boise *Statesman*, who from 1869 to 1872 "slept in the front room of the upper floor" of a frame building. In Donaldson's version the burglar was in the room and rummaging "in a heap of papers and trash" when Reynolds sat up in bed and said to him, "Excuse me, pardy, but who are you?" The burglar said, "I'm a robber. Ain't you skeered?" Reynolds said no, not to speak of, and asked what the man was looking for. "'Money,' said the thief, vigorously thrusting about." Reynolds told him he had lived in the room for six years and had found no money. He then presented his gun and told him to get, and the man said you bet, and vanished.

Rourke found the practical jokes brutal. They often are, in all lands. Mark Twain loved as well as any man the practical joke, if it was not on him. With great relish he tells of "General Buncombe" who was "shipped out to Nevada in the invoice of territorial officers, to be United States Attorney." Because Buncombe was a pompous gasbag some of the men plotted to teach him a lesson. Into Buncombe's office rushed a wildly excited man, who said he was a rancher in the Washoe valley. Above him on the mountainside was another ranch; the spring thaws had moved it down in an immense landslide and it had completely covered the lower ranch. The lower belonged to a man named Hyde, the upper to one named Morgan. Hyde wanted Morgan prosecuted for trespass, and he wanted him to be forced to take his ranch back where it belonged. A judge and defense attorney and spectators were all on hand for the trial, as well as legislators and Indians. Such a strong case was made for Morgan that Buncombe banged his law books and "quoted from everything and everybody, poetry, sarcasm, statistics, history, pathos, bathos, blasphemy, and wound up with a grand war-whoop for free speech, freedom of the press, free schools, the Glorious Bird of America and the principles of eternal justice!" But the judge thought it a profanity "to meddle with the decrees of heaven" and gave the case to Morgan: "Heaven created the ranches, and it is Heaven's prerogative to rearrange them, to experiment with them, to shift them around. . . ." Buncombe left in a rage but that evening begged the judge to pace back and forth for half an hour and consider the matter. The judge did, and at last concluded that the ranch still belonged to Hyde and would be as good as it ever was, if only he would dig it out. It was then that "the fact that he had been played upon with a joke had managed to bore itself . . .

through the solid adamant of his understanding." He fled from the territory.

Most of the camp editors used humorous items, gathered locally or from distant places, to lighten their dull reports of national and territorial affairs. The Boise *News* lifted the following from a Cincinnati paper.

A Negro soldier who had been in a battle was interviewed. "Were you in the fight?" "I had a little taste of it, sa." "Stood your ground, did you?" "No sa, I runs." "Run at the first fire, did you?" "Yes, and would hab run sooner, had I knowed it war coming." "Why, that wasn't very creditable to your courage." "Dat isn't in my line, sa. Cookin's my perfection." "Well, but have you no regard for your reputation?" "Reputation's nuffin to me, by de side ob life." "Do you consider your life worth more than other people's?" "It's worth more to me, sa." "Then you must value your life very highly." "Yes sa, I does, more dan all dis world; more dan a million of dollars, sa; for what would dat be worth to a man wid de bref out on him?"

An Iowa editor had been fined two hundred dollars for hugging a girl in church. The *Avalanche* ran comments from different newspapers. "Cheap enough! We once hugged a girl in church ten years ago and it has cost us a thousand a year ever since.—*Y. America*. That's nothing!—We hugged a girl in school some twenty-five years ago, and have had to support her and the family ever since.—*Tioga Dem*. Come to Salem, Oregon; we have hugged a dozen and it ain't cost a cent.—*Review*. That is cheap! Our editor paid five dollars to hug an actress a short time since.—*Or. Sentinel*. Talk of hugging by the wholesale! Just come here among the Mormons, if you want to see big business done in the 'bundling line'.— *S. L. Vedette*." The editor then commented: "If this nonsense isn't sufficiently run into the dirt, it has gone far enough. . . . From a land of fifty women to a man, to where the 'situation' is just the reverse, the contrast afforded ought to let us out. There is no necessity for a 'prices current' on the article in Owyhee. In the winter time one is in luck to be able to hug a stove-pipe, while wood is $12 per cord."

Most of the editors seemed to love stories of marriages that didn't pan out. In Montana were Joseph Wiggs and his Betsy, who had come west to make their fortune. For a while they occupied love's young dream but then a handsomer man made eyes at Betsy and soon she refused to shake the fleas out of the blankets before making the bed, or to help yoke the oxen in the morning, and by the time they reached booming Alder Gulch they were not speaking. They wanted a divorce and were told to post notices and a miners' jury would hear the evidence and decide the case. So the husband posted this notice: "Betsy and me has agreed to split blankets and rustle on different trales. She will take one and me tother. A miners meeting is hereby called for nex Sunday on the flat just abov Nevada to here our storeys and giv us splittin papers, everybody cum." Just about everybody in the camp turned out. Joe told his side of it, and Betsy told hers, but they couldn't agree on the division of the property. A young miner known as Jeff Davis called out to the husband, "Say, pard, what will yer take fer the gal and the stuff she claims?" Joe said he would take two hundred dollars for everything except the oxen and wagon, and he would throw the woman in. "Done," said Jeff, "if the meetin' will give yer both dividin' documents." The miners' court granted the dividing documents, and Jeff led them all to the nearest saloon to set up the drinks.

In Silver City a prosperous gentleman from the country stopped at a hotel and got into conversation with one of the boarders there. As told by the *Avalanche* the boarder said, "Will you have a cigar, sir?" and the stranger replied, "Don't mind if I do." A cigar was passed to him, "also the one which our boarder was smoking for the purpose of giving a light. He carefully placed the cigar first handed him in his pocket, and took his knife out and cut off that end of the lighted one which had been in the mouth of his generous friend, and commenced smoking, saying, 'It ain't often that a man from the country runs afoul of as clever a fellow in the city as you are.' "

The editors loved to pick up tidbits of information, true or false, for their readers, preferably with a humorous cast. Tonopah's *Bonanza* reported a disastrous mistake. A

well-known young man of Goldfield bought a pair of gloves for his girl and at the same time his sister bought a pair of hose for herself. The clerk in wrapping got the items mixed and the explosion came when the sweetheart opened her package and found that it contained a long pair of silk stockings. She blushed. Then she opened a dainty little sweet-scented note and read the following tender lines: 'I am sending you a little present. Oh, how I wish that no other hands than mine would ever be permitted to touch them after you put them on. But alas! Scores of fellows may touch them when I am not at your side, . . . I bought the longest pair I could get, and if they are too long you may let them wrinkle down. A great many girls I know wear them slipped down a little. I want to help you put them on and see how they fit your dear self when I call Tuesday night. You can clean them easily, dear, with benzine if you leave them on till they dry.

Though that sounds as fishy as a Mark Twain joke, the boys at the bars probably choked on their whisky when they talked about it.

The next story has appeared in a number of books. A young miner, despairing of finding a white wife, married a squaw. The ceremony was performed by a rough old justice of the peace who pronounced them man and wife with these words: "Here stands a white man and a squaw, married under the white man's law: up the hill and down the level, kiss your wife, you ugly devil." The miners decided to charivari the couple, and this they did not only with tin cans and cowbells but with a huge trough twenty feet long and four feet deep, across which they stretched lariats dipped in resin. A sapling pole served as a bow on this enormous fiddle, which also had strands of rope, resin-dipped. When the bow was scraped across the fiddle by a half-dozen husky men a sound was made "the like of which was probably never before heard on earth." The cans and bells and fiddle made such an appalling racket that at last the haggard bridegroom came forth and surrendered. He thought the vigilantes had come for him.

During these years there was much abuse in some of the newspapers of the Mormons. Toponce, who freighted from Salt Lake City to the Montana camps, says the Mormons tried to convert him to their beliefs. One day he told a Mormon named Byrnes that Moses was a grafter. Byrnes in a rage demanded an explanation. Toponce told him that while Moses was up in the mountain after the commandments Aaron and the children of Israel turned in their gold and melted it down for a golden calf. Was that true? Yeah, Byrnes said. Moses then came down and with an ax knocked the calf to pieces. "What became of all those gold chips?" Byrnes said he guessed it was turned back to the people, but told to find where in the Bible it said that, he searched and found nothing. So in another rage he jumped up and down and cried, "What will I do? I'll tell you what I'll do! If I don't find a full explanation of what became of that gold I'll leave the Church!" Years later Toponce met him in Ogden and asked, "Did you find out who got the gold?" and the man said, "No, and I left the Church."

There were many miners whose housekeeping was not above suspicion. One of them had a dog that he called Two Waters but no one could discover why he had given the beast such a name. Then one day a miner remembered the time when he had been invited to dinner and had complained that the dishes and utensils looked as if they had never been washed. The old miner had said to him, on that occasion, "They're jist as clean as two waters can git them."

One time when directing a man Kit Carson pointed to a distant peak and asked him how far away it was. The man thought it was ten or twelve miles. Kit said, "When you see me next time tell me how far it was." It had taken the man two days' riding to reach it. In the clear dry mountain air persons from the East thought everything in the distance was much nearer than it actually was. Toponce tells of two men who started across a valley. They thought it was three or four miles across but after five hours of traveling it looked as far across as ever. "Finally they came to a little stream about three feet across. One of the men sat down on the ground and began taking off his boots. The other said, 'What are you doing that for?' The man said he was going to wade the creek. The other said, 'Why, man, you can jump it.' 'Yeah,' said the man with his boots off, 'it looks that way but it might be a mile across.'"

The *Avalanche's* impious readers must have loved this story. At a certain crossroads in Alabama stood a small grocery, or whisky shop, before the War, where "bust head" and "chain lightning" were dealt out to the unwashed, at the small sum of five cents a nip or twenty cents a quart. The proprietor of this place was Bill Sikis, who among other pets had a crow, black as the ace of spades. This crow had learned among other things to repeat quite distinctly the words *damn you,* which he heard frequently in the shop. A violent fight in the shop frightened the crow, which flew away to the woods, never to return.

About three miles distant was an old meeting-house, used on the special occasions when a circuit rider happened to come through. The crow took up its abode high in this building, and was there one day when a huge crowd assembled, in which was an old lady who walked with crutches. She gave her whole soul to the preacher's thundering denunciation of all grades of sinners, and was thinking what a noble man he was when from high above a dreadful croaking voice uttered the words:

Damn you!

The preacher and congregation looked aghast at such profanity, and each peered into his neighbor's face in vain to detect some sign of guilt. Quiet was at length restored, however, and the sermon proceeded, but ere ten minutes had elapsed, the ominous "damn you" again electrified and startled the audience and just as the preacher cast his eye upward to search for the delinquent, the crow flew down from his perch, and 'lighting on the Bible, calmly surveyed the terrified crowd, as he gave another doleful croak: "DAMN YOU."

The effect was electrical. Giving one startled and terrified glance at the intruder, the preacher sprang through the window, carrying sash, glass and all with him, slitting his coat tail into kite tails and thus rendering it useless as a checkerboard, and set off at a break-neck pace through the woods, closely followed by the horror-stricken congregation, who had piled out of the building, pellmell, after him. In the general scramble the old lady with crutches had been knocked down in the church, where she lay unable to rise, and on observing her, the crow flew down beside her, and cocking up his eye at her very knowingly, again croaked: "DAMN YOU!" The old lady eyed him savagely for a few moments, and then burst forth in reckless defiance: "Yes—and *damn you,* TOO! I had nothing to do with getting up this Methodist meeting, and you know it!"

Entertainment

Nothing could have been duller for the people in northern climes than a snowbound winter. Most of them lived in dugouts, shacks, and shanties that it was impossible to keep warm in cold weather, and they did indeed, as the *Avalanche's* editor said, hug the stovepipe and try to crawl into the oven—if they had an oven. Because the miners were not coming in with pokes of dust to drink and gamble the saloons and gaming halls were pretty empty. Everybody was waiting for spring and good weather.

The *World* said that some of the residents of Idaho City, shuddering at the thought of a long winter, had proposed to establish a literary club and have weekly lectures, as was the custom in "large towns of Nevada Territory." Societies of various kinds were formed in different camps, including literary and debating groups; and profound subjects were debated, such as, Resolved, that gold has been more of a curse than a blessing; or, That the love of a woman has more influence on man than love of gold. The lectures were usually flamboyant rhetoric; what their windy insincerities did was to highlight the hell-roaring obscenities and brutalities of the camp's life. The editor of the *Black Hawk Mining Journal* in Colorado was touched: "It was profuse in eloquence, in pathos, in high and noble thoughts. Three or four word pictures were introduced which were as fine as the tomes of the classics, ancient or modern afford." The one who produced the eloquence was "not ashamed of being a *man.*"

There were groups of players or single celebrities like Montez who went from camp to camp. The traveling players put on anything from *Hamlet* down to the most unabashed tearjerker, in such famous opera houses as Tabor's in Denver and Piper's in Virginia City, Nevada. In Piper's, Drury says, *Hamlet*

was most often presented. When Edwin Booth told the stage manager the kind of setting he wanted for the 5th act he was told that "he should have a practicable grave." He was astounded at the thoroughness in executing this promise.

Entertainers in a Utah mining camp

A section was sawed out of the stage floor, a couple of Cornish miners did valiant pick-work, and that night the yokel grave-diggers shoveled some interesting specimens of ore onto the stage and were able to hand up Yorick's skull from the profound depths of a real grave. When Booth cried out, "This is I, Hamlet the Dane!" and leaped into Ophelia's grave to tussle with Laertes, he had the experience, unique in his long stage career, of landing on bed-rock.

what was to have been a murderous thrust, but which was only a feeble gesture that missed the villain by three feet, while a thin falsetto voice wailed, "Die, ye villain!" The villain was speechless. The audience exploded with profane jeers and hootings, while the villain gazed with compassion on the sad ineffectual creature who was supposed to be a picture of the outraged manly man.

Popular in the camps, of course, were the blood-and-vengeance melodramas. There appeared on a Boise stage the humilated and betrayed husband, and the triumphant seducer and villain. As told by one who saw it, in the climax of the third act the villain was to rush onto the stage, and the husband, waiting there, was to dispatch him, with the grace and finesse of any betrayed husband who knows what public morality expects him to do. The husband did rush on the villain, crying, "Ho! ho! base betrayer uv me sacred home, ruiner uv me fair name, prepare yerself to die!" He then yanked forth a property sword and made

When order was restored the villain walked to the front of the stage and said to the audience, "Ladies and gentlemen, I've been in this business twenty years. That was the damndest weakest 'die' I ever heard."

It was not often that a camp was favored with a combination as good as Oscar Wilde, passionate lover of the lily, and Eugene Field, passionate lover of the hoax. When it was learned that the visiting lecturer adored the lily, two of Denver's most enterprising madams adorned themselves and their girls with sunflowers and marched through Denver's business center. An outraged

In Tombstone

Irish policeman jailed them but a judge had to turn them free: They had been charged with meretricious conduct but his honor couldn't find that word in his law books. One of the newspapers came out with a poem. Then the one most esthetical turned to the Chief and in language not brief but quite exegetical, said, Why make all this bother about a big sunflower? Now isn't one flower as good as another? Suppose I should carry a shamrock or daisy, would that drive you crazy? Field got himself up in garments much like those affected by Wilde and was driven across Denver, as he gazed adoringly at a lily or waved wanly at the bug-eyed multitude. The reception committee and the flower of Denver society waited at the *Tribune* building, including that newspaper's business manager, who was also Field's boss. When the boss recognized the impostor he rushed at him furiously with a broom and knocked off his wig. Wilde was so wounded in his finer sensibilities that he sneaked out on an early train. The poem concluded: Oh, Oscar! Oh, Oscar! Pray take a hoss car and hasten to aid us; aid or we're busted. No daffydowndilly, no sunflower or lily in Denver is trusted.

The more sophisticated men in the camp loved not so much the tricks of the necromancer as the comments they drew from the yokels. When the cocoanut shell was broken and let fly a half-dozen canary birds, one solemn oaf said, "I'n see the birds musta been hatched in it but I doan see how the mother bird got in to lay the eggs." The more sophisticated also loved practical jokes— and Mark Twain, of course, played more than his share of them. Colton tells this story:

As I was sitting in the house of an old Californian to-day, . . . I felt something break on my head, and, staring around, discovered two large black eyes lighted with their triumph. It flashed upon me, that the annual egg-breaking festival here had commenced. The rules of this frolic do not allow you to take offence, whatever may be your age or the gravity of your profession; you have only one alternative, and that is, to retaliate if you can. You have not to encounter the natural contents of the egg—these are blown out; and the shell is filled with water, scented with cologne, or lavender; or more often with gold tinsel, and flashing paper, cut into ten thousand minute particles. The tinsel is rubbed by a dash of the hand into your hair, and requires no little combing and brushing to get it out. . . . When a liquid is used, the apertures are closed with wax, so that the belligerent may carry it about his person. The antagonist is always of the opposite sex. You must return these shots, or encounter railery, which is even worse. Having finished my chat, I bade my good old California friend, and his daughter, my egg-shell opponent, good morning; but turned into a shop, procured an egg or two, and re-entered the mansion of my friend by a side door, where I watched for my victim. A few moments brought her along, all unconscious of her danger. I slipped from my covert, and, unperceived, dashed the showering egg on her head. Her locks floated in cologne.

In the mining camps, as in the cow towns, people were shaken out of their boredom by the bullies, who, when well-liquored, had to fire their guns. This they often did in saloons. They would destroy the mirrors, use bottles of whisky and rum as targets, shoot out the lights, and sometimes ride in on their ponies. They were seldom or never punished. The explanation is that they were only boys having a little fun, and that the next day they came in and paid for all the damage. It is said that the saloonkeepers liked to have things enlivened, and thought the shooting was good advertising. Sometimes the advertising and fun were carried too far, as we shall see when we come to Slade.

Dancing   The principal forms of entertainment seem to have been drinking, wooing, gambling, and dancing. If there were no women in the camps the men danced with one another, and as Quiett says, if a miner could find an old crinoline costume he would put it on and become the belle of the ball. If no apparel made for women was to be had, the men who took the part of women would put a colored string around one arm, or wear a badge or mark. What times those rough men had, dancing with one another all night to the music of no more than an old fiddle! They preferred, of course, to dance with women, and welcomed the fandango and hurdy-gurdy halls, with their night loves of various hues and ages. But before the night loves came the men danced with one another, hopping or bounding over the rough boards or the bare earth with an energy that would scare the wits

The Tabor opera house

out of most dancers today. What ear-shattering music! What body odors! Half-drunk or wholly drunk, arms around one another, heavy boots playing bass to the shrill fiddle or the cacophonies of the organ, around and around they went, all night, and sometimes for a night and a day and half of the second night, before they fell to their

absence of ladies was a difficulty which was very easily overcome, by a simple arrangement whereby it was understood that every gentleman who had a patch on a certain part of his inexpressibles should be considered a lady for the time being. Those patches were rather fashionable, and were usually large squares of canvas, showing brightly on a dark ground, so the "ladies" of the party were

hard beds to dream their wanton dreams or moan in their nightmares.

Borthwick saw them in action:

It was a strange sight to see a party of long-bearded men, in heavy boots and flannel shirts, going through all the steps and figures of the dance with so much spirit, and often with a great deal of grace, hearty enjoyment depicted on their dried-up sunburned faces, and revolvers and bowie-knives glancing in their belts; while a crowd of the same rough-looking customers stood around, cheering them on to greater efforts, and occasionally dancing a step or two quietly on their own account. Dancing parties such as these were common enough. . . . men danced, as they did everything else, with all their might. . . . The

as conspicuous as if they had been surrounded by the usual quantity of white muslin.

In Bannack when Stuart was there the "dances were very orderly; no man that was drinking was allowed in the hall. The young people danced the waltz, schottish, varsoviane, and polka, but the older ones stuck to the Virginia-reel and quadrille." The Boise *Democrat* published a description of the quadrille, now as obsolete as bloomers:

We both bowed to both of us, then together, then the fiddle tuned, and the thing started; grabbed her femail hand, she squeezed mine, we both slung each other but she slung the most because I think she loved me for a little while; then we changed base clear across the room, jumped up and over so many times, passed each other twice times, then my dear and I dosed a doe and hopped home again. . . . then we two forward four, one ladies changed, we X over, turning around, twiced sash chayed sideways. I backed feller takes his gal back to place; right lady spin right gentleman, spin left gentleman, left gentleman spin left lady, all twist each other, do it over again, keep it up, all turn around, all turn the other, backward, sideways, each couple swing tother couple, cross over, back; all promenade to seats.

Matinee at Tabor's

to place, she dittoed; side couples to the right, to the left, side couples tother way, side couples turn ladies, ladies turn side couples, gentlemen turn side couples, all hands round, back again, first feller take opposite side gal, sling her around and take your own gal and tother feller's gal forward and back, twist both gals twice times, sling 'em to opposite feller, let him do the same as you did, and back again to places, light gentlemen balance to heavy ladies, duplicate, promenade all, gals in the center, fellers get hold of each others hands, bob up and down, arms over the ladies' waterfalls, ladies stoop, jump up and down, each

An Idaho editor told his readers that

The following beautiful description of waltzing is so true to life that we reproduce it. . . . "A group of splendid ones is on the floor, and lovingly mated, and gents encircle their partners' waist with one arm. The ladies and gentlemen closely face to face. They are very erect and lean a little back. The ladies lean a little forward. (Music.) Now all wheel and whirl, circle and curl. Feet and heels of gents go rip, rap, rap, rap. Ladies feet go tippety, tip, tippety, tip, tip. Then all go ripperty, clipperty, slipperty, flipperty, skippety, hoppety, jumpity, sumpity, thum. Ladies fly off by centrifugal momentum. Gents pull ladies hard and close—reel, swing, slide, look tender, look silly, look dizzy. Feet fly, tresses fly, all fly. It looks tuggity, haggity, pullito, squeezity, pressity, rubbity, rip. . . . The maidens tuck down their chins very low, or raise them exceedingly. Some giggle and frown, some sneer, and all sweat freely. The ladies' faces are brought against those of the men, or into their bosoms, breast against breast, nose against nose, toes against toes. Now they are again making a sound like georgy, porgy, deery, peery, didy, pidy, coachy, poachy. This dance is not much, but the extras are glorious. If the men were women, there would be no such dancing. But they are only men and so the dancing goes on by woman's love of it."

Bonnie are the hurdies O!
The German hurdy-gurdies O!
The daftest hour that e'er I spent
Was dancing with the hurdies O!

"The German hurdy-gurdy girls, dancing for a living at fifty cents a dance, had become an institution." Dimsdale, tubercular editor of the Montana *Post,* saw the hurdy-gurdy and described it. "As soon as the men had left off work, these places are opened and dancing commences." The hall has a bar at one end, with champagne at $12 a bottle, whiskey at twenty-five and fifty cents a drink. "The outer enclosure is densely crowded . . . with men in every variety of garb that can be seen on the continent. Beyond the barrier sit the dancing women, called 'hurdy-gurdies,' sometimes dressed in uniform, but more generally habited according to the dictates of individual caprice, in the finest clothes money can buy. . . . On one side is a raised orchestra." A dance costs a dollar in gold.

He describes a

first-class dancer. . . . There she stands at the head of the set. She is of middle height, of rather full and rounded form; her complexion as pure as alabaster, a pair of dangerous looking hazel eyes, a slightly Roman nose, and a small and prettily formed mouth. Her auburn hair is neatly banded and gathered in a tasteful, ornamented net, with a roll and gold tassels at the side.

She glistens and gleams with jewelry.

Her cavalier stands six feet in his boots, which come to the knee and are garnished with a pair of Spanish spurs, with rowels and bells like young water wheels. His buckskin leggings are fringed

at the seams, and gathered at the waist with a U.S. belt, from which hangs his loaded revolver and his sheath knife. . . . His long black hair hangs down beneath his wide felt hat, and in the corner of his mouth is a cigar, which rolls like the lever of an eccentric as he chews the end in his mouth. After an amazingly grave salute, "all hands round" is shouted by the prompter, and off bounds the buckskin hero, rising and falling to the rhythm of the dance, with a clumsy agility and a growing enthusiasm testifying his huge delight. His fair partner, with practiced foot and easy grace, keeps time to the music like a clock, and rounds to her place as smoothly and gracefully as a swan.

The dance closes with promenade to the bar, where the men pay; and after "scarcely an interval, a waltz, polka, schottische, mazurka, varsovienne, or another quadrille commences."

That all sounds sedate and proper enough; judges were there, Dimsdale says, and the "legislative corps, and every one but the Minister. He never ventures further than to engage in conversation with a friend at the door, and while intently watching the performance, lectures on the evil of such places. . . ." What did the newspaper editors think of them? The *Avalanche* copied from the Oregon *Herald* "The citizens of Vancouver have been annoyed lately by the large number of hurdy-gurdy and pretty-waiter girl whisky shops which have been opened in that hitherto quiet village." That was in August, 1866. In October of the previous year the editor began to tell his readers what he thought of hurdy-gurdies:

A brutal affair took place in the hurdy-gurdy house in Ruby on last Sunday night, . . . Two men, named respectively Gibson and Hanson, got into a fight about some matter, during which Gibson pounded Hanson in a shocking manner. Hanson, it is said, was so badly intoxicated as to be almost helpless, yet the crowd in attendance permitted him to be thus maltreated, and had it not been for the arrival of the Sheriff, there is but little doubt that a funeral would have been the result.

Six weeks later he was writing,

The buck hurdies, of Silver, on Thursday night, were having a howling time all to themselves—pistols cocked and flourishing. . . . When the rows commence the good citizens of that neighborhood run under their houses, and prospect holes are in great demand;—and after things quiet down, one can see the heads popping up all

A hurdy-gurdy house, Virginia City, Montana

round, in the attitude of trying to discover whether all's quiet on the Jordan.*

By April of the next year the editor seems to have got enough of them:

So far as our observation goes, these institutions are admitted public nuisances. They are prolific of more rows, fights and funerals than other common resorts. They have their chiefs and a luckless wight meets with ill usage in their hands. During the past week a man was brutally beaten with a pickhandle in the Silver Hurdy House. From expressions we've heard freely made, this beastly conduct has almost reached the point where endurance ceases to be a virtue. There is a disposition to respectfully inform the proprietors to suspend operations or try other localities or take the chances of an "unfriendly ebullition of temper."

———
* Jordan Creek.

By July of the same year he was reporting that

The buck hurdies—or those in the habit of visiting these nuisances—were on the rampage Saturday evening last. A friend who happened by relates that he saw one man on a table with six-shooter in one hand and a bowie-knife in the other, inviting an individual (equipped in like manner) to *jest* step out of the crowd till he'd show him how clean he could blow the top of his head off—"*c-l-e-a-n* off!"—while one of the barkeeps was on a counter flourishing a revolver and excitedly exclaiming, "I'll blow the brains out of the first man that shoots in this house." Another hombre was at the back end of the room, with six-shooter in hand, cocked, wanting to see the man who would fight him—he would shoot, he would; he was chief, he was; and if they didn't believe it, *jest* try it and see; he was bound to *spile* somebody's snoot, and he didn't seem to care whose. This may be very fine sport to the chieftains of a hurdy-gurdy house, but there are others in the community who refuse to "see it" and who dislike to be taxed to pay for new coffins on these special occasions. Look out, boys, or you may attend a tight-rope performance yet.

Sixty miles north, the *Statesman* had said two years earlier,

They are of no earthly use; no amusement to sensible man, and no benefit to the business of the place. The women are the lowest class of Swiss, chubby, black, can't talk English, live in the rear of the dance house on the cheapest kind of plunder. . . . They are imported into our country like slaves for the express purpose, paid a few dollars a month by an overseer, and then hired out to be cat-hauled round a room, drink whisky, and become the associates of a lot of drunken rowdies at four bits a dance. . . .

Now and then a hurdy-gurdy house was a good deal fancier than the one the editor had in mind in Boise. From the ceiling of the Bird Cage in Tombstone curtained boxes were suspended, into one of which a man could slip with the girl of his choice; and what they did there was nobody's business. A horrified newsman from the Arizona *Star* discovered that these love boxes were about fifteen feet above the floor. As for the hurdy-

gurdy, the "climax of obscenity" came around one o'clock. Meanwhile the men were "worked for all they were worth" in buying drinks. The girls of the place in striped or pink tights "or with dresses not of sufficient length to cover the hips danced the can-can through with barbaric vigor." Then men blacked over to represent Negroes, "clad in tights, with women's undergarments over them, sprang on the stage and vied with each other in the obscenity of their actions." It sounds much like the cheap burlesque joints of a later time. There seem to have been several ranks among harlots in Tombstone: the bird cage gals were the real aristocrats; next under them were those who banged the pianos; third were the dancers and drink pushers; and at the bottom were those out soliciting on the streets or "waiting at the door of their cribs."

Drinking   Whether fact or fiction a story says that a dirty-bearded miner stood one morning in the doorway of a saloon and watched a friend staggering along the street under a huge sack of flour. After squirting his juice and rumbling with astonishment he turned to a man and said, with infinite disgust, "Look at that crazy son-of-a-bitch—all stooped over under a ton of

Saturday night in a Denver Bagnio

Placerville, Idaho

Idaho City, Idaho

In Virginia City, Nevada

flour and I bet he ain't got a pint of whisky in the house."

Along with the bully boys, gamblers, bankers, and editors, saloonkeepers, Mark Twain said, occupied the very highest social level in the camps. Indeed, "The cheapest and easiest way to become an influential man and be looked up to by the community at large, was to stand behind a bar, wear a diamond-cluster pin, and sell whisky. I am not sure but that the saloonkeeper held a shade higher rank than any other member of society. . . . To be a saloonkeeper and kill a man was to be illustrious."

This Magnificently Fitted, Elegantly kept Saloon is supplied with the Finest Liquors and Cigars. The patronage of the thirsty is respectfully solicited.

Sometimes the editor put in a good word for his favorite bistro, as when the Bannock* *News* said: "We have the evidence of our own palate as well as the word of connoisseurs for saying that Boston & Cody keep bully liquors and cigars." Nothing in the newspapers received more advertising than the saloons. Drury cites the Carson *Appeal* to show how an editor lends a hand:

Lyman Frisbie has just received an invoice of the best potables that ever were concocted. . . . The whisky is of the most humanizing and exalting character; the brandy is of the choicest flavor and most amiable propensities; . . . the rum is of the quality which none but the most reckless of families are willing to be without; and as to the wines and ales, they are simply vinous and melted nectar, fit for the gods, the goddesses, and the general public.

Byers of the Denver *News* again and again calls the attention of his readers to a white two-story building with the fancy name of Criterior Hall. "If there are any so credulous as to believe that we have no good living in this country, we advise them to drop in at the Criterior presided over by the prince of landlords, Ed Jump." The food and liquors there, he said, "fairly rivaled Delmonico's. . . ."

Of the saloons, as an old-timer said, "they was good ones, bad ones, and worse ones." The good ones were usually closer to the banks, the worse ones to the lairs of the fancy women. Those that

could afford to be were wondrous elegant, with their huge mirrors and polished mahogany bars, brass spittoons and rails, and murals of naked women and innocent angels. Some of them were of immense size: the Long Tom in Columbia, when 15,000 hell-roarers jammed its streets, reached from Broadway all the way to Main Street, and at each of its two doors had two armed guards. Some had sixty-foot bars with as many as twelve bartenders and a consumption of six barrels of whisky a day. At the height of Goldfield's glory one saloonkeeper saw seven million dollars pass through his cashier's wicket in two years. That was his estimate anyway. It included various sums that he stored for patrons. When it was rumored that the banks were going broke so many men rushed to him with their funds that "It piled up so fast I couldn't get in the safe, so I sacked it and dropped it on the floor. It piled up there so I couldn't move around." At Crook Avenue and Main Street in Goldfield (Nevada) there was, in Beebe's opinion "for several years probably the most celebrated guzzling and gaming four corners in the world"—the Northern Saloon, the Palace, the Mohawk, and the Hermitage; and not far from them the Phoenix, Mint, Oriental, Combination, Silver Dollar and Last Chance. Beebe thinks that in only two years Rickard's Northern grossed more than $7,000,000, including its gaming tables. Among the more famous saloons in Dawson during the Klondike rush were Rickard's, Wyatt Earp's, and Boozer Brown's. The Chloride in Pickhandle Gulch in Candelaria was another famous one; and

The Crystal, most "elegant" bar on the Comstock Lode, still open.

_____
*The Montana camp name was spelled "Bannack."

Crystal bar, Tombstone, still open

as Emrich points out, there were "Buckets of Blood over the whole West, but the two famous ones remaining are on Park Street in Livingston, Montana, and on C Street in Virginia City, Nevada."* The Montana bucket was for a while pouring riches into the purses of Tex Rickard and Soapy Smith, before they headed for the Klondike. Not far from it Madame Bulldog had her saloon. Weighing 190 stripped (and she was often stripped, for she liked to move around free and easy when she acted as her own bouncer), she is said to have put the formidable Calamity Jane through the door.

Editor Drury says that by "authentic count in 1876 there were exactly 100 retail liquor dealers in Virginia City; 37 in Gold Hill; 7 in Silver City —a grand total of 144 in the Comstock sector. Ten wholesale liquor dealers and five breweries helped out. The Delta, the Silver Palace, El Dorado, Palace, Capitol, the Sawdust Corner—these were among the best known." The Gold Hill *News* boasted that Virginia City and Gold Hill "have more saloons to the population than any place

in the country"; and on the Sabbath "we are entertained with a horse-race or a fight between a bulldog and a wildcat."

In all the saloons the bartenders were among the most enterprising men in the camps. As was the custom, in payment for drinks he took a pinch of gold dust from a poke and managed to let a little of the gold fall on to the bar, which from time to time he wiped off. Before closing he would go outside and get mud on the soles of his boots, and on returning would step carefully around over the floor behind the bar, allowing the mud to gather up the gold. The stories say that he might wash out as much as twenty-five dollars' worth on a good weekday, and one hundred dollars on a booming Sunday. In the Klondike Wilson Mizner, a fast-buck man, gave Jack Kearns (later to be Jack Dempsey's manager) the job of weighing out gold dust, and instructed him in ways to make a neat profit on the side. After hiding some thin strands of syrup in the Kearns mop of hair Mizner told him that when in profound thought or in moments of extreme agitation

---

* Whether the present Bucket of Blood is on the original site is a matter of dispute.

Virginia City, Nevada

it might occur to him to run through his hair the fingers he used to pinch the dust. One Mizner admirer swears that at the end of a good day's run a Kearns shampoo sometimes panned out as much as a hundred dollars' worth.

In some camps the saloons set up the first drinks free when the miners came off work, and the first drink in the morning was free to all comers. Some of the men, Drury says, "took advantage of this, . . . and went to twelve or thirteen saloons for their morning nips, but as a general thing, I think, the respectable ones didn't extend

their routes to more than seven places." Borthwick was impressed by the "confidence placed in the discretion of the public—namely, the mode of dispensing liquors. When you ask for brandy, the barkeeper hands you a tumbler and a decanter of brandy, and you help yourself to as much as you please: the price is all the same; . . . and in the case of cocktails, and such drinks as the barkeeper mixes, you tell him to make it as light, or stiff, as you wish. This is the custom even at the very lowest class of grogshops. They have a story in the States connected with this, so awfully

Jerome, Arizona

old that I am almost ashamed to repeat it. I have heard it told a thousand times, and always located in the bar of the Astor House in New York; so we may suppose it to have happened there.

A man came up to the bar, and asking for brandy, was handed a decanter of brandy. . . . Filling a tumbler nearly full he drank it off, and, laying his shilling on the counter, was walking away, when the bar-keeper called after him, "Saay, stranger! you've forgot your change—there's six-pence." "No," he said, "I only give you a shilling; is it not a shilling a drink?" "Yes," said the bar-keeper; "selling it retail we charge a shilling, but a fellow like you taking it wholesale we only charge sixpence."

What was the quality of the liquor they sold?

I have not tasted it [Greeley wrote], but the smell I could not escape, and I am sure a more whole-some potable might be compounded of spirits of turpentine, aqua fortis, and steeped tobacco. Its look alone would condemn it—soapy, ropy, tur-bid, it is within bounds to say that every pint of it contains as much poison as a gallon of pure whisky. And yet fully half of the earnings of the working men . . . are fooled away on this abominable witch-broth and its foster-brother to-bacco. . . .

Some of it was undoubtedly that bad, but the quality varied widely. In some saloons they would add to alcohol water, burnt sugar, a chem-ical or two, and call the concoction either whisky or brandy. In a Montana camp, some of the drinks served, according to Dimsdale, were Tan-gleleg, Forty-rod, Lightning, and Tarantula-juice, all of which were pretty deadly. In many saloons a person could buy sangerees, mint juleps, sherry cobblers, and various other cocktails. Bowles found that

More French wines are drunk in California, twice over, than by the same population in any part of the eastern States. Champagne is mother's milk, indeed, to all these people; they start the day with a "champagne cock tail," and go to bed with a full bottle of it under their ribs. At all the bar-rooms, it is sold by the glass, the same as any other liquor, . . .

That most of the miners were a hardy breed who could be killed with nothing less than a gun of large bore, or a headlong fall three hundred feet down a mine shaft, is a part of legend if not of fact; and those who write of their drinking

contend that they were able year after year to consume enormous quantities of tarantula juice, bug juice, Taos lightning, and tangleleg with little damage to their organs. Sometimes the stuff they drank was nothing more than a poor brand of alcohol into which had been mixed about everything a barman could think of, including burnt sugar, flavoring extracts, cheap whisky and rum, and plug tobacco. It was said that the light-ning struck a man stiff with the first gulp; that the forty-rod fetched him to his knees after he had taken a few amazed and faltering steps; that the tangleleg so confused his senses that he could no longer tell which leg to put forward, and so flopped them around in tangleleg fashion as he tried to walk; and that the tarantula juice gave him the notion that the top of his skull was about to take off and soar. These potent ego-fortifiers were not mincingly measured into a jigger, as nowadays, but were poured in eager gurglings from huge barrels. It was in Butte, a writer says, that the Shawn O'Farrell was invented, which is now known as the boilermaker—a tumbler of raw whisky with a glass of beer to wash it down be-fore the man choked to death.

Food might be scarce in the camps but there was usually an abundance of drink. Almost thirty years ago we published a story which at that time we thought had its origin in Idaho City but which probably has, like so many legends, no traceable source. It was a very tough winter in the camp, with snow eight feet deep and the temperature below zero, when a small packtrain of ten bur-ros came in sight. The thirsty miners made a rush for the beasts and stripped the packs off; amazement and horror filled their faces as the appalling truth stood revealed. Nine of the beasts were loaded with whisky and one with flour. A man turned to those around him, and his voice and face must be imagined as he cried, "What in the name of the Almighty does the fool think we're gonna do with all that flour?"

Huntoon's at the corner of Main and Wallula advertised in the *News*: "Rum, old Jamaica, in original packages from the Hudson Bay Com-pany's Store, in Victoria. . . . Scotch whisky, gen-uine, & Glenlivet. . . . 3,000 gallons Old Bourbon, Rye and Wheat Whisky. . . . French Brandy— 1,000 gallons fine cognac, brandies, genuine 'Sa-zerac,' Godard, Pelievoisin, Otard, Depuy & Co.

. . ." That is a typical advertisement in camp newspapers.

Colton observed that "Quite a sensation was produced among the gold-diggers this morning by the arrival of a wagon from Stockton, freighted with provisions and a barrel of liquor. The former had been getting scarce, and the latter had long since entirely given out. The prices of the first importation were—flour, two dollars a pound; sugar and coffee, four dollars; and the liquor, which was nothing more or less than New England rum, was twenty dollars the quart. But few had bottles: every species of retainer was resorted to; some took their quart cups, some their coffeepots, and others their saucepans; while one fellow, who had neither, offered ten dollars to let him suck with a straw from the bung."

On the Comstock, says Lewis, after most of the whisky had been drunk that would be drunk there, and the Lode was panning out fast, there were still a hundred saloons, with 75,000 gallons of hard stuff and twice that much wine and beer gurgling down throats every year.

The heavy drinking led, of course, to heavy fighting, often with fists, sometimes with guns or knives. One Tuesday morning, said the Boise *News,* "Several parties were found in the streets . . . some with fractured skulls, some with bunged eyes and swollen faces. . . . puddles of blood were distributed over so large a district that it was almost impossible to locate the fight. . . ." Drunkenness led some of the men to do fantastic things. McIlhany saw a man named Bob Love pick up a huge decanter in the Arcade in San Francisco, look at the handsome mirror above the bar, and say, "Jack, that's a fine looking glass" and shatter it. He then said goodnight and walked out. The next day he returned, asked how much the mirror was worth, and on being told that a new one would cost $500 laid $500 in gold on the bar. "Does that square us?" "That settles everything, Bob. Come again." The Boise *News* reported a "terrible stampede" in a hurdy-gurdy house:

Several frightened individuals came past Boston and Cody's corner running for dear life . . . exclaiming as they ran, that four or five men were killed up town. We . . . took a look at the wreck. The sash-doors were completely carried away, and the pavement strewn with broken glass. A simultaneous rush of the crowd for the street when the firing began had carried everything before it,

bursting windows and doors, and tumbling heels over head in the wildest confusion.

So commonplace a thing was hardly newsworthy.

"We're going to get drunk, by gum, by gum, in Shorty's saloon tonight; we'll guzzle our ale before it's stale and dance until broad daylight!" We can imagine them singing such words in the Howling Wilderness Saloon on the Mother Lode. And we can see some of them the next morning in the condition of Pat Mahaffey, who, according to the Idaho *World,* "came home the worse for liquor. Turning sick, he sat by the stove and vomited into a box in which his wife had placed several goslings. After a bit, Pat looked down and saw them and roared: 'God a'mighty, wife, when did I swaller them things?'"

Horace Greeley wrote, after an appalled survey of conditions in Denver, "More brawls, more fights, more pistol shots with criminal intent in this log city . . . than in any community of equal numbers on earth." Eastern greenhorns were easily led into superlatives. Still, it takes superlatives to describe such conditions as those when San Francisco had 537 places selling liquor, and temperance societies and homes for drunkards were trying to care for the more hopeless cases. Alcohol and drunkenness became, in fact, a very grave problem in nearly all the camps. San Francisco had its 537 drinking places by 1853. In the previous year the Sons of Temperance were in the field, and penitent miners were taking to water all over the Sierra foothills. By 1856 the Placerville newspaper gave its readers the "Toper's Soliloquy":

To drink or not to drink, that is the question; whether 'tis nobler in the mind to suffer the slings and arrows of outrageous thirst or take up arms against the Temperance League and by besotting frighten them? To get drunk—to sleep it off no more. To get drunk without a headache, and to walk when drunk—'tis a consummation devoutly to be wished. To get drunk—to sleep in the street; to sleep! Perchance to get "took up"—ay, there's the rub! And thus the Maine Law doth make sober men of us all; and this, the ruddy hue of brandy, is siclied o'er with the pale cast of water —to lose the name of Drink!

The conditions Helper found in the Diana Saloon in San Francisco, or Marryat, when looking over the camps, could have been seen in almost any bonanza camp. Helper says the saloon was filled "at all hours" and that "one-fifth of them [were] drinking because they have been lucky, and the other four-fifths because they have been un-lucky."

The walls of this saloon were covered with paintings and engravings of "nude women in every imaginable posture of obscenity and indecency." Marryat found that a "vast quantity of liquor" was daily consumed in San Francisco;

From the time the habitual drinker . . . takes his morning gin-cocktail to stimulate an appetite for breakfast, he supplies himself at intervals throughout the day. . . . saloons line the streets, and you cannot meet a friend, or make a new acquaintance, or strike a bargain, without an invitation to a drink, which amounts to a command; . . .

The *Avalanche* concluded that there were six types of drunks to be found in the saloons:

The first is ape-drunk. He leaps and sings and yells and dances, making all sorts of grimaces and cutting up all sorts of "monkey-shines" to excite the laughter of his fellows. The second is tiger-drunk. He breaks the bottles, breaks the chairs, breaks the heads of fellow-carousers, and is full of blood and thunder. Of this sort are those who abuse their families. The third is hog-drunk. He rolls in the dirt on the floor, slobbers and grunts, and going into the streets makes his bed in the first ditch or filthy corner he may happen to fall into. He is heavy, lumpish and sleepy, and cries in a whining way for a little more drink. The fourth is puppy-drunk. He will weep for kindness, and whine his love and hug you in his arms, and kiss you with his slobbery lips, and proclaim how much he loves you. You are the best man he ever saw, and he will lay down his money or his life for you. The fifth is owl-drunk. He is wise in his own conceit. No man can differ with him, for his word is law. His arm is the strongest, his voice the sweetest, his horse the fleetest, his town the finest, of all in the room or land. The sixth and last animal of our drunken menagerie is the fox-drunk man. He is crafty and ready to trade horses and cheat if he can. Keen to strike a bargain, leering round with low cunning, peeping through cracks, listening under the eaves, watching for some suspicious thing, sly as a fox, sneaking as the wolf. He is the meanest drunkard of them all.

For Borthwick, one of the more sensitive observers,

Drinking was the great consolation for those who had not moral strength to bear up under their disappointments. Some men gradually obscured their intellects by increased habits of drinking. . . . But, though drunkenness was common enough, the number of drunken men one saw was small, considering the enormous consumption of liquor. The American style of drinking is so different from that in fashion in the Old World, and forms such an important part of social intercourse, that it certainly deserves to be considered one of the peculiar institutions of the

Portable bar.
Leslie's Newspaper, 1879.

country.* [Of empty bottles he said that] the enormous heaps of them, piled up in all sorts of out-of-the-way places, suggested a consumption of liquor that was truly awful.

Some visitors hoped to do something about it. After saying that it "is almost inconceivable what an excitement was produced upon nations and individuals" by the discovery of gold in California, Woods, a minister, hoped that his book might be a warning to those who planned to come to the diggings. He tells with sadness of the case of a "merchant of education and refinement" who "within three months of the time he arrived in the country became a subject of *mania a potu,* and died in the streets of San Francisco." The *Avalanche* told its readers that when back in Iowa a man was invited to speak at a temperance meeting he replied:

When drunkards to the right of us, drunkards to the left of us, drunkards in front of us, blunder and stagger, your mission is God-like. When physicians stop giving whisky and alcoholic medicines to their patients—when editors practice temperance as well as preach it—when young men have the courage to say "No"—when preachers stop the practice of giving wine to pledged temperance men at the sacrament—when temperance becomes as fashionable as drunkenness now is, cold water will be at a premium.

The same newspaper printed a little homily for those struggling with the demon:

A young wife, within a few months after her marriage, was horrified one night on going to the door to let her husband in to find him intoxicated—so much so that he could hardly stand. . . . Friends were in the drawing-room expecting him every minute. With a woman's quickness and instinct, she hurried him to his chamber, with loving words put him to sleep, made some excuse for the mutual benefit of friends and servants, watched over him through the weary hours of the night, and the next morning when he awoke, she had no harsh reproof for him, no withering rebuke, but tenderly alluded to the facts of the case and kissed away his confusion. There it dropped. He never tasted another drop; raised a family of sons who are responsible and successful merchants in New York and Philadelphia, and dying, left a name for mercantile integrity which any man might envy.

It was not only habitual drunkenness in the camps; it was often prolonged bouts of drinking as a way of demonstrating vigor and manhood. Dame Shirley's twelfth letter is a fine description of what she called a

perfect Saturnalia. . . . I believe that the company danced all night; at any rate, they were dancing when I went to sleep, and they were dancing when I woke the next morning. The revel was kept up in this mad way for three days, growing wilder every hour. Some never slept at all during that time. On the fourth day they got past dancing, and, lying in drunken heaps about the bar-room, commenced a most unearthly howling;—some barked like dogs, some roared like bulls, and others hissed like serpents and geese. Many were too far gone to imitate anything but their own animalized selves. . . .

Nearly every day, I was dreadfully frightened, by seeing a boat-load of intoxicated men fall into the river, where nothing but the fact of their *being* intoxicated saved many of them from drowning.*

She writes her sister of a Christmas celebration at Indian Bar that was a four-day drunk; it petered out but picked up fresh energy after New Year's Day.

The carryings-on during the nights in Silver City, said the *Avalanche,*

would discount the illustrated paper known as *Day's Doings* in New York. For instance, a posse of the most quiet and respectable young men in town got on a bender the other night. After getting pretty well "set up" they adjourned to an attic where they divested themselves of most of their clothing and then the deviltry commenced. Scalp dances and war whoops were performed in the most approved Indian style. Some quiet citizens, who were passing by, heard the revels of the Bacchanalians and went up-stairs to learn the cause. They were instantly seized, stripped and compelled to join the Bedlamites in their drunken revels. Before dispersing, all hands joined in a shirt-tail foot-race through the principal streets in town. This open air exercise cooled down the wretches, or no telling what depredations might have been committed.

To suggest the power of alcohol in the camps possibly there is no better instance of the things that happened than the threat of General Thomas

---

* Drinking habits seem to have changed in much of the Old Country.

* She expressed regrets to her sister for such an unpleasant letter "But I am bound, Molly, by my promise, to give you a *true* picture (as much as in me lies,) of mining life. . . ."

NEW STORIES **LIBRARY** STARTLING ADVENTURE

Entered According to Act of Congress, in the Year 1892, by Street & Smith, in the Office of the Librarian of Congress. Entered as Second class Matter at the New York, N. Y., Post Office, March 21, 1889. Issued Weekly. Subscription Price, $5.00 Per Year. March 31, 1892.

159.   STREET & SMITH, Publishers.   NEW YORK.   31 Rose St., N. Y. P. O. Box 2734.   10 Cents

GENTLEMAN JOE THE GILT-EDGED SPORT

By Jos. E. Badger Jr.

THE GILT-EDGED SPORT HAD DRIVEN A DAGGER THROUGH HAND AND CASH, PINNING THE GAMBLER FAST TO THE TABLE.

F. Meagher, acting governor of the Montana Territory. When the legislative assembly proved to be balky in passing the laws he wanted, he said the bills would be passed or he would by martial law take all whisky away from the legislators and from Virginia City.

Gambling A historian has said that in the early days of the West the profession of gambling "enjoyed an esteem equal to medicine, law, and banking." That was Mark Twain's view. But we must bear in mind that by professional gambler is meant not the suckers who sat at the tables and lost their money but the proprietors of the gaming halls. Many of these men moved with the gold rushes and set up in the new camps as they were established; a few of those in the Klondike got their start on the Mother Lode or Comstock Lode.

For many of the common men, with little knowledge and less gumption, losing their gold dust was the chief form of entertainment. A tale Glasscock dug out of the files can almost stand as the record of thousands of them. A miner named Spindle had been doing well in his diggings, so well that a professional called Eagle-Nose began to spy on him and covet his savings. He managed to lure him into a game, and of course let Spindle have what he thought was luck, to put his suspicions to bed, if the simple fellow had any. Spindle was allowed to win again and again; he must have been thinking that this was a hell of a lot easier way to get rich than shoveling gravel into a rocker. The moment came, as in such affairs it always does, when Spindle had four kings and a queen, and was allowed a glimpse of an ace on the bottom of the deck in Eagle-Nose's hands. He never suspected, poor fool, that the ace had been put there for him to see; and so bet by bet he was raised, until all that he owned and had won was in the middle of the table. Was he trying all the while to conceal his glee, while telling himself over and over that there could not possibly be more than three aces against him? No doubt. In the showdown when four aces were spread against him and he said there was still an ace in the deck, and was handed the deck to examine and could find none, he must have thought his eyesight was failing him. Poor devil, he left the hall and in the dark of night went to a precipice and jumped off. Eagle-

Nose didn't do so well either: he was caught slipping an ace and was shot dead, as all slippers of aces were in bonanza camps, when the eye and the draw were fast enough.

In 1866, J. E. Wharton tried to describe what he called the gambling hells of Denver:

Wine, liquors and cigars of the rarest vintage and the most costly brands garnished the splendidly furnished bars. . . . Around the walls of the room were ranged the tables of the gamblers, each temptingly displaying its piles of new and shining bank notes, besides the implements of the nefarious trade, and presided over by a smiling demon, under whose blandishments there lurked a heart that considered all men his prey, . . .

Faro and monte, he says, were the favorite games.

Richardson stopped off in Denver on his way west.

Gaming was universal. Denver and Auraria [facing Denver across Cherry Creek] contained about one thousand people, with three hundred buildings, nearly all of hewn pine logs. One third were unfinished and roofless, . . . One lady by sewing together corn sacks for a carpet and covering her log walls with sheets and tablecloths, gave to her mansion an appearance of rare luxury. . . . Chimneys were of sticks of wood piled up like children's cob houses and plastered with mud. . . .

The Denver House was a long low one-story edifice, one hundred and thirty feet by thirty-six feet, with log walls and windows and roof of white sheeting. In this spacious saloon, the whole width of the building, the earth was well sprinkled to keep dust down. The room was always crowded with swarthy men armed and in rough costumes. The bar sold enormous quantities of cigars and liquors. At half a dozen tables the gamblers were always busy, day and evening. One in woolen shirt and jockey cap drove a thriving business at three-card monte, which netted him about one hundred dollars per day. Standing behind his little table he would select three cards from his pack, show their faces to the crowd, and thus begin:

"Here you are, gentlemen; this ace of hearts is the winning card. Watch it closely. Follow it with your eye as I shuffle. Here it is, and now here, now here and now," (laying the three on the table with faces down) — "where? If you point it out the first time you win; but if you miss you lose. Here it is you see," (turning it up;) "now watch it again," (shuffling). "This ace of hearts, gentleman, is the winning card. I take no bets from paupers, cripples or orphan children. The ace of hearts. It is my regular trade, gentlemen — to move my hands quicker than your eyes. I always have two chances to your one. The ace of hearts. If your sight is quick enough, you beat me and I pay; if not, I beat you and take your money. The ace of hearts; who will go me twenty?"

Three-card monte seems to have been one of the most popular games; others were stud and draw, roulette, chuck-a-luck, wheels of fortune, faro, blackjack, dice, as well as some with pretty fancy names, according to Parkhill, at least in the Denver area. Faro, he says, was the "most popular game in western gambling houses," for the reason that the percentage in favor of the house was usually only one and three-fourths, and because it was believed that this game could not be rigged. Beebe says that in the Stage Drivers Retreat in Columbia there were "143 faro-bank games in progress at one time."

Borthwick saw the gambling on the Mother Lode. Spanish monte, introduced by Mexicans, was

played on a table about six feet by four, on each side of which sits a dealer, and between them is the bank of gold and silver coin, to the amount of five or ten thousand dollars, piled up in rows. . . . At either end of the table two compartments are marked on the cloth, on each of which the dealer lays out a card. Bets are then made by placing one's stake on the card betted on; and are decided according to which of those laid out first makes its appearance, as the dealer draws card after card from the top of the pack. It is a game at which the dealer has such advantages, and which, at the same time, gives him such facilities for cheating, that any one who continues to bet at it is sure to be fleeced. Faro, which was the more favorite game for the heavy betting . . . is played on a table the same size as a monte table. Laid out upon it are all the thirteen cards of a suit, on any of which one makes his bets, to be decided according as the same card appears first or second as the dealer draws them two by two off the top of the pack. Faro was generally played by systematic gamblers, . . . while monte, from its being apparently more simple, was patronized by novices. There were also roulette and rouge-et-noir tables, and an infinite variety of small games played with dice, and classed under the general appellation of chuck-a-luck.

Faro game. Houseman is in straw hat

He points out that gamblers were the professionals who

laid out their banks in public rooms, and invited all and sundry to bet against them . . . and any one who did business with them was no more a gambler than a man who bought a pound of tea was a grocer. . . .

As for the suckers,

The Mexicans showed the most admirable impassibility. I have seen one betting so high at a monte table that a crowd collected round to watch the result. After winning a large sum of money he finally staked it all on one card, and lost, when he exhibited less concern than many of the bystanders, for he merely condescended to give a slight shrug of his shoulders as he lighted his cigarita and strolled slowly off.

Sonora, California, had, says one historian, more gamblers, drunks, whores, and gold nuggets than any camp of its size. Frank Marryat, an Englishman, arrived there in September, 1851, and left his impressions of it:

It was dark when we entered Sonora; and as the habits of the people here are nocturnal, the evening may be said to have commenced as we alighted. It certainly had commenced, for Greenwich Fair might be spoken of as a sober picture of domestic life compared to the din and clamor that resounded through the main street of Sonora. In either side were gambling houses of large dimensions, but very fragile structure, built of a fashion to invite conflagration. . . . In most of these booths and barns the internal decorations were very glittering; chandeliers threw a glittering light on the heaps of gold that lay piled on each monte table, while the drinking bars held forth inducements that nothing mortal is supposed to be able to resist. On a raised platform is a band of music, or perhaps Ethiopian serenaders, or, if a Mexican saloon, a quartet of guitars; and in one house, and that the largest, is a piano, and a lady in black velvet who sings in Italian and accompanies herself, and who elicits great admiration and applause on account of the scarcity of the fair sex. . . .

"The gambling-saloons" in California were, Borthwick says,

Faro in Bingham, Utah

very numerous, occupying the most prominent positions in the leading thoroughfares. . . . thronged day and night. . . . on entering . . . one found a large well-proportioned saloon sixty or seventy feet long, brilliantly lighted up by several very fine chandeliers, the walls decorated with ornamental painting and gilding, and hung with large mirrors and showy pictures . . . a dozen or more tables in the rooms, each with a compact crowd of eager betters around it, and the whole room was so filled with men that elbowing one's way between the tables was a matter of difficulty. The atmosphere was quite hazy with the quantity of tobacco smoke, and was strongly impregnated with the fumes of brandy. . . . The people composing the crowd were men of every class . . . respectable-looking men . . . rough miners fresh from the diggings, with well-filled buckskin purses, dirty old flannel shirts, and shapeless hats . . . Mexicans wrapped up in their blankets smoking cigaritas, and watching the game intently from under their broad-brimmed hats; Frenchmen in their blouses smoking black pipes; and little urchins, or little old scamps rather, ten or twelve years of age, smoking cigars as big as themselves, with the air of men who were quite up to all the hooks and crooks of this wicked world (as indeed they were), and losing their hundred dollars at a pop with all the nonchalance of an old gambler. . . . dirty, squallid, villainous-looking scoundrels, who never looked straight out of their eyes, but still were always looking at something . . . could have made their faces their fortunes in some parts of the world, by "sitting" for murderers, or ruffians generally. . . . some wretched, dazed, woebegone men . . . the glassy unintelligent eye looking as if it saw a dim misty vision of everything all at once; the only meaning in the face being about the lips, where still lingered the smack of grateful enjoyment of the last mouthful of whisky. . . . Appearances, at least as far as dress was concerned, went for nothing at all . . . and frequently the man to be most courted for his dollars was the most to be despised for his looks.

The "take" in the larger halls can be imagined when we realize the implication in Buffum's statement, that in San Francisco in its beginnings the Parker House hotel leased for $200,000 a year; and that just one room in the El Dorado used for gambling paid a rent of $65,000 a year.

Were these gambling halls honest places? Some unwary writers have thought they were. Possibly they had been deceived by the many tales of honesty in the mining camps, such as that of Jim Dyer in Idaho, who always left his door unlocked so that a hungry passerby could enter and cook a

meal. Jim says he never had a thing stolen except his sourdough, in which he took special pride; after that he bought a slop jar and after putting his dough in it kept the jar under his bed.

In the camp gambling halls "every false and dishonest trick," as Helper says, was employed to fleece the unwary and the ignorant. In California camps which Helper visited there were two housemen at a table, who exchanged secret signals, "one of them a lank, cadaverous fellow, with a repulsive expression of low cunning, full of hypocrisy and deceit . . . who was formerly a Baptist preacher in Connecticut." He thought the public officials as dishonest as the gamblers. Knower heard through the muslin partition that separated his bedchamber from the next room the owners of a gambling saloon discussing their problems, and learned that in roulette the ball could be stopped on red or black. If $20 was bet on the red "the tapper would bet $10 on the black, and they could not make the red lose without making the black win."

It was Kelly's conclusion that

The miners never thought of sitting down to dawdle over an honest game to pass the evening; they would not give a cheese-paring for the dull, stupid monotony of fair play: the excitement of cheating in card-stealing, card-dropping, packing the deal, or defrauding the pool, constituted, according to their standard, the main interest of the amusement. Merit never was rewarded to mere skill in play; but when a fellow won by a gross fraud, a shout of approbation complimented his knavery. As a necessary consequence, serious rows and bloody encounters sprang out of those debasing scenes. On one occasion, at the national game of "poker," I observed a player slily dropping kings into his lap, as opportunity offered, until he assembled all the male portion of the royal family in a cluster; and then, with the full confidence of an all but invincible hand, he substituted ounces for dollars, bragging half-a-dozen to begin with. An opposite competitor covered the amount, and advanced an extra half-dozen, on which the other further improved, doubling the sum . . . when, to his utter mortification and discomfiture, four aces were displayed to his astonished gaze.

Wilson Mizner, con man of the Klondike, swore that he had never known what real crookedness was until he went to Alaska. To show what double-dealers the ordinary run of gamblers were he set up a poker game and dealt from a pack of fifty-two aces. Each player bugged his eyes at the five aces in his hand and wondered how he could get rid of the extra one. It is easy to imagine the infernal glee playing over the Mizner face as he slyly watched them.

One of the first questions that might occur to a gambler is, What was the record pot or stake? It seems not to be known. It apparently is a fact that Sam Bonnifield lost a $150,000 pot to Louis Golden one night in Dawson. The two men were professional gamblers. But the most spectacular gambling was really in mining stocks—as when the cold and ruthless William Sharon of the Bank of California got a tip on a new vein in the Comstock, and buying at bargain prices every share of Belcher stock that was offered watched it climb to $1,525 in three months. Emrich, who seems to have searched all over the map, says the largest wager ever made was the Territory of New Mexico, which was signed away by its governor, at pistol point, to match the raise of ten thousand head of cattle and a big ranch. In Texas, of course. The same author tells us that in the Texas House in Leadville a hundred thousand dollars changed hands in a night. He quotes an unnamed writer: "At the Santa Fe hotel, enormous piles of silver weighed down the tables, and frequently ten thousand dollars changed hands in ten minutes."

What was the character of the gamblers? Just about every kind under the sun. Quiett quotes Jafet Lindeberg, one of the founders of Nome:

Say what you will, the gamblers are the best developers of a mining country. In my long frontier life I have never seen a time when a gambler refused a man or woman who asked him for money. And I have seen men in actual need refused by plenty of other people. Those who make money at the tables can be induced to turn it over quickly enough for grubstakes because they figure it's just so much velvet, and when they lose the money to the house, the proprietor figures about the same way. So he is generous with it. There have been more prospectors grub-staked to sink holes in apparently impossible places, more rich strikes brought in, by men who were grub-staked by gamblers than by any other class of people. Everybody jumps on them and criticizes them, but, after all, most of them are damn nice and if I was broke I'd rather ask a gambler for a lift than anybody else.

Some of them cherished the superstitions common to gamblers, one of which said that a woman in a mine would lead to bad luck. In Mexico it is said that miners fled the premises when a woman came in, and then had a priest with holy water purify the underground workings. Another common superstition concerned candles or lamps. If they went out three times or fell off the walls it indicated that some well-dressed pomaded lecher was busy with a miner's wife. Animals in mines were looked on as friends, endowed with mysterious powers. Rats, bats, and reptiles were believed to have a sixth sense that warned them of floods, cave-ins, and fires; and rats were fed and pampered to insure their presence. Mules and burros were also thought to have a sense of imminent danger. Such superstitions were, of course, found in the Peter Sapps and their friends, rather than in men who made a profession of gambling.

Some of the gamblers, according to those who saw them, were men of exquisite manners and politeness; other were bully boys who were conscious of the guns and daggers hanging from their belts. Wild Bill Hickok, who made his living at card tables, was sitting at a poker game when a bully, who seems to have lost to Hickok in a previous game, slipped up and shot him. Donaldson gives a typical scene after the bullies had drawn their guns:

Joel B. Oldham kept an immense gambling house and saloon on Main Street, Boise. . . . The door stood wide open, and passers could note the extent of the bar and see the games of chance in full operation. My wife and I walked past one evening when the place was so crowded that tobacco fumes made it difficult to see past the door. Suddenly, oaths and shots rang out, and a fight was on! More shots came, and instantly the crowd in the saloon fell on their faces and crawled out the door. It was one of the most ludicrous sights I ever saw; miners, great hulking brutes, scrambled through the dust until protected by the adobe walls and then stood up and looked about for better cover. The banker had quickly grabbed all coin in sight and crawled out behind the first rush. When the confusion quieted, we looked in and saw the bartenders lying prostrate beside the huge timber that ran about the bar for a fortification.

Those who studied the nonprofessional gamblers in mining camps perceived that many of them were obsessed—obsessed in the way of the miserable wretches who in Nevada's saloons today feed a five- or six-hundred dollar paycheck into dollar machines and then in a sweat light a cigarette and amble outside. Delano observed that "a great many miners spent their time in riot and debauchery. Some scarcely ate their meals, some would not go to their cabins, but building large fires, would lay down, exposed to the frost; and one night, in the rain. Even after the monte dealer had cleared nearly all out who would play, the game was kept up by the miners themselves in a small way, till the fragments of their purses were exhausted. There were two companies at work near me, who, when I first went there, were taking out daily in each company, from one hundred to one hundred and fifty dollars. This they continued to do for more than two weeks, when it seemed as if the gold blistered their fingers, and they began a career of drinking and gambling, until it was gone. Instead of going to work on their claims again, they were seized with the prospecting mania," and took off in search of richer fields.

Like Delano and other visitors, Colton studied the men in the gaming halls.

They were playing at monte; the keeper of the bank was a woman, . . . There was no coin on the table; the bank consisted of a pile of gold, weighing, perhaps, a hundred pounds; and each of the players laid down his ounce or pound, as his means or courage permitted. The woman, on the whole, appeared to be the winner, though one man, in the course of half an hour, took ten pounds from her yellow pile. [Some of the miners dig out gold] simply and solely that he may have the wherewithal for gambling. This is the rallying thought which wakes with him in the morning, which accompanies him through the day, and which floats through his dreams at night. For this he labors, and cheerfully denies himself every comfort. All this is the result of habit. A Mussulman looks upon gambling as a species of larceny,—as a crime which deserves the bastinado. I saw a Turkish cadi at Smyrna sentence a man to thirty-nine lashes for having, as he termed it, *swindled* another out of fifty dollars at faro. Give me a Turk when there is a rogue to be caught or a crime punished.

It seems appropriate to bring these dreary pages on gambling to a close with a brief picture of the boys imitating their elders. So much gold and silver was dropped from the gaming tables that enterprising youngsters entered the halls the morning after to pan the litter. An old-timer who was in Columbia has said, "After a night's play the discarded decks of cards would be three inches deep all over the floor. Gold dust and coins falling into that were lost until we boys found them." After finding them, "we boys" gambled just as earnestly as those who had sat at the tables the night before.

CHAPTER II

Girls of the Line

They have been called whores (AS. hōre), harlots, prostitutes, trollops, strumpets, courtesans, drabs, hussies, harridans, sluts, chippies, wantons, fancy women, painted ladies, ladies of the shadows, frail sisters, soiled doves, vile libels on their sex, girls of the red light, and girls of the line, the last being the most widely used term in the West because the cribs and parlors nearly always stood in a line up and down the street. Books have been published on the history of prostitution and a great many on sexual practices in different ages and areas but we know of none that makes a study of the tendency in the human male to glamorize the soiled doves. As Caroline Bancroft says in *Six Racy Madams,* "A petite madam who turned into the darling of every masculine author dealing with Denver's scandalous red-light district—that was Mattie Silks," whom we shall look at in a moment. Mattie has appeared in a lot of books and occupies a whole paperback.

We are not going to glamorize the darlings in this chapter; on the other hand we think there is some truth in what Beebe says:

The position of the madame in the nineteenth-century American West might well be the subject of learned monograph or doctoral thesis were the tastes in scholarship in universities more robust and realistic. As much of a personage in any frontier community as the sheriff, the town banker, the Wells Fargo route agent or the parson, her personal character was as various as the personal character of humanity everywhere. Her archetype would be a completely worldly wise woman, far gone to be sure in moral obloquy and a shuddering and detestation to the respectable female element of the community, but proverbially generous, usually witty and the object of an affection among the menfolk that was about equally compounded of familiarity and chivalry. The rougher the frontier and the newer the outpost of civilization, the more exalted the estate of the pioneer madame. As urban civilization advanced

and women of a "respectable" nature became more frequent to the scene, the lower became the madame's station in the social scale. But in the beginning she was half procuress and half amiable heroine to the frontier mind.

You will observe that he is talking about the madams—that is, the owners and managers of the parlor houses or brothels—and not about the lone harlots who did business in the cribs and the streets. Mattie Silks and Jennie Rogers are two of the most famous madams; Julia Bulette is possibly the best known of all the crib girls. The madams were more intelligent or they wouldn't have been madams.

As Miss Bancroft says, there are authors who have made darlings of a few of them; during their time there were men who tried to reform them and make angels of them. Otero tells of seeing the girls of a dance hall backed into a corner by two "clerical-looking gentlemen" who converted the girls to a holy life, or so bored them that they pretended to be converted. Sadie, Big Hattie, Careless Ida, Lazy Liz, and Nervous Jessie were all wiping at their eyes when Otero saw them, and promising to be baptized the next day, which was Sunday. If as males we look at the doves with eyes not misted over with notions of Mother and Sister and Mary we see them as a pretty stupid lot, though they were not so much immoral as amoral. It is true that a lot of girls back East were given glowing accounts of opportunities in the West by procurers and pimps, and on arriving in the land of gold found no opportunities, except the cribs, or marriage to toiling and penniless miners. The harlots, like the miners, wanted riches, and they joined in gold rushes from camp to camp, until, too old and dirty and ugly to be desired by any but a sot, they killed themselves, or died in poverty alone. Now and then a girl was in the business for vengeance—Jennie, for instance, who, according to Hunt and Draper, "had endured bestial treatment from a philandering husband and in revenge against the entire male sex had established herself in the house of Madame Clara Dumont and was bent on wrecking every man she could." But most of the miners were too tough to be wrecked by a Jennie.

Then as now many of them were foul in their speech and murderous in their passions. In the Klondike rush, says one who was there, most of them were "experienced, hard, bold-faced, strident harpies with morals looser than ashes." Parkhill quotes Laura Evans: "I used to be mean —never let a customer get away while he still had a penny in his pockets. That would have been against our religion." Were they cruel? "At Leadville I had a beau named Arthur. At one party I thought he was dancing too much with another girl, so I knocked him, head first, through a plate glass door, and his head got stuck in the plate glass and like to cut his throat. When Arthur and I got mad at each other we'd fight with knives, and I've got scars where he cut me up. I loved that man." So many of them have said, "I loved that man," no matter how low he was as pimp

and gambler and liar, that we must assume that masochism was a strong force in their motivations.

Over their pimps, or over a matter as trivial as birthplace, the madams and the crib girls fought one another with the fury of female cats, using any weapon they could lay hands on, including sadirons and guns. Hall calls a fight between Mollie May and Sallie Purple "The Battle of the Painted Ladies": these two madams had, in the same line, elegant houses of sin, which in itself was enough to make them deadly rivals. Somehow they got into a feud over their birthplaces, and whether Connaught or Tipperary was the more desirable one became the issue to be settled. The two madams and their partisans tried to settle it with guns, and apparently put on such a lively show that Leadville's newspaper the *Democrat* drily informed its readers that "Both parties are resting on their arms and awaiting daybreak to resume hostilities."

An astonishing percentage of the madams were Irish. In Nevada City, California, Mary Mahaffey is said to have been enormously popular with the gamblers, including a few pranksters who, on hearing that a greenhorn was infatuated with Mary, thought to have some fun. They told her

A typical row
of cribs.

to accept his proposal of marriage, and to make a good story of her virginal state and her wish to devote her life to husband and children; all of which she did, filling the simple lubber to over-flowing with happiness. A wedding was arranged, a phoney clergyman was engaged, and the enraptured yokel dug up every cent he could for a sumptuous wedding feast that would include only the finest food and liquor from San Francisco. After everything had been eaten and drunk the hellcat bitch in Mary came to surface, twisting her lips and filling her eyes with gloating as she aimed a shotgun at her astounded "husband" and told him, "Now, you goddam stupid bastard, git!" Not all the doves had such a happy working alliance with evil as Mary had but there is no reason to think that most of them were not much like her.

Many of them were violent, cynical, and bitter; what kind of end did they come to? It was pretty sad, according to most of the writers on the subject whom we have read. Otero, who was on the scene in many a camp, says it was no uncommon sight to see gamblers and pimps and dance-hall owners at a stage station, to meet girls coming in, who back East had been promised high wages and a good life. "But when they reached their destination, they would find themselves forced to accept a life of debauchery, or be thrown into the street in a strange town, there to starve to death among the riff-raff." New Mexico's ex-governor liked to lay it on pretty thick; there were surely men in most camps willing, even eager, to marry the girls, and though few of them were Hollywood dreamboats life with them might have been better than debauchery or famine among the riffraff. It is probable that most of the girls who came out were pretty stupid, who had unrealistic dreams of fabulous wealth and position. Otero may be closer to the facts when he goes on to say that at about thirty-five or forty most of the crib girls were through, and that many of them then took to drink or destroyed themselves with poison. Miss Bancroft says that in handling their girls (called boarders) the madams "had to have an expansive maternal instinct as well as the ability to discipline. Many of the girls were moody and frequently depressed enough to try the laudanum route to suicide—a real dampener to business. (Laudanum was a liquid opium that could be bought by the quart at any drugstore

and was used as a pain-killer and tranquilizer by the tablespoonful) . . . ." Some of the girls "were temperamental and wanted to change houses repeatedly. . . . Many more were fundamentally unstable. . . ." Professor Greever says that in Deadwood "many years later a county official inspecting the undertaker's records of the 1880's was appalled at the number of girls who had taken their own lives but had been listed in the local newspapers as victims of pneumonia or mountain fever." The only one known to defy the "syndicate" "was a singer named Inez Sexton; given the alternative of rustling the boxes [in a theater] or shooting herself, she walked penniless out of the theater, took a hotel room, and told the owner her story. He spread it, the church people arranged a musical to benefit her, and with the proceeds she left the Black Hills forever." Unusual also is the story of Queenie, a taxi dance girl, as told by Miller. When her man accused her of burning him, that is, of giving him a venereal disease, she was so cast down that she said to the delivery boy, "You won't have to bring me any more groceries," and killed herself. The one who told Miller the story said that Queenie had been a camp angel and that for her funeral the camp declared a holiday. "That's how much the town people thought of Queenie—how she'd always nursed them, took care of sick wives and their kids—all that. But you know what? Queenie's man didn't attend the funeral."

Even most of the madams, some of whom were wealthy in their prosperous years, seem to have died destitute. The one in California camps, known as Madame Moustache because of the hair on her upper lip, was typical of most of them; she had a house in the gold rush years and took a fortune from the suckers. She then left California for a while but returned to it, and died poor, alone, and forgotten.

The red-light district, named for red lights or curtains in the windows of some of the cribs and houses, was known in most of the camps as "The Row" or "The Line." Miss Bancroft points out the unusual fact that in a lot of camps the Line was on a street parallel with the main street, and only about a block distant—in Denver, Holladay, later, Market—Street, and Larimer; in Cripple Creek, Myers Avenue and Bennett Avenue; in Leadville, State Street and Chestnut. In Virginia City, Nevada, it was D Street, a short block downhill from the famous C Street. In Silver City, Idaho, it seems to have been on Jordan Street where it met Long Gulch; and in De Lamar, Idaho, it was down Jordan Creek from the main camp, in what was known as Tough Town, with many saloons around it. The principal madam there was Jennie Mitchell. Emrich says that after the Barbary Coast lost its lure the largest red-light area in the West was in Butte, with more than a thousand girls of the line at one time. Of famous, or notorious, madams he gives Little Gertie, Highstep Jennie, Bulldog, Lattice Porch, Rowdy Kate, Blonde Marie, and Crazy Horse Lil in Bisbee's Brewery Gulch, who robbed her patrons of what the girls left in their purses. "Timberline was tall, crippled Slanting Annie walked with a tilt, and Madame Featherlegs" rode horses with "her long red underwear flapping about her legs like chicken feathers."

Most of the red-light areas were too commonplace to become famous. An exception was Leadville's, where the Winnie Purdy house is said to have had a financial angel in the town's bank. Another was Myers Avenue in Cripple Creek, which was lifted into fresh prominence after Julian Street took a horrified look at it and published an article about it. The camp's leaders were so incensed that they bombarded the magazine's editor with telegrams, and when those left him unmoved there went over an AP wire the following words: TONIGHT THE CITY COUNCIL OF CRIPPLE CREEK COLORADO APPROVED UNANIMOUSLY CHANGING THE NAME OF MYERS AVENUE TO JULIAN STREET. And Julian Street it still is.

The parlor houses and cribs were sometimes on the same street, but there was, says Bancroft, "a great distinction between the two. The cribs were single operations run by prostitutes in business for themselves. . . . a bedroom with a door and one window fronted on the street. A kitchen-living room was to the rear, and a privy stood out back, often on an alley." The crib girls charged from 25¢ to $2 depending on their age and attractiveness. Usually the charge was $1 to which was added the profit on beer sold and perhaps a tip. If they prospered, they moved up in the world to the extent that they would buy a small house and have their names engraved on the glass transom of the door. . . . In the city the better crib girls built up a clientele that would follow them to a brick building with four or more apartments where each girl ran her own business.

Instead of numbers on their cribs they placed names, at least in such camps as Cripple Creek and Virginia City, Nevada, where the much-glamorized Julia Bulette was number one.

In the parlor houses the girls paid the madam for their board and room. Each girl had a "handsomely furnished bedroom," and a trunk that not only housed most of her clothes but served as a safe, and the girl who made change from that safe in the presence of her "patrons" sometimes paid for her imprudence with her life, as Julia Bulette did. Bancroft says that an inventory of the Jennie Rogers house in Denver revealed that "all the bedrooms had enamel or brass beds, dressers, commodes, slop jars, rockers, straight chairs, rugs, lamps, lace curtains, and some even had writing desks. In the top houses the girls lived well." They were expected to dress "in the latest *haute couture*," each girl having seven or eight evening dresses and several afternoon costumes. Her dresses "were often priced higher than for 'good women' even without the extra markup designed for the madam. The girls seldom saved any money for the future" but such madams as Jennie Rogers and Mattie Silks "made fortunes despite heavy expenses." Among the heavy expenses was the payoff in some camps to the police; in Denver the alliance between law enforcement and prostitution became a national scandal. Says Miller:

During any number of occasions when I was looking through those old police files, I would see where evidence had been removed, or blurred out, or inked over—all sorts of things; and I would see, too, where pictures of arrested persons had been ripped out. . . . Some of those policemen owned their own stable of girls, some of whom worked in the cribs, and others in the parlor houses.

Some of the more enterprising madams advertised what they had to offer, in the little *Red Book* which was so small that gentlemen could carry it in a vest pocket. Blanche Brown offered "Lots of Boarders. All the Comforts of Home"; Minnie Hall told the unhappy husbands that she had "30 Rooms, Music and Dance Hall, Five Parlors and Mikado Parlor, Finest Wines, Liquors and Cigars, 20 Boarders, and a Cordial Welcome to Strangers." Belle Birnard also welcomed strangers with boarders "Strictly first-class in Every Respect." For a while Denver's madams had their own blue book, which was of course a red book, a trade publication and guide. It is said that when respectable folk of the town objected to the brazen advertising some of the madams put signs in their windows, one of which said with jolly frankness, "Men Taken In and Done For." There is a story, which Bancroft dismisses as a legend, that the light-loves were asked to wear a yellow ribbon in their hair, so that respectable folk would know them for what they were, and that the madams thereupon dressed their girls in yellow from their hair to their slippers. Fact or legend, Miller quotes an old-timer: "Instead of just yellow ribbons, . . . Mattie and Jennie Rogers got their girls complete outfits of yellow dresses, yellow gloves, yellow hats, yellow parasols—and I think gold slippers, too." When not busy welcoming strangers the girls often played cards, a favorite game being panguingue, Miss Bancroft says, "because a girl could lay down her hand, retreat to her bedroom for a date, return a short time later, and enter the game again without a hitch."

In the earlier camps Indian women were often delivered into prostitution by their men, who would have sold their mothers for a drink of rum. Richardson says that in New Mexico in 1859

There were only one or two American ladies in the Territory; though the number has since increased. Many native women were mistresses of the white residents by the consent, even the desire, of their degraded husbands. Chastity is practically unknown among them, but they possess all the other distinctive virtues of their sex. These poor creatures, utterly devoid of personal purity, willing to give or suffer anything to obtain jewelry and silks, are uniformly tender and self-sacrificing, ready to divide their last crust with the hungry, and deny themselves every comfort to nurse the sick and minister to the wretched.

We shall now present a few of the more glamorized crib girls and madams, beginning with Molly B Damn, who came to the Coeur d'Alene area in March, 1884, an "uncommonly ravishing personality," says William Stoll, who was an attorney in that area. We are gravely assured that she was a woman of culture and refinement who quoted Shakespeare, Milton, and Dante. "Here was a medley of contradictions: her blue eyes would one moment melt with tenderness and sympathy, and in the same instant her feminine poise

would be submerged in the basest and most profane vulgarity. With one hand she would rob the unwary; with the other she would give liberally to charity, or nurse the sick." Three months after her arrival there came to the camp "a bluff, red-shirted braggart of a miner" with a thousand-dollar poke, his earnings from months of hard labor. He went from saloon to saloon showing off his treasure and by the time he was drunk he entered Molly's place. Stoll says that whether Molly rolled him or other persons under her direction was never known; it was his belief that if anyone but Molly had robbed the miner that person would have been hanged.

The next morning Red-shirt got up a small outfit on credit and took to the hills. He "marched out of town and up the trail like a man." Three months later one of his friends came to the camp to say that Red-shirt was flat on his back, dying. Molly said, "Take me to him." Stoll says the man's response to her was angry words: "He's broke, I'm broke, and you know why!" But he consented to take Molly to the sick man, and for hours they trudged through forests and up mountains and down. They found the sick man semiconscious; all night Molly doctored him with the only medicines she had, whisky and quinine. For days and nights she nursed him until he was able to sit up and eat a little, and then she returned to the camp. Stoll says "she was never the same afterward. . . . Within a year she was found, one morning, dead in her bed. The day of the funeral the miners knocked off from their diggin's and sluice boxes, saloons closed their doors, blinds were drawn in business-house windows, and a procession of hundreds followed her remains to a burial ground on a sunny hillside."

In another chapter we introduce the San Francisco harlot known as Belle Cora (her name was Arabella Ryan). She was a strumpet of the baser kind, whose story has an interesting postscript in Volume XXX of the *Overland Monthly*. This story says that at the time Cora was in jail, waiting for the verdict of the vigilantes, a woman was practically kidnapped and taken to Belle, whom she found dressed in black silk and lace—a woman "below medium height and inclined to be stout. She had dark lustrous eyes and her complexion was snow-white and of satin smoothness." The kidnapped woman, Maria, expected no mercy; she felt herself doomed from the moment Belle tried to get her to drink wine, and then tea, both of which Maria believed to be poisoned. Maria was to be a witness at Cora's trial. The story says that Belle told her that if she would testify that Marshal Richardson at the moment he was shot had a gun in his hand she would leave the house alive. Finding Maria stubborn, Belle summoned two men, one of them armed with a knife; this man offered Maria a thousand dollars to disappear. Convinced that she had to accept the offer or die, Maria agreed, after being told that she would find the money in a certain place. She went to the district attorney with her story and was asked to say nothing about it until the day of the trial. Cora was acquitted, even though Maria testified that at the moment Richardson was shot she was only three or four feet from him and he was unarmed. Maria says she also testified at the second trial, under the vigilantes. She reveals it as her opinion that Belle told her lover Cora that if he did not kill Richardson she would, because in some manner Richardson had offended her. Whether Maria's story is fact or the senile imaginings of an old woman we do not presume to say.

Our version here of Martha's story is Jackson's. She had a saloon in Columbia and a back room for card games; and also in back was the apartment occupied by Martha and her brand-new husband John Barclay. After observing how well he did as a gambler she had suggested to him that as a team they could both do better. One afternoon a John Smith, full of bad liquor, entered her saloon and ordered a drink. Smith, a correspondent had written his San Francisco paper, was "much esteemed" but was sometimes "under the influence of liquor." Smith deliberately or accidentally knocked a pitcher off the bar. That started things. Martha cussed him out, and Smith, losing his head, seized her and shoved her down to a chair, and was giving her a good slapping when Barclay entered, gun in hand. He shot Smith. Though Barclay was thought to be, as a gambler, as respectable as any other businessman, he had made the mistake of marrying a whore of low quality. Columbia's citizens felt terribly upset, and J. W. Coffroth, a state senator, saw a chance to show off his forensic powers. Before he was done pointing out the many virtues Smith had been endowed with the camp was in a lynching mood. Men rushed at once with sledges and

crowbars to take the jail apart and bring Barclay out, and they marched him down the road to a good spot for hanging. There, jury, defense attorney, and judge were chosen, with Coffroth prosecuting. The defense lawyer is said to have done his best but he was no match for Coffroth's vocabulary or wind; and when the sheriff came gallantly to the scene to save the prisoner he was knocked out and bound, and Barclay was swung high. It must have been an ordeal for sensitive persons, if any were present, to see the poor devil seize the rope and draw himself up a few inches and hang there, knowing that when his strength failed him he would have to loosen his grip and fall. It was even more hideous when his executioners grabbed the rope and pulled him up a few feet and let him fall, repeatedly, in an effort to break his grasp; or when a miner managed to get close enough to strike across his hands with a pistol barrel until the bones cracked and blood gushed, and useless arms fell at his sides. A reporter for the *Gazette* made a neat package of it: "Drawing up his legs, the prisoner gave a few convulsive kicks and then hung straight in the red glare of torches and bonfires which cast a horrid flush upon the scene."

About nine o'clock one Sunday night, in Idaho City, a racket was heard in the house of a Mexican harlot named Belle Roqueraz. A pistol exploded, and a man burst from the front door crying, "I'm murdered!" with the long blade of a knife stuck deep in his belly. He ran to the front of the Cincinnati saloon and called to Jonathan Duncan to help him. Duncan pulled the knife out, and the wounded man, one Edward Baldock, fell down as if dead. A doctor came and dressed the wound, observing meanwhile that the only sign of life was a convulsive tremor running through the man's frame. At about the same time Doctors Hogg and Harris were summoned to the harlot's house, to attend a man by the name of Ben Bloomer, who was found to have received severe blows on the head, perhaps from a heavy revolver.

Next day when an inquest was held over Baldock it was learned that he formerly had been Belle's man, and had been invited, while in a distant camp, to come live with her again. On arriving he learned that the fickle one had changed her mind. Saturday evening he demanded admittance and was told to get. The next day he showed his friend Duncan a large knife which he carried, and Duncan showed him a fine pocket pistol, which he borrowed. Thus armed he went from saloon to saloon, drinking and telling the world what a dangerous man he was; and about nine that evening he again pounded on Belle's door. Again denied he kicked the door in, and found Bloomer putting wood into the stove. Baldock advanced toward him, gun in hand, and struck Bloomer heavy blows across his skull. Seizing the gun with his left hand, with his right Bloomer drew a knife and plunged it deep into Baldock's side. In the next moment he wrested the gun from him, and Baldock went staggering off the premises, crying in a loud voice that he had been murdered. He was right about that. Bloomer was acquitted on the grounds of self-defense.

A writer in the Dillon *Tribune* says that as late as 1930 there was a headboard in the original cemetery on the hilltop north of Bannack (Montana) which bore this legend:

NELLIE PAGET
age 22
Shot April 22
1864

Back in Illinois was a young man named Howard Humphrey, who had a girl named Helen Patterson. Though Helen had promised to become his wife she was bored, and in 1862 she headed west with a sister and her husband, after promising to return to Howard within a year. Howard said he would wait. He must have been a man who could have given Job lessons in patience, for according to the story he waited not seven or twenty-seven years but fifty-seven! Then "white-haired and crippled by age" he decided to go west to see what had happened to his girl. She had entered one of the dance halls in Bannack, and because attractive girls were as rare in that area as ten-pound nuggets she was enormously popular. Howard's dull honest face must have become dimmer and dimmer in memory as she danced with Plummer's road bandits. Before long the bandits were hanged or chased out but there were plenty of hoodlums around, one of whom got the idea that Helen Patterson, now known as Nellie Paget, was born to be his. We don't know what she did to arouse his jealously but anyway he shot her. The Dillon writer believed that

Howard actually made the long journey west and searched the area for her. That was in 1917 and by that time she had been dead over fifty years.

Miss Bancroft says that Colorado's Silver Heels is *"all* legend." She may be right. As the story appears in a number of authors it says that when in 1861 a smallpox epidemic swept the camp of Fairplay this surpassingly lovely harlot nursed the sick when most of the people were fleeing; and that at last she was stricken, and on observing that the disease had destroyed her beauty she disappeared. Some writers go so far as to say that years afterward she returned, heavily veiled, to kneel by the graves and weep; and most of those who have written about her love to say that Mount Silverheels, named for her, stands above the cemetery and all that is left of a once-booming camp. A full professor tells most of the story as though it were all literal fact. In a recent booklet on racy madams Bancroft, telling of smallpox in Salida, declares that a madam, Laura Evans, told a doctor to furnish uniforms for her girls and they would all nurse the sick.

One very pretty girl, later given the name of Silver Heels Jessie, was sent to nurse the minister's wife. . . . The minister was so grateful to the fine young nurse that he wanted her to stay on as housekeeper and companion to his wife. "No," the pretty girl answered. "Now that my job is done, I'll be on my way back to Miss Laura's on Front Street." The minister was flabbergasted. It was he who had been most fanatic in leading a crusade to shut down Front Street. That was the end of that particular crusade.

Fairplay and Salida are about sixty miles from one another: the skeptical reader may wonder if the two heels are the same blend of fact and legend, or if there is no fact in either tale.

We reserve the most famous of the crib girls for a later page and now present a few madams. Jennie Rogers was only one of many on Denver's Market Street, but according to Parkhill she was "far and away the most beautiful" of them all, and according to Bancroft, "the most spectacular madam Denver ever had" and the "most beautiful madam of the 1880's in that area—"a tall, willowy brunette, nearly six feet tall" but with "a nice bosom." She was a daredevil horsewoman, and from 1884 to 1909 "the undisputed queen of the underworld in Denver," though some writers "have placed the crown on Mattie

Silks' curly head." Jennie's name seems not to be known: Parkhill says her first name was Sara Jane, but according to one obituary she was Leah J. Tehme, though when she spelled it herself it was "Leeah." Her death certificate gave her as the daughter of one James Weaver. She married a doctor but found him dull, and so ran away with a steamboat captain named Rogers. In St. Louis while helling around, one of her male friends was the chief of police, and after Jennie set up shop in Denver the chief now and then made the long journey out to spend a few nights with her. Emerald earrings were her trademark. She envied Mattie's house and would have given her soul, or what was left of it, for a house as elegant. Parkhill says the chief decided to take care of that but Miss Bancroft rejects the story, which is one of a politically ambitious millionaire, the skull of a human female secretly buried in his backyard, a rumor of murder, and blackmail and payoff. Jennie had a few faces, male and female, put on the facade of her house, which can be seen in the Parkhill and Bancroft books, and these have led to endless speculation; but whether she put in stone a story of something that happened or thought a few stone faces would make her house envied above Mattie's nobody seems to know.

One of the chief problems in the lives of most of the madams was boys friends. Jennie had more than her share of trouble that way. After the chief passed out of her life for reasons unknown, she married a barkeep at the Brown Palace and for reasons of her own set him up in business in Salt Lake City, five hundred miles from her house of joy on Market Street. Dropping in on him unexpectedly she caught him in amorous dalliance and took a shot at him but hit only one of his good hugging arms. Later, when asked why she shot her man, she is said to have cried, "Because I love him, damn him!" words which are now accepted as a classic expression of grief from a brokenhearted whore. That one shot was enough: before long the barkeep took off in full flight and put what he thought was a safe distance between him and the willowy brunette.

She wasn't so willowy by that time; like most of the madams she ate well and was getting fat. With money which she got in ways strictly dishonest she built her house of mirrors, which some **admiring writers** have called fabulous. Miller

calls it that and tells his readers that it "contained the astonishing number of twenty-seven rooms, including four parlors, a ballroom, a kitchen, a wine closet, sixteen bedrooms"—and—perhaps this is the fabulous part—only one bathroom. The parlor to the right of the reception hall was lined with mirrors, and the furnishings included Oriental rugs and crystal chandeliers. At the height of her popularity Jennie's house was "patronized by many big names. During the 1880s and early '90s the legislature was meeting at successive addresses on Larimer and Market Streets, never more than a block or two away from her house." A druggist whose shop was only a block from Jennie's has said in the *Rocky Mountain News*, "Each afternoon about three o'clock the august lawmakers would retire to Jennie Rogers' Palace of Joy on Market Street and there disport themselves in riotous fashion. Nothing was thought of that sort of thing in those days. . . . Men took their liquor neat and women took what they could get their hands on."

"Tempestuous and emotional, she had her melancholy days. It was then that she turned away from people to her great love for horses." This great love "led to the most long-time love of her life for a man, John A. Wood," a big cab driver twenty-three years old. It seems that this John is also the simpleton she set up in business in Salt Lake City; Miss Bancroft surmises that she sent him so far away because she didn't want him snooping around when the police chief was visiting her. After John fled from the Mormon zion he set up shop in Omaha, and Bancroft assures us that Jennie and Jack yearned for one another across the distance. In any case, she wrote to him, he returned to Denver, and on August 13, 1889, they were married. He still set out the drinks in Omaha while she grew rich off the earnings of her girls, but their visits to one another were "frequent" until John died at the age of thirty-eight.

When old, and sick with Bright's disease and other ailments, Jennie went to Chicago and fell for the blarney of an Irishman twenty years her junior. She thought her marriage to him was on the level; learning that it was not and that he was "too sharp an operator even for her" she returned to Denver and reopened her house; but she was sick and tired and soon followed her dog, which we are told she had buried in a gold casket. She was laid "peacefully at rest beside her great love," with an excellent view of the Bancroft lot in Block Three. Thirty-six claims were filed against her estate, one of which came from the Irish crook in Chicago, who managed to get $5,000 out of it. Of the $200,000, the three heirs got only $30,000, the remainder going to lawyers and deadbeats. Which seems to suggest that not even the madams, the brainiest women on the Line, were as bright as they are sometimes thought to be.

The House of Mirrors eventually became a

Mattie

Buddhist church but not before the enterprising Mattie Silks bought it and set up her business there. The reader who wants in detail the story of the most famous of all the mining-camp madams is referred to Parkhill. Elsewhere we tell about her six-gun duel with Katie Fulton. Except for that, and the grand assault she made on the snobbish Potter Palmers, her story is tediously like Jennie's. Mattie, who said, "I was a madam from the time I was nineteen years old," and, "I never worked for another madam," had had houses in such hell-roaring cowtowns as Abilene and Hays City before she came to Denver in 1873, with a husband who had given her the appropriate name of Silks. George Silks found the Far West unfriendly to his deeper instincts and soon took off, to be seen in those parts no more. Cortez D. Thomson, whose wife had died in childbirth, leaving him a daughter, was looking for a woman who would take care of a man who liked to spend large sums gambling but was willing to do a little pimping on the side, and with the incredible low taste in men which all harlots seem to have Mattie took him to her blonde plump bosom. She also took his daughter as her own, and it is gravely said that the girl was never to know how Mattie earned her living.

Both Cortez and Mattie had wild tempers and impulses to violence. Before meeting Mattie, Cortez had fallen so hard for a madam known as Lillie Dab that when she showed him the door he made a futile effort to kill himself. His humiliation he seems to have worked out on Mattie, for it was his habit to knock her down and kick her around the room; and when at last she filed (but later dismissed) a suit for divorce she alleged that he assaulted, struck, bruised and beat her in a most cruel, barbarous, and inhuman manner, causing her great physical pain; and that the night of March 13, 1891, he did knock her down, black her eye, beat her face and body, and kick her in a most cruel and brutal way while she was on the floor. She said she could endure no more of it; but when he came with the same old blandishments and lies and promises she again hugged him to her ample bosom and murmured through bruised lips that he was her man. But Mattie was not the gentlest of soiled doves; she now and then got so drunk that she was hauled into court, where she paid her fine, a Denver newspaper said, "like a little woman."

In her old age a newspaper heard that Mattie might be close to death and sent a reporter to interview her. Max Miller gives two pages of these notes. They tell us that Mattie was "a most unusual woman with humor and detachment about a phase of life she had once known and left—and with humor and detachment about the whole of life which she was soon to leave." To this reporter, a young woman,

She defended calmly but without emotion—the life she had led. And she said at the first, "I went into the sporting life for business reasons and for no other. It was a way for a woman in those days to make money and I made it. I considered myself then and do now—as a business woman. I operated the best houses in town and I had as my clients the most important men in the West."

What an autobiography she could have written! She said:

"I never took a girl into my house who had had no previous experience of life and men. That was a rule of mine. Most of the girls had been married and had left their husbands—or else they had become involved with a man. No innocent young girl was ever hired by me. . . . Some of my girls married the customers. And—" and this was the only part of the interview where Mattie Silks became the least bit defiant—"my girls made good wives. They understood men and how to treat them and they were faithful to their husbands. Mostly the men they married were ranchers. I remained friends with them, and afterwards with their husbands, and I got reports. So I knew they were good wives."

But the thing that most delights us with Mattie is the way she put herself off on the Potter Palmers as a lady of unimpeachable character. The story, as Parkhill tells it, runs like this: the businessmen in Denver wanted someone to build a railroad to their town and persuaded a tycoon to come out and look it over. He looked over more than the town. Wanting him to feel at home the Denver men took him to the Silks parlor, and he was so smitten by what he saw, especially Mattie, that he came up with an idea. Would she like to be his traveling companion for a few weeks to the Pacific coast? Knowing the Cortez temper, Mattie demurred. But Cortez also had an idea. He proposed that she tell the businessmen that she could not get away because of a mortgage of five thousand on her house, but that she could get away all right if they could find five thousand

Cortez Thomson

somewhere. They found it. She handed it over to her pimp, so that he could also have a little fun, and took off with the tycoon; and it is said that he opened his purse wide and told her to help herself. He seems to have got his money's worth; after a tour of the coast he suggested that they play man and wife for another month or so; and of course he mentioned his friends the Palmers, who had a sumptuous summer residence in Wyoming. We can imagine that Mattie played her part well, because the Palmers, who are said to have snubbed Lady Peale, an actress, accepted into their home without question a madam who carried on her business at the same address for forty-two years, and was, according to Miss Bancroft, "the most stable and consistent high-class madam Denver ever had." No doubt Mrs. Palmer would have died of the most ghastly humiliation if, looking into the laughing eyes of the plump blonde, she could have read her thoughts.

Mattie took her man on a vacation to Europe, and then ran up to the Klondike to see if that gold rush had any need of her talents. Worn out at last by the heavy duties of a gambler's life, Cortez lay down and died, and Mattie, who had al-

Said to be Julia, in the Bucket of Blood, Virginia City, Nev.

This seems more likely to be Julia

ways had a taste for the legal side of things, took up with a saloon bouncer named Ready, who so loved diamonds "that he had a big one set in a front tooth." She married him in 1924, when she was seventy-six. She had long ago bought for $14,000 Jennie's House of Mirrors, and had had her name as "M. Silks" set in tile in the entrance; and she then became "undisputed queen of the Red-light District," though another madam, Laura Evans, consumed by envy, said that Mattie was strictly third-class, and her house was an Old Ladies Home. Wealthy, and full of such honors as a madam can come to, she died at eighty-one, and was laid away in respectable Fairmount Cemetery at the side of Cortez, under the name of Martha A. Ready.

Through the mists of legend it is possible to see most of the female adventuresses pretty clearly, from Lola Montez to Mattie Silks, but so much

nonsense has been spun around the name of Julia Bulette, such a bright halo has been set above the head of this crib girl, the most famous or notorious of them all, that it takes a bit of doing to penetrate the fog of masculine embellishment and see the woman.* We have tried to do it, with the help of Marion Welliver, editor of the *Nevada Historical Society Quarterly* and assistant director of the Society.

Like so many of the harlots this woman probably gave herself a name that she thought was more elegant than her own, quite as Mattie Silks persisted in spelling it "Madame" instead of "Madam," because she thought it was "tonier." Typical of the way writers have glamorized Julia is Emrich, for though he admits that she is lost in legends she was for him "dark-eyed, tall, slim, beautiful, but beyond mere physical beauty possessed the deeper feminine characteristics which

* The copyrighted photo used here of what is thought by some to be a painting of Julia is used by permission of Claude and Jane Weddle, San Clemente, California. It has been said she was born in Liverpool under the name of Smith; by others that she was a Creole; and by a Nevada historian that she was "Born a miscegen in London, England, in 1832. . . . the blood of the negro coursed in her veins." The painting of a woman above the Bucket of Blood bar in Virginia City, Nevada, pointed out to tourists as Julia, is certainly not Julia but it might be her maid, if she brought one West. A photo was found of what seems to be a mulatto with the name Julia written on its back; if there was a maid possibly she took the name of her mistress; or this may be another Julia. There is another painting purporting to be Julia, by Ben Christy; and the *Nevada Centennial Magazine,* 1964, has one which is attributed to the *Enterprise.*

endeared her to the men—generosity, compassion, sympathy, understanding. Her personality as well as her wit and humor must have been vibrant and living to dominate Virginia City as she did." He says she offered her male visitors "good conversation, taught them to recognize fine wines, . . . served dinners distinguished by rare and unusual delicacies." Without citing a single source he does his best to make the legends imperishable: no holiday or parade, he gravely informs his readers, was complete without Julia "riding enthroned on the gleaming brass and silver engine of Company No. 1." If fire threatened the camp she was there "with food and coffee for the fire fighters," and now and then manned a pump. If the town was threatened by Indians "she chose to remain with the men to nurse them, rather than to seek safety with the other women in Carson." He writes it down as fact that the "men of the town worshiped her," that the "good ladies of the place detested her"; and he declares that when she was made an honorary member of the engine company and designated number one it was "the highest social honor possible in Virginia City." And he really pulled all the stops when, as late as 1949, he said that Julia's grave "on Flowery Hill had been carefully tended, its white picket fence repainted from time to time. All other graves in the cemetery are neglected; weather-beaten picket fences lean awry, and headstones have crumbled. Alone, respected and honored by Virginia City, the grave of Julia Bulette, a 'lady of the line,' marks the cemetery."

What are the facts? Visible from Virginia City there is a white picket fence on the mountainside. We talked to the oldest old-timers there and they all told us the same story, that to attract tourists enterprising men set this fence there; but the fact, a historian has told us, with an admonition not to reveal names, is that about 1920 the fence was moved from the actual grave so that it would be visible from the town, and that the fence standing over the actual grave was at that time photographed. In any case, thousands of tourists every year take the difficult road up the mountain, as we have done several times, to look at the fence and the earth inside it. The historian who is our informant claims to know where the actual grave is. This area on the mountainside, says another Nevada historian, was "reserved and laid off for the exclusive use of the firemen for

Virginia City." She says the name on the headboard was spelled "Jule," but another historian says the marker in 1920 had the name as "Julie."

By one writer she was compared to Cherry Melotte in Rex Beach's *The Spoilers*. Oscar Lewis says she has been more written about "than any other Comstock woman. . . . Julia, who in the words of one writer, 'caresses Sun Mountain with a gentle touch of splendor,'" whose cottage or crib became "an oasis of elegance in a community of tents." It all sounds much like Stuart Lake's "biography" of Wyatt Earp. Such house as she had stood on the corner of D and Union streets, and could hardly have been more than a lumber shack of two small rooms. At the time of her death she was not giving elegant dinners but was taking her meals in the crib of Gertrude Holmes, which stood next to hers in the north. Beyond Gertrude's were the cribs of Elizabeth Hayes, Martha Camp, and Catherine Thompson. It seems to be true that Julia served as a kind of mascot to the fire company and that she had a place of honor in a Fourth of July parade. But the story that she loved to ride through the camp in her own splendid carriage, with a crest of four aces and a lion couchant, Dr. Effie Mack calls "too fantastic."

Though Lucius Beebe puts his story of her in a small book that he called *Legends* he says she "lived briefly and breathlessly to find herself the toast of the richest mining community on earth . . . a humane and compassionate strumpet who was tolerated by Father Manogue, . . ." Her "rococo premises" became known as Julia's palace, "the cultural center of the community." To say that a strumpet's cottage was the cultural center in a camp with thousands of people and its own opera house is to bring the legend into full view of even the least critical judges. But Beebe, a gallant man when it came to legends, goes on to say: "Only at her table were rough manners banished amidst the service of fine wines instead of whiskey and skillfully prepared French dishes instead of the beef and biscuits of the town's beaneries. She brought airs and graces where comparative barbarism had reigned and the miners accorded her an homage that elsewhere would have been the prerogative of a great lady." Beebe cited no souces, knowing that legends have none. He says that when "she took the air in the afternoon it was behind a matched pair of bays in a Brewster carriage"; that she wore diamonds and the "richest furs ever yet seen on the Comstock"; that she had flowers on her table and served vintage champagne; and that she was "wealthy even by Comstock standards and able in her own person to disdain the favors of any but the most agreeable and affluent customers." He says she was "undoubtedly Creole," though the *Enterprise* said right after her death that she was a native of London.

Whatever she was and wherever she came from it is a fact that she was found murdered on her bed Sunday morning January 20, 1867. It seems probable that her jewels and furs had caught the attention of men with larceny in their souls. It seems to have been a rather sensational murder. January 22 the *Enterprise* told its readers that

The most cruel, outrageous and revolting murder ever committed in this city was that of Julia Bulette on Sunday morning. She lived in a little house by herself, near the corner of D and Union streets, in a thickly settled neighborhood, and within a stone's throw of the station house. The murder was probably committed about five A.M. but it was not discovered until nearly noon, when the body was nearly cold and stiff in death. At eleven o'clock a Chinaman who was employed to make fire, sweep, bring in the wood, etc., came into the house as usual, kindled a fire and left, thinking she was asleep, as he could see her covered up in bed. About half an hour afterward, a woman who lives next door came to call her

to breakfast, and discovered her to be murdered. . . . She was found lying on her left side, with a pillow over her head and face, the bedclothes beneath her head being saturated with blood. Her throat was lacerated with the marks of finger nails, and the blood-suffused and distorted countenance, together with the writhing position of the body, showed conclusive evidence of strangulation. . . . There were two small wounds on her forehead. . . . and the back of her left hand was somewhat lacerated in her struggle to free herself from the grasp of the fiend who had her in his power. . . . the murderer took a set of furs worth $400, two gold watches and chains, and several pieces of valuable jewelry, even taking the earrings from her ears. . . . It certainly is to be hoped the murdering villain may be captured and eventually adorn the end of a rope. His victim was known as Jule Bulette, and was a native of London, England, whence she emigrated, when quite young, to New Orleans, and thence to California, in 1852 or 1853, where she lived in various cities and towns until April, 1863, when she came to Virginia. She is said to have married a man by the name of Smith—from whom she afterward separated—and has an uncle and a brother still living in the State of Louisiana. She was thirty-five years of age, belonging to that clan denominated "fair but frail," yet, being of a very kind-hearted, liberal, benevolent and charitable disposition, few of her class had more true friends. Julia Bulette was some time since elected honorary member of Virginia Engine Company No. 1, of this city, in return for numerous favors and munificent gifts bestowed by her upon the company; she taking always the greatest imaginable interest in all

Engine Company No. 1 in 1861

The murderer of Julia

matters connected with the Fire Department, even on many occasions at fires working at the brakes of the engines. She was still an honorary member of the company at the time of her death, therefore it was deemed eminently just and proper that she should be buried by the company.

That was on-the-spot reporting by a famous newspaper. We may assume that the one who wrote those words knew her by sight, and was familiar at least with the exterior of her cottage, which was only a short distance from the *Enterprise* building. He says only that it was a "little house": if it was the cultural center of a huge mining camp, it is strange indeed that such a remarkable fact was overlooked in the story about her.

The man who was hanged for her murder fifteen months later was tried under the name of Milleian, but in a pamphlet published by his attorney in 1868 his name was given as Jean Marie A. Villain. The surname may seem wonderfully pat for a brute who strangles a woman but originally of course it meant only a feudal serf. According to this Frenchman's simple-witted confession two other men entered the cottage and

killed her, while he, posted as a lookout, fell asleep. But a lot of Julia's things were found in his possession.

The trial opened June 26 before a jury and a packed courtroom. In summing up, the prosecution said that Virginia City had seen blood run like water but was now stricken with horror by a deed "more fiendish than ever before perpetrated on this side of the snowy Sierra." Though Julia had been a woman of "easy virtue" for her acts of charity "hundreds in this city have had cause to bless her name. That woman probably had more real, warm friends in this community than any other person." All that is only average exaggeration in criminal trials. Beebe says her body in a coffin with silver handles was borne out of St. Mary's, "a concession permitted by the Church because of Julia's notable charities and good works," but historian Welliver thinks she was taken from the fire station. The builders of the legend would have us believe that the funeral and burial were an event of extraordinary splendor, but the *Enterprise* reported that

Owing to the disagreeable storm prevailing, and the very muddy state of the streets, the procession was not so large as it would have been under more propitious conditions. Company No. 1 to the number of about 60, preceded by the Metropolitan Brass Band, marched on foot. The funeral was also attended by 18 carriages filled with friends of the deceased. She was taken to Flowery Hill Cemetery to the east of the city, where in her lonely grave her good and bad traits alike lie buried with her.

Those who have seen these mountains in mid-winter storms, and who, in warm dry weather, have driven up the mountain to the area where she was buried, can easily imagine how the beasts attached to carriages struggled in the mud on those steep slopes, and how eager the plodding members of the band must have been to get back to the saloons on C Street. The legend says that on its return down the mountain the band played "The Girl I Left Behind Me."

Milleian was defended by Charles E. DeLong, a noted trial lawyer, and later the first U.S. minister to Japan. DeLong did everything he could for his client at the trial, and after the conviction appealed the case so many times that a year passed before a gallows was erected on the mountainside to the north, above the Jewish cemetery.

Tourists are told this is Julia's grave. It isn't

This is, but the fence is no longer there. The girl is holding the original headboard

Various newspapers described the scene. Here are a few details from the Carson *Appeal* of April 25, 1868:

Before 8 o'clock this morning little clumps of men might have been seen about the Courthouse, and by 10, B street was packed from side to side. A delegation from the Sisters of Charity visited the doomed man as early as 7 o'clock, and remained with him most of the morning. We were unable to gain admittance to the cell, on account of the density of the crowd about the door. . . . Among the vast multitude assembled in front of the Courthouse, at 10 o'clock, we noticed quite a number of ladies, and also a great many of the class to which Millien's victim belonged. About 9½ o'clock a perfect torrent of humanity poured into the city from all directions. . . . Men came by wagon loads, in carriages, buggies, on horseback and afoot. . . . Some time before the prisoner was brought from the jail, every available housetop, balcony, door and window along B street, from Taylor to Sutton avenue, was crowded with the curious. [When the condemned man was taken in a carriage down the street] the crowd swayed to and fro for a moment, and then surged down B street, jostling and running—men, horses and wagons in the wildest confusion . . . vast multitudes preceded and followed the doomed man. The streets leading out to the Geiger Grade were thronged with people as far as the eye could reach, and along the road beyond the Sierra Nevada works the side of Cedar Hill was black with people. . . . Women, with children in their arms, hair disheveled and flying in the breeze, could be seen hurrying across the bleak, burnt hills to catch a distant glimpse of the horrid sight.

On the scaffold Millien gave thanks, presumably in French or through an interpreter, to various persons, including "all the ladies who had visited him during his confinement." He neither confessed nor denied his guilt. In his cell he had written a talk to be read from the scaffold; in it he was bitter toward the United States, sentimental toward France, and brutally unfair to the attorney who had done everything in his power to save him.* He betrayed "Not the least tremor, not the quivering of a muscle, nor the rush of blood or the lightest paleness." After hanging for about thirteen minutes the body was taken down.

A thrice-told part of the tale is that during his imprisonment the good women of the camp took him delicious things to eat and drink, as in California other good women, bewitched by a criminal, took drink and food to an especially vicious killer named Vasquez. An ironic conclusion to the whole sordid thing in Virginia City is found in a question a Nevada historian sent us: "Did you know Mark Twain arrived back in Virginia City to give a lecture and found the hanging giving him competition?" It is safe to say that Mark Twain didn't like the competition at all.

We now return to the girls. Cock-Eyed Liz, a Colorado madam, is reported to have said, "A parlor house is where the girls go to look for a husband and the husbands go to look for a girl." To the question, "How many husbands did the girls find?" the answer is, "Quite a few." Professor Greever, who seems to share a common male view that practically all the light-loves were abducted into whoredom and held there at pistol point, says: "The hurdy-gurdies, poor but pure girls who entertained by dancing for hire because they had to do so, often married the men who patronized and made fine mothers."* No doubt more of the poor but pure girls found husbands than those in the parlor houses and cribs, though the better houses had what the madams proudly called ballrooms, where the energy and enthusiasm of most of the male partners, if not of the female, were about what Charles Teeter saw in his journey overland:

As soon as the sun had sunk below the western horizon a spot was selected suitable for the purpose, around which a long rope was stretched, when the fair ladies of the train were led forth to occupy this inclosure. Soon, however, the music struck up, and the dance commenced and such a hopping, skipping and whirling as the company displayed was enough to frighten away every wild beast that prowled around us, and the bold Indian too.

A common attitude of crib and parlor girls, and surely of many of the hurdy-gurdies, is illustrated by a story Drury told. On the Comstock Porky

---

* This talk appeared in English in the *California Police Gazette*, May 2, 1868. See also the *Enterprise*, April 25, 1868.

* Possibly Professor Greever based his opinion on W. J. McConnell, first Idaho head of a vigilance committee and territorial governor. McConnell's statement was copied in his journal by Charles Teeter, who himself was in the Boise Basin in the early sixties and saw the girls—"strange as it may sound to those of modern times, these girls were pure women, who simply did the work they bargained to do, and when their contracts expired, most of them married men whose acquaintance they had made while pursuing their vocation. . . . The poor girls, and they danced only because they were poor, had kind hearts and wonderful patience and forbearance." That's the way McConnell saw them and apparently the way Teeter saw them in that area. Most of the girls were German.

and Big Nell lived together and managed a restaurant. People—he does not say what kind—persuaded them to marry, and they were then "so ashamed they had to leave town."

Of all the persons on the scene in those camps, who left an opinion on this matter, we like best Mrs. Mathews, a no-nonsense woman and a shrewd observer, as anyone can discover by reading her book. She says: "Sometimes a good citizen, wealthy and respectable, marries his wife from some of these corrupt houses, and he seldom ever regrets his choice. He builds her up to be respected and respectable. I have heard of several cases. More of such men would make Virginia City better." One of the oft-told stories, sometimes fact, sometimes fiction, and always heavy with sentiment, is of the harlot who marries and is then pestered by men who think she is still for hire. Jean Davis found a version of the story in Montana. Liz was a dance-hall girl and of course she was very beautiful, as in this legend she must always be. She met a handsome young man named John and they fell in love. When one evening a drunken miner tried to force his attention on Liz she seized a shotgun and "blasted him to eternity." After that feat she was known as Shotgun Liz. The following year she was found dead "in her dingy room above the dance hall," and John remained in the camp "to be near the grave of the one who was dearest to him."

The story of Red Stockings has been told by a number of writers; our version here is Miss Bancroft's: "The most colorful character of California Gulch's underworld was a mysterious twenty-year-old girl. Her charm and cultivated Boston accent captivated all the better class of men in the settlement; while her prettiness and habit of wearing red ribbons in her dark hair complemented by flashing red stockings over her trim ankles attracted the others. She called herself Nellie but never told her last name." She rode a spirited horse, accepted men in a log cabin, and "told her story to a confidant or two." Born to wealth, she had been sent to Europe to polish her education and had had an affair with a French officer. Rejected on her return by the people of her social level she had run away with a gambler and eventually became a harlot in the Gulch, which at last she left with a hundred thousand dollars (rumor said) to become "a fine wife and mother in Nevada."

The love of a good handsome man for a beautiful whore eager to change her ways has many versions, one of which belongs to Forbes Parkhill; he carries melodrama a league or two beyond where Shakespeare left it in his sad tale of Romeo and Juliet. Lois, younger sister of Lillis Lovell, a leading madam in Creede, fell head over garters in love with a young Denver businessman. He was just as madly in love with her. When again and again he begged her to marry him she wept on his shoulder and said in heartbroken whispers, "You must not marry a Market Street girl! It would ruin you, and I love you too much to wreck your life." If the two idiots had really meant it they could have left Denver and gone far away to another city. As it was, the young man left town on a business journey, and while he was gone Lois drank poison and died.

That much burden on credulity is as much as any author should ask of a reader, but there is more to come. The man returned after a week or two and went at once to the red-light house, again to beg Lois to become his wife. Told by a Negro janitor that she was dead, and at peace in a grave down on the river, the horrified lover asked the janitor to show him the spot. There he said he wished to be alone a few minutes, and when alone he stretched himself out on the mound of his dead Lois and blew his brains out.

Of the stories of prostitutes who changed their ways and became good wives and mothers (and there can be no doubt that some of them did) we find the one of Joe Plato as plausible as any other. Joe had a ten-foot ledge next to the Sandy Bowers claim on the Comstock. One night when drunk he is said to have given half of it to a San Francisco harlot. On learning that he had a very rich piece of the lode he hastened across the mountains to find her, hoping to buy her half for a song; when she proved to be too foxy for him he proposed marriage and was accepted, and they both rushed back across the Sierras, praying that a claim-jumper would not be sitting on their ten feet. When Joe died in the seventies his wife then had all of it and it was a fortune. It is said that she then married a rich and prominent San Francisco businessman and became "an ancestress of note." Alas, none of us knows how many whores and bandits are in our family lines, nor should care.

Of the tales of harlots-turned-respectable that don't put too much strain on the skeptic, that of Cock-Eyed Liz, a Colorado madam who got one eye knocked out in a brawl, is as good as any we have found. In 1897 when she was almost forty-one, a character named Alphonse Anderline, known as Foozy, right out of the blue asked Liz to marry him. She did. Liz gave her age as "over twenty one" and he gave his as thirty-five. They "lived happily ever after," says a Colorado historian. "That's a real switch for you. Imagine the leading madam of the town becoming a respectable citizen, openly living as faithful wife in her former palace of joy! But that's what Foozy and Cock-Eyed Liz did. They let the girls go." Foozy, handy with tools, enlarged the building, so that they could rent an apartment for income. Today the building is the Edwards Apartments. Which reminds us that now and then a whorehouse became a church building, or a museum, or a private residence. Jennie Mitchell's in an Idaho camp was moved, picket fence and all, to become the home of a ranch family on the slope of Spring Mountain.

The two girls with Liz were not harlots.

The Jennie Mitchell house.

CHAPTER 12

Sports*

Wild game was so abundant, including fowl, in all the more northern camp areas that the miners could have had plenty of good meat if they had not been so obsessed with digging gold out and rushing to the nearest gambling table to squander it. "Wild geese," Colton said, "prevail here in the greatest abundance; every lagoon, lake, and river is filled with them. They fly in squadrons, which, for the moment, shut out the sun; a chance shot will often bring two or three to the ground. The boys will often lasso them in the air. This is done by fastening two lead balls, several yards from each other, to a long line, which is whirled into the air to a great height. In its descent the balls fall on opposite sides of the neck of some luckless goose, and down he comes . . ." But if hunting was a favorite sport there is little record of it in the books written by those who saw the camps.

A few Fourth-of-July celebrations were so spectacular or violent that editors took note of them —as when barrels were filled with powder, cans, bottles, and scrap metal and exploded; and now and then an exhibitionist drew a crowd, as when Yankee Driggs, to raise funds to start a tinshop, walked back and forth for thirty-six hours on a plank eighteen feet long, raised a foot and a half above the earth. And in areas of deep snow there were winter sports. In Silver City one winter riding on snowshoes was popular. The *Avalanche* says that

At almost any time during the day can be seen some one, on long Norwegian snowshoes, coming down our mountain sides with the speed of wind—oftimes running at the rate of a mile a minute. Riding thus is quite an art, and it is amusing in the extreme to see a company of novices attempt it for the first time. They go to the top of a hill, adjust their shoes, make ready and

start; and by the time they begin to run with considerable speed they are almost sure to lose their center of gravity, and immediately some can be seen standing on their heads, others turning somersaults, and, in fact, performing in the most wonderful manner all kinds of involuntary gymnastic exercises; after which, from the summit of the hill downwards, may be observed a line of arms and legs like fence-posts sticking up through the deep snow. It's rare fun—for spectators.

Occasionally there were footraces or horse races. One day in Nevada's Virginia City when men were footracing on C Street a woman who had been watching from a window leaned out and hurled a challenge. A few moments later she came dashing out, stark naked and even barefooted, and took off down the hill, with not a single gentleman in pursuit of her. In the Denver area horse racing was popular, as it still is. An extraordinary race there seems to have been between Rocky Mountain Chief and Border Ruffian; contributions were called for to make up a winner's purse, and a "staggering $95,000 was taken to the Clark and Gruber Mint, melted down, and formed into a huge gold nugget." This immense ball of gold was taken to the track grounds and suspended from a cable, so that it could freely turn in the wind and shame the sun.

Most of the sports were brutal, but no more brutal than they would be today, if allowed: "prize-fighting" has become almost a parlor game, and a rooster fight, if put on and discovered, would set the humane societies marching all over the land. Rooster pits seem to have been popular on the Comstock Lode but hardly known in many camps. There, William Sharon of the Bank of California, and other wealthy men with a taste for blood, kept roosters for fighting and wagering. Toponce tells an amusing yarn about it. Sharon

---

* Readers with delicate sensibilities would do well to skip this chapter.

and some others had fighting cocks in a pit when a man slipped up to Toponce and pushed into his arms a rooster that was stone blind from too many duels. Giving the man a dollar for the bird Toponce then put him in the pit. "He ran round and round in the pit and the other roosters thinking he was running after them got excited and one after another flew out of the pit, leaving my rooster the sole survivor and $80 in gold on the floor." Curious, Sharon picked up the cock and examined him. "Look!" he cried. "That damned Mormon cattleman has licked us all with a blind rooster."

There were dogfights but the cultivated taste of gentlemen preferred fights between bulldogs and wildcats—and these seem to have been more popular on the Comstock than elsewhere. At the Alhambra Theater on a Sunday afternoon occurred a fight between

an eastern fighting bulldog named Turk and a 42-pound wildcat, a vicious brute, for $100 a side.

. . . In due time the cat was introduced upon the stage and was immediately followed by Turk; but at the first kiss of the dog, the cat took to the audience. . . . The first bound of the cat took it upon the piano of the orchestra. . . . turned a back handspring. The next leap of the "varmint" was at the contrabass, and both player and instrument went down instanter with broken heads. The cat . . . then sprang out among the audience [and there was a] grand and lofty tumbling, leapfrog, and such-like feats of dexterity. . . . Time: shortest on record—1:59.

The cat was not usually so lucky. The *Enterprise* reported another fight:

The great wildcat and bulldog fight came off at the Opera House last night according to programme and was witnessed by a full and excited house. . . . There was rigged up on the stage a large cage. Toward the roof of the cage, placed so as to perfectly light up the interior, were eight gas jets. All being in readiness, Fight No. 1 took place. The smallest cat was let into the den and shortly afterward Mr. Gage's large white bulldog,

July Fourth in Cripple Creek

Footrace, a popular sport; the left leg of one racer and the right of the other are tied together

"Hero." He went after the cat "thar and then." The cat stood his first charge and then began to want to leave. After a few wild plunges about the cage the dog got a square hold on the beast and promptly killed it.

The second battle was won by a cat. A fresh dog was put in: "Over and over they rolled, fighting so rapidly that it reminded one of a big bunch of firecrackers exploding. . . . The fight lasted over 20 minutes. . . ." The dog killed the cat, and the multitude in the House, composed of women as well as men, cheered.

Fistfighting had to be popular in camps so brutal—and it *was* fistfighting then and not clowning to deceive the spectators. A part of the report in the Boise *News* of the King and Heenan fight in England reads as follows: 1st Round: Heenan closed in, grasping King by the neck and giving him an old fashioned hug, . . . 2nd Round: King forced the fighting as soon as he came to the scratch, and Heenan closed in again. After hugging King some time he threw him. 3rd Round: After hard exchanges Heenan again hugging King, and threw him on the ropes with awful violence. 4th Round: King came up with visible marks under his eye. Heenan again attempted the hugging game, . . . 5th Round: King got in a tremendous blow on Heenan's ear and another on his temple, when the latter again

hugged and threw King violently. 7th Round: King gave Heenan a rattler on the nose, drawing a stream of blood. . . . 16th Round: Heenan came to the scratch with a dreadful eye. . . .

Up in Alaska a big brute of an Irishman stood in a ring scowling round him, for it had just been announced that his opponent was in a hospital, and this Irishman wanted to fight. So he bellowed out to the world that he would lick Paddy Ryan if the coward was present and dared to come forth. As O'Connor tells the story, Paddy thereupon began to roar, and "shedding his foliage like an autumn oak" he rushed into the ring, stripped down to his red winter underwear. At the word go, he went like a bull for his enemy; both landed hard blows that shook the frozen north and both fell flat. "It was the only double knockout in Nome's prize-fight history."

In the other Virginia City was another Irishman, named Hugh O'Neil. In the *Post,* October 8, 1864, he inserted the following:

A Challenge

EDITOR POST:—Hearing a great deal about fighting in and about your place, and occasionally my name mentioned among the most common, I only have to say that one or any of the fighting fraternity can find me ready to fight any one of them for five thousand dollars ($5,000) in gold, on or after the 10th day of October, 1864. Any reasonable amount will be found at Dance & Stuart's store in Virginia, as a forfeit for the whole amount.

HUGH O'NEIL

A week later there appeared in the *Post* the following:

A Card

EDITOR POST:—In answer to Hugh O'Neil's challenge in your columns, I have only to say it was not my intention to seek further honors in the P.R. If such had been my desire, I would have gone East—but if O'Neil must have a fight, and means business, although he is some fifty pounds heavier than I am, and would have to come down that much, according to the rules of the ring; yet if he will call upon me, I will make two matches with him, catch weights. One, a glove fight for $500 a side, and a ring fight for $1,000 a side, open for $5,000. Money ready at any moment. If this offer is not accepted within a fortnight, I will not receive or accept any further challenge, as it is my wish to follow my business undisturbed, and to live as a private citizen. Yours, &c., JOHN C. OREM

For some incredible reason, the first legislature of the Territory of which Virginia City was a part had made it unlawful for persons to fight "to the terror of the citizens of the Territory," but a little later boxing was legalized, if gloves were used.

O'Neil accepted the terms, with what must have been a look of contempt down his Irish nose: although Orem called himself the middleweight champion of the United States he was not even a welterweight. He is said to have stood about five feet and a half tall and to have weighed 138 pounds. O'Neil was several inches taller and about fifty pounds heavier. A log building was erected behind a saloon on Wallace Street, complete with bar and a small loft for the ladies. For weeks the two gladiators soaked their hands in anything they could find that would toughen them, and wore skintight buckskin gloves for the fight. It took place on January 2, 1865, and ran for 185 rounds, but they were only one-minute rounds, or until one of them was knocked down or fell to a knee.

Barsness says that O'Neil entered the ring in green, with the Irish harp and his name "embroidered thereon." Orem was in black, with a girdle round his waist on which there was "a single star on a black ground with a red border. Round his waist were girdled the stars and stripes with the eagle and bearing the motto, 'May the best man win.' "

It must have been quite a spectacle—a man weighing less than 140, but cool and fast, against a big scowling lout who was determined to knock his head off. It is said that 36 times during the 185 rounds Orem deliberately sank to one knee to end the round, and that O'Neil, frustrated and infuriated, then tried to fall on him and crush him. Unable to land a hard blow on an opponent who was never where O'Neil thought he was, the giant a number of times seized Orem by the neck and simply flung him down. The account of the fight says that Orem was knocked flat at least fifty times, and O'Neil eighteen. Orem worked on his opponent's eyes, hoping to close both of them; and O'Neil flailed around him, hoping no doubt to knock the nimble little fellow out of the ring and halfway across the camp. In the 167th round Orem was caught napping and slugged behind the ear, and from then on his footwork was pretty erratic; but the giant was so nearly blind by fight's end that his friends had to haul him home in a wagon. The fight was a draw and all bets were off. Of course they had to fight again, some months later, and this time Orem won on a foul. To the first fight Dimsdale gave nearly the whole front page of his *Post,* and concluded with the pious thought that America and Ireland "may shake hands. . . . Cowardice or anything really unmanly cannot be named in connection with this fight."

A more dramatic and terrible fight occurred in Butte between a man named Ward, a carpenter, and Gallagher, a miner. Both men were big and powerful and full of git-up. In the 48th round Gallagher's left arm was broken between the wrist and elbow, and until the end of the fight in the 105th round the arm hung helpless at his side. As a consequence of it he took terrific punishment; one writer makes the astonishing statement that for 97 rounds he was knocked to the floor in each round. In the 98th Gallagher managed to land a tremendous blow with his good right hand, and for the other seven rounds Ward was dazed. He went down for good in the 105th and died the next day.

There was bull-fighting in some of the California camps. In Columbia Borthwick saw posters which announced that one Señorita Ramona Perez would slay a bull with a sword. This celebrated bull killer was no lady but a man "who entered the arena very well got up as a woman, with the slight exception of a very fine pair of mustaches, which he had not thought it worth while to sacrifice. He had a fan in his hand, with which he half concealed his face, as if from modesty, as he curtseyed to the audience, who received him with shouts of laughter—mixed with hisses and curses. . . . The señorita played with the bull for some little time with the utmost audacity, and with a great deal of feminine grace, whisking her petticoats in the bull's face with one hand, whilst she smoothed down her hair with the other." After slaying the bull "she" ran out of the arena, "kissing her hand to the spectators, after the manner of a ballet-dancer leaving the stage."

But perhaps the most brutal of all the sports were the combats of bull and bear. Horsemen went into the forests with lariats to catch bears, and one of these scenes Colton saw and described:

As soon as bruin felt the lasso, he growled his defiant thunder, and sprung in rage at the horse. Here came in the sagacity of that noble animal. He knew, as well as his rider, that the safety of both depended on his keeping the lasso taught [sic], and without the admonitions of rein or spur, bounded this way and that, to the front or rear, to accomplish his object, never once taking his eye from the ferocious foe, and ever in an attitude to foil his assaults. The bear, in desperation, seized the lasso in his griping paws, and hand over hand drew it into his teeth: a moment more and he would have been within leaping distance of his victim; but the horse sprung at the instant, and, with a sudden whirl, tripped the bear and extricated the lasso. At this crowning feat the horse fairly danced with delight. A shout went up which seemed to shake the wild-wood with its echoes. The bear plunged again, when the lasso slipped from its loggerhead, and bruin was instantly leaping over the field to reach his jungle. The horse, without spur or rein, dashed after him. While his rider, throwing himself over his side, and hanging there like a lampereel [sic] to a flying sturgeon, recovered his lasso, bruin was brought up again all standing, more frantic and furious than before; while the horse pranced and curveted around him like a savage in his death-dance over his doomed captive. In all this no overpowering torture was inflicted on old bruin, unless it were through his own rage,—which sometimes towers so high he drops dead at your feet. He was now lassoed to a sturdy oak, and wound so closely to its body by riata over riata, as to leave him no scope for breaking or grinding off his clankless chain, . . .

The "baccaros" then brought a wild bull from a herd,

as full of fury as the bear. Bruin was now cautiously unwound, and stood front to front with his horned antagonist. We retreated on our horses to the rim of a large circle, leaving the arena to the two monarchs of the forest and field. [For a little while] Neither turned from the other his blazing eyes; while menace and defiance began to lower in the looks of each. Gathering their full strength, the terrific rush was made; the bull missed, when the bear, with one enormous bound, dashed his teeth into his back to break the spine; the bull fell, but whirled his horn deep into the side of his antagonist. There they lay, grappled and gored, in their convulsive struggles and death-throes; We spurred up, and with our rifles and pistols closed the tragedy; and it was time: this last scene was too full of blind rage and madness even for the wild sports of a California bear-hunt.

Some of the advertisements presented a bull as one of the worst killers ever to leave Spain, and the bear as an enraged grizzly that would demolish the bull on sight. Jackson, who lived not then but in the twentieth century, says the nearest the bull had ever been to Spain was the San Joaquin Valley, and the bear "was nine times out of ten a mangy, ailing specimen remarkable chiefly for its obvious anxiety to find an exit from the ring and leave bull and audience far behind." He says further that if a bear did knock a bull silly with one blow the miners demanded their money back and ran the promoters out of the camp. One fight was described by a newspaper man whose name is unknown: "Yesterday's hugely advertised bull-and-bear fight proved a distinct disappointment to our citizens. So anxious was the bear to get away that he made straight for the plank walls of the arena, which he clawed frantically in his attempt to scale them." A doctor gave the beast a good caning to drive it back, and the reporter surmised that since the doctor was heavily bearded "the bear looked up, saw his antagonist for the first time, muttered, 'Et tu Brute!' and gave up."

Though Jackson may have been right in most instances he seems not to have been right in all of them. As Borthwick approached Moquelumne Hill he found rocks and trees along the roadside, and a little later the camp itself, covered with placards:

WAR!   WAR!   WAR!
The celebrated Bull-killing Bear
GENERAL SCOTT
will fight a Bull on Sunday the 15th inst.,
at 2 P.M. at Moquelumne Hill

The Bear will be chained with a twenty-foot chain in the middle of the arena. The Bull will be perfectly wild, young, of the Spanish breed, and the best that can be found in the country. The Bull's horns will be of their natural length, and *not sawed off to prevent accidents.*' The Bull will be quite free in the arena, and not hampered in any way whatever.

The arena was about a hundred feet in diameter and inclosed by stout boards ten feet tall. Roundabout were tiers of seats. "The scene was gay and brilliant," the miners dressed in red, white and blue shirts, under "hats of every hue." Red and blue French bonnets, gay Mexican blankets, and Mexican women in "snowy-white dresses" helped to make a colorful scene. Gen. Scott was on a chain, and inside his cage in the center of the arena; in two pens were two bulls. The bear, a grizzly, according to Borthwick, that weighed 1,200 pounds,* had already slain several bulls, whose horns had been dulled. The "General was rolled out upon the ground all of a heap . . . floundered half-way round the ring at the end of his chain, and commenced to tear up the earth with his forepaws." One of the bulls was poked and prodded into the arena and "did not seem to like the prospect." Neither did the General, apparently, for he at once dug a hole and lay down in it. The bull

was a very beautiful animal, of a dark purple color marked with white. . . . He stood for a moment taking a survey of the bear, the ring, and the crowds of people; but not liking the appearance of things in general, he wheeled round, and made a splendid dash at the bars, which had already been put up between him and his pen, smashing through them with as much ease as the man in the circus leaps through a hoop of brown paper.

It sounds as if this bull was not of the mangy, frail kind. He was again prodded and driven into the ring, and by this time "he had made up his mind to fight." Head down, he charged, and the bear seized the bull by the nose. This seems to have been the General's principal offensive trick;

_____
* If so, it was a very large bear.

the bear not only seized the nose but held the bull and "embraced him round the neck with his fore-paws." He even "shook him savagely by the nose" until there wasn't much of the nose left. The bear then hurled the bull down but the bull escaped and retreated; but after looking at his enemy a few moments he charged again. Again the bear "took the bull's nose into chancery" and again the bull broke free; he was again retreating when the bear seized one of his hind feet and was "dragged round the ring." The spectators were shrieking their delight when the bull, his nose and lips a "mass of bloody shreds" decided to lie down.

But he was instantly forced to his feet and back to the fight. By prodding him and waving red flags the hustlers in the ring worked him up to another rage and he charged again. This round "was fought more savagely than ever" but most of the fight was gone from the bull, and a good deal of his blood. A purse of $200 was taken up and the second bull was sent in, and one of the "finest fights ever fit in the country," as one overjoyed man put it, was soon in progress. But even two bulls, attacking at once, were no match for a bear whose deep pelt and thick hide the horns could not pierce. The bulls were shot to put them out of their pitiable plight, and Gen. Scott withdrew, triumphant, into his cage. Borthwick tells us that the next bull sent against him was lucky enough to drive a horn between ribs and into a vital spot.

Helper also tells of having seen posters announcing bull-and-bear fights but in this case they would be chained together and would fight until one had killed the other. Determined to "moderate or diminish the sin as much as possible" Helper prepared himself by going to hear a sermon by a preacher who "assumed a sanctified visage" on Sunday but whooped and yelled as loud as anyone else at a brutal fight. A number of persons who saw these fights left descriptions of them; we think Helper's is one of the best.

By this time the whooping, shouting and stampeding of the spectators attested that they were eager and restless to behold the brutal combat; and an overture by a full brass band. . . . The music ceased; the trap-door of the bull's cage was raised, and "Hercules," huge, brawny, and wild, leaped into the center of the inclosed arena, shaking his head, switching his tail, and surveying the audience with a savage stare that would have intimidated the stoutest hearts, had he not been strongly barred below them. His eyes glistened with defiance, and he seemed to crave nothing so much as an enemy upon which he might wreak his vengeance. He contorted his body, lashed his back, snuffed, snorted, pawed, bellowed, and otherwise behaved so frantically, that I was fearful he could not contain himself until his antagonist was prepared. Just then, two picadors—Mexicans on horseback—entered the arena, with lassos in hand. Taurus welcomes them with an attitude of attack, and was about to rush upon one of the horses with the force of a battering ram, when, with most commendable dexterity, the piccador who was farthest off lassoed him by the horns, and foiled him in his mad design. As quick as thought, the horseman from whom the bull's attention had been diverted, threw his lasso around his horns also; and in this way they brought him to a stand midway between them. A third person, a footman, now ran in, and seizing his tail twisted it until he fell flat on his side; when, by the help of an additional assistant, the end of a long log-chain was fastened to his right hind-leg. In this prostrated condition he was kept until the other end of the chain was secured to the left fore-leg of the bear, as we shall now describe.

Running a pair of large clasping-tongs under Bruin's trap-door, which was lifted just enough for the purpose, they grasped his foot, pulled it out, and held it firmly, while one of the party bound the opposite end of the chain fast to his leg with thongs. This done, they hoisted the trap-door sufficiently high to admit of his egress, when out stalked "Trojan," apparently too proud and disdainful to vouchsafe a glance upon surrounding objects. He was a stalwart, lusty-looking animal, the largest grizzly bear I had ever seen, weighing fourteen hundred pounds.* It was said that he was an adept in conflicts of this nature, as he then enjoyed the honorable reputation of having delivered three bulls from the vicissitudes of this life. It is probable, however, that his previous victories had flushed and inspired him with an unwarrantable degree of confidence; for he seemed to regard the bull more as a thing to be despised than as an equal or dangerous rival. Though he gave vent to a few ferocious growls, it was evident that he felt more inclination to resist an attack than to make one. With the bull, the case was very different; he was of a pugnacious disposition, and had become feverish for a foe. Now he had one. An adversary of gigantic proportions and great prowess stood before him; and as soon as he spied him, he moved backward, the entire length of the chain, which jerked the bear's foot

and made him rend the air with a most fearful howl, that served but the more to incense the bull. Shaking his head, . . . casting it down, and throwing up his tail he plunged at the bear with a force and fury that were irresistible. The collision was terrible, completely overthrowing his ponderous enemy, and laying him flat on his back. Both were injured, but neither was conquered; both mutually recoiled to prepare again to strike for victory. With eyes gleaming with fire, and full of resolution, the bull strode proudly over to his prostrate enemy, and placed himself in position to make a second attack. But now the bear was prepared to receive him; he had recovered his feet wild with rage, and he then appeared to beckon to the bull to meet him without delay. The bull needed no challenge; he was, if possible, more impetuous than the bear, and did not lose any more time than it required to measure the length of the chain. Again, with unabated fierceness, he darted at the bear, and, as before, struck him with an impetus that seemed to have been borrowed from Jove's own thunderbolt; as he came in contact with the bear, that amiable animal grappled him by the neck, and squeezed him so hard that he could scarcely save himself from suffocation. The bull now found himself in a decidedly uncomfortable situation; the bear had him as he wanted him. Powerful as he was, he could not break loose from Bruin. A vise could not have held him more firmly. The strong arms of the bear hugged him in a ruthless and desperate embrace. It was a stirring sight to see these infuriated and muscular antagonists struggling to take each other's life. It was enough to make a heathen generalissimo shudder to look at them. . . . The Spanish ladies—they laughed, cheered, encored, clapped their hands, waved their handkerchiefs, and made every other sign which was characteristic of pleasure and delight. The contending brutes still strove together. Hercules quaked under the torturing hugs of Trojan. Trojan howled under the violent and painful perforations of Hercules. But the bear did not rely alone upon the efficacy of his arms; his massive jaws and formidable teeth were brought into service, and with them he inflicted deep wounds. . . . He seized the bull between the ears and nostrils, and crushed the bones with such force that we could distinctly hear them crack! Nor were the stunning butts of the bull his only means of defence; his horns had been sharpened expressly for the occasion, and with these he lacerated the bear most frightfully. . . .

Finally, however, fatigued, exhausted, writhing with pain and weltering in sweat and gore, they waived the quarrel and separated, as if by mutual consent. Neither was subdued; yet both felt a desire to suspend, for a time at least, all further hostilities. The bull, now exhausted and

---

* We suspect that both Borthwick and Helper were deceived in the weights.

panting, cast a pacific glance toward the bear, and seemed to sue for an armistice; the bear, bleeding and languid after his furious contest, raised his eyes to the bull, and seemed to assent to the proposition. But, alas! man, cruel man, more brutal than the brutes themselves, would not permit them to carry out their pacific intentions. [The two beasts were goaded with spears until they again rushed at one another.] During this scuffle, the bull shattered the lower jaw of the bear, and we could see the shivered bones dangling from their bloody recesses! Oh, heaven! what a horrible sight. How the blood curdled in my veins. Pish! what a timid fellow I am. . . . Shall I tremble at what the ladies applaud? . . . neither the bear nor the bull could stand any longer. . . . they dropped upon the earth. . . . while in this helpless condition the chain was removed from their feet, horses were hitched to them, and they were dragged without the arena. . . .

A second bloody contest followed.

CHAPTER 13

Camp Angels

In an environment where the need of them was so great one would expect to find many women ministering to the destitute and the sick but if there were outstanding camp angels their names, with few exceptions, have not come down to us. But we must bear in mind the fact that mankind has never found the good people half as memorable as the bad: for every tinhorn gambler like James Butler Hickok or Wyatt Earp who has been made a national hero it is possible that a lot of Nellie Cashmans have been forgotten. Spurning the few persons who actually were angels of mercy and whose names have come down to us, many writers have participated in building legends, as with Silver Heels, or in transforming harlots into paragons of virtue. There can be no doubt that a few of the harlots had big mother hearts, for they were human females, after all; one of them we shall try to give a portrait of in this chapter. But of women, if they existed, who were not prostitutes, yet were devoted to the relief of suffering, our search of the records has turned up very few names.

Mrs. Mathews tells of a Mrs. Beck, of Virginia City, Nevada, who worked twice as hard as any other women she knew, yet was constantly on the go taking medicines or food or clothes to destitute families. "I think she gave from $600 to $700 every year, . . . from her own pocket, besides as much more that she collected. Then there was the time she gave to running around and hunting up cases. She would sit up many nights making over clothes for children. . . . I think she spent one-half of her time in the interest of the poor." Mathews gives several pages to Mrs. Beck's charities but we never found her name in any other book or in the Virginia City papers that we examined. It sounds as if Mrs. Beck has far more claim on our attention than the killers and blackguards who fill so many pages in the books of history, yet she and her kind are forgotten but for a casual page by a Mrs. Mathews.

There were men who were camp angels, and now and then a name appears in a paragraph or a page: "An extraordinary person who made a great contribution to the Klondike was Father Judge, the Roman Catholic missionary, whose name was blessed by everyone in Dawson." He came not with gold pan and pick but with medicines and bandages, and the day he arrived "half the town was sick with scurvy, pneumonia, typhoid," so he put up a hospital tent and then two more, and "night and day he labored for the sick." He was urged to spare himself, "for he was a frail man, but he refused to listen to advice, continued to work unsparingly. When he could get away from his routine at the hospital, he begged food and supplies, blankets, and grave-clothes, . . ." and the people of the town pitched in to solicit funds and a hospital was built. In the middle of the Klondike fever Father Judge died, at the age of forty-nine, worked to death, some of the people said. "When he died, it was said that everybody in Dawson went to the funeral, and everybody wept."

Nellie Cashman, also a Catholic, was born in Ireland, and after coming to this country went West with her sister, who married, had five children, and died, leaving the children for Nellie to care for. Nellie's chief, if not only, biographer is John P. Clum, who knew her well when he was mayor of Tombstone, and she had the Russ boardinghouse on the corner of Fifth and Tough Nut streets Before writing about her, Clum talked with Mike, oldest of the five children and at that time president of two Arizona banks, and with other persons who had known her; but his long essay about her is uncritical and almost reverent. He wished that he had gotten information from

Nellie, the glamorized painting

her own lips, and so do we, when he saw her for the last time at Fairbanks, Alaska, in 1908.

Clum established the *Daily Citizen* in Tucson in 1879 and met her there: "Her frank manner, her self-reliant spirit, and her emphatic and fascinating Celtic brogue impressed me very much. . . . she was about my own age (28)." Two years before that she had been in the gold rush to British Columbia, and from time to time afterward joined in stampedes with men. Like Calamity Jane and many others adventurous women of the West she preferred the company of men to that of women, and was happy when traipsing with them over rugged trails and sharing the discomforts of their rude camps. Clum says she had learned that scurvy was a curse in the camps, and averted an epidemic of it in British Columbia by having potatoes shipped in, with priority over all other freight, including whisky. "And so it

came to pass that no matter where Nellie went in later years there was sure to be someone about who knew Nellie's record in the Cassiar District."

In 1878 she came down from Canada; paused to look over Virginia City and Pioche in Nevada, both in their decline; and hearing of a strike in southern Arizona headed for Tucson. She was soon over in Tombstone, which was established in 1879. If any miners were in want, or sick, "Nellie was in her element." When in Tombstone a miner fell down his own shaft and broke both legs and was found "in a most pitiable condition" Nellie raised five hundred dollars to take care of him. It is said that when she asked for a contribution it was given, and when she arranged a benefit, tickets were bought. Any idea that she was impulsive and addle-pated vanishes, says Clum, "before the facts of the major deed undertaken" by her in 1880, "involving physical, moral, and financial responsibilities over a period of years." She "consecrated some of the best years of her life to a service, which, in the fullness of her religious zeal, she doubtless believed a divine influence had placed in the pathway of her life, not only as a privilege but as a sacred duty." We would surmise that her motivation combined both the religious and maternal emotions. In her own unorthodox way she was a nun. As Clum says, it was but natural that all her most unusual undertakings were promoted by her within the framework of her religious faith.

Typical of the stories told about her is the one in 1884 related to five men under sentence of death for murders committed at Bisbee. The execution was to take place in Tombstone at the Courthouse, at one o'clock March 28, 1884. Because the murders had been unprovoked and cold-blooded, and the killers were chronic criminals, many persons wanted to witness the hangings. A "brutal-hearted, mercenary group leased an adjacent vacant lot and erected a grandstand overlooking the courtyard and prepared to sell standing room thereon to all who were willing to pay a substantial fee." The more enlightened citizens were shocked but unwilling to do anything about it. Though only two of the men were Catholics Nellie offered to be Mother Confessor to all of them, and thereupon the other three became Catholics. The woman must have had a way with her. On learning that their hanging was to be the occasion of a Roman holiday the men asked

Nellie to do something about it, and she told them not to worry at all, that not a single soul would watch from the grandstand when the men were swung from the earth to meet their Lord and Creator. In view of the intense feeling in the community against the men, Clum is probably right in saying that Nellie's resolve to wreck the grandstand was a test of her "tact and courage."

The first thing this "daring and strategic leader" did was to clear the field so that she could act. To the chief of police and a few others she suggested that in view of the hostility in the town it would be a good thing if all the people were to retire before midnight. This suggestion "met with enthusiastic approval." Nellie then asked some of her most robust miner friends to appear about two o'clock, quietly, with tools; and the grandstand was taken down and carried away. "Another thing that oppressed the doomed men" was the rumor that their bodies would be sent to a medical laboratory. Nellie told them not to worry about that: she would post a guard over their graves and they would rest in peace. After the men were hanged and buried, "two old prospectors with their blankets and coffeepot and frying-pan strolled away from Nellie's restaurant and disappeared. . . ." For ten nights they slept by the graves.

In the same year, says Clum, she saved the life of E. B. Gage, superintendent of a mining company. The miners had presented certain demands and then struck, and when Gage proved to be adamant they decided to kidnap and hang him. Nellie got wind of it. At ten o'clock the night of the planned hanging she drove in her buggy to the Gage home, just south of the camp; and when she had him on the seat beside her she drove slowly away until she was beyond the town, when she laid the whip on her beasts and fled with him at top speed to the nearest stage station. Why she did not merely tell him what was planned, so that he could get a horse and flee, Clum does not say. One suspects after a while that the ex-mayor is not allowing the implausible to get in the way of his stories.

In a lighter vein he tells of the empty lot close by the Russ house, and of boys who gathered there to feud and fight, in imitation of their elders, then as now. Hearing their horrible shrieks and profanity Nellie would rush from her house and separate them and lecture them and try to

In 1925, not long before her death

shame them; but to show that her heart was big and in the right place she would then bring to each of them a generous piece of pie. They were not stupid boys. It did not take them long to see the inevitable association between howling curses and pie, and so thereafter they waged sham battles that were even more horrendous than any that had gone before. United States Marshal George A. Mauk, who was one of the boys, and who told Clum the story, said the sham fights were an unqualified success.

Another story on the lighter side concerns Sam Lee, her Chinese cook. One day, feeling a call to visit his people in Hong Kong, he asked Nellie to let him take a portrait of her with him, so that one of his friends over there, a distinguished artist, could make a painting of it. He promised to bring the painting to her and he was as good as his word. The likeness of her in these pages is a photograph of that painting.

In the summer of 1884 the cry of gold was heard in Lower California, and Nellie joined a small group going from Tombstone, which included Mark A. Smith, later a United States Senator from Arizona. It is said that she saved the lives of the men, for the reason that she had more stamina. After the men had prowled for gold in the desert wastes until only a gallon of water remained, with no sign of water or living thing in sight, Nellie said she would go forth alone and find water. It is alleged that she said an angel would guide her. What man, sitting down in the hot sands to wait for a woman to save him, can say that an angel did not? In any case she came to a Catholic mission, had goatskin water bags filled and packed on burros, and with a Mexican guide set off to find the future senator and his companions. Again and again she put on such manly garb as overalls and went off with prospectors. And then came the Yukon.

An Arizona newspaper reported on November 20, 1897, that Nellie was "preparing to organize a company for gold mining in Alaska. Her many friends in Arizona will wish her success, for during her twenty years residence in the territory she has made several fortunes, all of which have gone to charity." If Nellie had known that the Yukon rush demanded much more in fortitude and courage than any rush before it she probably would not have been daunted at all. She went, and established a store in Dawson, in front of which Clum photographed her June 23, 1898. He says he was in a darkroom changing the film in his camera when a voice out of the gloom asked him for a contribution to the local Sisters' Hospital. The brogue told him it was Nellie. He learned that she had climbed the Chilkoot Pass* and he read in the Yukon *Sun* that "Miss Cashman is the pioneer woman in this country and is widely known for her good deeds." She had a small grocery shop and was grubstaking prospectors. An English miner is said to have entered her store one day to ask where the Union Jack was, since Dawson was in Canada; and Nellie took him outside and pointed to the Stars and Stripes,

above which, "in neat arrangement were several British postage stamps bearing pictures of the Union Jack." "There's my flag and there's yours," she said, and they both had a hearty laugh.

Clum tells a story that he heard from his son-in-law, who served as a witness for Nellie when she filed on a claim. The oath said that the person swearing it had filed a claim at a certain time on a certain spot; but when Nellie took the Bible in her left hand and raised her right hand, she said, "So help me, God, I was never there," and kissed her thumb instead of the Bible.

Clum seems to have asked a few persons who had known her what they thought of her. A Wells, Fargo man wrote him: "Nellie Cashman was one of the most wonderful women I ever met. She was unique. Though she seemed to prefer to associate with men, there never was a spot on her moral character. I have always regarded Nellie as a most remarkable and admirable woman." A Seattle businessman who knew her in Alaska wrote: "I think the principal lure of the far-north mining camp to her was that she might be near the prospectors and miners to whom she was doctor, nurse and missionary. Nellie was a devout Catholic and spent much of her time with the Sisters and working with and for the Sisters' hospital."

In her last years her nephew Mike tried to persuade her to leave the North and accept a comfortable home in a warmer clime, but her heart was where the miners were. She even voyaged several hundred miles down the Yukon and then fought her way to Coldfoot, the camp in Alaska farthest north. There she staked the last of her mining claims, and in that area she sometimes journeyed for seventeen days at a stretch by dog-team. She was at Coldfoot in 1924 when she became seriously ill; she was taken a thousand miles to the hospital at Fairbanks. She seems to have been obsessed with the notion that her Coldfoot claim would yield a fortune, for she wanted to go back up there; but friends persuaded her to go to Victoria, where she was placed in the Sisters' hospital. Like all dauntless persons she found it hard to realize that she was close to her end. She died January 4, 1925, and was buried in the Cath-

---

* See the Klondike in "Other Gold Rushes."

olic cemetery there. Monuments to her compassion and charity are scattered over the west.*

Some readers may be startled to find Calamity Jane placed among camp angels; we hasten to assure them that we have put her there only after reading everything about her we could find, after studying many photographs of her, and after giving a good deal of thought to the authentic facts of her life. She wasn't the angel of mercy that Nellie was. She probably was hardly an angel at all compared to many whose names have not come down to us. But there can be no doubt that for all her faults she had a big heart and a wonderful courage, and that when sickness was around her she pitched in as nurse, if sober enough to do it.

Few if any persons on the American frontiers have been the subject of such extremes in opinion or of such abysmal nonsense. Robert J. Casey says, "Long ago somebody asked me, 'Who was Calamity Jane?' And the answer was simple. I had the legend pat, complete with names, dates, addresses and bibliography. Today if anyone were to ask me the same question the answer would still be simple: 'I don't know.' " Mr. Casey found that "one school of thought" said she was a blend of Florence Nightingale and Elizabeth of Hungary. Another school, "just as bigoted," thought she was a "female bum." In some measure she was both. She claimed to have been, says O'Connor, "army scout, bull-whacker, pony-express rider, Indian fighter, frontier Florence Nightingale, camp follower, and mule-team driver. . . . The All-American tomboy. . . . The little girl from Missouri who could outshoot, outride, outdrink, outcuss, outfight, outchew, and outlie any of her male companions." And all that merely exaggerates her prowess.

The best of the books about her, the most detached and scholarly and the most unaffected by the legends, is *Calamity Jane* by Roberta Beed Sollid.* In her Preface Mrs. Sollid says,

No career is so elusive to the historian as that of a loose woman. Calamity Jane was that sort of woman, and known details about her life are hard to find. Like most prostitutes and drunkards she left little behind in the way of tangible

Calamity preferred male attire

evidence. . . . She was all but forgotten by her contemporaries until, for some unknown reason, sensation-writers and historians began to take an interest in her.

_____

* A church house in Tombstone, for instance, that since 1882 has been used as a rectory. According to a letter to us from Father Doyle it served only as a temporary church until a permanent building could be erected. The present church was built in 1947. Father Doyle says, "No doubt Nellie took part in raising funds for the church completed in 1882, and in 1923 she presented the late Bishop of Tucson a check for $1,000 for repairs on the Tombstone church."

* Published by the Montana Historical Society. Mrs. Sollid seems to have searched all the newspapers and county and other records of the time and we hereby acknowledge our indebtedness to her exhaustive researches.

She cites five motion pictures that pretend to be based on her life, in which with Hollywood's genius for miscasting such persons portray her as Jean Arthur and Yvonne de Carlo!

Present-day opinion about Calamity Jane varies from the frequently-met belief that no such person existed to the complete acceptance of all the fantastic stories about her. . . . People of the older generation in Deadwood were not pleased to know that further research was being made on such a person. . . . They thought her of no con-

sequence and not at all important in Deadwood history.

Nevertheless, the sensation-mongers have called her The West's Joan of Arc, The Black Hills' Florence Nightingale, Lady Robinhood; and Edward L. Wheeler, after he invented the plains terror he called Deadwood Dick, with iron-cast shoulders, limbs like steel bars, and a mask over his face through which "there gleamed a pair of orbs of piercing intensity," plus a "wild sardonic

It was said that she was a crack shot

laugh," or, for a change, "a terrible blood-curdling laugh," sent his hero in pursuit of (among many persons and things) Calamity Jane. Scared silly, she seems to have hidden in the hills, while "grim and uncommunicative" there roamed through the country "a youth in black, at the head of a bold, lawless gang of road-riders, who from his unequalled daring has won and rightly deserves the name of—Deadwood Dick, Prince of the Road"—who, when last heard of, was giving lessons out in space to the astronauts. But in sixty-four volumes she would be the wife of Dick, this woman who took and shed husbands, nearly always without benefit of clergy, with no more thought about it than for a plug of Climax when she sank her teeth into it, or a bottle of whisky after she had drained the last drop.

Before putting before the reader the real Calamity it is best to expose some of the more preposterous legends—and none is sillier than the deathless romance she has in a number of books with Wild Bill. Her marriage to Hickok never existed in fact, and a document brought forth which purports to validate it in 1873 is a clumsy forgery. O'Connor points out that

As recently as 1958 a magazine article asserted that she was the mother of his child, that "as an 18-year-old girl she fell in love with Wild Bill Hickok the moment she set eyes on him. She saved his life, married him, gave him a divorce so he could marry another woman, and then became his mistress while posing as his partner." This account (in the *American Weekly,* June 1, 1958, by the *much-respected Homer Croy*) also stated that the marriage ceremony——

and so on and on through the maudlin and tiresome inventions that don't even have the virtue of plausibility. The defenders of Wild Bill, a tribe almost hysterical in their idolatry, say that Hickok, a very fastidious man, would never have considered even an intimacy with a woman for whom a bath was a weekend chore, when water was at hand. In her last alcoholic years she wrote or had a woman write for her four or five pages which she called the story of her life* and which, to raise money for drinks, she hawked from door

_____
* Given in full by Sollid.

Calamity Jane at Lewistown, Montana, 1899.

to door. Though some writers have accepted it as gospel we cannot be sure that there is a single fact in it. Of Wild Bill she makes only brief mention: "I started for Fort Laramie where I met Wm. Hickok better known as Wild Bill, and we started for Deadwood, where we arrived about June." She didn't even know that his first name was not William. Says Sollid, "During her active life, as far as is known, she said little that would indicate any special affection for him." This romance reached its glowing apogee in her death, and no matter what his syntax we'll let a South Dakota poet laureate tell it: "By a strange hand of fate, the Judge of the Universe called her before the bar of omnipotent justice on the same day of the month, twenty-seven years later, that her consort Wild Bill, beside of whom she requested to be buried, met his death." Young in his *Hard Knocks* goes the poet one better: "she passed away on August 2, 1906. On the same day and month and same hour Wild Bill was assassinated thirty years before."

Hickok died August 2, 1876, and Calamity August 1, 1903. Possibly she had a girlish crush on the tinhorn gambler and man-killer; the breathless romancers say that when dying she whispered her wish to be buried at his side, as indeed she was; and not long before her death she stood by the fence around his grave and was photographed.

What was her name? We're not even sure of that, for the simple reason that she was illiterate and apparently was not sure of its spelling. Sollid calls her Martha Cannary; others spell it Canary. In her "autobiography" Calamity gave it as Marthy Cannary and said she was born May 1, 1852. The year of her birth has never been settled, or the letter *E.* which appears as her middle initial on her headboard—"Mrs. M. E. Burke." In regard to what she looked like you can find about anything you want, in spite of the fact that some very good photographs of her, from young womanhood to old age, survived.* For the romancers she had hands of creamy whiteness, dainty feet, and a perfect figure; and piercing black eyes and raven hair. For another writer she had hair the color of copper and large brown eyes. For still another writer she was "tall, thin, and stooped" and had light brown hair and blue or blue-gray

eyes. One of Estelline Bennett's friends said she was a "pretty dark-eyed girl" but the *Rocky Mountain News* said her eyes "emit a greenish glare." A man who knew her said she was six feet tall, but in 1878 a newspaper said she was "a little creature," and another paper (of the many Mrs. Sollid examined) thought her "a cross between the gable end of a fire proof and a Sioux Indian." "Fire proof" are the words in Sollid's book.

Mrs. Sollid sums it up:

Close observation of the existing pictures of her show that she had dark hair and high cheek bones. Her figure was slender as a young woman, but stocky during her later life. Shortly before her death she seemed to be slim again. In no picture is her appearance striking or even very attractive. She looks plain and mannish, in fact, whether dressed in her buckskin suit or in a dress." In her "elegant leather suit," and indeed in every photograph that shows her in masculine attire, she looks more like a man than a woman, for she had a large strong face, much stronger than General Custer's and Wild Bill's, who also appear in the Sollid book with her.

What did she do? In the way of labor and hell-raising she did just about everything that men did. When in March, 1876, General Crook was ordered to discipline the Indians, there "is positive evidence," says Sollid, "that Calamity Jane was with the troops on this expedition." Whether she ever did any scouting we don't know but she would have us believe in her "autobiography" that she served with half the generals and was in half the Indian battles of that time. General Dawson said that in the later years of her life she was no longer able to distinguish between the stories that were true and those that were false. She had never wanted to. Whether she was a psychopathic liar with a criminal mind or merely one of those happy enthusiastic souls for whom truth is a dull and pretentious thing that deserves no hearing we also do not know. But it is true that she lied like a trooper from her earliest appearance on the stage of history to the day of her death and that what we know about her is only what was written about her in newspapers, and in diaries and other books by persons who knew

_____

* Mrs. Sollid reproduces twelve, possibly thirteen, photos of her, including the one in her coffin.

her. That she preferred the work of men and scorned the work of women there can be no doubt.

She freighted—the *Sidney Telegraph* reported that she had arrived from the Black Hills and had been promoted to "assistant wagon boss." She was a bullwhacker—"Mr. Charles Fales, who is considered an authority on the subject of bullwhacking, recalled seeing her as a whacker off and on over a period of time. . . ." Sollid quotes an old mule skinner, "I worked with Jane on three freight trips with a bull string. She took her place as any man would, and did her share of the work with the best of them. But when night came, if any of the boys wanted to go to the brush, she was always willin' to pull off a pants leg."

Apparently she drank with the best of them and the worst of them and in the later part of her life became a chronic alcoholic who, like those sunk in alcohol, was willing to do about anything for a drink. She was familiar with the inside of many a jail. Just like a man in his cups she got into many a fight. A Miles City paper reported that "Calamity Jane has been heard from again. This time she bobbed up serenely at Rawlins, got drunk and knocked a frail sister out of time for which she was arrested and fined ten dollars and costs." In another town she swore out a warrant for a man named Steers, who at that moment was serving as her husband; she said he stabbed her with a butcher knife and bashed her on the face with a rock. He was fined and hit the trail but after a few miles decided that it would be for him a dreary world with Calamity; and so he returned to her and then they both hit the trail, with Martha carrying the pack. They were soon in another drunken brawl and on her way to find a policeman came to a saloon and as usual couldn't resist it. She had time to throw some stones through a plate glass window before she was hauled away. In Laramie she must have been well known for her destruction of plate glass, for the *Boomerang* had this item in 1887: "To say that the old girl has reformed is somewhat of a chestnut. She was gloriously drunk this morning and if she didn't make Rome howl she did Laramie. Her resting place is now the soft side of an iron cell."

Says Sollid after giving the above: "Two weeks later this gentle creature turned up in the capital city of Wyoming in a very dilapidated condition." In Livingston a "spicy charge was made against" her when she and a man named Townley were hauled in for fornication. Though her guilt was as plain as the ravages of alcohol in her face a jury turned her free. In Lead an officer had to drag her from a saloon and load her into a hack, and against him, said the *Black Hills Daily Times,* she used vile language. She spent more and more hours in various jails as she approached her end, for with every drink she could buy or beg she was digging her grave. In the year of her death a Bozeman newspaper told of the time when a bunch of tenderfeet began to "chaff her" at a bar, whereupon she drew her six-gun and told them to dance. "They did other things that she commanded. Calamity Jane was not a person to be trifled with. The manner in which she shut up that saloon was powerfully convincing." Mrs. Sollid uncovered a few long-dead editors who once fled from her wrath, one of whom "climbed upon a convenient desk, sprang through the skylight, ran nimbly across the adjacent roofs, jumped through another skylight and hid in a friend's office." When almost at the end of her trail she was riding across the hills one day and accosting every man she met with the question, "Got a bottle?" Finding none with a bottle she settled for a chew of Climax plug "that would have made a Kentuckian ashamed of himself."

Was Calamity Jane a whore? A number of writers have said she was. Otero, who saw her in Arizona or Hays City, says she was "regarded by the community as a camp follower, since she preferred to ply her well-known profession among the soldiers rather than among the teamsters, freighters, herders and hunters." Parkhill, without citing any source, says that at one time she was a harlot in the house of Madame Moustache. We have seen the statement of a mule skinner that for the boys she was always willing to pull off a pants leg but if that makes her a whore are all this nation's single and married women whores who are equally accommodating? The dictionary says a whore is one who has unlawful sexual intercourse, especially for hire. No one has shown, and so far as we know, no writer has asserted, that Martha took pay from the soldiers and teamsters. If she yielded because she liked to, or to please them, or for any other reasons besides pay, she was no more a whore than the men who asked her to, and not half the whore that a lot of Hollywood characters are, whom Madison Avenue elevates to the level of national heroines. Mrs. Sollid says that she was known as a camp follower "and at the age of sixteen was so dissolute that she and others of her class were ordered to leave town." She seems to base that categorical statement on McClintock's *Pioneer Days in the Black Hills.*

It seems to be a fact that she lived with a man and passed him off as her husband any time it suited her whim. She is said to have complained, "I have been called the common law wife of King, Conors, Wilson and Dorsett." In her five pages about herself the only man she claimed as a legal husband was a Clinton Burk (or Burke), whom she says she met in 1884 in Texas and married in 1885, and by whom she claimed to have had a daughter; but Mrs. Sollid dug up three newspaper items which prove that she was a long way from Texas during the months she claims to have been there. And though she said she lived with him many years, in 1886 she was charging a man named Steers with beating her up, and according to a Cheyenne paper she said in 1887 that she was married to the man. But we haven't the space here to set down all the known details about her "husbands" and common-law men,* who have nothing, in any case, to do with the question whether she was or was not a harlot. It has been said by some writers that for a few days or weeks she was now and then an inmate of a brothel. When broke she may have gone to one to raise some whisky money but we cannot imagine Martha Cannary in a whorehouse, peering from behind a curtain and soliciting passersby whose whole soul and meaning was in the outdoors world where the men were—on horseback and wagonback, on the plains and trails, in the gambling halls and saloons.

We suspect that the best way to determine what she did or did not do is to determine what she was. She is reported to have been angered by a statement that she was a horse thief, a highway woman, a three-card monte sharp, and a minister's daughter; and to have said that they were all false, and especially the last one. Mrs. Sollid sums her up this way:

Calamity Jane's violent personality certainly contained some characteristics that endeared her to the rough and ready companions of the early boom days. Her spasmodic generosity; her brusque, rough sympathy for the underdog; her failure to be impressed with riches, fine clothes or "society"; her lack of pretense or affectation of virtue; her complete rejection of simpering femininity; such attributes were esteemed on the frontier. She was a symbol of rough virtue

and ignored any censure of her unconventional frontier ways.

The wife of Dr. McGillycuddy, who knew Martha, reports his impression of the woman when he saw her nursing a boy stricken with smallpox: "He spoke of her kindness to anyone in trouble, as well as her utter recklessness and lack of morals, expressing his opinion that there was much more real feelings in the hearts of the average rough human beings of the West than in those of more cultured centers." McGillycuddy himself wrote to an editor twenty-one years after her death: "Jane was a healthy girl of an affectionate disposition, and naturally had many husbands, license shops and preachers being scarce in those days, . . . She was not immoral, she was unmoral, she lived in accordance with the light given her, and the conditions and the times." Estelline Bennett, one of the first white women in Deadwood, and a woman of culture and sense, said of her: "More than anyone else who lived in the twentieth century, Calamity Jane was symbolic of old Deadwood. Her virtues were of the endearing sort. Her vices were the wide-open sins of a wide-open country—the sort that never carried a hurt." Buffalo Bill knew her well. He says: "Only the old days could have produced her. She belongs to a time and a class that are fast disappearing. . . . Calamity had nearly all the rough virtues of the old West as well as many of the vices. . . . She was one of the frontier types and she has all the merits and most of their faults."

Many opinions similar to Cody's could be given but we think he sums it up pretty well. We who were born on the frontier and grew up on it do not find it hard to understand this woman: we knew women long ago who were much like her—the kind who were more at home in a saddle than in a kitchen, who were more familiar with a rifle and an ax than with beauty aids, and who preferred the vast clean world out of doors and the way of life of men. Martha was almost as much

---

* For this difficult and mixed-up matter see Sollid, or Clarence Paine, whom she cites.

a part of the plains as the sagebrush and the coyote, and it was the tragedy of her last years that the frontier had passed away and a new era had come, in which she could find nothing that was part of her. The difference between her time and today is suggested by the wish of Deadwood people to let her be buried and forgotten and nothing more be written about her.

Two old-timers of the Black Hills, Jessie Brown and A. M. Willard, in *The Black Hills Trails*, try to sum it up: "She was a strange mixture of the wild, untamed character of the plains and mountain trails, and generous kindly womanhood. But under the rough exterior there beat a heart so big and friendly as to be without measure. Brave, energetic, unfettered, kind, always on the line of action, with helping hand ever turned to the poor and unfortunate, . . ." Possibly that overstates it but the story is told of the time when she saw a mule skinner brutally treating a mule. "When she told him to stop, he flicked off her hat with his mulewhip. She pulled her guns and ordered him to put the hat back." It is said the skinner didn't obey at once but when he looked at her eyes he picked up the hat and set it on her head.

Mrs. Sollid gives a whole chapter to Martha's labors as a nurse; she starts off with a statement by Brown and Willard but says that the "most colorful picture of the woman's work" in the smallpox epidemic year of 1878 is to be found in Bennett's *Old Deadwood Days*:

. . . everyone of them was remembering those days in '78 when Calamity Jane alone took care of the smallpox patients in a crude log cabin pest house up in Spruce Gulch. . . . Smallpox was the most dreaded scourge of the frontier town. Usually people died because of dearth of nursing, of facilities for taking care of the sick, and bad sanitary conditions. For the same reasons, it spread with terrifying speed. . . . All that a town like Deadwood in '78 knew to do for smallpox patients was to set aside an isolated cabin and notify the doctor.

The doctor found a half-dozen very sick persons in the cabin and said he would return after supper. "No one offered to go with him, but when he went back he found Calamity Jane there." He asked her why she was in the cabin and Miss Bennett reports her as saying, "Somebody's got to take care of 'em. They can't even get 'emselves a drink of water. . . . tell me what to do, Doc, and I'll stay right here and do it." He told her she would probably catch the disease and she said that was a chance she would take. "She came unscathed through the long smallpox siege and most of her patients lived. Dr. Babcock believed that without her care not a one of them could have pulled through. She never left the pest cabin during those hard weeks except to make hurried trips down to Deadwood for supplies that the grocers gave her."

If that is the unexaggerated truth of it we can say that in all our reading about the mining camps and their people we found no instance that excelled it in courage and devotion. The late Stewart H. Holbrook was skeptical of most of the tales about Calamity, as he had need to be, but he gave her "full credit for her service to the sick." So far as we have been able to learn Estelline Bennett was a good witness. Mrs. Sollid cites other writers who knew Calamity, and who support Bennett's view of her as a nurse. There is Dee:

In the year 1878, eight men came down with smallpox, they were quarantined in a little shack. . . . Calamity had volunteered to care for these men, of whom three died. She would yell down to the miners in the gulch below for anything she needed, and throw down a rope by which to send supplies. They would bring her what she required to the foot of the hill and she would haul them up hand over hand. Her only medicines were epsom salts and cream of tartar. When they died she wrapped them in a blanket and yelled to the boys to dig a hole. She carried the body to the hole and filled it up. She only knew one prayer, "Now I Lay Me Down to Sleep." This was the funeral oration she recited over the graves. But her good nursing brought five of these men out of shadow of death, and many more later on, before the disease died out. Think of a trained nurse these days nursing eight men, working as cook, doctor, chambermaid, water boy, and undertaker, with the duties of a sexton thrown in. If anyone was sick in camp, it was "send for Jane"; where Calamity was, there was Jane; and so she was christened Calamity Jane.

That may be only a legend. Where the Jane came from nobody knows; Mrs. Sollid thinks it possible that "jane" was already a synonym for woman, as skirt, twist, and femme have become in a later time.

After telling of her nursing in the gulch another of Sollid's sources says: "The woman in her was bound to come out in one way or another. She turned the same trick over at Pierre, where a settler family was practically abandoned in a quarantine for black diphtheria. Calamity broke the quarantine." One of her pallbearers, "whose account seems impartial," says: "She was a fine nurse. There wasn't anything she wouldn't do for anybody and whenever she had any money it went just the same way, easily and for the first fellow that asked for it." That's not the way of the professional whore, except with her pimp. Mrs. Sollid says the only extant edition of the Deadwood *Pioneer* "contains an article which shows that Calamity gained a reputation as a nurse soon after she first arrived in Deadwood." Indeed, the newspaper headline was "Calamity Jane as Nurse," and the article said, "The man Warren, who was stabbed on lower Main St. Wednesday night, is doing quite well under the care of Calamity Jane, who has kindly undertaken the job of nursing him. There's lots of humanity in Calamity, . . ." That was in July, 1876, two years before the smallpox epidemic in the gulch.

In the remarks at her funeral, twenty-five years after the epidemic, her service as a nurse had not been forgotten: "How often amid the snows of winter did this woman find her way to a lonely cabin of a miner who was suffering from the disease of those times and who felt sorely the need of food and medicine. When the history of this country is written too much can not be said of the results of this woman's labors in helping you to build and to complete the work you had undertaken. As I think of her labors and voluntary sacrifices I hear the voice of Christ as he said: 'Even as much as ye have done it unto the least of these my little ones, ye have done it unto me.'"

Funeral orations delight in exaggerating the merits of the deceased but the evidence presented by Mrs. Sollid convincingly supports her statement that "Calamity could be called upon for aid whenever anyone was sick and in need of help. If she were sober, she would go immediately to them and take care of their needs. If she were drunk, she would sober up and be ready to care for the sick the next day." It does not seem likely that any of her critics could match her record of service to the least of these my little ones.

In some who write about the early American West there is a tendency to think most of the women were angels, and a few of them were fallen angels through no fault of their own; or a tendency to view all of them with cynical skepticism. The newspaper editors were on the whole a cynical lot with a deep vein of sentimentality that was forever on the watch for stories about women that presented an image of virtue. Typical of such tales is that taken from a New York newspaper and reprinted in the *Boise News* in Idaho City, May 14, 1864. It is the story of a beautiful girl who married a handsome young man who was soon called to the war, and of a libertine who at once began to woo and propose elopement to the wife. When he offered her ten fifty-dollar greenbacks to settle her affairs she accepted his proposal, and soon afterward sent him a note:

I have to inform you that circumstances beyond my control will prevent me from fulfilling my engagement to elope with you to-night. I expect my husband home on furlough soon, to spend Christmas and New Years, when we shall enjoy a hearty laugh at your discomfiture. Meanwhile, I shall keep the money as a Christmas present for him, and when this cruel war is over, it will come handy to assist him to start in business. Yours "tenderly"

P.S. When next you undertake to play the libertine, you would do well to beware of a soldier's wife.

We don't know what the present generation thinks of such a woman but we have no doubt that most of the miners would have laid their poke of gold dust at her feet.

PART III

Crime and Justice

Some Problems in the Camps

Women For some the chief problem was homesickness, and for others it was disease, or inadequate clothing and bitter cold, or fires, or prejudice and hate; but we suspect that, taken all in all, the major problem was the lack of what some writers have called the civilizing influence of the female sex. Most of the camp editors were conscious of this preoccupation with women and wrote about it in different ways, or reprinted articles that appeared in the Eastern press. Thus the *Avalanche* took from the New York *Sun* these extraordinary observations:

Some enthusiastic Frenchman once declared the human leg to be the most philosophical of all studies. "Show me the leg," said Gautier, "and I will judge the mind," and it does seem quite as natural that the limb should indicate the disposition as the shade of hair should indicate the temperament. What sloth, for instance, does an obese limb betray! What a shrew is the possessor of a limb like a walkingstick! But what a gentlewoman is she of the arched instep, the round ankle and the graceful pedestal, swelling to lightness! What a dogged obstinacy the stumpy leg with the knotted calf exhibits! What an irresolute soul does these lanky limbs betray! How well the strong ankle intimates the firm purpose—how the flat ankle reveals the vacant mind!

Young men about to marry—observe: The dark girl with a large leg will become fat at thirty, and lie abed reading novels till midday. The brunette with slender, very slender limbs, will worry your soul out with jealousy. The olive-skinned maid with a pretty rounded leg will make you happy. The blonde with large limbs will degenerate at thirty-five into the possession of a pair of ankles double the natural size, and afflicted with rheumatism. The fair-haired damsel will get up at five A.M. to scold the servants, and will spend her nights talking scandal over tea. The light, rosy girl with a sturdy, muscular, well-turned leg, will be just the girl for you. If you can find a red-haired girl, with a large limb, pop the question at once. The short lady should always possess a slender limb; the tall lady should possess an ample

one. These are the rules to observe in making your choice!

How seriously the miners took such advice is not known; most likely then as now it depended largely on prejudice, and the man devoted to a red-haired mother accepted that part as gospel, and the man whose father had the lanky legs of a Lincoln yet was not irresolute threw the paper aside. The editor of the Boise *News* burst forth with observations of his own:

All men who avoid female society have dull perceptions, and are stupid, or have gross tastes and revolt against what is pure. Your club swaggerers . . . declare female society stupid. Poetry is insipid to a yokel. . . . I can sit for a whole night talking to a well regulated, kindly woman. . . . One of the greatest benefits a man may derive from a woman's society is, that he is bound to be respectful to them. . . . Our education makes us the most eminently selfish men in the world.

Though some of the prospectors, and some of the miners and gamblers and bummers, were professional women haters, most of the men were strict believers in bad women and good women, and the good women were all good and the bad women were all bad. Many of them cherished sentimental memories of mother or sister or childhood sweetheart and would bristle and turn red with outrage if they heard coarse language in the presence of women.

A woman to a mining-camp [said Bancroft, who was a pretty sentimental soul] brought the odor of Araby, brought the sunshine of Eden. The atmosphere was mellowed by her influence; the birds sang sweeter for her coming, the ground was softer to sleep on, the pick was lighter, and whiskey less magnetic. Fair was the form of her, radiant her presence, thrilling her touch. Her dress was as the drapery which shrouded the mysterious holy of holies, and sacred was the hem of her garment.

To all of which it is customary to say, "amen." Said Ferguson: "I never knew a miner to insult a woman. . . . Every miner seemed to consider himself her sworn guardian, policeman and protector, and the slightest dishonorable word, action or look of any miner or other person, would have been met with a rebuke he would remember so long as he lived, if, perchance, he survived the chastisement." Ferguson was there and looked around him but he overstates the case.

It became a cherished custom in the camps to wash for any woman who appeared in the diggings a pan or two of sand and gravel, and to present to her, with an elaborate bow and a clumsy compliment, the gold washed out. In the richest diggings such a gift was sometimes worth a hundred dollars or more. A story is told of Henry Thomas Paige Comstock that he washed a panful for the Mormon wife, whom we met in a previous chapter, and slyly salted it with a double handful of gold dust from his pockets.

Haskins says that when, at a little distance, a group of men were looking at a young woman, one of them said she was not half as pretty as his Matilda Jane in Jersey. The others were so angered by the remark that they "gave the youth to understand that they would have a settlement with him for expressing such an opinion" but he was saved from a flogging when, after another look at the girl, he made amends. Mark Twain tells how in "a certain camp" the news spread that a woman was coming. The miners discovered where she was hiding and yelled, "Fetch her out!" The amazed husband appeared and said his wife was ill—they had been robbed of everything they had by the Indians; but the miners continued to yell, "We gotta see her, fetch her out!" So the husband fetched her out, and they waved their hats and cheered, and they touched her dress and listened to her voice "with the look of men who listed to a *memory* rather than a present reality"; and they took up a purse of $2,500 in gold and gave it to her, and swung their hats some more and "went home satisfied."

Helper tells of a married woman at whom a miner gazed so earnestly that she was embarrassed. Had she done him some injury? "Oh, no, madam. I assure you you have not harmed me in the least. But pardon me; I have been in the mines for the last two years, and it has been so long since I saw a lady, that I must own my admiration

of you has compelled me to be somewhat rude in my scrutiny of your charms." If he actually spoke to her in such stilted language she had a right to feel alarmed.

Fariss and Smith in their history of Plumas County (1882) say that when a Mrs. Bancroft came down the mountain trail to Rich Bar "down dropped pick and shovel from the hands of the miners, who had not seen the face nor heard the soft voice of a lady for many weary months, and her progress along the trail was watched with eager eyes for several miles." Shinn quotes a writer who said of the early San Francisco, when there were only fifteen women among the thousands of men, that "Women were queens, children were angels," and adds: "Bearded and weather-bronzed miners stood for hours in the streets to get a glimpse of a child at play. . . . A New England youth of seventeen once rode thirty-five miles, after a week's hard work in his father's claim, to see a miner's wife who had arrived in an adjoining district. . . . 'and, father, do you know, she sewed a button on for me, and told me not to gamble and not to drink. It sounded just like mother.'" Dimsdale, that outstanding Montana editor, thought that "The absence of good female society, in any due proportion to the numbers of the opposite sex, is likewise an evil of great magnitude; for men become rough, stern, and cruel, to a surprising degree, under such a state of things."

One old-timer said that when the first woman came to camp it was something he could never forget as long as he lived: "Everybody quit work; there were six or eight thousand of 'em. They built decorated arches over the street and marched four miles down the road with a band of music to meet her and escort her back to camp. By the time they got back the town was jammed full of miners that had come in from miles around to get a glimpse of the woman." Edwin Bryant, journeying over California, had one night as host "a gentleman of intelligence and politeness [who] made apology after apology for his rude style of living, a principal excuse being that he had no wife. He inquired, with apparent earnestness, if he could not send him two pretty, accomplished, and capable American women, whom they could marry; and then they would build a fine house, have bread, butter, cheese, and all the delicacies, luxuries, and elegances of life in abundance."

But it was not all chivalry, if we are to believe Mrs. Mathews, who had a pretty sharp pen. Having lost her purse she was trying to find a hotel room in Sacramento, telling a clerk that she would find a friend and pay the next morning. The clerk said, "I will pay for your bed if you will share it with me." For her "it seemed as if an adder had stung me. One moment my amazement held me spell-bound. . . . I made one bound across the room and reached my satchel. *Villain!* said I, 'do I look like such a person that you dare thus to insult me?' The man said no, but 'you know you have no money.' 'But I have this,' said I, leveling a pocket derringer at his head. . . . 'Oh! I beg your pardon, lady; I did not mean to insult you!' said he, cringing before the pistol and backing out the door."

In a California camp a woman named Eliza was in the saloon and hotel business with a man, whose name seems not to be known. When the place was closed for debt the man transferred it to the woman, who reopened it. When the sheriff came to arrest them for fraud the man shouted to the woman to shoot him, which she did her best to do but missed him. The man then fired on the sheriff, wounding him, and the sheriff's deputies fled, leaving the wounded man to fight it out with two assailants. When it was all over the man and the woman lay dead, and the sheriff, still alone, staggered off to the courthouse.

Neither women nor marriage was taken seriously by a lot of the men. O'Connor tells of a wedding ceremony performed in the Yukon by a French trapper who sang these lines: "We have no preacher and we have no ring, it makes no difference, it's all the same thing." The bridegroom is said to have responded with, "I'll love and protect her, this madam so frail, from those sourdough stiffs in the Koyukuk Trail!" McIlhany tells of a man in Marysville, with a wife, son, and daughter, who said one day to his friend the sheriff, a single man, "My wife thinks more of you than she does of me. I'll take the boy, you take my wife and daughter, and give me $500." When the wife raised no objections the sheriff accepted the offer.

Mrs. Mathews—to return to her—had a boardinghouse in Virginia City, Nevada, and her narrative of her troubles makes delightful reading.

A woman came to her because her wealthy husband had kicked her out. She was an alcoholic. He told Mrs. Mathews that if his wife remained with her six weeks and did not take a drink he would pay for her care; but when at the end of six weeks Mrs. Mathews asked for her pay the husband said he had changed his mind. He had thought it over and decided that he had had enough of her. The poor abandoned wife wept day and night and at last sued for money to support her "but the case was finally decided against the poor woman and she was left destitute."

And it seems that sometimes a wife was a liability because so many men coveted her. Marryat tells of

. . . Crockett, who had a very pretty wife; but the possession of this luxury, so far from humanising Crockett, appeared to keep him in a continual fever of irritation; for he was jealous, poor fellow, and used to worry himself because there was ever a dozen or two of hairy miners gazing in a bewildered manner at Mrs. C; but, if report speaks truly, the bonnet and boots of a "female" had been successfully exhibited in this region at a dollar a head (a glimpse of them being thought cheap even at that price), surely therefore Crockett might have excused the poor miners for regarding attentively the original article when presented gratis in the shape of a pretty woman.

The evidence suggests that most of the miners would have married just about any woman who was not disease-ridden or repulsively old; and quite a number of them, of course, married harlots. For celibacy, wrote the Reverend Colton, was a "misfortune, with which it seemed wicked to trifle. It is too selfish for pity and too selfish for mirth." Such high-falutin' philosophy would have abashed most of the men; what they wanted was a woman, and when a single "lady" arrived in camp she didn't last long, if marriage was her wish. Woods says that a woman "soon after her arrival here lost her husband. Before he had been dead a week, she received three proposals of marriage."

Sometimes a woman-hungry man took the law into his own hands. A famous case is that of William Douglas in Montana's Gallatin Valley, a rancher who had hired as housekeeper a young woman named Alice Earp. He pressed his suit so ardently that, perhaps afraid of him and repelled by him, she decided to go home. When he was not spying on her she managed to board the stage for Virginia City, but finding her gone Douglas set out on a fast horse and overtook the stage about noon. Jim Delaney was the driver that day, and Alice was sitting in the driver's seat with him. Douglas rode up and ordered Delaney to stop, which he did, and then told Alice to come down off the seat. She refused, saying that she was going home. Douglas said if she didn't return with him he would send her "home in a grave," and drew a gun. Again refusing she huddled against Delaney, whereupon Douglas shot her and she fell forward into the boot. At that moment Dave O'Brien and a companion rode up, unarmed, and sensing what had happened, O'Brien told Douglas to get off his horse and surrender. Asked if he was an officer, O'Brien is said to have replied, "Officer be damned! It doesn't take an officer to arrest a skunk like you. Get down." Douglas handed over his gun and surrendered. Alice died the next day and a crowd of angry men met the stage; court was in session, and at once Douglas was tried and sentenced to be hanged. "Douglas was hanged right at the jail door. . . . A rope was passed over the beam and a noose dangled about five feet from the ground. The other end of the rope was hidden in a high upright box in a corner of the stockade large enough to conceal the executioner. To the end of the rope in the box was attached a 250-pound weight. . . . Without a tremor" Douglas stepped to the little platform under the noose "and faced the crowd defiantly. He had refused to see a priest or a minister."

But not all the women-hungry men were so violent in their passions. Sometimes they tried to lure a mate with verse in the local newspaper:

There is Fowler and Fischer and Wallace
    of yore,
There's cows and there's chickens and
    many things more;

But none like your Perry that sells lager
    beer,
His tender heart is breaking each day
    in Pioneer.

So now, my dear Fanny, if you will
    incline
To join me in wedlock just drop me a
    line;

And great expectations with me you will
    share,
Not to mention the sour-krout and
    oceans of beer.

In such a manner wailed a lonely bachelor in Montana.

Now and then it was real love, or what passed for it, even among the Chinese. Drury tells of the hilarious case when "He Hawk, the Celestial Romeo" found that his ladylove had been kidnaped in Virginia City and hidden. Knowing that an open effort to recover her would meet with violence "he commenced operations secretly. His first object was to ascertain the exact whereabouts of his Que Nong." He learned that she was "in a hell pit in Chinatown," down under a trapdoor and an armed guard. It looked bad for He Hawk until a "friend opened a way by which a terrible legal vengeance could be wreaked on his rival, Ah Fon, . . . who claims to have bought and to own the woman." So a warrant was got out and a constable and deputy sailed forth to rescue the damsel; but while the officer was hustling the trembling lass to safety, Chinatown came to life and the "air was thick with stones, brickbats and other missiles," as well as gunfire. When at last He Hawk was ushered into the girl's presence, Que Nong's face became rosy with love and the two were immediately married. He Hawk was told that he had "authority of 'Melican law to protect his Que Nong" from Ah Fon and Hop Sing and the other villains who had hidden her deep in the hell pit.

Muriel Wolle dug up the story of Clifford Griffin and his Seven-Thirty mine in Colorado. Back East when the girl he was to marry was found dead, Griffin, brokenhearted, came to Colorado but the discovery of a rich mine did not assuage his grief. Though he became the wealthiest man in the camp he withdrew from the company of other men and spent his lonely hours playing his violin or chiseling out of solid rock a mausoleum. It was his habit, Wolle says, to play in the evenings for the other miners down below him; and one evening after playing extremely well he went to the tomb he had prepared, entered it, and shot himself. The miners found in his cabin a request to be buried there, and they not only buried him but set a granite shaft above him.

Apparently it is still there on the top of Columbia Mountain.

Another unusual love story is that of James S. Reynolds, publisher of the *Statesman* in Boise, whom we shall meet again in the chapter entitled "Feuds." According to Donaldson, who knew him, Reynolds was a widower when he came to Boise, but not long afterward he met a young woman and her mother while trying to find a teacher for the public school. He became enamored of the girl and against the advice of his friends married her. "The outcome was bad—too much mother-in-law and too much hidden past in the girl's case. After a few months of married life his wife ran away and took with her everything portable." Reynolds sold his newspaper and began a long search for his wife. When "I saw him in San Francisco in 1875, he described to me how he found the girl and her mother in an out-of-the-way place, one winter night. Certainly a tragic scene was enacted there. Jim considered himself fortunate in escaping with his life; two of his predecessors in the woman's affections were not as fortunate." Because of his wish to spare the feelings of the living, that cryptic account of it is all that Donaldson left to us, but the reader can imagine for himself what the euphemism, out-of-the-way place, stands for.

Aside from the few trueloves and false loves and devoted wives, there were, of course, the adventuresses, and in most of the camps, plenty of whores. Both of these will be presented in separate chapters. Of the soiled doves it is enough to say at this point that they were the only women whom most of the miners ever put their arms around. As wooers, most of the miners were as unprepossessing and as ineffectual as Peter Sapp, and with pokes as empty of gold dust. Reading advertisements in the newspapers Peter and his friends learned that wine, women and song "are supposed to make life palatial, and while nothing of an improper character is permitted we can furnish all three any night." What Peter and his friends wanted was a wife. He had heard that the "fancy girls at the bottom of the street were useful, but the whang-bang of their professionalism left a man restless for a real woman." In his diary a man named Rogers, who had left a wife behind and whose long winters were a "torture of morbid longing," recorded "dreams of his wife each night—a diary which stopped short the day his wife arrived."

Some of the harlots married miners and made good wives. Some laughed at marriage. A story is told of a miner in the Klondike who became so smitten with a dance-hall girl that he offered her weight in gold if she would marry him. Determined to get an extra five thousand dollars as a bonus for himself, her pimp attached a few pounds of lead to her underwear. The bridegroom was too drunk to wonder why he had so underestimated her weight, and too heartsick to care, after she fled with her pimp to the States.

Closely associated with their hunger for women was their hunger for mail. In northern Idaho when the mail stage arrived on its run between Lewiston and Florence, a man in the waiting multitude jerked his hat off and said, "Gentlemen, uncover in the presence of the United States mail!" It is said that every miner yanked his hat off and stood as solemnly as he might have stood before the flag. His reverence was probably for the letters from home, or for the "gallant expressman," who, Bret Harte wrote, "knew everybody's Christian name along the route, who rained letters, bundles and newspapers from the top of the stage, whose legs frequently appeared in frightful proximity to the wheels, who got on and off while we were going at full speed, whose gallantry, energy and superior knowledge of travel crushed all us other passengers to envious silence." A different view of the United States mail was taken by the *Alta California*, July 13, 1855:

The present Post Office system is the most outrageous tyranny ever imposed upon a free people. It forbids us from sending letters by such means of conveyance as we may prefer without paying an odious and onerous tax to the government. A private individual cannot carry letters, because it would interfere with the government monopoly, and so the Post Office charge must be paid, whether the service is rendered by it or not. . . . the Post Office system, so far as California is concerned, is a humbug and a nuisance.

There were of course private carriers who solicited subscribers; in California they charged $1.50 per letter in summer, $2.00 in winter. One of them wrote in his memoirs, "There was at this time no means of weighing gold dust; there were no scales to be had, until finally we got a thimble.

It would hold just $4 worth of gold dust." These private carriers were a hardy lot:

That intrepid and untiring mountain express-man, Capt. Singer, has been perambulating the mountains during the recent snow storm. He was of course obliged to traverse huge drifts of snow, before which the heart of any other man would have sunk in despair. But the Captain never surrenders. His last trip used up three pairs of boots, besides innumerable snowshoes. He deserves great credit for his untiring efforts to furnish the mountain boys with news from the regions of mud below.

Harlow tells of one Enos Christman "whose luck as a miner was uniformly bad." With only seven dollars to his name he asked for his letters, found he had four, for which he owed $8, and was given credit for a dollar. One hopes the letters were worth that much. How many letter expresses there were seems not to be known. Berthold and Wiltsee as a hobby collected the names of 446, to which Harlow's research added a hundred; and just as his book went to press he heard that W. W. Phillips had 775! There was a great deal of rivalry between the large companies. In 1853 on the occasion of a presidential message to the Congress, Wells, Fargo and Adams "Both sought the best horses to be found in that section of the country, sometimes agreeing to pay $100 for the use of a particularly fast horse for no more than four or five miles—for the relays were no farther apart than that."

Knower found that homesickness among the miners was the worst sickness of all. Some of them "were married men, separated a great distance from their wives and children. . . . I can recall several instances where I have known them to lie down and die from despair." If that was true, the miners who pined away and died must have been Peter Sapps who simply found it impossible to save enough gold to take them back home. The Reverend Daniel Woods wrote in his journal, "This is a day to be remembered. *Letters from home!* If any one would learn the full significance of these words, let him pass ten months in California without one word from his loved ones, . . . " Having had no mail for seven months he was overjoyed to learn that eight miles across the mountains were three letters for him. Though "suffering greatly from blistered feet" he made the long difficult climb, only to find three letters from politicians soliciting his vote!

Henry Page, in a letter to his wife, expressed what most of them felt for the mail carrier:

But with what heartfelt solicitude is he expected! Without him, the miner would be shut out from the world, and next to the trader, who furnishes the means of sustaining life, the Express man is of the most importance. His approach is sometimes heralded by a wandering miner, when the news spreads along the banks and bars for miles. The Express will be here in three days, in two days, to-morrow. Little else is thought of. "Hope he will bring a letter for me—and me—and me, too," goes from mouth to mouth, and then follows the oft repeated tale, "I haven't had a letter from home in—months." "I'll give five dollars for one." "Yes, ten. . . ."

Shaw describes a typical scene:

As the lengthening columns swayed and wriggled sometimes a half-mile in length, great anxiety and impatience were often manifested by persons wishing to get to the all-important window of the postoffice. Rugged miners who had not perhaps for a year heard a word from home, and anxious merchants whose fate depended upon their letters and invoices, seeing no hope of approaching the office for hours, would offer liberal sums to buy out some fortunate one in the line. From five to twenty dollars were average prices, but fifty and one hundred dollars were often paid for a good position near the window. The expression of countenance of those paying highest rates when forced to leave the window without a letter, was a study beyond description.

We can easily imagine that it was. Absence doesn't always make the heart grow fonder, but possibly extremely high postage more than forgetfulness caused a lot of heartache, though it appears that on some lines postage was collected on the receiving end. Zamonski and Keller tell of some tricks the miners used to get out of paying it. One trick was to return a letter and say it was not for him and he was sorry he had opened it by mistake. Another was to ask the clerk to open a letter and read a few lines, and then say, craftily, "I can't be sure, read on a little." After feeling that he had the gist of it the miner would say, "Nope, it ain't mine." We can hardly blame the men who felt that some letters were not worth five dollars!

The Weather Almost the only camps devastated by floods were in California. Such streams as Dear Creek, swollen from melting snows and heavy rains, sometimes roared down from the mountains, carrying on their turgid breasts uprooted trees, miners' shacks, and at last the foundations of theaters, boardinghouses, and saloons; but Nevada City, like all other camps razed by flood or fire, rebuilt with astonishing speed, and where it had fifty saloons it then had seventy-nine, and more brothel cribs than a man could count.

Sacramento, of the camps in California, suffered most, from both flood and fire. Richardson saw schooners sailing "through the principal streets. A friend assured me that one night, upon returning home in a boat, he found a cow in his drawing-room, and tied her to the hall banister, lest the flood should take her up stairs before morning." Delano looked at the flooded city from Table Mountain, "where I could command an extensive view of the valley, . . . [and] estimated that one-third of the land was overflowed. Hundreds of cattle, horses and mules were drowned . . . and Sacramento City . . . was almost entirely submerged. A small steamboat actually run up its principal streets. . . . The number of dead carcasses of animals, which floated down and lodged as the waters retired, produced a most loathsome effluvia, . . ."

Cold was a more deadly enemy, not in California camps but in those high on the mountains in Colorado, Idaho, Montana, British Columbia, and Alaska. As Hogg points out, cold made men quarrel violently on provocation that would have been ignored under other conditions. A man whose eyes swelled until they looked like the eyes of toads; who, sweating, paused to rest and found the sweat frozen in a few moments and his garments frozen to his hide; or who, sick with dysentery, took his trousers and underwear down to void and after a few minutes was so nearly frozen that he found it almost impossible to stand up—such a man was ready to murder for no reason at all.

That was up in the Yukon, where if a fire was allowed to die during the night everything froze solid, except the living occupants. Huge icicles, called glaciers, formed on the ceiling and around the walls; and on waking in the morning a man was not surprised to learn that the cold had made icicles of his breath across the blanket under his chin. Farther south the winter of 1888 was a stinger long remembered. As Dick d'Easum tells it, "Horses froze solid standing up." Lafe Griffin of Boise went out to find his livestock and "huddled over a sagebrush fire all night. His horse froze to death. A greyhound across the river froze in a strawstack." It reached fifty-four below in some Idaho areas, and probably was colder than that in parts of Montana and Canada.

When Alex Toponce was panning gold in Colorado he fell into a creek one bitter night, and before he could reach a cabin and a fire his legs froze till they "felt like sticks of wood." Two companions in the cabin "got my boots off by ripping the outside seam of the boots clear down to the soles. We had a tub that had been made by sawing a whiskey barrel in two." The tub they filled with snow, poured two pails of water over it, and forced Toponce to sit with snow banked around his legs up to the knees. "My legs were frozen solid. . . . When the melting snow began to draw the frost out of my legs . . . the pain was so great that I could hardly stand it." But the two men held him and poured whisky down him. "I could not keep from yelling with pain. The frost as it was drawn out formed a film of ice around my legs as thick as a pane of glass." Toponce tells of a Montana winter when they were unable to save their cattle and had to shoot them, "as their horns would freeze and burst off from the pith; their necks would freeze and their legs, so that they couldn't move at all." Herds of buffalo froze to death. "In the spring you could walk for miles on buffalo skeletons, your feet never touching the ground." His mules were fed on cottonwood bark and *buffalo flesh*.

Other natural enemies were snow, winds, and fire. On the Rocky Bar trail in Idaho, and across the Sierras, there were trees blazed, to mark the maximum snow depth, almost forty feet above the earth. But snow was friendly enough when it did not isolate camps and reduce them to famine. In Idaho City they formed the Sliding Club and challenged all rivals: "we shall lay back in our chair, and melt our frozen ink, and be in readiness to make a record of as many broken heads, legs and arms as possible."

Winds were sometimes as deadly as cold. Mark Twain's description of a Washoe zephyr became famous: "a soaring dust-drift about the size of the United States set up edgewise. . . . a Washoe wind is by no means a trifling matter. It blows flimsy houses down, lifts shingle roofs occasionally, rolls up tin ones like sheet music, now and then blows a stage-coach over and spills the passengers; and tradition says the reason there are so many bald people there [Virginia City and Gold Hill] is, that the wind blows the hair off their heads while they are looking skyward after their hats."

It was the way of mining camps to exaggerate both the good and the bad. The Washoe zephyrs that poured in howling furies down over Mount Davidson and into Virginia City were given all that was coming to them in the camp's newspaper and in the tales that the miners wrote to the folks back home. These gentle zephyrs were said to pile hats and drunks in huge drifts all the way down Six-Mile Canyon, and even to pick up mules and carry them wildly braying over the hills and out of sight. A man called a man a liar on B Street, his words were swept downhill to C Street, where a man who heard them knocked down a man standing a few feet from him and bit his nose off. A man in a C Street saloon wondered aloud how Julia Bulette was, in her crib a street below, and Julia, clinging to a window ledge, with her body straight out in the wind, shouted up the hill, "Why don't you come help me?" Up in Idaho (possibly it was the same zephyr) a man arose one morning to find all his chickens blown against a barn wall and held there. To measure the velocity of winds he secured a logchain to the top of a fifty-foot pole set in the earth. When the wind blew the chain out parallel with the earth it was a pleasant breeze, all right; when it began popping links off it was a fair gale. When the chain was torn loose and sailed away like a kite a man had to get busy and tie his house down.

Fire   The story was told of an Eastern greenhorn who took lodgings in a bonanza hotel and after looking things over asked what the tenants could do in case of fire. Without even looking at him the clerk said, "Jump out your window and turn left." That was about all a person could do in most of the camps, though a few of them had firefighting equipment. In Arizona and New Mexico mud-bricks were used but up north practically all the camps were built of wood, and some of them were burned to the ground

again and again. Sometimes a fire started in a brothel because of fights there—in Dawson when a soiled dove hurled a lighted kerosene lamp at another soiled dove she started a conflagration that destroyed a good part of the town. Sometimes they were started by criminals so that in the confusion and panic they could loot the buildings: the first number of the San Francisco *Herald,* June 1, 1850, warned its readers that gangsters would burn the town if they could. Bancroft says that "Two or three attempts to fire the town were sometimes made in a single night."

The Boise *News* after pointing out that "our city [Idaho City] is built almost entirely of pine" said that a fire would leave nothing but ashes. The editor then told the story of a friend who in California had

accumulated quite a fortune, and placed it all in a store full of goods. There were two reigning belles in the city where he lived; everybody acknowledged their superiority over all the rest of the ladies in town. Our friend had retired to his neatly fitted room, in his own building, up stairs, over his store, and was meditating and cogitating over things mercantile and matrimonial, and arguing with himself which of the two ladies he would marry on the following morning, for such was his wealth and standing in society that he knew he had but to ask either, and she would answer "yes," when the cry of fire reached his ears, and before he had time to dress himself his store was in flames, and in half an hour he was not worth a dollar.

No other camp suffered more disaster from fire than Idaho City: all around it, up and down the ravine at night there were pine torch fires to give light to the toiling miners. A black headline in the *World* said, IDAHO CITY AGAIN IN ASHES. Practically the whole camp was burned that time, the loss being estimated at a million dollars. The editor had just completed an essay on the second anniversary of the great fire of May 18, 1865, when the whole camp was destroyed, when he heard a cry of fire, and a few hours later looked on the ashes of a blackened camp.

Of the 1865 fire he had written: "A great fire Thursday night laid nearly the entire town in ashes, sweeping away the fortunes of very many of its business men, and leaving innumerable families homeless and destitute. . . . A narrow, scattered remnant of buildings on the outskirts of the town alone remain to mark the boundaries of

Idaho City." The fire when first seen was "massive sheets of flame from the roof of a dance house." There was a wind from the south—one of the Washoe zephyrs, no doubt—and in a

few minutes the adjoining buildings, including the Sheriff's and, Probate Judge's office, Wells, Fargo & Co.'s Express Office, the Idaho Saloon, . . . and Totman's saloon was also on fire. . . . It swept furiously up the range of buildings on Main and Montgomery streets, carrying every building before it. Taylor's Exchange, the City Hotel, Magnolia Hall, Harris's drugstore, the Umatilla market, followed in rapid succession. . . . as the flames swept through some of the two-story buildings they seemed to gather strength and unquenchable ferocity. The east side of Main street was next on fire. Mix's drugstore, Robertson & Co.'s Jewelry store, Charlton & Langworthy's brokers' office, Fred Bell's saloon, Lauren's saloon, and other's on the east side and on the west side the Washoe saloon, Prior's saloon, Peff's bakery, Gans and Bros bowling saloon. . . . The Forrest Theater, opened for the second night only, and which at the beginning of the fire was crowded . . . to witness the Potter troupe in Shakespeare's "Romeo and Juliet"—was soon floating in flaming brands through the sky.

The wind changed, and the fire, having destroyed nearly all of the northern part of the camp, "now seemed to part and turn in two fiery circles, to the right and left." One part of it

came furiously down High Street. Here a great many new buildings had recently been put up for residences. . . . These soon disappeared. . . . the whole town presented a fearful sight. The forked tongues of fire seemed to lick the very clouds; dense volumes of black smoke rolled in streams down the Basin; the whole city from bank to bank was a sea of boiling flames. [The Catholic Church, standing high on a hill, was saved.] To sum the whole up, the town from the Jenny Lind Theatre on the north, to Powell's on the south, has entirely disappeared. . . . The origin of the fire is not known, but it is generally supposed to be the work of incendiaries. Several attempts were made during the week to fire various portions of the town.

A week later the editor said that attempts to fire the town "were made on four occasions during the week previous to the conflagration" by persons consumed by "love of plunder." From fire the editor's attention was now diverted to three cold-blooded murders.

Charles Teeter had a business in the city and was one of those who saw the fire and wrote about it:

An artist's notion of the fire in Marysville, California, 1851

Men that were worth thousands when they opened their eyes that morning were literally ruined before they closed them again. Yes, fortunes that men had toiled for through long years were swept away in an hour. There was no insurance on the property. . . . The fire broke out about nine o'clock in the evening near the central part of town, and as the city was compactly built out of pine lumber, thoroughly seasoned and full of pitch it burned like a torch, spreading in every direction; it crossed street after street with such irresistible force that nothing could stay its progress.

Such a fire as it was few men ever saw; people that were six miles away at the time declared that it so lighted up the Basin that they could see to read common print without difficulty, although the night was dark. As soon as the fire started, George Dwight hastened up town and entered Mr. Craft's store, and was accosted by that gentleman in this wise, says he, "I have got use for you, Mr. Dwight." "All right," says George, "I am at your service; what would you have me do?" "I want you," says Mr. Craft, stepping to the safe and unlocking it, "to take charge of this dust until morning." The safe door was then thrown open disclosing over 100 pounds of gold dust, nicely laid away in buckskin purses, with owner's name stamped on each purse. . . . The contents were then shifted into a sack, which being securely tied, was shouldered by George Dwight, who without a word disappeared in the darkness and was seen no more that night. . . . A few minutes later his store and its whole contents were enveloped in an ocean of flame. . . . every building was swept away except a very few around the outskirts of the place. A bed of hot ashes and cinders covered the ground except in the middle of the streets. Over this burned district could be seen hundreds of men wandering hither and thither, evidently trying to locate the particular spot that they had so recently occupied in the transaction of their business. . . .

. . . every miner for miles around knew just what had happened and so rushed to town as fast as their legs could carry them, many of them getting there in time to carry out what they could pack on their back. As the fire spread in every direction . . . the crowd kept just in advance of it, and as fast as the proprietors were forced to abandon their stores by the intolerable heat, the crowd would rush in, grab whatever they could, and get out with it in double quick time. To deposit these goods in the center of the street was useless, as they very well knew, for the streets being narrow, the flames extended across them in sheets . . . and so these men would continue their march, loaded down as they were, until the city was left behind and their cabins reached. Dozens of men passed by our store during the fire. . . . Some would have a sack or two of flour, some a side or two of bacon, some a five gallon keg of molasses. . . .

Borthwick was in Sonora when it burned, and left a description of the fire. It began about one in the morning, and a half hour later the camp was "utterly annihilated." He had hoped to help save the furnishings where he lodged but found the landlady "*en deshabille,* walking frantically up and down, and putting her hand to her head as though she meant to tear all her hair out by the roots." A waiter, no less out of his wits, was "chattering fiercely." Nobody in that house tried to save anything. The flames spread so fast that one man hadn't time to snatch his gold watch from under his pillow, and many barely escaped with their lives. It was thought to save a part of the camp by tearing down a row of houses and exposing bare earth: that men could have imagined they had time for such a project indicates that they were all beside themselves. In the distance, as was usually the case when a camp burned, a multitude stood or sat, naked or half-naked, amidst piles of goods, watching in the brilliant night as rows of buildings vanished with the speed of sheets of paper in a bonfire. Then darkness fell and there was only the hot odor of it. At daylight they saw that the "whole city of Sonora had been removed from the face of the earth. The ground on which it had stood, now white with ashes, was covered with still smouldering fragments, and the only objects left standing were three large safes belonging to different banking and express companies, with a small remnant of the walls of an adobe house." Prowling in the ashes were "unprincipled vagabonds" staking claims where houses had stood, knowing that under some of the buildings there were pockets of gold.

On April 15, 1856, *Richelieu* was playing in Hangtown's new theater when the cry of fire emptied it. A young Englishman who was present said that in the camp the "flames boomed along like a sea in a tempest; the terrified inhabitants running two and fro to save their property or their lives. The whole town, a mile in length, was consumed. A few fireproof safes stood out in relief, as dark monuments, and some men and many animals burnt to a cinder." That fire even consumed the equipment of the fire company! Georgetown, one of the lustiest of the California camps, almost melted and disappeared during the heavy rains of the spring of 1849, when miners could neither work nor sleep because of the kind of incessant downpour that nearly destroyed Lewis and Clark and their men on the mouth of the Columbia. After its long bath of flood and mud it rebuilt itself with stronger timbers, and watched it all vanish in flames in the summer of 1852.

The winter of 1848-49 was one of the severest in California's history, with snow thirty feet deep in the mountains. When the spring rains came, over a hundred inches all told, and sudden warm weather melted the snows, the valley was a sea over most of the distance from Marysville to San Francisco Bay. Thousands of tons of merchandise freighted to the mines went gaily off in the torrents of snow waters and mud, as the Yuba, the Bear, the American and nearly every other river in the area overflowed their banks and joined their waters. Some men took off in the flood on makeshift rafts, hoping to overtake their property; and the few with two-story structures carried their belonging to the upstairs and a day or two later took off in rowboats from the highest windows. The adobe buildings melted and flowed away; the frame buildings became rafts for beasts and people. The brand-new city of Sacramento, gateway to the mines, went almost completely under, or off in the floods. By April, after the water levels had fallen many feet, its streets were avenues of deep mud, where forlorn residents waded to their knees to see if anything had been left. As if such disaster were not enough to rebuke the lust for gold, cholera struck next; then other floods came, other diseases, and within two years the city was devastated by fire five times. In that of November 2, 1852,

Twenty-two hundred buildings, [says Helper] with other property, valued at ten millions of dollars, were completely reduced to ashes. The wind was blowing very hard at the time the fire commenced, and the roaring of the flames, the rapidity with which they spread, the explosions of gunpowder, as house after house was blown up, formed a scene rarely excelled in terrific grandeur. Men, women and children ran to and fro in the greatest confusion, excited almost to frenzy, in the effort to save their lives and effects. Within six hours after the fire first broke out, more than nine-tenths of the city were swept into oblivion, and the people were left to sleep on the naked earth without any shelter but the clothing they had on.

But the will to live, and to live there, would not

Creede, Colorado, after destruction by fire

be put down. Sacramento was tough. It built and burned and built again, and by 1854 was the capital of the new state. That was less than seven years after Jim Marshall ran babbling to Sutter with gold in his hands.

San Francisco, as everyone knows, has had its share of terrible fires. They began almost in the moment of its founding. Its first, on Christmas Eve in 1849, consumed only about fifty homes, the ramshackle buildings in its path being pulled down. On May 4, 1850, three hundred houses went up in flames. It was believed that this fire was set by criminals, who then looted shops. Water at this time sold for sixty dollars a cartload. In June, a month later, a third fire did damage in the amount of three million dollars; and in September there was a fourth fire. It was the fires set by criminals that were one big reason for the rise of the vigilantes.

Perhaps no camp had more spectacular fires than Virginia City, Nevada. On the Comstock so many miles of tunnels ran under the city that it was not unusual when a segment of the town fell into darkness and disappeared. In the summer of 1863 the whole mountain shuddered in the depths of its tunnels, and then roared, as a part of it collapsed into the Mexican mine. The whole structure "to a depth of 225 feet dropped with a force which swept rock and timbers like straw into the adjoining galleries of the Ophir," a part of which was obliterated. "Later a great collapse at Gold Hill wrecked the upper levels of the Imperial, the Empire and the Eclipse mines. Earth, rock and timbers weighing hundreds of thousands of tons settled into the depths with one mighty crash." When the camp was not collapsing into its cavernous belly it was burning.

A drunk lighting his pipe knocked a kerosene lamp over in a lodging house and started a fire that burned a hundred buildings in an hour, and another thousand in the next hour or two. When stores of powder began exploding, terrified people seized a few possessions and fled over the ridge to Gold Hill, a mile distant. Stores of exploding dynamite shook not only the camp but the whole mountain behind it and the miles of tunnels in its depths. Residences, brothels, saloons, hotels, business houses—all were made of wood and all went up in flames. The Ophir mine had a pile of iron wheels in its yard, and so intense was the heat that the wheels were fused into a solid mass.

Parts of the fire fell into mine shafts, and timbers deep down began to burn. The panic and frenzy were so extreme and uncontrolled that one writer has thought that "Probably no such frantic pandemonium ever accompanied the burning of a city." When the last ember had died more than two thousand of three thousand buildings were ashes, and two-thirds of the people in the bonanza city were homeless; but here, as in other camps, the people pitched in with amazing energy and fortitude to rebuild, even though right after the fire they "wandered through the debris of Virginia City . . . with such a look on their faces as men and women wear when they gather around the coffin to look upon one who in life was very dear, but who is gone forever." Drury saw the camp destroyed in the fire of October 26, 1875. He saw the *Territorial Enterprise* building "utterly destroyed" but he saw the town come back strong, with the International Hotel, "sumptuously furnished with mahogany furniture, ceiling-high mirrors and magnificent chandeliers," becoming the center of the new life, not a brick or board of its six stories standing today.

The most horrible of all the mining camp fires was possibly that in the Yellow Jacket mine on the Comstock, in April, 1869. It opened into the Crown Point on one side and into the Kentuck on the other. In the miles of tunnels under the camp there were millions of feet of timber, a thousand feet deep in the main shafts, and in all the drifts and crosscuts. In the vast ore chambers that had been stoped out the huge beams were set on one another to great heights, with heavy plank floors in these immense structures at every six foot level. All this plus the gases made a perfect setup for a fire. When smoke and gases began to boil up out of the shafts, and men coming up in the cages were found to be unconscious, the wives and mothers were lost in wild frenzies. In those awful depths the cages were overloaded and men clung to them all the way around like huge insects; and other men running wildly in the roaring heat were falling into shafts and disappearing. Those at the surface were able to raise and lower the cages only two or three times. Then the wives and children became so mad with fear and grief that they had to be seized and held, back from the shaft mouths.

A lighted lantern was set on the floor of a cage, with a message to those below: "We are fast subduing the fire. It is death to attempt to come up from where you are. We shall get you out soon. . . . Write a word to us and send it up on the cage, and let us know where you are." As if it made any difference! The cage was stopped at various levels, on its descent eleven hundred feet, and on its returning, but there was no message and the lantern was out. After five hours a cage descended to a certain level and every man but one known to have been working there was found dead. Only in wives and mothers was there hope that any might survive, though firemen and miners from above still descended and fought in the depths. It is impossible to conceive their heroic struggle in the infernos of smoke and gas and boiling waters. Clothes were burned off some of them, and flesh off fingers to the bones. By nine in the evening the fires were again rising; and by two in the morning, nineteen hours after discovery of the fire, only thirteen bodies had been brought out. When, after three days, the fires below seemed to rage with greater fury, shafts were covered over with planking, blankets, and wet earth. When, after two more days, the shafts were opened, fires began to rage anew, and so the shafts were again sealed. Ten days after the fire first began forty-one bodies had been recovered, and the camp's hospitals were filled with men who had been gassed, burned, and maimed. After thirty-nine days a few more bodies were brought up but fresh air rekindled the fires; some of the drifts were then sealed and abandoned, and when opened three years later some of the stones in them were found to be red hot.

Tombstone's first disastrous fire is said to have been caused by a barrel of whisky whose quality offended the owner of an Allen Street saloon. Resolved to turn the whisky back to the wholesaler, he tried to determine how much remained in the barrel, but on thrusting a gauge in to measure it he lost the gauge. A barkeep with a lighted cigar in his mouth came over to offer advice and help. In a few moments the saloon was on fire.

Disease During one of the overland journeys a man suffered terribly with an aching molar, and because there was no dentist in the camp a man named Ferguson, who tells the story, volunteered to pull the tooth. "So I got the captain seated on the ground, with his head between my knees, got out my lance (jackknife) and commenced chopping and digging away around the gum of the tooth. The women all ran away as soon as I commenced to mutilate the patient's mouth." For the extraction Ferguson used some kind of crude tool, possibly a pair of pincers or pliers, that he got "hooked on at last." With a final wrench and jerk, he fetched the tooth out, and the patient, after spitting out a "few mouthfuls of blood," pronounced it a fine job.

That was about par for surgery on the frontier, and "doctoring" for illness was not even that good, though a surprising number of the pioneers carried around their own materia medica. Because of malnutrition, especially the lack of vitamin C, scurvy was one of the most dreadful of the diseases. Buffum says that at least half the miners known to him came down with it: "Many, who could obtain no vegetables, or vegetable acids, lingered out a miserable existence and died. . . . I noticed its first attack upon myself by swelling and bleeding of the gums, which was followed by a swelling of both legs below the knee, . . ." Some of those afflicted with scurvy put their faith in sour pickles as a remedy, but these were very expensive, costing in the California camps as much as eight dollars a pint bottle. Emrich says that "The only, yet most certain cure, for scurvy was a diet of raw potatoes soaked in vinegar." Shaw, on the other hand, put his faith in onions, and paid as much as a dollar for a small one.

Cholera, or what they called cholera, was a common sickness on the overland journeys and in the camps. Page says that in Sacramento it struck so hard that gambling dens were closed, streets were empty, and all business came to a standstill. A kind of hospital was set up at last but most of the able-bodied men were busy carrying away the dead. Bancroft's estimate is that five hundred, including seventeen doctors, died in that camp, and that thousands fled. "Uncounted numbers died on the roads outside the town, and throughout the state there must have been a loss of five thousand more, for in Hangtown alone one authority says that the mortality was seven hundred men."

The chief symptoms of this dreaded disease were intense pain, vomiting, purging, cramps, and cold chills. After a while the vomiting ceased, but in a few minutes, suddenly and without warning, the sick man was gasping for breath, his eyes bugged out, his tongue protruding, and in no time at all he was dead. Delano had what he took to be cholera: "I felt strangely. Was there a change in the weather? I could not get warm. I piled on more clothes. I felt as if I was in an ice-house," with chills creeping along his back, his legs drawn up, his head under the bedding, and his body shivering all over. He had intense thirst, and a wish to have ice water run continuously down his throat. Then he grew steadily warmer, until he seemed to be living fire; and all the while he gulped cold water as sweat burst out all over him.

There were innumerable sad scenes, as persons died far from home, their deepest wish being to see once again their loved ones. Colton says,

As I was passing this morning one of the little huts sprinkled around the skirts of Monterey, my steps were arrested by the low moans which issued from its narrow door. On entering, I found on a straw pallet a mother whom disease had wasted to a mere shadow, but whose sufferings were now nearly over. She did not notice my entrance, or anything around; her eyes were lifted, fixed, and glassed in death. A slight motion drew my attention to another corner of the hut, where I discovered, in the dim twilight of the place, a little boy lying on a mat, whom I supposed asleep; his young sister was near him, and trying to cross his hands on his breast. She did not seem to notice me, spake not a word, but went on with her baffled task, for the hand which she had adjusted would roll off while she was attempting to recover the other. Now and then she stopped for a moment and kissed the lips which could return none, while her tears fell silently on the face of her dead brother.

In Idaho the *Statesman* was moved to anger:

A scene was witnessed on the streets of Idaho that will long be remembered by all that have one drop of human sympathy in their composition. The county is bankrupt and the County Commissioners being unable to make further provisions for the indigent sick, the occupants of the hospital were this morning thrown out of doors upon the mercy of the public. It was a sickening sight: no home, no friends, no bed to sleep in, no place to go to stay hunger or shelter them from the cold.

According to the editor the county was bankrupt because of dishonest officials:

For shame! Oh! for shame! . . . when it is remembered how many tens of thousands of dollars have been and are still being wantonly squandered. . . . Through one subterfuge or another the different county officers have contrived to gob-

ble all the money that has been collected. . . . the wholesale plunder. . . .

In many cases homesickness undoubtedly hastened death. Ferguson wrote in his journal, "you know him, he sickens and dies, and no one knows whence he came. His friends never get tidings of his fate, and not unlikely an aged mother is looking for his return even unto this day." Because of malnutrition, insufficient bedding and clothing, shacks, dugouts, and lean-tos for shelter, and a longing to be back home, thousands perished. Thousands like Peter Sapp lay desperately ill, back in a dank dark dugout, with only a clumsy homesick friend to nurse them, and lived, but were never as strong again. Delano tells of a fiddler who fell ill but aroused himself sufficiently to "draw the bow to 'Auld Lang Syne,' . . . with tears trickling down his care-worn face," for neither wife nor children would ever see him again. "He gradually grew worse and died the last of January, and was buried on the hill-side above the cabin, leaving, as the only memento of his former existence, his violin, which long hung against the rude wall, and a half-written letter to his wife, which she probably never received. Many cases equally as melancholy came to my notice."

Pests   Rats were a problem on all the frontiers of the West, those intelligent, bold, and acquisitive creatures that invaded camps and made off with everything they could carry or drag, including countless items for which they had no use at all. In San Francisco they were unloaded from ships from all over the world, and spread north, east, and south. The mining camps drew mice and rats by the thousands from miles around, against whose ravening hunger only food in tin pails or cans was secure. Cats became the favorite animal next to the burro. One freighter, on his way to Bannack, Montana, saw on a farm near Denver a tabby with seven kittens and bought all but one kitten for $2.50. He hauled them hundreds of miles and sold them in Bannack for more than forty times what they cost him. Mice of various shapes and sizes swarmed everywhere, and were such a scourge in the wooden buildings, to which they had easy entrance, that mice droppings in the food was a commonplace of life. We lived in early life in log shacks and saw the droppings in flour, oatmeal, beans, rice, and in all food not protected by glass or tin. The miners must literally have eaten tons of it, and the odor of mice and rats in the shacks and shanties must have been the principal odor.

Wasps, which many of the miners called yellow jackets, had their nests hanging everywhere in thickets along streams, and many an unwary miner who inadvertently struck a nest was suddenly covered over by the furious creatures. He was then a sad sight for a while, with forty or fifty swollen and infected stings over his arms and face. Marryat says swarms of infuriated wasps destroyed everything they attacked. As for fleas, he says they not only abounded in "the skins of every beast you kill, but even live on the ground, like little herds of wild cattle; and ants are of all shapes and sizes, and stand up savagely on their hind legs and open their mouths, if you only look at them. The wasps attack any meat that may be hanging up, and commence at once cutting out small pieces, which they carry home, and it is astonishing the quantity they will carry away with them."

It was fleas in the warmer climate; in the Klondike and some other areas it was huge mosquitoes in such dense swarms that often a man could not see the sights of his gun for the hundreds of mosquitoes standing along the barrel. It was sand flies and deer flies and the huge black horsefly, the probe of which is as sharp and painful as the stab of a large needle—the only bloodsucker that put horses in absolute frenzy. In Arizona, New Mexico, and southern California it was the scorpion and the serpent, and the deadly spines of desert plants; and throughout the Rocky Mountains it was the rattlesnake and poison oak, and the wolf and the grizzly, which we shall look at under sports. But compared to many areas of the world the American West was and is relatively free of both poisonous animals and plants.

Union vs. Rebel   It may strike a reader as strange to put the Civil War among camp problems but it was there in amazing strength and bitterness and led to some of the most brutal killings.* Just before the war, and during it and for years afterward, some of the camps were divided into two hostile and implaca-

ble factions—the Democrats (Rebels) and Republicans (Unionists). The bitterness was especially virulent in Idaho City and Boise City. Thomas Donaldson, who was in the thick of it, says "It would have done your heart good [if it happened to be a Republican heart] to have seen the Rebels' faces when the news came of the election of General Grant to the Presidency, in 1868. Consternation does not express it." When certain Rebels passed him before the election, he says they hissed at him under their breath, "Old villain! Old radical humbug! Somebody ought to do him! . . ." But the day news came of Grant's election

I walked down Main Street and I felt good! On all sides the old Rebels saluted me! "Howdy do, Gov.! Why, howdy!" An hour before they would have cut my throat, had they dared. They were an abject lot, abject but treacherous. My, but I did a sight of secret but inward jollifying that week!

. . . The humor of the thing began when the Rebels, along in the fall of 1868 and spring of 1869, started to flee the country. They fled in the night on stages, in wagons, or on horseback, and some, I fear, on foot. Men, women, and children were in a mad scramble. Arizona caught some, Nevada more, and Utah, unfortunately for her, caught quite a dump of them. Their departure was a godsend to Idaho.

On Alder Gulch, in Montana, the first boomtown was called Nevada City; the second was named Varina, for the wife of Jefferson Davis, but a judge who despised the rebels saw the name and turned purple, and said that by the living God the name would never shame his records. He changed it to Virginia City, and Virginia City it was.

But in no camp was the bitterness more extreme or prolonged than in Idaho City, and no others editors were quite as virulent in their hatred of the Union as those of the *World*. On December 24, 1864, it reprinted from another source an article which began:

The re-election of Lincoln would add four years more to the reign of social immorality and vice, by which private morals and virtue have nearly perished from our country. The depravity of manners, the scandalous indeceny and the obscenity of Lincoln's own daily conversation, seems to have fallen like a fatal epidemic upon the people. Never before did vice of every description appear

* See Chapter 26, "Feuds."

with such front and shamelessness in all our citizens and towns. Washington, that under all
other administrations, was the seat of manners,
of refinement, and of social culture, has, under the
present administration, become a den of vulgarity,
indecency and vice. [Lincoln's papers, said the
editor] are remarkable documents, and years
hence will be looked upon with wondering awe,
and many secret thanksgivings by readers that
they did not happen to live in an age so dark
and barbarous.

Lincoln, he told his readers, was not a great man,
and did not have on the people's affections the
hold of Clay and Jackson. A month later he was
reprinting from the Paterson, New Jersey, *Register* the statement that Lincoln's election was a
fluke, and that the people "cannot now escape the
odium of having put a mere joker, an uncouth,
uncultivated flatboatman, a buffoon, into the seat
which Washington honored."

As Plautus said, *Lupus est homo homini.*

Race Prejudice

In most and possibly in all the mining camps of the West there was in "whites" hostility to the South Americans, Mexicans, Chinese, Indians, and Negroes. The California legislature in May, 1850, imposed a license of twenty dollars a month on all foreign miners; this forced thousands of them to abandon their claims and seek employment or return to their native lands. When the foreigners in Sonora decided to resist the tax they came close to being lynched. There and elsewhere most of them left the areas, and many of them, driven by thoughts of vengeance, took to robbery and murder. It was out of such a situation that the Murieta legend developed.

When the miners heard of new discoveries and rushed to new areas most of the Chinese moved with them. Sometimes they walked for hundreds of miles, as from the Comstock to Idaho. Hearing that they were on the way the *World* wondered what was to be done with them: "Had none of this race ever been permitted to settle like vermin on the mineral lands of California, the mines of that State would have still been open to the industry of thousands of now broken, penniless men." Irishman O'Meara then quoted from the *Avalanche*: "If their family cognomen only contains something like a Big O with a fly-speck over the right upper corner; or should it contain a vague indication of having originated in medieval times in the land where lager beer is the most popular beverage; or in any way convey the impression of being a lineal descendant of some of those ancients who were forbidden the use of pork. . . ."* To which O'Meara, the editor with the O and the flyspeck, retorted in a fury:

The editor of that paper was a Know Nothing years ago, and hates adopted citizens and foreigners with hearty hatred yet. The above is simply a fresh running of his old sore. . . . Yet he hankers after patronage from the Irish, the Germans and the Jews, whom he so heartily despises and wantonly insults. . . . he is for dollars; no matter how basely or disreputably he gets them—he simply wants dollars. He would beg and entreat an Irishman, a German or a Jew, for subscriptions and advertising, and yet plot in a midnight, secret wigwam, with others of his dark-lantern brethren, to proscribe, disfranchise, and degrade below the political status of the negro, that same Irishman, or German or Jew, from whom he had received business patronage. The fellow calls us a "bigoted wretch". . . .

From which it is plain that "race prejudice" was not confined to hostile feelings between whites and nonwhites.

The people far from the scene, then as now, loved to offer severe censure and lofty advice, especially in regard to the treatment of Indians. Some of the mining-camp editors struck back at them. The *Avalanche*, for instance, quoted from the Oregon *Herald*: "Let benevolent people in the East send out their protective Indian agents; they have a right to do so; but, in nine cases out of ten, we will guarantee they will accomplish no good; and some of them will be shot and scalped by these same Indians whom they may try to benefit. . . . Our belief is that the true policy of the Government to pursue towards the Indians is *extermination*, instead of protection." The *Avalanche* was in an area where Indian raids and brutal massacres had become common; O'Meara up in Idaho City was striking out on all fronts, and never with more fury and venom than when the Irish were under attack. He called the *Harper's Weekly* an "infamous sheet" that contained "some most infamous slanders upon the Irish in America. It speaks of them as 'ignorant foreign-

---

* The *Avalanche* editor who wrote these words was probably named J. L. Hardin or Joseph Wasson. See under "Feuds" the fury with which some of the camp editors denounced one another.

ers,' as persons in whom 'ignorance and vice prevail.' " James Harper "approved the anti-Irish and anti-Catholic riots and murders" when he was mayor of New York.

One big reason for "race prejudice" is given by Carl I. Wheat in his essays on Dame Shirley's Letters when he says that for many Yankees "a man who could not speak English was a monstrosity." It does not fall within the scope of this book to try to give the causes of race prejudice but because it was a powerful force in the mining camps, and nothing so far as we know has been written about it except a paragraph here and there, we are going to give an overall view of it, setting forth the principal reasons why the Indians, the Mexicans, and the Chinese were disliked, and the forms taken by the persecution of them.

There is little to be said about Negroes in the camps, for the reason that very few Negroes were there. The kind of trouble they would have had if many of them had been present is suggested by Beidler:

I was Deputy Marshal under G. M. Pinney in Helena on the day of election when the colored folks first voted. There was bitter feeling. The territorial legislature had passed an act prohibiting colored people from exercising the right of franchise conferred on them by the Constitutional amendment. Marshal Pinney received orders from Washington to see that the colored vote offered should be cast.

Hearing a shot fired, Beidler rode toward it and came to a Negro dead on the sidewalk and a white man with a smoking six-shooter. Beidler

An Alaskan mother

asked who had killed Nigger Sammy, the janitor, and was told it was the man with the gun, "a powerful brute." The man had said to Sam, "You nigger son of a bitch, are you going to vote today?" Sam said, "I don't know, boss. I ain't doin' nothin', Marsa, an' I ain't got nothin'." Sam threw his coat open to show that he had no weapons, and, according to Beidler, the man then shot him. Beidler then jumped the man and "held him up with his own six-shooter." A crowd gathered, and some wanted to hang him, some wanted to let him go. Beidler called two powerful men to his assistance and it took the three of them an hour and a half to take their prisoner two blocks. Beidler says that on reaching the jail "I hadn't enough clothes on me to dress a china doll." The murderer of the Negro broke from jail and fled.

There are in the records a few other brutal murders of Negroes but they are no more brutal or less justified than hundreds of murders of whites. Jean Davis dug out of a Montana paper the story of a man named Reed who married a half-breed French-Crow woman who, after bearing him a child, became enamored of a big Negro cook, who worked for Reed. Finding out about it Reed chose a spot and told the Negro to dig a grave there. This the Negro did, without suspicion, for he had dug other graves; but the moment the grave was dug Reed shot him. He then forced his wife to take the shovel and pile the earth upon her lover.

Among the reasons Indians were disliked by whites were these, that they were filthy, and lazy, and wild and cruel and murderous.* Dame Shirley remarks in a letter to her sister that in California it was said that Indians were so unclean that if a squaw "should venture out into the rain, grass would grow on her neck and arms." Those who have inspected Indians on their western reservations and have looked into their huts and shacks can only stand appalled at the filth. To say that this is only in part their fault is true. As for the Dame, few persons in California saw so much in the red people to admire. In some areas, as Shaw says, "They were often driven to such an extremity for food that they ate every kind of insect and every creeping thing however repulsive. Snails, lizards, ants, and even lice I

have often seen eaten with apparent relish. . . . wasps taken 'fresh' from the comb were considered a great luxury." Perhaps they were and are: no prejudices on earth have less reason under them than those in regard to food. White people ate those parts of the viscera of beasts that they were accustomed to eat, and thought the Indians filthy and degraded if they also ate other parts.

Delano thought "Their habits are filthy—frequently in the extreme. I have seen them eating the vermin which they have picked from each other's heads, and from their blankets. Although they bathe frequently they lay for hours in the dirt, basking in the sun, covered with dust." They shortened their hair by burning it off, or by laying it across a piece of wood and sawing it off with a clam shell. They

are taught from necessity to be watchful and wary, and to look upon all men as enemies whom they do not know to be friends. Being in a state of perpetual warfare, they hold it to be a virtue to steal from those . . . whom they do not regard as friends. They do not steal from their own people, and during my residence with the Oleepas, I never saw a quarrel; and I firmly believe that nine-tenths of the troubles between the whites and Indians, can be traced to imprudence in the former. . . . revengeful feelings are instigated, and being unable to distinguish between the innocent and the guilty, it being their custom to visit the insult of an individual upon his tribe, they take vengeance on the first white man they meet. . . .

On Feather River in 1850 "It had become common to charge every theft of cattle on the Indians. A party of miners missed several head of oxen, and a cry was raised that the Indians had stolen them. Fifteen men started out, well armed, swearing vengeance. Proceeding to a rancheria . . . they found a few bones, which they considered proof positive." They surrounded the huts, killed fourteen Indians and demolished their shacks. On their way back to camp they found the oxen that they thought had been stolen. "Had the Indians, under similar circumstances, killed fourteen whites, an extermination warfare would have ensued." In Grass Valley a war between red men and white was barely averted when a white man named Holt forced a squaw into his shack and for two days sated his lust on her. The en-

---

* On the troubles of California miners with Indians, Kelly's *A Stroll through the Diggings of California* is excellent.

A "squawman's" house and wife

Another "squawman" and his family

raged Indians went forth for vengeance and, as was their custom, killed the first white man they came to, who happened to be a good man and one of their friends.

The red men *were* lazy, by white standards, but not by their own. "A foreigner," says Colton, "may be induced to work for money, but not a Californian [he means Mexican], so long as he has a pound of beef or a pint of beans left. Nor is it much better with the Indian: take from him the inducements to labor which rum and gambling present, and he will refuse to work for you." So would a lot of the whites. "The blanket, which he wore last year, will answer for this; his shirt and pants can easily be repaired; his food is in every field and forest. . . ."

Contempt for Indians was not true of all white men in the West but it was true of most of them; and here is Mark Twain on the Goshoots, "a silent, sneaking, treacherous-looking race; taking note of everything, covertly, like all the other 'Noble Red Men' that we (do not) read about, and betraying no sign in their countenances; indolent, everlastingly patient and tireless, like all other Indians; prideless beggars—for if the beggar instinct were left out of an Indian he would not 'go,' any more than a clock without a pendulum; hungry, always hungry, and yet never refusing anything that a hog would eat, though often eating what a hog would decline . . . a people whose only shelter is a rag cast on a bush to keep off a portion of the snow, and yet who inhabit one of the most rocky, wintry, repulsive wastes that our country or any other can exhibit."

Captain James Brownlow, in a letter to the Knoxville *Whig*, got some things off his chest, and the *Avalanche*, bugged by Indian atrocities in the Owyhee area, was happy to reprint them: "I think I have become pretty well acquainted with that poor Indian generally known as Lo. I think if J. Fenimore Cooper could spend two or three weeks with me in the Sacramento Mountains he would cease writing the life of Lo in the shape of cheap novels, or if he could only see his 'lovely Indian maid' wading through the entrails of a coyote or a lizard as long as one's arm, and winding up with a glandered horse-head for dessert he would come to the conclusion that there was more hyena than romance in her nature."

Bryant found the red people objectionable for other reasons. When celebrating all night, he said, they would shout and howl like wild beasts; they would fix a scalp on a pole and dance round and round it with such wild insane screeching and yelling that persons five hundred yards distant found it impossible to sleep. If ill, they entered their dugouts where a fire was burning, with only a small hole in the roof; and on shelves dug into the earth walls the Indians lay with sweat pouring from them. "I incautiously entered one of these caverns . . . and was in a few moments so nearly suffocated with the heat, smoke, and impure air, that I found it difficult to make my way out."

There have been published thousands of stories dealing with Indian cruelties. Browne tells of the end of a Dr. Titus: wounded, surrounded by Apaches and knowing that capture was inevitable, he shot himself. If he had been taken alive he would have been hung by his heels from a tree and a slow fire would have been kept alive under his head. Of another white man, Barnes says, "We recovered his body. The diabolical wretches! —they had stripped him entirely of his clothing, dug out his eyes, torn off his scalp, opened both his breasts, took out his heart and entrails—a tribute to his bravery—cut off both his feet, cut his head nearly off, and otherwise disfigured him. . . ."

Most of the overland emigrants despised and feared the red men because of their attacks. Ferguson tells of his agonies as he traveled for days with almost nothing to eat and with an inflamed arrow wound; and says of a companion that how he "stood it to ride as he did and live, has always been a mystery to me. He had four arrow wounds in his body, and was red all over with inflammation, and swollen as full as his skin could hold, and so weak he could hardly sit on his horse, but he bore it all without a murmur." After the most dreadful sufferings this man died and was the "sixteenth of our party killed by Indians."

The attitude of the mining camps toward the red people is presented by the *Avalanche*:

Our country and adjacent country are infested with these savages. They go where they please, steal what they wish, and to add spice to their worthless lives take the lives of good citizens, and slosh round with impunity. . . . For over three years the daring miner has been trying to develop this upper country. He has been met and surrounded by danger at all times. To-day he is liable to be shot or robbed if he sticks his head out of his cabin, or goes prospecting, or travels the oldest thoroughfares. . . ."

The Indians had, of course, their side of it, and it wasn't a pretty side either. They naturally and inevitably hated the miners because they swarmed by the thousands over their virgin and beautiful hunting grounds and devastated them. They chased away the deer, the rabbits, the game birds; they muddied the streams until the fish choked to death; and they gouged hills and meadows down to bedrock in their feverish insane search for a material that the Indians had never valued. Stuart says that when a group of miners were trapped on a river island by a band of Sioux and all of them killed, the red men "finding the buckskin bags filled with the gold dust, and not knowing what it was, emptied it out on the sand and took the sacks."

The Indians were as ruthless with the Chinese as the palefaces, regarding both yellow and white as poachers on their lands, as in fact they were.

A Chinese known as I-John, who lived in Silver City, was the lone survivor of about ninety Chinese marching up from Nevada. "Troops who investigated the incident found dead Chinese scattered along the road for several miles. The soldiers collected the bodies and buried them in a common grave."

The white men not only invaded their lands; they sometimes raped their women, and for their own sordid gain they debauched Indians with whiskey and rum. As the whites kept coming from the land of the rising sun, first in small groups, then by hundreds, by thousands, the wiser Indian chiefs knew that their people would have to kill the white people or drive them back to the lands they came from, or eventually they would have no lands at all. How right they were is revealed in their wretched existence today on some of the poorest land in the West. In this matter, as in all things and in all times, neither white nor Indian could see any side but his own, and the hatred of the one by the other is be-

An old Indian left to die

These are five Southwest Indians taken captives after their part in the massacre of a wagon train. Three of them were hanged.

The Apache Kid, a deadly killer in the Southwest
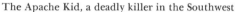

yond the power of words to exaggerate. Of the Indians on an Arizona reservation in the eighties Tassin says they were a "lot of dirty, thievish, half-starving Indians, wards by name and victims by fact, of a blind, misdirected governmental policy, brutalized instead of civilized by their contact with the few white outcasts who have taken refuge among them."

The Chinese the whites despised for somewhat different reasons. At the end of 1848 only seven Chinese were registered in what was to become the state of California, but three and a half years later twenty thousand of them had poured in. Word must have been continually sent back to the old country, for they kept coming. Various proposals were put forward for the limitation of their number, most of them pretty absurd. The Chinese laborer was "beaten and robbed," says Jackson; "he was at the mercy of anyone who felt like fleecing him"—and fleece him they did. And stab him they did, as witness: "I was sorry to have to stab the poor fellow; but the law makes it necessary to collect the tax; and that's where I get my profit." A Mexican bandit is said to have suspended six Chinese by their queues, from trees, and cut their throats, to watch them die for the fun of it.

One reason the white men disliked them was

The two lads, white and Negro, are two of Geronimo's captives

their lazy way of placer mining—scratching, the Americans called it. Borthwick, who saw them in the placers, says that their rather hit-or-miss indolent way reminded him of women—that they "handle their tools like so many women, as if they were afraid of hurting themselves."

Another reason for white irritation was the racket the Chinese made. An irate Montana editor says that when their gambling began after dark, "greenbacks, opium, and joss-gods" were the stakes. "We don't know and don't care how many years they claim to have been infesting the earth, and only wish they would go to bed like decent people and stop playing their infernal button game of 'Foo-ti-hoo-ti,' so a fellow could get a nap."

To a considerable extent the dislike of the Chinese was exported from California in the various gold rushes. Virginia City had heard about them, away up in Montana, and when a coach-load of them arrived in the summer of 1865, Thomas J. Dimsdale, editor of the *Post,* made plain his contempt by giving to the Chinese such names as Ho-fie, So-Sli, Lo-Glung, Ku-Long,

Whang and Hong, and by saying that all the mice in the area had fled. Another editor said he didn't mind "hearing of a Chinaman being killed now and then," but admonished his readers not to kill them "unless they deserve it, but when they do—why kill 'em lots." Because in his opinion all Chinese were kleptomaniacs "the prospects for an early Chinese funeral are brightening." The editor of the *Madisonian* mixed admiration with contempt:

Chinamen are heavy on the pack. While the heathen is apparently physically deficient, he can carry a load that would disgust [not dismay or astound but disgust] the boss mule of a pack train. One was noticed going down the gulch with two large rolls of blankets, a sack of rice, a couple of hog's heads, a lot of heavy mining tools, a wheel-barrow and a hand-rocker swinging to his pack-pole. It was a mystery how that Chinaman managed to tote that weary load along so gracefully, and not grunt a groan.

In the Montana area, as in all others to which the Chinese migrated, this humble people seldom tried to stake a claim but were content with the tailings that the white men left behind. It took them forty years to recover from the camps in the Virginia City, Montana, area the gold which white men in their greed and haste had shoveled out of rocker and sluice box. And it may be that one reason for white irritation was the patience with which they toiled and the stamina with which they endured. It has been said that a cause of irritation was the Chinese refusal to be assimilated but in this they had no choice. Their women sometimes became whores to the white man's lust but intermarriage was rare.

In Helena a "Committee of Ladies" gave notice to the Chinese that they must at once "suspend the washing or laundry business," and the editor lent a hand by putting the ultimatum in rime:

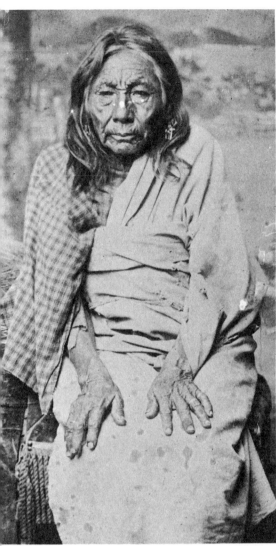

A Cheyenne squaw

An Apache squaw

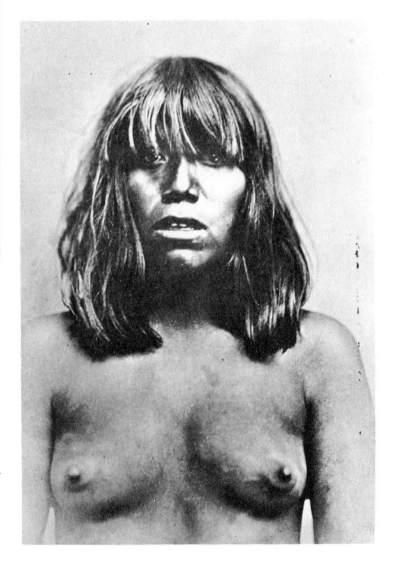

Chinamen, Chinamen, beware of the
    day
  When the women shall meet thee
    in battle array.
Ye hopeless professors of salsoda and
    soap,
  Beware of the fates that await ye;
No hangman's committee with
    ladder and rope,
  But the ladies are coming to *hate*
    ye.
Ye almond eyed leather faced
    murthering heathens!
  Ye opium and musk stinking
    varments,
We will not object to your livin'
    and breathin'
  But beware of the washing of
    garments!

Down in Tonopah, Nevada, the Chinese were
told to pack up and get out. When their leaders
begged for time to gather their possessions some
of the white men said yes and some said no; and
some of those who said no smashed through locked
doors, and after robbing the helpless Chinese
brutally beat them. One old man, Wing Sing,
was beaten to death and hidden in the brush, and
by that time most of the Chinese were a mile or
two down the road, plodding along in the dust on
their way to Sodaville and points beyond.

As though they did not have enough enemies
among the whites they sometimes turned on one
another, as in the feud between the Yan-Wos
and the Sam-Yaps on the Stanislaus River in California. Whatever the cause, the Sam-Yaps felt
compelled to save face, and so sent a challenge
to the Yan-Wos:

There are a great many now existing in the
world who ought to be exterminated. We, by
this, give you challenge, and inform you before
hand that we are the strongest, and you are too
weak to oppose us. We can therefore wrest your
claim, or anything else from you, and give you

An Alaska Indian

notice that it is our intention to drive you away before us and make you ashamed of yourselves. . . . You won't stand like men; you are perfect worms; or, like the dog that sits in the door and barks but will go no further. If you won't accept the challenge, we tell you, by the way, to go and buy lots of flour, and paint your faces; then go in your houses, shut the doors and hide yourselves, and we'll kill every man of you that we come across. Shame! Shame!

It sounds like the blustering of small boys half-scared out of their pants. Next, of course, the Yan-Wos had to save face or paint themselves with flour and hide. They seemed to have had quite a time getting their courage up. Some writers have said there were nearly two thousand Yan-Wos and Sam-Yaps saving face; and that both sides had blacksmiths who with loud banging forged spears and swords and various man-killing contraptions for the big face-saving war that was to come. The Sam-Yaps had some guns, but a Chinese, at least in those early days, was as likely to shoot himself as the enemy. For ten dollars a day and whiskey the Yaps hired some miners to teach them how to shoot, and on the day of the great battle possibly a hundred shots were fired. In all the whoop and holler, thrusting and banging, only four were killed and four wounded. As a disgusted San Francisco reporter said, it was a lousy fight, because "so few were killed." But a lot of faces were saved, and after the banging was over, the warriors went back to their digging.

It is a way with people to find customs annoying that are not their own. The *World's* editor took a Sunday walk to the graveyard, and

witnessed a strange scene enacted there by a half-dozen Chinawomen, who were engaged in the ceremony of furnishing provisions to the dead celestials. Baked pork, fruits, apples, and rice were placed at the foot of one of the graves and a large number of wax tapers, joss sticks, and so forth, were lighted and placed around it. Cups of tea were then poured upon the ground, and an immense quantity of fancy paper was burnt. The women were laughing and apparently joking during this time, and occasionally they would prostrate themselves at the foot of the grave and bow their heads to the ground three or four times in succession. After going through with these performances, one of them deliberately pulled out her handkerchief, and, selecting a good seat on a box, she commenced singing, at the same time covering her face with her bandana. The song changed to wailing and crying in which all of

them joined finally, and they kept up an unearthly din for about half an hour. We can hardly believe that their exhibitions of grief were genuine as we could see no trace of tears. They may have been hired mourners; and if so, that accounts for the businesslike manner in which they went at it.

The persecution of Indians by the whites often easily matched in brutality that of the red man's assaults on the whites. As Gard points out in his *Frontier Justice,* the "scalping of Indians by whites had been common east of the Mississippi, where, in some sections, it was encouraged and rewarded by public bounties." Like a bounty, that is, for the ears or front feet of a coyote. Some of the ministers "urged them to hunt them with dogs as they did bears." In a few mining camps —Denver and Central City are instances—money was raised by subscription, so that twenty-five dollars could be paid for each scalp brought in. Over in Deadwood a little later, where, as Mrs. Tallent, first white woman in the Black Hills, says, men had to be resolute who entered in the face of Indian wrath, the bounty was as high as two hundred dollars, and Indian heads were sold at auction.

Many a white man proposed complete extermination of the Indians, a fate that would have been more merciful than the degradation that fell to the lot of most of them. The *Alta* told its readers that "an armed body of Americans" had "publicly organized themselves in the village of Sonoma, for the avowed purpose of exterminating the Indians in this valley and burning the ranches and lodges. . . ." They were "not unnaturally reckoned by everybody in the West," said Bowles, as "'pesky sarpints.'" The government "is ready to assist in their support, to grant them reservations, to give them food and make them presents; but it must and will, with sharp hand, enforce their respect to travel, their respect to lives and property, and their respect to trade throughout all this region. And if this cannot be secured, short of their utter extermination, why extermination it must be." The idea that government can enforce respect is loud in the land as we write these words. In the view of Bowles the Indian was "false and barbaric, cunning and cowardly, attacking only when all advantage is with him, horrible in cruelty, the terror of women and children, impenetrable to nearly every motive but fear, impossible to regenerate and civilize."

When in the mining camp of Weaverville the butcher was killed by Indians, the sheriff took after them with a posse of seventy men, and massacred every one of a hundred and fifty-six, except three small children. According to Bancroft the California Indians didn't have enough gumption to unite against their white enemies—

hence when now and then one of them plucked up courage to defend his wife and little ones, or to retaliate on one of the many outrages that were constantly being perpetrated upon them by white persons, sufficient excuse was offered for the miners and settlers to band and shoot down any Indians they met, young or old, innocent or guilty, friendly or hostile, until their appetite for blood was appeased.

Even poison was used against them. The *Enterprise* reported that fifteen or twenty Indians had been poisoned on East Walker River: "It appears that a ranchman in that vicinity sprinkled strychnine over the carcass of a fat ox that died of some disease, and that the Indians cut up and eat the animal. . . ." A story put out said he had set the poison for coyotes, and this may be, but he fled the country. The editor went on to say:

We say bully for the ranchman mentioned above, and hope he *has* put a quietus on fifteen or twenty of the Lo family. . . . No doubt many leading men and narrow minded philanthropists will hold up their hands in religious horror at this mode of treating the "poor Indian." Those holy humanitarians doubtless would control the "noble red men" by moral suasion. . . . We are not certain but that we should enjoy the spectacle of Big-foot and his co-friends in the hair-dressing operation upon some of those sanctimonious, cheese-brained craniums that uphold the murderous savage.

The *Avalanche* reprinted the above and in the same issue quoted Mark Twain from a California paper: "It is funny, the absurd remarks people make about the far West, and the wild questions they ask about it when they are discussing the Indian difficulties. It is humiliating to me to consider how high an opinion we have of our importance out there . . . and then to discover how little some people know about us."

The whites took delight in short-changing the red man because of his ignorance. Knower says he was short-changed by a merchant, and after complaining about it found the merchant glad to apologize, "saying that he had weighed it in the scales that he used when he traded with the Indians." The Indians were also cheats but like all cheats didn't like to be cheated. At first in their innocence they would offer a handful of gold in payment but on learning that they were being robbed, they would go, says Buffum,

in a party of ten or twelve, and range themselves in a circle, sitting on the ground, a few yards distant from the shop; and then in a certain order of precedence, known to themselves . . . they proceed to the counter in rotation, to make their purchases, as follows: placing on the palm of the hand a small leaf or piece of paper, on which is perhaps a tea-spoonful of gold dust, the Indian stalks up to the dealer, and pointing first at his dust in hand, and then at whatever article he may desire, gives a peculiar grunt—Ugh!—which is understood to mean an offer; if the dealer shakes his head, the Indian retires, and returns with a little more gold dust

until his offer is accepted. But if "they should purchase half-a-dozen hard biscuits for a tea-spoonful of gold, and want several dozen, they will return with one tea-spoonful more, obtain six biscuits and retire, and then return again, and so on until they have obtained the desired quantity."

Most of the white merchants were eager to sell the Indians all the whisky and rum they could pay for—and often the liquor was chiefly water mixed with whisky or rum, into which had been stirred tobacco and cayenne pepper to give it color and potency. In some old diary Hulbert found the tale of a young squaw who had "swallowed nearly a pint of the pop-skull dispensed in the near-by grogshop under the name of whiskey." Among those who tried to save her life was a "large, obese, crosseyed Indian, earnestly engaged in native performances." In nothing but a breech-cloth, his eyes bugged out and "revolving in an unearthly manner" he leaps and stamps and grunts and bellows, all the while pounding his chest and "swelling like a puffball"; and with his mouth to hers blowing into her his bad breath and spittle. He fills his mouth with water and sprays her with it. He heats a knife in a fire and does wild antics with it. He beats and thumps himself, screeches and bellows, puffs and swells and gags, and massages and manipulates the poor

A very old squaw

dying thing all in vain. The white man's fire-water triumphed over her and over thousands.

If it wasn't whisky it might be rape. The *Statesman* told the story of Bruneau Jim, a young Indian with two young Indian women, who was accosted by two white men. At once the white men strove to rape the girls, and when Jim interfered, one of the men seized his arms and the other with a knife stabbed him over the right eye. When the Indian broke free and ran the white men "violated the squaws, but not without difficulty, as the marks on the ground show." The two women took Jim to a grove of willows on the Boise River and nursed him until he died.

There is some prospect of identifying the fiends who committed this outrage. . . . The Bruneaus have been peaceable always, they are intensely poor and are reduced to the number of about two hundred. They were about to perish from starvation last January, and asked to join the Boise agency. They were brought in, and have ever since conducted themselves entirely peaceably, some of them doing odd jobs about town for a slight pay in grub or old clothes. There is no extenuation to be mentioned for the perpetrators of this diabolical outrage and murder.

Beebe found in the California State Library some notes left by an incredible blackbearded giant named Aaron Ross, "who took a dim view

An Eskimo child

Patchy, captured as an Indian baby

of crime and low life generally and Indians in particular." When in old age Ross dictated some of his recollections he said:

I told them where the Peg Anne were camped and that night they went up and ran off a lot of the Injun's horses and went over the mountains with them. Then they killed two night herders on the prairie and came into town and told what they had done. Some of the men, cowboys and others, had one Injun named Crowtaw in jail but couldn't get hold of him as he had a big knife and stood them off; so they came down to the stable and wanted me to come up and get him. I went up and went into the jail and he came at me with a knife, and I took hold of him, ketched him by the hand and took the knife away and throwed him and held him until the rest got hold of him and then I let go. They took him out and hung him and then we went around town and caught two more and somebody shot them and killed them. Then the soldiers made a fuss and said we'd have to bury them. We put a rope on them and dragged them down to the river and throwed them in and let them go. Made good Injuns out of them.

Perhaps no massacre of Indians by whites exceeded in the whole West in brutal senselessness that under Colonel Chivington. After six or seven hundred Arapahoes and Cheyennes surrendered at Fort Lyon and were turned away by the commander there because he didn't have food for them they encamped on Sand Creek. Colonel Chivington, with 750 armed men and a howitzer battery, marched against them, even though more than half the Indians were women and children and it was blizzard and bitter cold that morning of November 29, 1864. He is said to have given the order, "Spare no one," and no one was spared. Lieutenant Cramer said afterward that "No Indian, old or young, male or female, was spared." No prisoners were taken. Some historians have said that many of the bodies were brutally mutilated; and even Kit Carson, who despised the red people, was shocked.

As Quiett and so many others have said, "This inhuman slaughter roused the Indians all over the plains to concerted attacks on the whites. They interfered with the Union Pacific Railroad then building west, tore up telegraph lines, burned stage-stations. . . . They burned and pillaged farmhouses and murdered white men everywhere. . . ." The Chivington massacre led, of course, to the massacre of Custer and his men. And it led to chronic massacre in the Owyhee region.* After a group of Indians including two women and two boys were killed, the *Statesman* wrote,

Indians, old and young, took delight in wearing castoff clothing of whites.

_____

* For long graphic accounts of massacres in the Owyhee area see the *Avalanche* of March 30, May 4, 11, 25, 1867.

Let vengeance and extermination be visited upon all the hostile tribes until there is not a redskin left between here and California and Nevada, with a disposition to shoot travelers from behind the sage brush. We don't believe in Borobola-gha and Pocahontas stories about the virtues of the "Noble Red Man," especially of the Shoshone variety, and think the more of them are killed the better. At the same time it is to be hoped that no outrage upon innocent and peaceable Indians, who take no part in robbery and murder, will be committed to bring disgrace upon the community.

A reading of the records discovers stories of "good" Indians—stories of heroism and pathos that relieve the dismal bloody history of white-red relations. The Nez Perce Indians did not go on the warpath when thousands of gold seekers overran their lands. The reason, McConnell thinks, was the friendliness they had felt for the white people ever since Lewis and Clark and their men went down the Clearwater, and because of the Christian spirit that the Reverend Spalding and his wife had made familiar to them while teaching among them. Of Indian heroism Edwin Bryant gives a touching incident from a California newspaper. After a Walla Walla Indian volunteered to help Colonel Frémont he found himself

pursued by a party of the enemy. The foremost in pursuit drove a lance at the Indian, who, trying to parry it, received the lance through his hand; he immediately, with the other hand, seized his tomahawk, and struck a blow at his opponent, which split his head from the crown to the mouth. By this time the others had come up, and with the most extraordinary dexterity and bravery, the Indian vanquished two more; and the rest ran away.

Of pathos we like the well-worn story of the note which one white man sent to another: "This Indian is a drunkard, a liar and a notorious old thief; look out for ⊕m." The Indian thought the white man's note was a witness to his good character and was enormously proud of it.

Some whites had a more civilized view of the common hatred of "foreigners" but though M'Collum thought the Chileans "quiet and inoffensive," he had nothing good to say "of the wild Indians—kindness toward them would be mostly labor lost; they seem almost incapable of appreciating the offices of humanity. . . ." Greeley put his finger on it when he said

the Indians are children. . . . Any band of schoolboys, from ten to fifteen years of age, are quite as capable of ruling their appetites, devising and upholding a public policy, constituting and conducting a state or community, as an average Indian tribe. . . . I have learned to appreciate better than hitherto, and to make more allowance for, the dislike, aversion, contempt wherewith Indians are usually regarded by their white neighbors, . . .

Theodore Roosevelt saw the matter in this light:

As for the whites themselves, they too have many and grievous sins against their red neighbors for which to answer. They cannot be severely blamed for trespassing upon what was called the Indian's land; for let sentimentalists say what they will, the man who puts the soil to use must of right dispossess the man who does not, or the world will come to a standstill; but for many of their deeds there can be no pardon. They utterly despised the red man; they held it no crime whatever to cheat him in trading, to rob him of his peltries or horses, to murder him if the fit seized them. . . . Their deeds of terrible prowess are interspersed with deeds of the foulest and most wanton aggression, the darkest treachery, the most revolting cruelty; and though we meet with plenty of the rough, strong, coarse virtues, we see but little of such qualities as mercy for the fallen, the weak, and the helpless, or pity for a gallant and vanquished foe.

Father de Smet, the famous missionary to the Flatheads, put it well:

Poor unfortunate Indians! They trample on treasures unconscious of their worth, and content themselves with fishery and the chase. When these resources fail, they subsist upon roots and herbs; whilst they eye with tranquil surprise the white man examining the shining pebbles of their territory. Ah! They would tremble indeed could they learn the history of those numerous and ill-fated tribes that have been swept from their land, to make place for Christians, who have made the poor Indians the victims of their rapacity.

When Bret Harte was writing for the Eureka newspaper a band of Indians was slaughtered by whites on Gunther Island, and Harte wrote a blistering editorial about it. An enraged mob marched against the newspaper shack; Harte held them off with two old pistols until a friend could bring help from an Army post. But that ended Harte's career in Eureka.

"I am no believer in the Noble Savage," Richardson said, after his long journey overland and. through the West. "If he ever existed outside of Cooper's romances, he was long ago extinct. The Indian is cruel, bloodthirsty and treacherous; but he often behaves quite as well as the Pale-face." Sometimes a white man was actually impressed by an Indian, as Colton was by a chief over seven feet tall, whose "long black hair streamed in darkness down to his waist. His forehead was high, his eye full of fire, and his mouth betrayed great decision. His step was firm; . . . He entered the court with a civil but undaunted air." Rickard, it seems to us, is fair to both sides when he says that the Indians "had ample justification for fighting the early trappers, prospectors, and settlers, however much he, and we, may sympathize with those that opened up the Great West." As for the Apaches, they "had no reason to be friendly to the Americans, any more than to the Mexicans, whom they had hated and fought for three centuries; they had been treated by the whites with the same cruelty that they themselves had practised. . . ."

The whites and Mexicans disliked one another not only because of differences in skin and habits but because their nations had recently been at war. A "rooted hatred" existed between them, Marryat says; and the Americans called the Mexicans greasers, "which is scarcely a complimentary sobriquet, although the term 'greaser camp,' as applied to a Mexican encampment, is truthfully suggestive of the filth and squalor." When a tax of $30 a month was levied on Mexicans, in the Sonora area in June 1850 they banded with the Chileans and refused to pay it, and the surrounding camps, expecting attack, prepared for battle. By July the whole area came close to civil war when four Mexicans were discovered "piling brush upon and burning the bodies of two American miners." Before judge and jury, Shinn says, the Mexicans swore that they had happened to find the corpses and were cremating them because cremation was their custom. But they were found guilty; a rope was thrown over the limb of an oak and they were brought forth to be hanged. At that moment three judges happened to come along, and their appeals to the miners were so persuasive that the Mexicans were returned to jail. A day or two later nearly a hundred armed citizens were marching through the camp; and 300 armed miners from the Green Flats Diggins,

where the corpses had been found, marched over to see the

punishment of the murderers. Every knife and revolver in every camp within a radius of a dozen miles was strapped to some stalwart miner's side, . . . When rumors came of an uprising in a Mexican camp three miles distant a sheriff and a posse went over to the camp and arresting 110 Mexicans marched them to Sonora but they were released the next day. Tuesday was the beginning of the famous trial. Fully two thousand excited and armed men were in the streets of the town.

Since there was no evidence at all against the four Mexicans they were acquitted and the mob began to disperse, though before the trial ended there were rumors of "new outrages and murders. . . . This tendency to despise, abuse, and override the Spanish American may well be called," says Shinn, "one of the darkest threads in the fabric of Anglo-Saxon frontier government."

Writers have pointed out that the Chinese asked no more than to work the tailings and dumps, and affected a mock humility. This the Mexicans refused to do: they had been in this land long before the white men came, and only recently had seen a part of their country wrested from them. When a heavy tax and other penalties were assessed against them—an oppressive and outrageous law, as historian Hittell has pointed out —Mexicans had a choice of starvation or flight. Dame Shirley, whose letters some writers have thought temperate, reported that on slight pretext Mexicans were whipped and sometimes hanged. Some visitors to the area were shocked by the severe measures adopted in an effort to drive all Mexicans out. When a Chilean (Mexicans and South Americans were all the same to the whites) named Concha had claims staked in the names of some peons, and whites came over and drove them out, Concha got a warrant from the alcalde in Stockton and with fifty armed peons killed two Americans while trying to arrest them, and seized thirteen. After the whites managed to free the captives they tried the Chileans, found them guilty of murder, and hanged three of them. The others had their ears cut off or were flogged and then driven out. This is known as the Chilean War.

By the early fifties the whites were made aware of the fact that more and more Mexicans were becoming road bandits, and were preying on those who refused them the right to dig for gold. Many of the thieves made a regular business of it, stealing horses, running off the cattle, robbing saloons and shops and unwary travelers, and building up gangs.

As for the Chinese, many of the whites, including now and then a judge, simply refused to accept them as human beings. "They are a harmless race," Mark Twain said, "when white men either let them alone or treat them no worse than dogs; in fact, they are almost entirely harmless anyhow, for they seldom think of resenting the vilest insults or the cruelest injuries. They are quiet, peaceable, tractable, free from drunkenness, and they are as industrious as the day is long. A disorderly Chinaman is rare, and a lazy one does not exist."

The trouble started in San Francisco, their port of entry, and spread over the whole West. In San Francisco speeches were made, societies formed, and whereases by the dozen drawn up, all to discourage and stop the immigration of Chinese. Brace in his *New West* quoted from the *Bulletin* a letter that expresses the attitude of whites toward them: "A most inhuman scene occurred on Sunday evening at the corner of Green and Dupont Streets. A party of young scamps—the oldest not more than eleven years—attacked a peaceable Chinaman, without provocation, and beat him in an unmerciful manner. They pulled him down, beat and kicked him, and pelted him with stones until the blood ran out of his wounds. A large crowd stood around at the time, and none of them offered to interfere." One man pushed through, saw that it was a Chinese, and exclaimed, "Oh, it's nothing but a Chinaman!" The letter concludes: "Hardly a night passes but what a Chinaman is attacked by these young ruffians." Then Brace cites another letter: "Not long ago, a gentleman passing along Kearney Street, interfered to save a little Chinese boy from the attacks of a dog, whom half a dozen white-skinned scoundrels were setting upon him, that they might enjoy the precious sight of the agony of the screaming child." He cites the newspaper *Alta*: "Last evening, at the fire on Dupont Street, a crowd of Waverley Place loafers, and thieves, and roughs, who were being kept back from the fire by the police, amused themselves by throwing a Chinawoman down in the muddy street, and dragging her back and forth by the hair for some minutes."

Bowles estimated that there "are no fewer than sixty to eighty thousand Chinamen here [in the Coast States]. They never, or very rarely, bring their wives. The Chinese women here are prostitutes, imported as such by those who make a business of satisfying the lust of men. . . ." After telling of the four dollars a month tax imposed on them in California he continues: "Thus ostracized and burdened by the State, they, of course, have been the victims of much meanness and cruelty by individuals. To abuse and cheat a Chinaman; to rob him; to kick and cuff him; even to kill him, have been things not only done with impunity by mean and wicked men, but even with vain glory. Terrible are some of the cases of robbery and wanton maiming and murder reported from the mining districts."

Typical of the brutal and wanton assaults on Chinese is the notorious case in Los Angeles, October 24, 1871, when a gang of whites, having been told that a Chinese had $7,000 in his store, sacked and looted Chinatown, hanging fifteen Chinese and shooting three, according to the shocked Los Angeles *Star* the following day. Its headline cried, MURDER! TERRIBLE OUTRAGES! FIENDS IN OUR MIDST! Major Horace Bell, who was there, says one Bancroft account based on the version in the *Star* is almost wholly inaccurate because it represented the Chinese as the fiends, and their looters and hangmen as instruments of outraged virtue. Bell says twenty-one Chinese were lynched, and that the number shot was never determined.

The difficult time the Chinese had in Idaho's camps was typical of all the camps. By 1885 the persecution of them in the Territory had led to trouble with China, after a Pierce merchant was robbed and killed. Eight Chinese were seized as suspects, and of these five were held for the sheriff at Lewiston. On their way down the Clearwater River they were taken from the sheriff by a mob and hanged. China demanded an explanation. The Secretary of State called on the territorial governor for an investigation. The matter was patched up with China and by the following year Idaho was trying to drive all the Chinese out of it.

Donaldson, who came to the area in 1869 and later declined the office of territorial governor, says he never knew of a Chinese who tried to cheat but had known of hundreds of white men who cheated them. The Chinese were reluctant to go to court because "not one judge or jury in a hundred" would find in favor of them, even if all the evidence was on their side. If taken to court they sometimes replied to questions in such incomprehensible jargon that they had to be dismissed. Donaldson tells of a Mr. John who "paid four dollars for his license to work a claim, but he was forced to pay this more than once during the year because white loafers, hard up for cash, would procure blank licenses, visit the diggings under the guise of license collectors, and poor John would be threatened and bullied until he handed over four dollars—perhaps for the fourth time in a year." Donaldson says that "In a suit to recover money—I speak from experience—not one Idaho justice of the peace or one juryman in a hundred would have given judgment in favor of a Chinaman."

In the more important court cases a Chinese charged with crime would ask that Chinese witnesses be sworn

by the native method. Judge Lewis granted this concession one day in 1871 when sitting at Silver City; the case was that of murder, charged against a Chinaman. The interpreter, beginning his preparation to administer oath, explained that a Chinaman devoutly believed that if he violated the truth under oath, he would meet with dire punishment, continual decapitation, in the other world. Rube Springer, sheriff of Owyhee County, brought into court a live chicken and a tin platter. The Chinese witness cut off the chicken's head, and the blood was let drop on the platter. The witness then took several squares of red paper and, with a Chinese brush dipped in ink, wrote on the papers his oath to tell the truth. The bits of paper were then rolled into pellets, soaked in the chicken blood, sprinkled with incense, and burned with a taper. A Chinese explained to the fascinated judge that the burned papers went to heaven, and if when Hop Wo, the witness, got there it was found that he had sworn falsely, he would "hab mooch trouble. They chop, chop, chop, he head off, allee time!"

In 1884 a horse thief and murderer known as Doc Chafee was captured and lodged in the Silver City jail. Because there was a Chinese man in it and he detested the Chinese, Doc set the bedding afire. He got more than he bargained for. Before the fire was discovered and the jail door was opened both Doc and the Chinese were cremated. The citizens were less disturbed by the fact that they had lost a courthouse and jail than by the fact that they had been cheated of a hanging.

It should no longer be necessary to point out that "human beings" are much the same under their skins, no matter what their race or color. The infernal cruelties of Chinese and Japanese, and indeed of all peoples, are thoroughly authenticated in the records of warfare. The white persecution of Indians, Mexicans, and Chinese is a sickening part of American history, but if the matter had been turned around the record would be much the same. Now and then something happened that for a moment revealed the emerging humanity in the brute. When, for instance, the Chinese were allowed to mine certain areas near the Columbia camp and three Arkansas bullies jumped their claims, the white miners immediately chased the bullies out.

And there is the love story of Polly Bemis.* She was one of many Chinese girls shipped to San Francisco in the white slave traffic; from there they were taken to various mining camps by those who bought them, and Polly and several others were transported to the placer camp of Warren in central Idaho. Among the men there was Charles Bemis, said to have been an educated man from a distinguished New England family. A man who preferred gambling to the hard labor of a miner he had a hall next to a dance hall, where Polly was one of the dancers and harlots. Now and then she slipped over to the Bemis bedroom in back of the gambling hall to put it in order, and when she found herself in trouble with tough drunks she called to Bemis, who "never failed her."

It was a habit with some of the miners to come in with pokes of dust and ask Bemis to put most of it in his safe and not let them have it while drunk. There came one day a half-breed Indian who consigned most of his dust to Bemis but who by midnight was drunk and asked for his dust. Bemis said no, and soon afterward went to bed. About daylight the half-breed crawled through a window into the Bemis room and again asked for his gold; and when again refused lit a cigarette

---

* As it appears in most books it is as much legend as fact. Our account here is based chiefly on that left by George J. Bancroft, mining engineer, who knew both Polly and Bemis; the Bancroft essay is in the collection of his historian daughter Caroline. Sister Alfreda Elsensohn gives a version of the story in her *Pioneer Days in Idaho County*. Peter Klinkhammer, a friend of Bemis and Polly, took a number of photographs of them.

and said if the dust had not been returned by the time he finished the cigarette he would shoot one of Bemis's eyes out. Bemis was going back to sleep when the half-breed, as good as his word, aimed at a Bemis eye but missed it. The "ball entered the face under the eye and plowed through the lower part of the head to the back of the neck." Polly heard the shot and came running, and a few moments later summoned the marshal and a doctor. The doctor said Bemis would die but Polly nursed him back to health, and with a razor blade cut the bullet out of the flesh of his neck.

When recovered Bemis asked her to marry him, and after their marriage they bought a small tract of land at the bottom of the deep canyon of Sal-mon River's Middle Fork, where the winters are extremely mild; and there they planted an orchard and berry bushes, and Polly had cows and chickens and a garden spot. Bemis did a little hunting but chiefly he liked to read and smoke and play solitaire. What a strange pair they were! Polly had never been out of that canyon for more than thirty years, when, in 1922, her husband died at the age of seventy-four. She went to Warren once in a while, and once went all the way to Boise, to visit with Chinese families there; but she had almost completely forgotten her own language. In her last illness she asked to be taken to the canyon home and buried at the side of her husband.

Polly Bemis and her man

CHAPTER 16

Crime and Frontier Justice

In his *Frontier Justice* Gard says "there was little crime in the primitive camps. Men could pile gold-bearing gravel on their claims or leave thousands of dollars' worth of gold dust in their tents without worry." Ryan seems to have started the legend in his *Personal Adventures in California,* published in London in 1851: "The miners dwelt together in no distrust of one another, and left thousands of dollars' worth of gold-dust in their tents, while they were absent digging." Pointing out that Ryan wrote of the camps of 1850, "as a result of wide observation," Shinn then quotes from a number of persons who saw the camps and reached the same conclusion.

The miners needed no criminal code. It is simply and literally true that there was a short time in California, in 1848, when crime was almost absolutely unknown, when pounds and pints of gold were left unguarded in tents and cabins, or thrown down on the hillside, or handed about through a crowd for inspection. . . . Men have told me that they have known as much as a wash-basinful of gold-dust to be left on the table in an open tent while the owners were at work in their claim a mile distant.

Elsewhere he says: "The unwritten, unformulated law that ruled each camp was the instinct of healthy humanity to mete out equal justice to all. There was no theft, and no disorder; few troublesome disputes occurred about boundaries and water-rights." Other men who saw the early camps, in California, support Ryan's statement. We shall quote from a few of them.

William T. Coleman, who became the leader of the vigilantes in the San Francisco cleanup, says that on entering California he met a man carrying three thousand dollars in dust and nuggets. When asked if it wasn't dangerous to show off his wealth to strangers the man said "it was perfectly safe, the people were honest, or made

to be honest; that there was no room in the country for thieves, and that there was no such thing as highway robbery; . . ." That was in 1849, before the area was overwhelmed by what was called the "scum and refuse of the earth; scoundrels from the British penal colonies, rogues from Sydney and Van Diemen's Land," as well as professional criminals from the States.

Delano tells of meeting a stranger who told him

he had made fifteen hundred dollars in a short time, and taking out his purse, exhibited the money in gold coin. As it was heavy in his pocket, he arose, and going to my bed, which was spread out under a tree nearby, he turned a corner of the blanket down, and then put the purse under it, leaving it there till morning, without going near it again, apparently with as much unconcern as if it had been so many chips, although we were entire strangers. In the morning, after breakfast, and when he had harnessed his mules, he went to the bed, and taking his gold, jumped into his wagon and drove off, as carelessly as if he had run no more risk than in depositing his money in the vault of a bank.

In *A Frontier Lady* Sarah Royce says: "I had seen with my own eyes, buck-skin purses half full of gold-dust, lying on a rock near the road-side, while the owners were working some distance off. So I was not afraid of robbery; . . ." Much later, Stoll an attorney in the north Idaho area, decided that

life was safe, property was safe. My first month there I lived in a tent. I have often had in this tent as much as two tons of grub, something in a district so far removed from civilization as this, second only in value to gold. I have tied my tent flap shut in the early morning and returned late at night, day after day, but never have I lost so much as a toothpick.

He says a man named Hussey came to the camp with three thousand dollars to found a bank. "He had nothing but tent walls; he had no safe. Resources and deposits he kept in a suitcase, sleeping beside it unarmed. He was never troubled."

Borthwick even found everything safe in the gaming halls. When gambling was slack, the owners, he says, would walk away from their tables, leaving on them thousands of dollars in gold.

It was strange to see so much apparent confidence in the honesty of human nature, and—in a city where robberies and violence were so rife, that, when out at night in unfrequented quarters, one walked pistol in hand in the middle of the street—to see money exposed in such a way as would be thought madness in any other part of the world. But here the summary justice likely to be dispensed by the crowd, was sufficient to observe a due observance of the law of *meum* and *tuum*.

We suspect that the sharp eyes of the owners watched the tables.

Jackson, who was not there but who read what those had to say who were there, sums it up:

In the first year of the California gold rush there grew up a concept which found its way into songs and skits, into news stories, into comic books of the day, and finally into the public mind. Though a stereotype, it was essentially a stereotype with a reason behind it. This was the character of the "Honest Miner," unfortunate perhaps in his pursuit of riches, scrabble-bearded and ragged, reckless with his money, yet a man in whose eye a letter from home might start a tear, a rough and hardy frontiersman but one who would always stand by his "pard," who would impulsively adopt a baby in some "Roaring Camp," defend the rights of others, and trust implicitly in the honesty of his neighbor. This "Honest Miner" was to some extent the creation of the romantic fictioneer of a later time, but there was truth behind the highly colored portrait. . . . By 1850 . . . rascals were common in the gold region. No man's property or life could be called safe. The easy-going days of the "Honest Miner" were gone. . . . There were hard characters among the miners, plenty of them. Yet, no matter what else they did, there was one point at which they stopped, one line over which they refused to step. They did not steal gold. By common consent, honesty was the best policy.

To say that *none* of them crossed the line and stole gold is to say what was not true. Mr. Jackson did not wholly free himself of the stereotype. Elsewhere in these pages will be found men who became fabulously rich but who in the early years when they were miners spied on other miners and jumped their claims. Jumping claims, as we shall see in a moment, was one of the principal sports.

According to Dr. Knower, "Every thing was honorable and safe until the overland emigrants from western Missouri arrived there. They were a different kind of people; more of the brute order." But the ones who led to swift justice and vigilantes were not the lean lanky unwashed men from Missouri, but, as Major Bell discovered, the

hordes of men that were fugitives from justice elsewhere. They sank their identity under assumed names in the excitement of the times, . . . The "killers" of the early days were a queer study. Mighty good fellows to meet, many of them, without fearsome aspect or manner. . . . But each one of them was rather proud of having killed his man, or men. They liked to reminisce over these affairs. I have sat down in camp with these men and listened to their after-supper conversation. . . . often it would run like this: "Have you heard of Bill Magee recently?" "Oh, yes. The last I heard of him he killed Jake Sipes up on the Trinity, then went over into the Pit River country, staked a claim, found that a man had got in ahead of him so he killed him."

The innocent belief that all men were honest did not last long in the early camps, and if a sense of betrayal then drove human passions to an opposite extreme it must be admitted that the provocations were grievous and many. Professional criminals swarmed in from many parts of the world and were soon partners in crime with corrupt sheriffs, judges, and lawyers (whom see elsewhere). An instance of a courtroom scene that was a shameful farce is given by Bancroft.

Court was sitting when there was a knock on the door, and a man thrust his head in and asked if the defendant had been found guilty. When the foreman of the jury said he had been found not guilty the man at the door exploded with profanity, and laying his hand on his pistol butt said they would have to do better than that. When he returned a half-hour later he was told that the defendant had been found guilty. Bret Harte tells of a man poking his head into a courtroom to ask if the accused had been found guilty. The answer was no, "Very well," said the man. "Take your time but remember that we'll need this room to lay out the corpse." A notorious case of justice perverted is told by Dimsdale. A man in the Helena area named Daniels killed a man named Gartley with a knife. Because by that time the vigilantes had disbanded the civil authorities took charge of Daniels. With the defense in the hands of shyster lawyers, and Daniels openly swearing vengeance on judge and witnesses if convicted, a set of "extenuating circumstances" was conveniently found and Daniels was declared guilty only of manslaughter. Soon afterward he was reprieved. Hearing that he was repeating his threats against the witnesses, some of the former vigilantes captured him and hanged him that night.

Another thing that promoted swift and remorseless justice after the camps became fully aroused was the indifference to crime in people besides those hired to keep the peace or to punish the guilty. Beidler says in his dry way that on one occasion when those voting for hanging were to go up a hill, and those for acquittal were to go down, acquittal won because those "in favor of no hanging were a lazy lot of loafers and naturally went down hill and beat us." Sharing a cynical indifference to crime were many of the camp newspapers, though there were exceptions. Typical of reporting of crime, without anger, is the *Gold Hill News* (August 22, 1876):

Another disgraceful affair took place in the Delta Saloon* about three o'clock this morning. About twenty men, some of them county officers, were engaged at the hour named in drinking whisky and talking politics. The disputants merged from politics into personalities. Bad names were bandied and over a dozen revolvers were drawn and held aloft. Officer Curby ran in and attempted to quell the disturbance and arrest the leaders. He was threatened with death and shouldered out of the saloon. Chief among those concerned in this outrage was an official whose duty it is to prosecute those who offend against the laws of the State.

The early *Statesman* in Boise spoke out with great courage against crime and criminals and corrupt officials; and so did the *Bonanza* in Tonopah:

Our city is becoming rotten to the core. . . . There are too many lecherous loafers who live on the earnings of fallen women; too many hop fiends; too many yeggmen and fakirs; too many saloon bums and men of the never-sweat stripe. Crime in every form from murder down to rolling and pocket-picking is the result, and the law-abiding citizens are suffering therefrom. Strangest of all is the fact that this execrable ghoulish state

of affairs has only developed within the last month or so.

It may be that the editor was astonished when officers got the donation of an empty boxcar, and, packing in a hundred bums, shipped them on to another camp.

In camp after camp the harassed and exploited law-abiding citizens took the law into their own hands. The most widely quoted instance is Buffum's explanation of how Dry Diggings became known as Hangtown:

A Mexican gambler, named Lopez, having in his possession a large amount of money, retired to his room at night, and was surprized about midnight by five men rushing into his apartment, one of whom applied a pistol to his head, while the others barred the door and proceeded to rifle his trunk. An alarm being given, some of the citizens rushed in, and arrested the whole party. Next day they were tried by a jury chosen from among the citizens, and were sentenced to receive thirty-nine lashes each, on the following morning.

Never having witnessed a punishment inflicted by lynch-law, I went over to the dry diggings on a clear Sunday morning, and on my arrival, found a large crowd collected around an oak tree, to which was lashed a man with a bared back, while another was applying a raw cowhide to his already gored flesh. A guard of a dozen men, with loaded rifles, pointed at the prisoners, stood ready to fire. . . .

After the whole had been flogged, some fresh charges were preferred against three of the men—two Frenchmen, named Garcia and Bissi, and a Chileno, named Manuel. These were charged with a robbery and attempt to murder, on the Stanislaus River, during the previous fall. The unhappy men were removed to a neighbouring house, and being so weak from their punishment as to be unable to stand, were laid stretched upon the floor. As it was not possible for them to attend, they were tried in the open air, in their absence, by a crowd of some two hundred men, who had organized themselves into a jury, and appointed a *pro tempore* judge. The charges against them were well substantiated, but amounted to nothing more than an attempt at robbery and murder; no overt act being even alleged. They were known to be bad men, however, and a general sentiment seemed to prevail in the crowd that they ought to be got rid of.

At the close of the trial, which lasted some thirty minutes, the judge put to vote the question, whether they had been proved guilty. A universal affirmative was the response; and then

* Still in business.

the question, "What punishment shall be inflicted?" was asked. A brutal-looking fellow in the crowd, cried out, "Hang them." The proposition was seconded, and met with almost universal approbation. I mounted a stump, and in the name of God, humanity, and law, protested against such a course of proceeding; but the crowd, by this time excited by frequent and deep potations of liquor at a neighbouring groggery, would listen to nothing contrary to their brutal desires, and even threatened to hang me if I did not immediately desist from any further remarks. Somewhat fearful that such might be my fate, and seeing the utter uselessness of further argument, I ceased, and prepared to witness the horrible tragedy.

Thirty minutes only were allowed the unhappy victims to prepare themselves to enter on the scenes of eternity. Three ropes were procured, and attached to the limb of a tree. The prisoners were marched out, placed upon a waggon, and the ropes put round their necks. No time was given them for explanation. They vainly tried to speak, but none of them understanding English, they were obliged to employ their native tongues, which but a few of those assembled understood. Vainly they called for an interpreter, for their cries were drowned by the yells of a now infuriated mob. A black handkerchief was bound around the eyes of each: their arms were pinioned, and at a given signal, without priest or prayerbook, the waggon was drawn from under them, and they were launched into eternity. Their graves were dug ready to receive them, and when life was entirely extinct, they were cut down and buried in their blankets.

Innumerable instances are in the records of punishment as swift and remorseless as that. Though in certain areas lynching became necessary to get rid of the most vicious offenders and put others to flight, less severe forms of punishment were common. An instance typical of many happened in the same Hangtown. A deserter from the ship *Ohio* was caught making off with a poke of gold dust and a sack of silver dollars, and, found guilty, was sentenced to hang. But after those who opposed capital punishment had spoken for him it was decided to give him a hundred lashes, shave his head, and cut off his ears. All of which was done. One would think that a man with his ears gone and his back slashed into welts and furrows would have taken to his heels, but this thief, bolder than many, stole a mule and tried to escape on it. He was caught and tried again. The miners decided to flog him but after stripping off his shirt and looking at his

mangled back they told him in the jargon of the camp to git, and he lost no time doing it. Kip tells of a man caught stealing money whose punishment was a hundred lashes and the loss of his ears. These two stories may be about the same man.

After seeing some men punished Shaw thought that "Possibly a little less severity might have accomplished the desired end and object, but it was not so regarded at the time. I am sure there was no element of persecution or revenge present in the cases I have related, but the public welfare alone was considered." One of his cases is of a sailor who abandoned ship to flee to the diggings. He was caught making off with two bags of gold. His feet were "tied to the foot of a tree, and a doctor cut off his ears, from the stumps of which he bled freely while he received his flogging." Greever tells of a man caught while trying to murder who was beaten to death with pick handles taken from a barrel where the attack occurred. Taylor says that near Stockton he saw a man "whose head had been shaved and his ears cut off, after receiving one hundred lashes, for stealing ninety-eight pounds of gold."

Sometimes the guilty person tried to steal not gold but a woman. The *Statesman*, April 12, 1866, gave an instance:

Last Friday a young man from an adjoining county addressed a letter to the wife of a non commissioned officer at Fort Boise. He had never spoken to her, but professed to have fallen desperately in "love at first sight," and asked her to appoint a place of meeting. The lady gave the "billet doux" to her husband, who next morning quietly nabbed the gay Lothario and brought him to justice. Taking him before the insulted wife he made him get on his knees and ask her pardon; then, having previously removed his coat and vest, the husband applied the cowhide until the blood flowed freely.

Sometimes a man who was a law unto himself was aided by a conniving sheriff or constable. When John Springer, a deputy sheriff, discovered in Idaho's Silver City that a horse thief had escaped from his jail he headed for Succor Creek and the ranch of Nat Graves. He asked Nat if he had seen a suspicious character, and Nat said yes, he had, and one of his horses was missing. Nat said that when he overtook the suspected man he was halfway across a river in a boat. "When he seen me he jist rared right up and dove in head-first. That's what he done, the last look I had at him." The sheriff knew that Nat's last look at the man was along the sights of his rifle.

The swiftness of frontier justice dismayed some editors in the East and fascinated others; some editors in the camps kept their readers posted on Eastern attitudes: "An eastern paper is surprised," said the *Avalanche*, September 15, 1866, "at the swift retributive justice which is meted out to violators of law on this Coast. It thus sums up:— 'San Juan, Nevada, State robbery at 6 A.M., of $3,000; reward offered at 7 A.M.; robbers shot and all money recovered at 2 P.M.; coroner's inquest at 3 P.M.; funeral of the thieves at 5 P.M.'"

But justice for the most part was meted out, not by vigilantes and not by men who were a law unto themselves, but, according to Shinn,* by the miners' courts. These had no permanent officers, unless, as here and there in California, the alcalde plan was adopted; and they had almost no written laws, and usually kept no records. Any person who felt himself wronged could call a meeting. He could tell his friends, who would tell friends, until at last his wish would become known in a whole area; if the miners thought he had cause for complaint they would assemble. They would then choose a presiding officer and a judge, impanel a jury of six, or sometimes twelve, and call witnesses. Now and then all the assembled miners served as the jury. A few matters the miners' courts refused to consider; they would not, for instance, be a party to collecting debts. It was thought such a contemptible thing to be dunned for money owed that a miner who one night was awakened and dunned put on his clothes, went to his claim and panned out the sum demanded, put it in a pouch, and returning hurled it into the dunner's face.

After a crime was committed and the suspected person was captured

Short speeches were made for and against the prisoner; the presiding officer charged the jury "to do the fair thing accordin' to common sense"; and after a brief deliberation they rendered a verdict, and the presiding officer decided the penalty. If there was no jury, the case was submitted to the decision of the miners present, who also fixed the punishment. In small camps, where only thirty or forty men assembled, this was perhaps the usual system; but in larger camps the jury system prevailed. In one case a trial was had and the prisoner hung within an hour; it being a cold-blooded murder, with a dozen witnesses to the act. In another case the jury spent four days taking testimony and listening to arguments. When it was decided to hang the prisoner, the crowd guarded the town and the building where he was confined, having reason to fear an attempted rescue; they gave him four days to prepare for death, and then, on the ninth day after his arrest, hung him to a bridge in the presence of more than two thousand men, . . .

Hanging was the penalty for more crimes than murder, but "cases of whipping were very numerous." In Grass Valley in 1850 a mule thief got thirty lashes. At Hay Fork the first thief caught was "whipped by the entire camp, each miner giving a blow." A man who sneaked into the tent of a Chileno when he was absent to rape his wife "was whipped so severely that he was hardly able to drag himself out of the camp he had disgraced." A storekeeper was whipped because he used false scales to swindle the miners. In Calaveras County a thief was given ten lashes; when that night he stole again he was given twenty; and when in the next camp he stole a mule he was sentenced to fifty, but they found his back in such a dreadful condition from the previous punishments that the mule's owner dressed the wounds.

Typical of the richer camps was Jackass Gulch in California. Shinn says that a piece of earth ten feet square often yielded $10,000 from the surface dirt. No claim larger than that was allowed there. After 1851 these were some of the rules adopted and enforced:

First, That each person can hold one claim by virtue of occupation, but it must not exceed one hundred square feet.

Second, That a claim or claims, if held by purchase, must be under a bill of sale, and certified to by two distinguished persons as to the genuineness of signature and of the consideration given.

Third, That a jury of five persons shall decide any question arising under the previous article.

Fourth, That notices of claims must be renewed every ten days until water to work the said claim is to be had.

Fifth, That, as soon as there is sufficiency of water for working a claim, five days' absence from said claim, except in case of sickness, accident, or reasonable excuse shall forfeit the property.

---

* His book on government in the camps is the standard.

There were sixteen articles. Under no circumstance was any person allowed to hold more than one claim, and he must labor on it at least one day in every three during the mining season, though several persons were allowed to concentrate their toil upon any one of their claims. Thirty days absence from a claim during the season resulted in loss of it "without remedy." Article Thirteen prohibited persons from serving on committees of arbitration or on juries, who were not American citizens or whose citizenship was in doubt.

After a while many of the camps had a sheriff, sometimes with deputies, but as often as not the sheriff was a scoundrel in the pay of criminals. Legal fees varied but might be three ounces for the alcalde or chief official, two for the sheriff, and one for each juror. Sometimes the jurors refused to render a verdict until they were paid. Sometimes to the fees was added as much whiskey as the court, jury, and witnesses could hold. Sometimes a man whose claim was jumped did not bother with judge and jury or miners' court but gathered some friends to go with him to chase off the claim-jumpers.

Claim jumping, ranging from the ignorant miner who thought a spot had been abandoned, to organized gangs that used blackmail and force, became, in some areas, almost as common as gambling and drunkenness. In early San Francisco it was so bold and brazen that some of the facts seem fantastic. If an owner found a jumper, or, as likely as not, several men, all armed, on his property, and refused to be bluffed, he was sometimes beaten within an inch of his life and thrown off. In some instances he was killed on the spot. Those who had property outside the town limits could hold it only by policing it day and night. One jumper fenced Union Square in the heart of the camp and tried to hold it at gunpoint. By 1854 the principal owners of property had been forced to establish an association for the protection of their rights.

Claim jumping, according to Professor Paul, under cover of some alleged prior right, became a regular kind of legal blackmail, in which the hope of the "jumpers" was not that they might win, but rather that they would be "bought off." Part of the trouble here was inadequate enforcement of the rule that a man must continue actively to develop his claim in order to retain title to it. Carelessness in describing and marking boundaries increased the chances for chicanery. Then again, there was the danger that the discoverers of a district would try to "rig" the rules in such a way as to monopolize the promising ground.

That was part of it but only a part of it. Often the jumpers simply took possession of a claim, when the owner was absent, and if he had any improvements on it, they pretended they had put them there, or they pitched them off the property. Some of the jumpers organized themselves into gangs that were much like criminal gangs of our time. One, calling itself the Merced Mining Co., staked out and fortified a number of claims on the Mother Lode, and then attempted to take by force a claim belonging to Colonel Frémont. Frémont hastily threw up fortifications, and there was open war until troops were sent in.

The *Statesman* reported, August 27, 1864:

We heard of one case where a man was building a fence around his lots and a jumper came along and commenced building just inside, claiming the same lot. Another where a man was in actual possession of his lot and getting material to build a house, and taking advantage of a temporary absence his lot was jumped and fenced. It is rumored that a number of persons of the border ruffian stamp, have within the last day or two signed a pledge to see each other through in their efforts to gobble up whatever town property they can, that they have in their wisdom determined to respect no fence, in fact nothing but a house and an actual resident on a lot.

Whether it was a mining claim, properly posted, or a lot or lots in a camp, or an entire ranch made no difference to the jumpers. Bancroft found that in Oregon the forcible possession of land and property belonging to others became so outrageous that a man named McDonald organized a gang and seized some of the best lands.

So bold had the claim-jumpers become that they openly avowed their purpose to resort to violence and murder if necessary . . . on several occasions settlers had been shot from thickets, and notices embellished with a death's-head and coffin were found in the post-office directed to citizens in various parts, threatening them with assassination if they refused to vacate their land. Forty of the settlers now banded, armed, and started in pursuit of the claim-jumpers.

On catching them they shot them because "hanging was not their forte."

Shinn gives the remarkable Scotch Bar case in northern California. After a rich discovery was made two parties laid claim to it. Bitter feelings compelled both sides to recruit armed volunteers until every gun in the camp had been pressed into service. After the opposing forces drew up in battle array and began shooting at one another the hundreds of men who had been spectators decided that the thing had gone far enough. They forced a settlement, without lawyers and courts, that for intelligence and fairness is said to have been without a rival in the Western mining camps.

Mrs. Davis found in a Montana newspaper a story of a gang that "had been organized in Idaho and Nevada" and moved to Montana for easier pickings, "a sinister, desperate band of frontier desperadoes." Finding a spot that looked good to them they served notice on its two miners that if they hadn't vanished from the area by sunset of the next day their fate would be "sudden death." The astonished men carried the news to a half-dozen camps roundabout, and about twenty men, well armed and with provisions for a long siege, "portholed one of the cabins not far from the claim of the two men. . . . At dusk a dozen of the claim jumpers appeared," but had barely had time to look round them when the one who appeared to be the leader fell dead. When the claim-jumpers opened fire on the cabin the men in it rushed outside, and when three more of the gang fell the others took to their heels. That way of handling professional criminals, and Abe Slocum's way, are a lot simpler and easier than a long series of appeals, and a higher court's reversal on an absurd technicality.

Like Jedediah Smith Abe was devoted to both his Bible and his gun. Hearing that a gang of jumpers intended to drive him off his claim, he kept his rifle within reach as he toiled. As the gang approached, Abe picked up his rifle, and when one of the men ordered him off he pointed out the boundaries of his claim and said they were welcome to stake claims beyond his. The leader of the jumpers said they would settle for half of it. The story says that Abe thereupon gave them a little sermon about God and peace and the rights of all men. That over with, he told them that if they intended to put a pick in his diggings they

should first prepare themselves to meet the Almighty, because as sure as his name was Abraham Slocum he would put one of them beyond all thoughts of gold.

Pointing out that claim jumping was a hazardous way of life, the editor of the *Avalanche* went on to say: "Take the history of frontier towns and settlements generally, and more shooting scrapes, murders and expensive litigation have grown out of disputed titles to lots and lands than from any other one source." There was so much litigation in some of the camps that it clogged the courts. From 1860 to 1865 just one mine on the Comstock, the Ophir, was involved in thirty-seven suits, being the plaintiff in twenty-eight of them. J. Ross Browne, who visited this bonanza camp, told the readers of *Harper's* in January, 1861, that "Nobody seems to own the lots except by right of possession; yet there was trading in lots to an unlimited extent. Nobody had any money; yet every body was a millionaire in silver claims. Nobody had any credit, yet every body bought thousands of feet of glittering ore." Everything underground, he said, "was silver, and deeds and mortgages on top." Eventually there were seventeen thousand claims, but half of the wealth taken out, and eighty per cent of the dividends came from two pairs of mines close together, the Crown Point and Belcher, and the Consolidated Virginia and California.

There were savage battles over disputed claims. Wells Drury was on the Comstock when "the most desperate battle over a claim was fought at the Waller Defeat shaft, between lower Gold Hill and Devil's Gate, on October 3rd, 1874." A few men were killed and an unknown number were wounded.

Whether that was a more savage fight than that over the famous Poorman vein in Idaho's Owyhee district no man knows. Two men, Hays and Ray, had sunk a shaft. A prospector named Peck found rich float a thousand feet south of the shaft, and after exploring the area became convinced that the mine a thousand feet north of him included his discovery. He promptly hid all the float he could find and left the area, because for some strange reason he thought it would be best to go away for a while. During his absence his float was found. When Peck returned, the finders of his float were busy building a fort, to defend themselves against Hays and Ray. That is the way

Rickard, a mining engineer, tells that much of it. We now switch to Richardson, who was on the scene.

The Hays-and-Ray Lode, as claimed and staked, was sixteen-hundred feet long. Other parties afterward claimed, for fourteen hundred feet, a lode which they called the Poorman, crossing the Hays-and-Ray at an acute angle, the two lines of stakes exactly representing the letter X. The Poorman party began to work their lode, not at either end, but at the very point of crossing the Hays-and-Ray; and there struck a "pocket" or "chimney" of ore of unprecedented richness—almost pure silver. Portions of it yielded sixty per cent of bullion—a result never before equaled in mining history. The Poorman owners, it was alleged, took out two hundred and fifty thousand dollars in two weeks. . . . The Hays-and-Ray proprietors claimed that the adverse party were reducing *their* ore. The "Poorman" not only denied this, but, with an armed force, drove off the Hays-and-Ray workmen from a portion of the ledge, and prepared for bloody warfare. And hence Fort Baker was erected.

The United States district court granted an injunction, restraining the Poorman party from taking out more ore, until both claims could sink shafts, and trace their veins, and a jury decide which owned the disputed mineral. This just and equitable decision excited so much feeling, that threats of tarring and feathering were even made by the friends of the discomfited claimants. But the belligerents finally compromised the matter and consolidated their interests. This was at least better than Nevada in early days, when in great quartz cases, one party and sometimes both, used to buy witnesses, jury, sheriff, prosecuting attorney and judge. The man in the play must have lived in a mining region before he learned the profound truth, that "Honesty is the root of all evil, and money is the best policy."

The conflict of interests between two other silver mines in this area, the Golden Chariot and Idaho Elmore, was not so easily settled. It is known in history books as The Owyhee War. According to historian Merle W. Wells, the trouble began when the Chariot "violated the neutral ground and broke through into the Idaho Elmore workings." There was tension on both sides and some shooting but "hostilities did not amount to much for a month." Then the Chariot's forces advanced on the Elmore in "an offensive marked by heavy firing that threatened to shatter and break up the timbering and bury the underground belligerents." When one side called for reinforcements toughs from Nevada were ready and the battle "assumed serious proportions."

By this time some persons wanted to know why the sheriff didn't stop it. "Just how the sheriff, all by himself, was going to stop a hundred armed men from firing on each other in underground workings was never proposed." The best the sheriff could do was to close the saloons. The Governor up in Boise then issued a proclamation and on arriving in Silver City found the situation so serious that he summoned troops from Fort Boise. Ninety-five soldiers with a brass cannon brought the leaders of the two mines to an agreement.

Dr. Wells makes the excellent point that some of the claimants chose to wage war instead of going to court because the cost of litigation consumed most or all of the profits. If the Owyhee War was fought to avoid the costs of litigation he wonders if they saved anything. The fighting cost the Elmore a third of the $600,000 taken out the first year, and the Chariot its entire first year's production. After the feud was settled the combined production of the two mines was more than $200,000 a month.

As Glasscock tells it, Jim Butler, discoverer of the Tonopah field, granted a hundred leases on his claims without a scrap of paper, but at last was forced to take a case to court. Judge Ray, before whom the litigants appeared, had, unlike most frontier judges, a huge pile of law books, though he seldom opened them. Key Pittman, later to become a famous Nevada senator, was sleuthing his way through a legal tangle when a man named Johns, a party in the case, shouted at Pittman, "You're a liar!" Pittman swung on the man's jaw, knocking him against the judge, who fell sprawling against the shelves of his books, bringing them all in an avalanche upon the three men. After the judge got his head out and surveyed the wreckage he allowed as how he ought to adjourn court, so that they could all refresh themselves at the Mizpah bar. That was the end of the case.

But the most spectacular crimes and the most shocking indifference to justice are found not in saloon brawls and claim jumping but in the raids on the coaches which carried the gold and silver to the mints and banks. Though hundreds of bandits were captured they had a pretty easy time of it. No plausible explanation has ever been made of the fact that Black Bart (see under

"Badmen"), always alone, was able to rob twenty-seven stages before he was taken. Of what use in those twenty-seven instances was the armed guard sitting with the driver? Jackson quotes Bancroft's ironic statement, "There seems to be a prejudice in some quarters against the profession of high-wayman. It has become the custom of our refined and discriminating civilization, when such a person is caught, to kill him; for which reason many good men have been kept out of the profession, and have in consequence fallen into evil ways." Jackson suspected that Bancroft had been reading De Quincey, who was opposed to murder because it often led to "robbery, drinking and Sabbath-breaking, and from that to incivility and procrastination."

Brace reported that

I have reached now the mountain region of California, where robbing is almost a profession. Not a week passes, sometimes not three days, in which we do not hear of a coach robbed or of teamsters or foot-passengers being plundered. No one resists. Your native Californian takes robbing as an Englishman without an umbrella takes a shower. It is unpleasant, he may grumble a little, but it belongs to a law of nature. A man is considered somewhat of a fool, who should be killed defending his property. I have often talked with the old settlers about this non-resisting habit. It does not arise from want of courage, for if a mountain Californian possesses any virtue, it is a reckless disregard of life. It seems to come from a low appreciation of money, where it is made so easily, and a cool, intelligent accepting of a fact, as a fact. They understand perfectly what a "six-shooter" means when presented to their face, and they hand over their "bullion."

By coach, Wells, Fargo transported nearly fifteen million in 1866. The bandits "had spies loitering around the stage offices, sometimes stage and express petty employees in their pay. If a stable boy swung a lantern in a circle back of the barn just as a coach was about to depart, a single horse's hoofs would be heard at a distance pelting away in the darkness, and a chance observer, had he 'been in the know,' would have guessed that that coach would be robbed only a few miles away." Harlow makes it sound awfully easy, and apparently it was. The stage robbers are bank robbers now, and they also seem to have an easy time of it.

Dimsdale saw a lot of it. He says:

Thus armed and mounted on fleet, well-trained horses, and being disguised with blankets and masks, the robbers awaited their prey in ambush. When near enough, they sprang out on a keen run with level shotguns, and usually gave the word, "Halt! Throw up your hands, you sons of bitches!" If this latter command were not instantly obeyed, there was the last of the offender; but in case he complied, as was usual, one or two sat on their horses, covering the party with their guns, which were loaded with buckshot; and one dismounting, disarmed the victims and made them throw their purses on the grass. This being done, and a search for concealed property being effected, away rode the robbers. . . .

According to the Gold Hill *News* it was commonplace for Wells, Fargo to offer substantial rewards for the capture of stage robbers. Some of the sheriffs offered rewards, as in Gold Hill for a murderer named Langan: "'Red Mike'—Under indictment for murder. He is a native of Ireland; is about 5 feet 5½ inches high, about 35 years old; sandy, almost red hair and whiskers; red face; had short whiskers on chin; good, high, strong forehead; a restless blue eye—looks away as soon as you look at him; rather quick of speech; quite solid looking. . . . I will give $100 for him in any shape you may bring him. . . ." On the same circular, sent far and wide: "Lafferty, alias Rosenbaugh—About 5 feet 5 inches in height; rather pale; an inveterate opium smoker; rather good looking; about 25 years old; heavy eyebrows; smooth face; light moustache. Will pay ten dollars for him—he is not worth that."

According to Drury, Wells, Fargo closed some of its mountain offices "it is alleged, because when a stage was robbed a jury couldn't be found in any of the outlying communities which would convict the highwaymen." It would be interesting to know what Ben Holladay, the terrible titan of staging, thought about that. Ben was once stopped in his own royal coach and told to put up his hands. After a few moments he said his nose itched and asked permission to scratch it. According to Harlow, the bandit said he would scratch it for him, and did, using the muzzle end of a Colt .45.

It was 1912, Beebe says, when a Wells, Fargo was boarded by bandits for the last time. On that day, the guard, a man named Troutsdale, was suddenly confronted by a bandit with an old-model Winchester, who demanded to know where the currency was. Figuring that he must be a green-horn if he didn't know where the money was kept, Troutsdale outmaneuvered him until he got hold of a wooden mallet, with which he killed the robber with a blow across his skull. After examining the gun to be sure it was loaded Troutsdale opened the door and looked out. There stood a confederate. The guard "blew what brains the man possessed all over the desert." Other men, lending a hand, stood the two dead robbers on their feet so that they could be photographed.

Nowadays a guard with courage to shoot a couple of robbers would have juries and courts against him, marching picket lines, outraged parole boards, and suits for damages from relatives. In that instance, fifty years ago, the Congress voted a cash award of $1,000 to Troutsdale because he saved a bag of registered mail. Theodore Roosevelt would have understood that. When, in 1903, he visited Butte, where feeling was bitter between rival copper interests, he had placed before him in a dining hall a huge tray with a cover on it. Removing the cover he saw a dozen six-shooters which had been taken from dinner guests at the door.

CHAPTER 17

Frontier Judges

"Thieves," said Shakespeare, "for their robbery have authority/When judges steal themselves." When looking at the judges on the American frontier and in the mining camps we have to put away the image of a detached, impartial, and very wise man, stiff with dignity and wearing a black robe, and see His Honor in eccentric garb, with a quid of tobacco in his cheek and a flask of whiskey in his pocket. They didn't all chew tobacco or drink with the jury, put their feet on their desks, swear at the witnesses, and rest judgments on whim and caprice instead of facts. But a lot of them did. We'll first look at some with minor frailties and then proceed to the autocrats and the crooks. It will be kept in mind that drinking and chewing and raising some hell were, after all, acceptable habits in the early West.

For a sobering moment we'll present Colonel Chivington, and then let that fairly civilized man, Walter Colton, alcalde of Monterey, open the subject. According to the newspapers a man named Kittery, in Denver, slipped up to a whore-house, smashed a window, and shot a drunken soldier who was trying to have some fun inside. An enraged mob gathered, and dousing the premises with turpentine set fire to it. When it was a sheet of flame they rushed to a dance hall and reduced it to ashes. They were then going to lynch the murderer when a man who hadn't yet taken leave of his senses brought Colonel Chivington to the scene.

Chivington addressed the mob, saying,

Two wrongs never can make one right. . . . We live in a land of Law—where every man has meted out to him equal and most exact justice. It would be just as cowardly for 10 or 20, or all of you men, to take him out from here, bound as he is, to hang him, as it was for him to shoot your comrade from the window. . . . if you should hang him without a trial, you would be just as guilty of a cowardly murder as he is.

Those were mighty strange words in a mining camp. Some of the men who heard them, and who were familiar with the antics of judges and lawyers, must have thought the stuff about equal and most exact justice howlingly funny. Whether the murderer was tried in a court of law, or lynched, we do not know.

As an alcalde Colton was a judge.

A Californian came into my court in great haste last evening, and complained that another Californian was running away with his oxen. Suspecting the affair had some connection with a gambling transaction, I immediately handed him a warrant for the arrest of the fugitive, when off he started at the top of his speed to execute it. In less than an hour he returned with his prisoner. I then asked the plaintiff if the oxen were his; he said they were. I asked him of whom he obtained them; he said of the man who attempted to run away with them. I asked him what he gave for them; this was a puzzler, but after hemming and hawing for a minute, he said he had played for them, and won them. I asked him what else he had won of the man; he replied, the poncho, and a thin jacket, both of which he had on. I then ordered them both into the calaboose for the night. The winner, who had apprehended the other, and who, no doubt, expected to get the oxen at once, looked quite confounded.

This morning I had the two gamblers before me: neither of them looked as if he had relished much his prison-couch. I made the winner return all his ill-gotten gains, oxen, poncho, and jacket, and then fined them each five dollars. The one who had served the warrant shrugged his shoulders, as if he had made a great mistake. . . . The next time they gamble they will probably settle matters between themselves, without a resort to the alcalde.

In the next case a California

mother complained to me to-day, that her son, a full-grown youth, had struck her. Usage here allows a mother to chastise her son as long as he remains unmarried and lives at home, whatever his age may be, and regards a blow inflicted on a parent as a high offence. I sent for the culprit; laid his crime before him, for which he seemed to care but little; and ordered him to take off his jacket, which was done. Then putting a riata into the hands of the mother, whom nature had endowed with strong arms, directed her to flog him. Every cut of the riata made the fellow jump from the floor. Twelve lashes were enough; the mother did her duty, and as I had done mine, the parties were dismissed.

A man

who had been absent some two years in Mexico, where he had led a gay irregular life, finding or fancying on his return grounds for suspecting the regularity of his wife, applied to me for a decree of divorce, *a vinculo matrimonii*. I told him that it was necessary, that on so grave a subject, he should come into court with clean hands; that if he would swear on the Cross, at the peril of his soul, that he had been faithful himself during his long absence, I would then see what could be done with his wife. He wanted to know if that was United States law; I told him it was the law by which I was governed—the law of the Bible—and a good law, too—let him that is without sin cast the first stone. "Then I cannot cast any stone at all, sir," was the candid reply. "Then go and live with your wife; she is as good as you are, and you cannot require her to be any better." He took my advice and is now living with his wife, and difficulties seem to have ceased. Nothing disarms the man like the conscious guilt of the offence for which he would arraign another.

We will let the Monterey judge conclude with a story about an assistant alcalde, at San Juan, who,

in reporting a case that came before him, states that one of the witnesses, not having a good reputation for veracity, he thought it best to swear him pretty strongly; so he swore him on the Bible, on the cross, by the holy angels, by the blessed Virgin, and on the *twelve* Evangelists. I have written him for some information about eight of his evangelists, as I have no recollection of having met with but four in my biblical readings.

Some frontier judges were held in such low esteem by those brought before them that they were cursed out. Greever says that at Hell Gate, Montana, there was a trial for trespass; when the defendant took a good look at the judge he burst forth, "Who in hell made you judge? You are a squawman with two squaws and your business, you son of a bitch, is to fill the country with half-breeds." The original story was told by Granville Stuart, who possibly was present. According to him, the courtroom exploded into a free-for-all fight after the defendant had addressed the judge, and "when the dust of battle cleared away it was considered a draw. But judge, jurors, constable, and prosecuting attorney" had all vanished.

Shinn says, and Gard in his *Frontier Justice* quotes with approval:

The men who were chosen justices of the peace in these mining camps were often eccentric and illiterate, but as a rule their honesty and good judgment were unquestionable. They had the full confidence of the people and were conscious of

Vanity Fair's "image" of a judge in 1873

the responsibilities of their office. In many a town of the mining region, the pioneers still remember their names with respect and still smile over their eccentricities. One of them, when dying, left all his money, after paying funeral expenses, to the boys for a treat; and it was duly spent in the saloons of the camp. Yet he is said to have dealt out justice with firmness and good sense; . . .

Since Shinn's book on government in the mining camps is the standard work in this field we'll go along with the view that most of the judges were able and honorable; but Mr. Gard presents a lot of evidence, and we found some sources which he may have missed, that support a different view. Drury, the Comstock editor who was in the thick of it, says:

It was bewildering to read the report of a trial in a Comstock court: "Judge Whitman contended that certain evidence was inadmissible. Judge Mesick read from decisions of the Supreme Court precedents clearly establishing its admissibility. Judge Leonard and Judge Belknap held that the evidence was pertinent, but Judge Seely took diametrically opposite grounds. Judge Rising thought that the precedents cited in support of the admissibility of the evidence were not in strictly analagous cases, and that the evidence should not be admitted in this case. Judge Aude then addressed the jury, urging the acquittal of the prisoner."

How familiar it sounds, all the way from our lowest to our highest courts today! No doubt what we must realize is that ignorance and prejudice and vanity have a lot of fun in even the most judicial minds. Soulé, in his *Annals of San Francisco,* presents Judge Almond, ex-peanut vendor, and asks us to imagine him "paring his corns or scraping his nails, while seated on a dilapidated old chair, with his feet on a highest level. . . . he might delay proceedings, in order to squirt tobacco juice. . . . " Once the judge formed his opinion, "and that sometimes occurred before even the first witness was fully heard," his mind was made up good and true, and "Nothing further need be said."

Stoll was one of the attorneys for the plaintiffs when Noah Kellogg, discoverer of the "largest lead and silver mine in the world," was sued by the two men who grubstaked him. According to him the judge and all the other attorneys chewed tobacco throughout the trial, even at the top of

their most impassioned oratory. Further, if one of them took a huge bite out of his plug he passed it around, and if any man refused it that was "an unpardonable breach of etiquette." The judge and the chewing lawyers (as well as witnesses and principals, no doubt) all spit their juice on the floor—"a cuspidor at this period would have attracted as much attention as a marble bathtub in the hotel"; and during recesses a Negro servant got busy and mopped up the mess. There was a crisis in the camp when one day the Negro said he was leaving; the judge with much eloquence persuaded him to promise to return, pointing out that if he were not there it would be impossible to hold court.

Gard quotes a writer named Kent Sagendorph:

Most justices hold court anywhere they happen to be at the time some judicial business pops up. I saw a justice delay a hearing because he was up on the roof of his barn laying some shingles. Another holds court in the village grain elevator, where he is the bookkeeper. And one fairly well known justice can invariably be found in a village poolroom, forever trying to make a combination shot for the side pocket, and nobody has ever seen him do it yet. He grumblingly leaves the game, sits enthroned in an ancient chair in the front window, and dispenses justice and sarcasm in equal parts.

Another justice I know of is a barber in a small community. Not long ago a prominent businessman was brought before him on a charge of speeding. The businessman was in a hurry, trying to get to a distant point to keep an important engagement. But, before hearing his case, the minion of the law kept the defendant cooling his heels for the better part of an hour while he shaved several stubble-bearded customers.

All of which suggest that some frontier ways are still alive in some areas.

It seems likely that some of them made a virtue of eccentricity. The Boise *News* reported that "Justice Walker fined himself $5 for becoming angry in court and swearing at an attorney." Such impartiality, an editor had written, probably had never occurred before. Oh, but it had, the *News* editor said, and quoted from the Nevada *Gazette:* "When Charley Mason was Justice of the Peace in Yreka, some 12 or 13 years ago, he was betrayed into blasphemy, while holding court, and fined himself ten dollars for contempt, and adjourned the court to drink out the amount at a convenient bar."

Gard gives no source for a hilarious incident. A Texas justice was known as Old Necessity because he was completely ignorant of the law. His only book, in court, was a mail order catalog bound in sheepskin. He "consulted it diligently before giving his verdicts." On an occasion when he put on his spectacles, consulted his book, and fined a man $4.88 the defendant loudly protested. His lawyer yanked him back to his seat and whispered in his ear, "Shut up, you damn fool! Be thankful he opened it to pants instead of pianos."

Gard also gives a Turner anecdote from his essay, "Prairie Dog Lawyers." When an attorney for the defendant learnedly quoted the law, the prosecutor howled at the judge, "That ain't the law." He slapped a ten-dollar bill on to the table and said to his opponent, "I'll bet you it ain't the law." The defendant's lawyer looked at the ten, the prosecutor, and the judge, and was silent; and at last the judge, who probably hadn't a ghost of a notion which one was right, said, "If you ain't got the nerve to cover it I guess you're wrong. I rule against you."

Some of the judges were, as Soulé said of the peanut vendor, simply incompetent. In the first issue of the first newspaper published in the mining camps of California Justice Barry of Sonora inserted in the *Herald,* July 4, 1850, these words, whose worst errors of spelling, grammar, and punctuation the printer corrected:

All persons are forbid firing off pistols or guns within the limits of this town under penalty; and under no plea will it be hereafter submitted to; therefore a derogation from this notice will be dealt with according to the strictest rigor of the law so applying as a misdemeanor and disturbance of the peaceful citizens of Sonora.

This justice had before him a Mexican whose trial had run for nearly two days. Barry wrote out his judgment:

Having investigated the case wherein—Barretta has been charged by an old Mexican woman named Maria Toja with having abstracted a box of money which was buried in the ground jointly belonging to her self and daughter, and carrying it, or the contents away from her dwelling, and appropriating the same to his own use and benefit, the supposed amount being over two hundred dollars; but failing to prove posittively that it contained more than twenty, and that proven by testimony of his owne witness, and by his owne acknowledgment, the case being so at variance

with the common dictates of humanity, and having bean done under very painful circumstances, at the time when the young woman was about to close her existence, the day before she died, and her aged mother the same time lying upon a bead of sickness unable to rise or to get a morsel of food for her self, and he at that time presenting him self as an angel of releaf to the poor and destitute sick, when twenty poor dollars might have releaved the emediate necessitys of the poor, enfeabled, sick, and destitute old woman, far from home and friends. Calls imperitively for a severe rebuke and repremand for sutch inhuman and almost unprecedented conduct, as also for the necessity of binding him over to the Court of Sessions in the sum of $500.00.

(Signed) R. C. BARRY, J.P.
Sonora, Nov. 10, 1851

Thomas Donaldson, onetime clerk of the Idaho Territorial Supreme Court, says of Judge Gilmore Hays of Silver City that he was a friend of the "famous colonel, John Doniphan," that he had "sat as judge more than once before Doniphan in Missouri," and that Doniphan had told him "that he attributed his success at the bar to the fact that he had never read a law textbook." The Idaho *World* complained repeatedly of the quality of territorial judges, saying that "Smith of the Lewiston district is said to be entirely incompetent"; that one named Parks "has never been a resident of the Territory, and failed to be here at the regular July term"; and that the quality and behavior of territorial judges was "not only a defiant neglect of the people, but is an insult." If that is so, the people are still being insulted and apparently don't mind it, for as we write these words in October, 1965, magazines and newspapers are telling us that the younger Senator Kennedy is trying to get a man appointed to an important judgeship who is incompetent, and that President Johnson is determined to elevate over judges of more experience and learning a man with invisible qualifications, for the very good American reason that his brother contributes huge sums to Democratic campaigns.

Joseph H. Jackson found another imperishable instance of Judge Barry's wisdom: "This was a gambling scrape in which T. Smith the monte deeler shot and wounded Felipe Vega. After heering witnesses on both sides, I ajedged Smith guilty of the shooting and fined him 10 dolars, and Vega guilty of attempting to steele 5 ounces.

I therefore fined him 100 dolars and Costs of Court, Costs of Coort 3 ounces." He continues: "H. P. Barber, the lawyer for George Work in silently told me there were no law fur me to rool, so I told him that I did not care a damn for his book law, that I was the Law myself. He jawed back so I told him to shetup but he would not so I fined him 50 dolars and comited him to gaol fur 5 days fur contempt of Coort in bringing my roolings and dissisions into disreputableness and as a warning to unrooly citizens to contredict this Coort." Judges Barry and Roy Bean should have known one another.

Like a lot of the frontier judges, Bean, and possibly Barry, had an unappeasable thirst for whiskey. In Placerville a judge adjourned court so many times during the trial of a claim jumper, to let lawyers, jurors, witnesses, and himself above all, refresh themselves at the nearest saloon, that according to a biographer of the discoverer of gold on Sutter's property, a "drunken lawyer addressed a drunken jury on behalf of a drunken prosecutor. A drunken judge having delivered an inebriated charge, a fuddled verdict of acquittal was delivered."

We have dwelt on race prejudice in another chapter; here we point out (what no literate reader needs to be told) that some of the frontier judges were guilty of it. Kip tells of a Mexican who, having lost a mule and found one that looked like it, was brought to court. He apologized, but the judge said that in addition to the apology there would be a fine of one hundred dollars, and fifty lashes. That was such outrageous punishment that an Englishman took pity on the Mexican, who said he had no money, and paid the fine. The Englishman then had some scathing things to say to the judge about the land of the free and the home of the brave, whereupon the angry judge offered him his choice of a fine of $50 or twenty-five lashes.

In June, 1860, a German couple at Carson City charged a Mexican with trying to break into their home to kill them. During the examination the woman swore that the Mexican had made indecent proposals to her, though her reputation was not of the kind to inspire confidence in her virtue. At the close of the examination the woman drew a pistol and thrusting it against the head of the Mexican killed him. Bancroft says that "neither judge, sheriff, nor the people made any

attempt to arrest her, but permitted her to return home in peace." It is hardly an overstatement to say that this was typical of treatment of Mexicans by "whites."

Indeed, the tax levied on foreign miners in an effort to drive them out of the diggings made criminals of many of them. The country had belonged to the Indians before the Mexicans came, and to the Mexicans before the whites began to pour in; and the whites for the Mexicans were trespassers and poachers and thieves. The Mexicans liked them little enough before the tax was levied; after that they hated them.

And the whites despised not only them but the Indians and the Chinese. In a Montana camp a Chinese girl made the mistake of doing an Oriental dance for the miners, the "full details of which," said a shocked newspaper editor, were unfit for publication. "Mary became drunk with excitement and stripped off her cotton trousers and nankeen chemisette. Her pantaloons contained a roll of $100 in greenbacks, a gold josh god, and other Chinese valuables, all of which Mary lost." When she brought a charge of theft against some of the men who had watched her dance, the court, says Barsness, had a "full two days of fun at the poor girl's expense." At last the judge said, "Mary, from China, stand up." He probably put on his most judicial face for his concluding words: "This court decides that you, while in company with white fellers, hereafter (if you want to keep your greenbacks) do *Keep your britches on.*"

A few of the frontier judges were men of delightful wisdom. The alcalde of Monterey was one. Another was Mark Aldrich, according to a story in the *Arizona Historical Review.* An officer had brought in a big, muscular, brutal Mexican. Aldrich kept in court a heavy leather strap, and his favorite punishment was 20 or 30 blows with the strap across the guilty man's naked back. But only half of them the first time. After fifteen had been given he would say, "You can come back tomorrow and get the rest of them," but by tomorrow the criminal was far from Tucson and riding fast, which was exactly what the judge wanted and expected.

According to Ramon Adams, Miguel Otero, onetime governor of New Mexico, was not a reliable witness. Otero tells a story of a judge named Benedict who had before him a murderer of the "greatest brutality, with absolutely no mitigating circumstances." The Judge said, "Jose Maria Martin, stand up." He told him he had had a fair trial and had been found guilty, and that "the court takes positive delight in sentencing you to death." He pointed out that Jose was a young man in good health, and the time was the spring of the year; soon the grass would be pushing its fresh green heads out, birds would be singing of love and mating, and flowers would be dressing up the hillsides. But Jose would be in a cell and would see none of it. On Friday, March 22, the sheriff would take him to an appropriate spot and hang him by the neck until dead. The judge said he would not ask God to have mercy on Jose's soul, for why should he ask an "All Wise Providence to do that which a jury of your peers refused to do?" But Jose did see the flowers and the grass and hear the birds sing again; the absurd thing called a jail did not hold him.

Some writers have thought they saw the wisdom of Solomon in a judge who heard evidence against a Negro named Sam Totem. One morning the judge saw Hop Lee and his friends marching along with a pole over their shoulders, from which hung the bound and bloody form of Sam. Hop told the judge that they were working their claim on Tennessee Creek when Sam ordered them off, declaring that the diggings belonged to him. They had used their picks and shovels in an effort to make him see reason, and then had bound him to take him before the judge. They wanted him punished as a claim jumper and lodged in jail.

Well, the judge knew Sam Totem and he had seen him working that claim. He ordered the Chinese to unbind Sam, and saw that he was a sad-looking case indeed. Holding court right there he told Hop to state again his side of it, and then asked the bloody Sam to state his. When this was done the judge said that in the absence of witnesses he would swear himself and testify. This he did. He said he had known Sam Totem for four years and knew him to be a truthful man. He knew that Sam had staked a claim on the creek, for he had seen him working there. As for Hop Lee, well, he knew that Hop put bogus stuff in his gold dust, such as spelter; so he found Hop and his friends guilty, fined them twenty-five dollars, and twenty dollars for the cost of holding court, and told Sam Totem to wash himself off and get back to his diggings.

Many of the judges were autocrats; whether they are thought to have been amusing or obnoxious may depend on a person's attitude toward judges. Kelly found that in San Francisco "judges sit on the bench attired like other men, . . . puff their cigars while laying down the law, on the enlightened principle of 'ex fumo, dare lucem.' . . . Law arguments under such a system are no longer dry and uninteresting, but flow smoothly along liberally lubricated with tobacco juice."

Bancroft, who was also there, tells about Judge Parsons. In 1851 the *Herald* supported those who wanted to drive the criminals out but five years later was a bitter opponent of the vigilantes. When supporting the reform party the editor tried to arouse the people to the lawless conditions, and by his strictures on the "masterly inactivity of the courts" brought down on his head the wrath of Judge Parsons, who asked the grand jury to declare the newspaper a nuisance and put it out of business. When Walker published his reply to the judge he was brought into court and fined five hundred dollars (a huge sum in those days) for contempt; and when he refused to pay it he was lodged in jail. "Four thousand people met and said the thing should not be; four thousand people condoled with Walker in prison, and requested Parsons to resign." The people's sense of injustice and outrage was so extreme that it caused the "irate judge to wither" and before long he resigned.

Up in Montana in a later period a man named Heinze, said by Professor Greever to have been "always selfish, usually unethical," for years fought the Montana copper companies. At the peak of his activities he had brought one hundred and thirty-three suits against Anaconda, employed thirty-seven attorneys, and spent millions in fees. One of the judges who favored him was "a saloon lounger and curbstone lawyer, cunning, shrewd, coarse, vulgar, of little education, indifferent to his opponents but stubbornly loyal to his friends. . . . In court he chewed tobacco, put his feet on a chair or a desk, looked out the window, dismissed a witness when he had heard enough. . . ." His "undignified, slipshod personal habits seemed to endear him to the lower levels of society, whose votes kept him in office."

On May 6, 1865, the *Statesman* in Boise reported the shooting in Idaho City of a man named Slater by a man named Murphy.

Murphy struck Slater in the face with his hand and immediately shot him with his revolver, the ball taking effect in the thigh. Murphy was about to fire the second time, when Slater asked him not to shoot again. Mr. Slater was said to be unarmed at the time. Murphy was arrested and taken before some court, we did not learn which, and discharged; the Judge remarking, as he dismissed the case, that "it was only a flesh wound."

Early Tucson had a German judge named Meyer, who by profession was a druggist. As told by Lockwood, the lawyer for a man arraigned before Meyer saw that his client was headed for a severe judgment, and so asked the court for a jury trial; whereupon the judge bristled and said to the lawyer, Oh, his client wanted a jury, did he? Very well, he would give his client two weeks on the chain gang, and the lawyer one week for disrespect of the court.

An article in the *Arizona Historical Review*

His Honor, Judge Roy Bean

says that the only two books in this judge's library were *Materia Medica* and *Fractured Bones*, both of which he consulted when perplexed by a difficult case. That seems highly unlikely, but if to make a good story we must find Judge Meyer poring over broken bones in an effort to decide a case of robbery or rape we will let it pass. The author says that this judge's favorite form of punishment was a work gang, and that he fairly rubbed his hands with joy when there appeared before him an accused man who possessed a skill that the town of Tucson could use.

A Tucson newspaper gave a picture of him in court. An electrician? "Vy, dis city in darkness has avaited your coming!" A mechanic? "Goot! Our tools vill now not vaste mit rust!" And so on. As told here, after sentencing the defendant to two weeks on the chain gang, and the lawyer to one week, Judge Meyer asked triumphantly, "How you like dat trial by shury?"

But the most famous or notorious of all the autocrats among frontier judges was Roy Bean, the law west of the Pecos. Born in Kentucky about 1825, he went to Mexico and was soon in trouble; fought a duel and broke jail; and in Los Angeles went into the saloon business with a brother. His brother was killed. Roy seems then to have had an affair with a Mexican girl (all his life he fancied himself as a devil with the ladies) which ended when a rival hung him to a tree. Fact or legend says that the girl managed to cut him down before he was completely strangled and that all his life thereafter he had a stiff neck. He is said to have set up a saloon business with another brother; by turns to have been a freighter, dairyman, and roustabout; and to have married and sired four children. When a railroad was being laid toward the Pecos, Roy had a tent saloon and moved along with the building, and by 1882 he had settled down with his saloon near the present town of Langtry. At the request of a Texas ranger he was appointed a justice of the peace in a lawless area, and thereafter for years won by election, except in 1886 and 1896. He lived in Langtry, which he swore he had named himself for the Jersey Lily, but which the Southern Pacific said had been named for a construction foreman.

Still fancying himself as a lady-killer he both wrote and telegraphed the Lily his undying admiration but got no reply until he told her that he had named a town for her. She then thawed enough to write him that though she could not make a personal appearance in the town which he had immortalized for her she would send twenty-five dollars toward a drinking fountain. To which he replied that "We ain't got no public park, and in the second place, if it's anything these hombres drink it ain't water." Later, according to her, the "bigwigs of the township" begged her to stop over long enough for a reception. This she did, by arrangement with the Southern Pacific Railroad. On January 4, 1904, the train stopped under the water tank, and, as reported in the San Antonio *Daily Express,* a lank character peered through a sleeper window and called to his pals, "There's the old gal now! Howd do, Lill?" Lill made a brief talk and ordered drinks for the crowd. She trudged "through the sage and prickly pear to the Jersey Lilly [sic] Saloon, cut a deck of cards, looked at the schoolhouse and promised fifty dollars toward repairing its wind damage."

The Judge, alas, never saw her; he had died March 16 in the previous year.

Horace Bell says Bean told him that he was the only man who could make anything out of the office of justice of the peace on the Rio Grande. He says Roy told him that his predecessor "came up to propose to me than I buy him out. He brought his commission along with him, his docket and all his papers, and dickered with me. So I gave him a demijohn of whiskey, two bear skins and a pet coon for the right, title, honor and emoluments of the office." Whether Bell was more interested in the facts or in a good story we cannot say. Roy, he says, held court in the bar-room of the Jersey Lily Saloon, using the bar as his bench and a keg of whiskey to sit on. That sounds about right. Roy's only lawbook was the 1856 compilation of the California statutes. When a man was brought before him, charged with the murder of a Chinese, Roy hunted through the codes and said he could find nothing in them

His Honor holds court. Lilly is the way he spelled her name

©N. H. Rose, 1927

about killing a Chinaman. So he turned the defendant free. When a man was killed by a Southern Pacific train and forty dollars and a revolver were found on him, Roy, sitting first as coroner, ruled that the railway was not at fault; and sitting as judge he found that the dead man had violated the law by carrying a concealed weapon, and so fined the corpse the forty dollars and confiscated the gun.

Bell says that on May 27, 1901, he read in a newspaper these words:

Judge Roy Bean, notorious throughout western Texas and many times the subject of magazine articles, also known as "The Law West of the Pecos," again distinguished himself last night by going through a Pullman car while the westbound Southern Pacific train was stopping at Langtry and, with a .45-caliber Colt in his hand, collecting from an eastern tourist thirty-five cents due for a bottle of beer. The tourist had bought the beer at the Judge's saloon but had rushed off without paying for it. Going through the car Bean peered into each passenger's face until he found his man when he said, "Thirty-five cents or I press the button." He was handed a dollar bill and was returned the correct change. As the judge left the car he turned in the aisle and said to the frightened passengers, "If you don't know what kind of hombre I am, I'll tell you. I'm the law west of the Pecos." The passengers thought it was a holdup.

According to some of his biographers Bean, illiterate and not-very-bright, was dominated by greed. After sentencing a horse thief to be hanged he discovered that the man had on him four hundred dollars in currency. He thereupon reopened the case, fined him three hundred dollars, and told him to get out of the country. His philosophy in regard to divorce was that if marriages he had made did not "take" he had a right to correct his errors—for a fee. Besides greed, a ruling passion was for publicity, and his genius for keeping himself in the limelight brought him to national attention as the law west of the Pecos.

So much for the autocrats. Bean was, of course, also a crook, as were many of the frontier judges. Bancroft says: "Courts of law were in bad repute in those days. Venality and corruption sat upon the bench in the form of duelling, drinking, fistfighting, and licentious judges. . . . It was not uncommon to see a judge appear upon the bench in a state of intoxication, and make to scruple to

attack with fist, cane, or revolver any who offended him. Two prominent magistrates bore the significant sobriquets of Mammon and Gammon." He quotes from an unnamed journal: "On the night of February 17, 1854, James P. Casey, William Lewis, James Turner, Martin Gallagher, and others entered the Mercantile Hotel, brutally beat the inmates, destroyed property, and committed other depredations." The police who interfered were overpowered and beaten; and when these criminals (two years later Casey was hanged by the vigilantes and the others banished) were freed by a corrupt judge and the mayor protested the "shameless proceedings" he was "insulted."

A San Francisco peanut vendor with the appropriate name of Almond decided in 1849 that there was an easier way to make a living. So by hook and crook he had another civil court established, with himself as its judge; and since his sole purpose was to get rich as fast as he could he made short work of windbags. Since his fee was an ounce of gold for each case he naturally wanted as many cases as he could jam into a day's session, so had no time for juries, the choosing of which brought no gold to his table, or for long-winded testimony from witnesses or eloquent tirades from lawyers.

Richardson says that on his way west he saw a "probate judge of the county lose thirty Denver lots in less than ten minutes, at cards, in this public saloon on Sunday morning; and afterward observed the county sheriff pawning his revolver for twenty dollars to spend in betting at faro." Burns in his *Tombstone* tells of an Arizona justice who lost his money in a gambling hall and thereupon arrested the proprietor, fined him fifty dollars, and returned to the game. Collier and Westlake say that "One justice ran a crap game in his courtrooms after hours, and, on the bench, whenever he asked for a drink of water; he was served whisky in a tin cup. He built a one-cell jail in the basement beneath his temple of justice, where he proposed to incarcerate the city's chief gambling-house proprietor who had refused to lend him money." They also declare it to be a fact that another judge who ran short of gambling funds went to a whorehouse, fined the madam fifty dollars and the girls ten dollars each and returned to the tables.

When in 1872, Judge Isham Reavis threatened John Wasson of the Tucson *Citizen* because the editor had revealed to his readers the nature of justice in the Reavis court, Wasson replied in his newspaper:

Reavis, a few words to you and your court. In common with the mass of the people of the Territory, we hold you and your court in the utmost contempt. We dare you to send along your contemptible warrant for our arrest for contempt of your contemptible self and court; but bear in mind if you do you will not be practicing upon any such as you have done in your district, who will submit through fear of your tyranny and disregard of personal rights and liberty. We dare you again on your contemptible sneak, to send your contemptible orders for our arrest. We promise the public a history of your tyrannical career in this respect, wherein the timid, poor and ignorant were outraged by your violation of personal rights and liberty. What became of the fines may also be considered. Your actions in the second judicial district justify the belief that you felt as the people did, that your court inspired frequent contempt. Your case will be continued upon the disagreeable facts as they come to hand and our space will permit. But send the warrant for contempt at any time, and we ask our readers to note your record as we give it space in this paper.

That ought to have sobered his honor for a few moments.

In Nevada County, Shinn says, a judge had a reputation as an honest but eccentric man. When two men, charged with stealing horses, were brought before him, friends of the two lawyers in the case, one prosecuting and the other defending, laid wagers on which attorney would make the best speech. The prosecutor did well with his side of it; the attorney for the defendants then called a witness and asked him what kind of character the defendants had had back East. It had been very good, the witness said. Hearing those words the judge shouted at the witness, "Good character, when they have been proved damned thieves? Your evidence is no good. Sheriff, take the thieves to jail."

On the Comstock there was "such wholesale corruption in the courts," according to such writers as Lyman, Lewis, and Greever, that the decisions of judges depended on the size of the bribe. When Judge Turner asked $10,000 for a decision agreeable to those who were willing to pay for it, the banks had all closed for the day and it was necessary to hustle around the saloons and shops and find the sum in gold pieces. It is said Mrs. Turner answered the door and seeing a part of the gold made a lap of her nightgown, into which all the gold pieces were then dumped, but that the gold was so heavy it tore the gown from her and left her standing naked in her pile of wealth. Lewis says hundreds of boundary disputes choked the Nevada courts, and that a judge was considered "reprehensibly lax only if he sold out to *both* sides. . . ."

Up in Idaho the *World* was telling its readers that

Wm Fagan, late a justice of the peace in this Precinct, ran off . . . and left divers debts, big and little, which he not only failed to liquidate, but lied awfully about, to deceive his victimized creditors into the belief that he intended to go no farther than Silver City [a hundred miles], to accompany his wife and two little ones that far on their way to California. But worse than this, the rogue either hid or destroyed his office record book, and consequently parties to suits, and those interested in judgments &c., are materially affected by his roguery. It is known that he was guilty of most shameful and even criminal acts in his official capacity, . . . He is a bad man and a very cheap rascal.

Still farther north it was about the same. Alexander McKenzie, representing a New York corporation, went to Alaska on a boat which also carried Judge Arthur H. Noyes. Soon after arriving, McKenzie had his men jump several rich claims and drive off the legal owners, with the judge and his henchmen collaborating. "All this was done before court was officially organized and on the flimsiest of pretexts without legal precedent or procedure. McKenzie's bond was fixed at $5,000, although one of the claims yielded $15,000 a day. . . . The Judge refused to hear appeals and would brook no interference." When writs were obtained from San Francisco His Honor paid no attention to them. The court there eventually thought the whole episode a "shocking record with no parallel in the jurisprudence of this country."

Parkhill says that when Soapy Smith (whom see) was brought before a Colorado judge, Soapy demonstrated for the court his sleight of hand. For the suckers he *pretended* to put within the wrapping around a bar of soap a five- or ten-dollar bill, and for a dollar each the men could guess which bar of a dozen or more had the bill. For the judge Soapy used a fifty-dollar bill and left a tiny piece of one corner barely visible. His Honor had no trouble at all choosing the right bar, and was so delighted that he exclaimed to Colorado and the world, "Why, this is no swin-dle! This is legitimate business." And after slipping the bar with the bill into his pocket he said, "The case is dismissed."

Did the lawyers prosper as well? We shall now take a look at them.

CHAPTER 18

Lawyers, Juries, and Jails

On his way west Richardson heard Brigham Young say in a sermon to the Mormons, "The Latter-day Saints are the happiest people in the world—the most industrious, the most peaceable among themselves. At least they would be, but for a few miserable, stinking lawyers on Whisky street, who for five dollars will prove that black is white!"

In his standard work on government in the mining camps, Shinn reminds us that "when Spain established her colonies, every one of them petitioned the king, in the most earnest and anxious terms, to allow no lawyers to sail for America; because they desired to live in peace and prosperity, freed from the malice of men and the malign presence of attorneys." Balboa wrote to the king, "One thing I supplicate your majesty: that you will give orders, under a great penalty, that no bachelors of law should be allowed to come here; for not only are they bad in themselves, but they also make and contrive a thousand iniquities."

After Winfield Scott Stratton (whom see) became a fabulously wealthy man, and discovered, Waters tells us, that most lawyers would accept any kind of case, even for professional crooks, he said, "I wish I were worth ten million and I would spend every cent of it putting every lawyer in Colorado in jail."

We who write these words have served on juries, and have many times sat in courts of law observing the procedure in civil and criminal cases. We have never observed an attorney in action whose primary interest seemed to lie in justice. It was in winning. But we cannot agree with Balboa that lawyers are bad in themselves, unless we are to assume that all people are bad. Before we look at some frontier lawyers in action it might be well to fix in mind a couple of facts. Few lawyers participate in writing the laws. If they take advan-

tage of every technicality and loophole which incompetent legislatures leave open to them they are merely exploring what they have to explore, unless all members of the legal profession are to take some kind of Hippocratic oath. The second fact is of weightier importance: the human male, denied the fields of honor and the battlefields, where in former times he worked off his natural aggressions, today must try to work them off in intellectual combat; and the courtroom is one of the few arenas left to him. It isn't indifference to justice that primarily motivates the lawyer; it is the human male's desire to triumph over his foe.

The frontier West had plenty of lawyers. To take Idaho as an instance: on the front page of the Boise *News* in Idaho City, July 2, 1864, were the advertisements of at least eighteen attorneys. The Idaho *World* March 25, 1865, published the names of forty-five who had been admitted to the bar since Idaho became a territory, or in a period of about two years.

After pointing out that the early San Francisco was overrun by convicts from Australia and other areas, who spent their nighttime "robbing miners in the dark streets and stabbing patrons of the crowded bars and gambling dens," and that they became so bold that most of the residents were afraid to leave their homes after dark, more than a hundred persons having been murdered in a few months, Gard says that the "few thugs arrested promptly hired shyster lawyers who usually kept them out of jail. It was almost impossible to find witnesses to a crime who dared to testify against the culprits." He cites Charles Cora, a typical thug, a petty gambler and machine politician, who catching Marshal Richardson unarmed shot him dead on the street. He was hurried off to jail and eventually brought to trial; but the "jury was fixed, the witnesses were rehearsed in perjury, and the proceedings were a farce"—so much of a farce, as we shall see when we come to the vigilantes, that the only way to deal with such murderers was to take them out and hang them. As the *Bulletin* pointed out, legal technicalities made a half-wit of justice, and called "into action the heated blood of an outraged community." Such lawyers as those employed by Cora, or by his mistress, a harlot, were, of course, out and out crooks. They even went so far at times, as many writers have pointed out, as to hire killers to do away with witnesses.

Some of the attorneys in the bonanza towns were a tough breed—much tougher than some of the polite sneerers we see in courtrooms today. In Tombstone was a lawyer known for his windy eloquence. One day when he was somewhere in the stratosphere, and his passages were a deep pur-

William M. Stewart, said to have been the ablest mining lawyer of the nineteenth century

ple, a jackass just outside the courtroom window gave off a tremendous braying. At once the other lawyer, who was to become a United States senator, leapt to his feet and objected. "The Court, please," he said, "it is contrary to judicial procedure when two attorneys burst into oratory at the same time!"

One of the tough lawyers was Judge Terry (whom see). One time, after a judgment had been rendered, and Terry told the court that he wanted his fee out of it, his partner was thought by someone to reach toward his weapons; and at once "fifty revolvers or bowie knives flashed in the hands of partisan spectators and participants." Terry once said that time and again during his years as an attorney he had an armed man stand behind him when he was in court to "prevent an attack in the *rear*."

Another tough one, and typical of frontier lawyers as a people, was William M. Stewart, who served for forty years in the United States Senate, and is said to have been the ablest mining

lawyer of the nineteenth century. We show him in action in a courtroom in "Brave Men and Cowards," and a photograph of him appears in that chapter.

Many stories have been told about him. One of them says that in the California mining camp of Rough and Ready he happened to be with a friend when he saw a mob dragging a victim to a tree. He drew his gun and shouted, "Stop right there! You have the wrong man! Release him instantly! Just what in heaven's name do you want to hang him for?" One version of this story says that the man was instantly released, whom Stewart didn't know from Adam.

Stewart's size, his bearing, his booming voice were all so impressive that the mob not only released the man from its clutch but turned to Stewart to listen. Stewart didn't know the man or what he was charged with but that made no difference; he was ready to defend him. He said they would try the man, then and there, before a jury. This they did, with Stewart defending him;

The legend says: Hawley Puckett Attorneys Law Notary Public

but when the man was found not guilty, the men who had been eager to hang him didn't like it at all, for their blood was up, as when they were promised a bear-and-bull fight or a shooting in a saloon. Casting around for a way to vent their emotions one of them said, "All right, by God, we'll hang the son of a bitch who said this man was guilty!" But that man, who had scented trouble, was already on a fast horse, hightailing it out of the country.

Somewhat like Stewart in his impetuosity and his bullying was Temple Houston, son of the famous general. As Horan tells it, the son wore his hair down on his shoulders, sometimes used a rattlesnake skin for a necktie, and in a courtroom employed just about any stratagem that occurred to him. One story says that to give judge and jurors a lesson in finesse in drawing weapons he snatched his .45 from its holster and fired several shots point-blank at jury and judge. They were blank cartridges but he scared the living daylights out of His Honor and twelve good men and true. This attorney, Horan says, became known as the "fightingest and greatest orator in the Territory."

Bancroft tells of two lawyers in Montana who "were driven from the territory by the exasperated people for attempting to clear the banditti when arrested"—that is, some members of the notorious Plummer gang. The vigilantes were tough on lawyers. In Colorado several attorneys put in a newspaper an advertisement which said they were closing their offices because the vigilantes had put them out of business. Zamonski and Keller, who spotted the advertisement, give their readers some of the reasons why the lawyers locked their doors.

In his account of the arrest and hanging of George Ives, one of the most brutal of the Plummer gang, Colonel Wilbur Fisk Sanders, himself a lawyer, and in this case the prosecutor, at the risk of his life, says of the three lawyers engaged by Ives and his friends, all distinguished in the area, that they were "somewhat remarkable characters, all of them out to thwart the administration of justice by all the means they could invent." Among their methods were preliminary motions to "quash the proceedings, to defer the trial, to continue the case, to delay the matter until the sense of offended justice had somewhat died out of the community." And first, last, and all the

time they strove to "manipulate the witnesses by such persuasions as they were susceptible to." All such methods are as commonplace in courtrooms today as the judge's black robe. The Boise *News* reported that when, over in Montana, an attorney failed to get his client acquitted before the vigilance committee he denounced the committee in language so abusive and violent that he was given fifteen minutes to get out of the area. "Other attorneys were required to pack up and leave. . . ."

Of the trial of Ives before a miners' court Dimsdale says that

Among the lawyers there was, doubtless, the usual amount of browbeating and technical insolence, intermingled with displays of eloquence and learning; but not the rhetoric of Blair, the learning of Coke, the metaphysics of Alexander, the wit of Jerrold, or the odor of Oberlin could dull the perception of those hardy mountaineers, or mislead them from the stern and righteous purpose of all this labor, which was to secure immunity to the persons and property of the community, . . .

It is true that no Coke was there, no Jerrold, but women were there and the miners were not hardy enough to be immune to them. Their tears and supplications and prayers saved the lives of two of the worst creatures ever to infest a mining camp.

The technical insolence, which so easily awes and abashes the unsophisticated, did not always work. Gard cites the case of a man who had killed his former wife and "had used technicalities to escape punishment." A mob "strung him from the same tree they had used earlier." When an emigrant couple "used legal loopholes to avoid conviction, two thousand enraged citizens gathered at the courthouse," took the man into custody, and hanged him. In that time, says White, lawyers

played the game of law exactly as the cheap criminal lawyer does today, but with the added advantage that their activities were controlled neither by a proper public sentiment nor by the usual discipline of better colleagues. . . . Indictments were fought for the reason that the murderer's name was spelled wrong with one letter; because while the accusation stated that the murderer killed his victim with a pistol, it did not say that it was by the discharge of said pistol; and so on. . . .

It was "formal legalism," White goes on to say,

that led in San Francisco to the vigilance committee of 1856.

The law was a game played by lawyers and not an attempt to get justice done. The whole of public prosecution was in the hands of one man, generally poorly paid, with equally underpaid assistants, while the defense was conducted by the ablest . . . men procurable. . . . To lose a criminal case was considered even mildly disgraceful. It was a point of professional pride for the lawyer to get his client free [isn't it still?]. . . . The whole battery of technical delays was at the command of the defendant. If a man had neither the time nor the energy for the finesse that made the interest of the game, he could always procure interminable delays during which the witnesses could be scattered or else wearied to the point of non-appearance. . . . Even of shadier expedients, such as packing juries, there was no end. . . . Of course the sheriff's office must connive at naming the talesmen; therefore it was necessary to elect the sheriff; consequently all the lawyers were in politics. . . . elections . . . to be sheer farce. . . . polls were guarded by bullies. . . . votes could be bought in the open market. "Floaters" were shamelessly imported. . . . five hundred votes were once returned from the Crystal Springs precinct where there dwelt not over thirty voters.

White was writing about San Francisco but the "conditions in the realm of criminal law" were much the same in all the larger camps.

Were there no honest lawyers in those areas in those times? There must have been a few, though in all our reading in this field we can call to mind only one who was so outstanding for courage and honor that writers have singled him out for tribute. This one was the fearless and incorruptible Colonel Sanders who, as Dimsdale says, was "the hero of that hour of trial" when the Plummer gang was liquidated. There was not a killer in the crowd of onlookers "but would have felt honored by becoming his murderer. . . . fearless as a lion he stood there confronting and defying the malice of his armed adversaries." Dimsdale also doffed his hat to Judge Byam who "will never be forgotten by those in whose behalf he courted certain, deadly peril and probable death." The courage of both attorney and judge is not easy to grasp by those for whom Plummer and his gang, and the terror they aroused in mining camps, is only something that they read about in books. And it was moral as

A typical frontier jury

well as physical courage; and, as Dimsdale says, in paying them tribute, "Physical courage we share with the brutes; moral courage is the stature of manhood."

It has been said that a court of law is no better than its judge. True it is that in ways both brazen and subtle a judge casts a powerful influence over a jury. That may be truer today than it was in the mining camps, for in those days, says Dimsdale, a "powerful incentive to wrong-doing [was] the absolute nullity of the civil law in such cases. No matter what may be the proof, if the criminal is well-liked in the community, 'Not guilty' is almost certain to be the verdict of the jury. . . . If the offender is a moneyed man, as well as a popular citizen, the trial is only a farce. . . ." Today the moneyed man would probably not cut much ice, for witness the outrageous sums that juries assess against corporations, believing them to be, along with the Federal government, a source of unlimited funds.

Marryat says that in California "I had the honour of shaking hands with a murderer quite fresh from his work, who had been acquitted a day or two previously by bribing the judge, jury, and the witnesses against him." Marryat and other English visitors were astonished and entertained by the "workings of democracy." In this instance Marryat points out that Americans seem to think " 'one man is as good as another,' and, as I heard a democratic Irishman observe, 'a d----d sight better!' "

In Idaho's Silver City the *Avalanche* told how a witness was successfully impeached:

Judge Scaniker, now of Boise City, is said to have successfully impeached a witness in Judge Crayner's court, in Stockton, Cal., by the testimony of the witness himself, thus: Question—Are you an American? Answer—Yes. Can you make a cigarita? Yes. And smoke it? Yes. And blow the smoke through your nose? Yes. Then, said Judge S. I submit to the Court whether or not I have impeached this witness. An American citizen that can roll a cigarita, smoke it, and blow the smoke through his nose will steal horses, and of course swear lies. The court sustained him.

Anyone who has watched the choosing of jurors in a courtroom is aware that on neither side are the lawyers interested in members of the panel who would make the ablest jurors. They seek those with prejudices that can be turned to account on their side, and ordinarily they prefer a stupid juror to an intelligent one. In California, Bancroft says,

Juries were summoned from the hangers-on about court-rooms, men fit for nothing else, scarcely able to live by their wits, and yet too lazy to work. Old familiar faces were they, blossoming under the genial influence of strong drink; old pensioners they seemed to regard themselves, as they did nothing but sit in the jury box, the same person sometimes serving several times in one day. Thus the courts had always at hand an acceptable, stereotyped jury of retired Peter Funks from the purlieus of Long Wharf, petty hucksters, perhaps, or sham bidders at Cheap John auction rooms.

"I remember," says Mark Twain, "one of those sorrowful farces, in Virginia (City), which we call a jury trial." A man was killed "in the most wanton and cold-blooded way" and of those who could have served on the jury and were better qualified—bankers, ministers, merchants—not a one was chosen, for they had read about the murder. "Ignoramuses" were called—a "jury composed of two desperadoes, two low beer-house politicians, three barkeepers, two ranchmen who could not read, and three dull, stupid, human donkeys! It actually came out afterward, that one of these latter thought that incest and arson were the same thing. The verdict . . . Not Guilty."

Every country, wrote the *Avalanche's* editor, "has some distinguishing characteristics. Among those of Idaho may safely be classed the sharp practice of Idaho juries, fast becoming proverbial among us. A notable instance of the kind recently occurred in the trial of a case before Justice Calloway at Centerville. The jurors returned into court with a verdict, which was refused to be announced before the payment of jury fees."

So much for juries. The jails were no better. They were indeed such a farce and a scandal that we present here what is, so far as we know, the largest photographic representation of them yet to appear in a single volume. In timbered areas the jails were small log structures, on which vigorous and determined prisoners spent their furies, often going through the roof, sometimes through a wall, sometimes under the foundation. Carson City, Nevada, was so scandalized as its jail repeatedly collapsed in one way or another, under the assaults of prisoners, that at last a huge beam was run the full length of the jail and securely

Coloma, California

Coulterville, California

FOUR CAMP JAILS

Bannack, Montana

Idaho City, Idaho

anchored at either end. The prisoners were chained to this beam. At one time it had as many as sixteen desperate men chained to the immense log, and we can imagine how they glared round them and with what curses they wrestled with the chains and padlocks.

A description of the Hangtown jail was left by a prisoner who was temporarily there—canvas roof and paper walls, its stalls filled with twenty horse-thieves: "I did as I had done before, coyoted out from 'neath the floor." Twenty years later, in 1869, Deer Lodge, Montana, was telling the world that it had "the most unretentive jail in America—an institution that was never known to hold a prisoner long enough to bring him to justice."

Many of the jails in California were built of stone, but the mortar could be picked out with a jackknife or even with a stick, and a stone pushed out of the wall. The mortar was so soft in some of them that a powerful man who didn't mind bruising his shoulder could have plunged against a wall and knocked half of it out. But even when made of logs the jails offered no impassable barrier to a determined man. In Yerba Buena, which was to become known as San Francisco, a prisoner, carrying the jail door to which he had been chained, appeared before the alcalde and demanded breakfast. Up in Oregon, the *Herald* told its readers how four highwaymen had escaped: "They were confined in a log jail, the entrance to which was from the top by means of an aperture about sixteen inches square. They

called for a drink, and the Deputy Sheriff reached a vessel containing water through the entrance to them; they seized his arm, dragged him through, gagged and pinioned him, and made their way out." If it was that simple, one wonders why they bothered with water and the deputy.

In Idaho, as in all the camp areas, stories of escapes were published. The *News* reported that "Crow, the desperate jail-breaker, has escaped again. . . . He is said to have escaped from all the State prisons on this coast, and all the county jails he was ever in except that of Boise County."

Surely one of the most spectacular escapes from a frontier jail was that of Red-Handed Mike, a "huge beast of a man," two hundred and forty pounds of ex-pugilist, "surly of temper and dangerous." He was arrested in the Coeur d'Alene area for a major crime, and when two men awaiting trial in the small jail heard of it they summoned the sheriff and told him that if Mike was put in with them they would kill him. Unable to find another building that would hold Mike the sheriff fetched a log chain, spiked it to a huge stump, and padlocked the other end of the chain to one of Mike's wrists. Giving him food and water and a blanket, he left him there for the night.

Or what he thought would be the night. At daylight, Mike, log-chain, and stump were gone. A half mile away in a blacksmith shop the sheriff found the stump and chain. Mike he was never to see again.

CHAPTER 19

Badmen

In the chapter on the vigilantes some of the worst of the "bad" men appear, and in the chapter on brave men and cowards, as well as elsewhere; but in this chapter we are going to look at a few of the more infamous ones who, more than most, may be typical of the badmen of the American West. There were a lot of them. "To attempt," Mark Twain said,

a portrayal of that era and that land, and leave out the blood and carnage, would be like portraying Mormondom and leaving out polygamy. The desperado stalked the streets with a swagger graded according to the number of his homicides, and a nod of recognition from him was sufficient to make a humble admirer happy for the rest of the day. The deference that was paid to a desperado of wide reputation, and who "kept his private graveyard," as the phrase went, was marked, and cheerfully accorded. When he moved along the sidewalk in his excessively long-tailed frock-coat, shiny stump-toed boots, and with dainty little slouch hat tipped over left eye, the small-fry roughs made room for his majesty; when he entered the restaurant, the waiters deserted bankers and merchants to overwhelm him with obsequious service; when he shouldered his way to a bar, the

shouldered parties wheeled indignantly, recognized him, and—apologized. They got a look in return that froze their marrow, and by that time a curled and breast-pinned barkeeper was beaming over the counter, proud of the established acquaintanceship that permitted such a familiar form of speech as: "How're ye, Billy old fel?"

Drury tells of one who pounded the bar in a C Street saloon and announced to the world that he was a ripsnorter and a tough customer to clean up after. He ordered whiskey and he wanted the strongest, for he had swallowed a cyclone for breakfast and he didn't want to be crowded. The bartender, says Drury, hunted for fusel oil "while rubbing the glass with red pepper. One by one the loungers near the stove quietly slid out of the side door, and when the bad-man turned around, the place was empty." Shaw says the "cowardly desperado would pull his revolvers around to the front so the handles could be seen, ruffle his hair, and, with fierce looks and terrible oaths, placing himself in front of his victim, would address him about as follows: 'Seen yer before, young feller! Can't call yer name. Oh, yes, Jones. Lemme tell yer, Jones, this yere's a bad place. . . . Don't tell me I lie (reaches for his pistol) —saved yer life—lend me a fiver.' "

A surprising number of men, all of them liars, seem to want the world to believe that they received a nod from Wild Bill, touched a saddle in which the Daltons rode, and actually were allowed to heft a Wes Hardin six-shooter. According to Berton, Clarence Andrews

once wrote that if all the men who claimed to have seen the shooting of Soapy Smith were laid end to end "the line would extend to the Equator and back again." . . . Even such a respected writer and jurist as the Hon. James Wickersham writes (in *Old Yukon*) that, in the summer of 1900 in Skagway, "in one of the banks a gentlemanly clerk named Bob Service was introduced smilingly, as a writer of poetry." The fact is, of course, that Robert Service never worked in a Skagway bank and, indeed, did not reach the north or write any Yukon poetry until 1904. Writer after writer makes the same mistake. In *Far North Country* Thames Williamson writes: "Service was in the Klondike during the fevered days of the gold rush," and Glenn Chesney Quiett falls into the same error in *Pay Dirt*. Actually, Service's most famous book of verse . . . was written before he ever set foot in the Klondike. Stanley Scearce . . . says that he met Service in Dawson in 1898 (years before the poet actually ar-

rived) and goes on to tell of his encounter with Soapy Smith in Skagway in the winter of 1899-1900, about eighteen months after Smith was killed.

All of which is to remind us that when writing of the badmen we are in the realm of legend. Deadshot Dick was born in the mind of a man named Wheeler and romped through more than thirty atrociously written and horribly printed dime thrillers. He was killed off so that his son could leap to the horse and keep the breathless saga going. For a lot of people Dick and his son were far more real than the President of the United States or their own father. Nickanora actually lived, and was, Drury thought, king of the stage robbers of Nevada. One of his favorite gestures was to give wild game and champagne dinners to distinguished people, "with all the grace of a Chauncey Depew or a Ward McAllister." He had his own gang; his total loot "must have amounted to hundreds of thousands." Time and again he was arrested but always freed, because he had so many accomplished liars to swear to alibis. Eventually he became known as The Spanish King, and the Wells, Fargo company became so sick of him that they sent an attorney to the mountain hideout to dicker with him. The story that came out of it said that, weary of fleeing, Nickanora agreed not only to rob no more stages but to protect them, for two thousand dollars a month. He dishonored the agreement, and after killing two of his pals following a robbery, apparently for their share of the loot, he vanished, presumably into Mexico.

Another psychopath (it is the fashion now to call them sociopath) was the Tennessee doctor, Thomas Hodges, of whom Bancroft says, with his usual lack of restraint, that he was "by far the most intelligent, accomplished and kind-hearted American gentleman who ever took the road to California." What he took was the name of a notorious criminal, Tom Bell, as a disguise, though inasmuch as he had had his nose smashed, so that the bridge lay flat with his cheeks, and he had only a button above his upper lip, disguise didn't do him much good. The Sacramento *Union's* Grass Valley correspondent solemnly told his readers that Bell "carries six revolvers and several bowie knives, and wears a breast-plate of thin boiler-iron around his body." Dr. Hodges decided in 1856 that the honest labor of a physician was not for him, and, growing fancy whiskers and long hair, he fixed his attention on the stage-coaches and their express boxes filled with gold. But when with five companions he undertook to rob the stage leaving Marysville, with $100,000 in the box, he had a bit of bad luck. A guard named Dobson, a rifle across his lap, was sitting with the driver, and when the bandits came up he got two of them, which was far above average for express guards. Inside the coach a man was shot in the head, another in both legs, and a woman was killed.

One of his impudent letters to the police said, in part, "But don't think for a moment that your vigilance causes me any uneasiness, or that I seek for an armistice. No, far from it, for I have unfurled my banner to the breeze, and my motto is 'Catch me if you can!'" He even invited the captain of police to join him for a while and learn some of the tricks:

Probably you hear a great many things, but you must know I am not guilty of every accusation that is alleged against me. For instance, some malicious scoundrel tried to saddle the murder at Frenchman's Bar on me, . . . although I am looked upon as a desperado and know that I could expect no leniency at the hands of the people should they ever catch me, still I am too proud to commit such an atrocious and cowardly murder as that was.

Truly Yours,
TOM BELL

A newspaper published the letter and Tom was on his way toward becoming a legend. But soon he was captured.

To a Mrs. Hood who had a small hotel that was one of his hideouts he wrote a letter that reeked of sentiment: "I am to die like a dog, and there is but one thing that grieves me, and that is the condition of you and your family. Probably I have been the instrumentality of your misfortunes." He had been so much the instrument that her three daughters had been dismissed from school. He wrote to his mother that he was about to make his exit; that he recalled her "fond admonitions" and that if he had heeded them he would not be "in my present condition"; and that he had "no one to blame but myself." In our own time, when it is the fashion to say that all crime is wholly due to en-

vironment, and that practically all criminals, even the professionals, are "sick," it is nice to find the doctor admitting like a man that his being hanged was nobody's fault but his own.

Very well, then, the doctor was a gentleman. Of the type that was brutal but not stupid, Geronimo and the Apache Kid were two of the most blood-chilling, the latter having had, one writer has said, "all the evil traits of a tribe notorious for diabolical outrage and unrestrained savagery. . . . southern Arizona and Sonora are dotted with crosses which cover the graves of victims of this ferocious renegade." He came from a people more notorious for their cruelty than even the Blackfeet, and in spite of the fact that he seems to have tortured as well as to have killed, and that Indian females which he took as mates he treated with the utmost brutality, binding them to trees when off murdering and hobbling them at night as if they were horses, there has been a tendency to "glamorize" him, as Billy the Kid has been glamorized, and to say that none of it was his fault. The Kid's murder of Sheriff Reynolds and his deputy is a Western classic of negligence in officers and of Indian cunning.

Another killer, typical of those who were both brutal and stupid, was Cherokee Bob, half Cherokee and half hell fiend, who hated the Union and Lincoln and Northern soldiers. In a theater with some of his henchmen two soldiers were killed and a number of them knifed. Bob then fled to Lewiston in Idaho, and from there to Florence, taking with him a brazen whore known as Cynthia, whom he is said to have won from a notorious gambler and killer. He said he would take his woman to a ball and they would treat her as a lady or settle with him; on protest from the respectable patrons the management ejected her. Bob then picked up one of his meanest henchmen, a creature named Willoughby, and armed to the teeth they went forth; but the two managers were ready for them. Willoughby was killed on the spot and Bob was shot several times, dying of his wounds a few days later. The two were buried in unmarked graves in the Florence (Idaho) cemetery and their bones are still there.

We will now present, in a little more detail, five of the most infamous badmen of the West. Juan Soto was as brutal and stupid as they come; Murieta was the typical Robin Hood of the sentimental legends; Black Bart was a "gentleman"

psychopath, like Bell; Joseph Slade, more of a
riddle than the others, seems to have been be-
deviled by impulses and passions beyond his control; and Soapy Smith can stand as the archetype
of the confidence man and swindler of all lands
and times. The five seem to us to include all
types of badmen who preyed on those taking gold
from the earth.

Jackson says that Soto "has been described by
many writers of the time, and all agree that he
was the most fearsome figure of an outlaw that
ever roamed California's hills." After one look
at his face no one is going to be eager to dispute
that statement. About three-fourths Indian, he
was a large man and powerfully built and "as
agile as a wildcat." But most likely it was his
face rather than his size that scared the daylights out of a deputy sheriff. Above his low forehead he had a huge mop of coal-black hair, and
each black brow had enough hair in it to make
a mustache. The same black hair, as you can see
in the likeness of him, covered his upper lip and
around each side and over his chin. It is true,
as Jackson says, that he had a "general expression
of animal ferocity" but a lot of criminals have
that. The thing about him that must have paralyzed the deputy, and was enough to put almost
any man to flight, was his crossed black eyes.
They both turned inward, one of them looking
down a little and the other up. When he rolled

Juan Soto

them he must indeed have been a fearsome sight,
for if the one looking down turned up or the
other turned down the other very possibly rolled
out of sight into the skull. It is said also that he
fell into insane rages and frightened even his
savage followers half to death.

Guards and miners waiting for the payroll, after Butch Cassidy had taken one, April 21, 1897

Harry Morse, if Jackson is right, was just about the greatest of the early Western sheriffs. Like some other fifteen-year-olds he came West with the Forty-niners, but finding that he didn't care for liquor and gambling or the uncertain future of a placer miner he went to San Francisco and by the age of twenty-eight was elected sheriff. He served in that area for fourteen years and was the chief instrument that put an end to the gangs of Mexican bandits. It was inevitable that he should have a showdown with Juan.

The most of what we know about that amazing meeting of badman and sheriff is based on the account given afterward by Morse, a reticent man, and by Sheriff Harris, of Santa Clara County, who during part of the fight was a witness. Jackson examined these records, and others; our brief recital here is based on his conclusions. Like all the other road bandits Juan and his assassins had their hangout far back in the hills, from which now and then they would ride swiftly out to rob and kill. They brought Sheriff Morse onto the scene when, in 1871, they burst into the store of a Thomas Scott, an ex-Assemblyman from Morse's own county; after looting the store and killing the clerk, they shot through a thin partition into a room where they had reason to believe Scott's wife and children were cowering. Such shooting was just ordinary fun for Juan and his brutes, as it was for countless badmen of the American West. During the shooting the neighbors had hidden but they had been peering out. When Morse came on the scene and the bandits were described to him he was convinced that one of them was Soto.

He invited Sheriff Harris and some of his men to come along, and with this party and three of his assistants, including a deputy named Winchell, he set out for the Panoche Mountains and Juan's hiding place. He came to a Mexican sheepherder, whom he had met on an earlier journey, and was able to drag from him the information that he knew where Juan and his men were hiding. Under what must have been coercion the herder agreed to go with Morse and point out the spot, if he would be allowed to turn back and vanish before he was seen. The Mexican led them to a ridge that overlooked a ravine or small valley, in which stood three shacks at some distance apart. In these, he said, were the sentinels; Juan and the main body

Sheriff Harry Morse

were far up a canyon. Whereupon he fled. Whether he told all he knew, or lied, hoping that the sheriff would ride into an ambush, we do not know.

The two sheriffs divided the party into three groups, one for each shack. Morse and Winchell rode toward one of them and on the way over met a Mexican. Morse asked him for a drink. The Mexican said all the water was in the shack. Leaving his rifle on the saddle and carrying only a pistol Morse followed the man, with Winchell at his side with a double-barrel shotgun. What they found in the shack was not water. Over by the wall across from the door was a table and behind it, facing the door, sat Juan Soto, with eight or ten men and women at the table around him. We are not told whether they were drinking or gambling or both. Jackson may be right in his surmise that Morse saw in an instant that the only way to save his life and Winchell's was to take the offensive, which he did, swiftly advancing, his gun drawn, and calling on the cross-eyed monster to put up his hands. Quite certain it is that in all his life Juan Soto had never put up his hands. He had no intention of doing so now. A sheriff less brave, less cool, and with less confidence in himself would have shot Juan when he refused to raise his hands at the second command. Jackson says the scene "comes through with the horrid vividness of a nightmare." Possibly it was not that for Morse, who had spent years running down bandits and who time and again had gone alone into the wilderness to find a criminal and bring him out.

Juan's eyes must have been something to look at. A pistol and a shotgun were leveled at him but whether he didn't draw then (he had two guns at his waist) because he was afraid to, or because he was momentarily paralyzed by insult and outrage, or because he was horribly fascinated by what he had thought could never happen, no one knows. But he didn't put up his hands. Morse then displayed incredible coolness. He would have been justified if he had shot the brute; what he did was to reach with his left hand for handcuffs, drop them on the table, and tell Winchell to put them on Soto. Winchell advanced to do it. He must have been trembling, for he was no Harry Morse. As he moved forward he took his finger off the trigger and rested the shotgun across his left arm. One wonders how he thought he would handcuff the fellow with one hand. He probably wasn't thinking. When he reached across the table to take one of Juan's hands that killer turned to look him full in the face. Before we too severely condemn Deputy Winchell for rushing blindly from the shack and taking to his heels we would do well to try to imagine what Winchell saw. In any case it was two ferocious black eyes staring at him while looking inward at the nose. The reticent Morse said only in his report that "The next second Winchell backed away and ran out."

One of the women, whom in his report Morse called a "muscular Amazon," now attacked the sheriff from behind. By all the laws of probability he should have been dead in another two or three seconds. Was Juan helpless with fascination? In the moment the woman seized Morse's pistol arm, a man grasped his other wrist, and Juan came up from his chair, a gun in each hand. Hollywood with all its silly Westerns has never done it better. Morse reported that in the moment of leap-ing up Juan called to his men to stand back, that he and he alone would kill the sheriff. With the superhuman strength that men sometimes have in such moments Morse tore his arms tree, and, firing as he went, got back through the doorway. His aim was high; the bullet went through the bandit's hat. Soto hurled himself across the table. Jackson says it is curious that "no one interfered" in what happened then and in what followed. It is even more curious that Winchell, armed with a shotgun, was nowhere in sight. Jackson may be right in guessing that the Mexicans stood back because they thought their Juan was a match for ten sheriffs. For ten like Winchell he might have been but against Harry Morse it was his last day on earth.

Moving swiftly back and firing now and then, Morse went for his rifle. Soto came after him, firing. He fired four shots, and according to Harris, Morse saw the move for each shot and hit the earth. Harris thought that four times Morse had been hit and had fallen. Jackson found this part of the duel a little incredible, and it certainly is. It may be that Harris didn't see clearly, for he was at a considerable distance and was racing forward. Or it may be that Soto, in fact, was not the deadly shot he was in legend. Besides, he was firing pistols, and the pistols of that time, no matter what the legends about Wild Bill and others, were far more noisy than accurate. When one of Soto's guns jammed he rushed into the shack and came out with three. This seems to indicate that he had suffered some impairment of his confidence. Morse had reached his rifle. Harris and his men were still coming on. Juan, hoping to run away and fight again, raced toward a clump of trees where a horse was saddled. The horse was excited by the shooting and got away from him. With a gun in his belt and in either hand Soto dashed toward another horse. Morse called to him, "Throw down your guns, Juan! There's been enough shooting!" What marvelous forbearance in a sheriff! When Soto still ran Morse fired; the bullet struck the bandit and spun him around. Whether he thought he was still going toward the horse or whether the shot renovated his courage is unknown but the big cross-eyed fellow now rushed down the path he had just gone up. He vanished into the shack and burst out after a few moments, his mop of hair wild, his eyes searching the world for the sheriff, a cocked pistol in each hand. The next shot went through his skull and brought him down.

From what copse or cave Deputy Winchell was peering out all the while the records do not say.

Bancroft spelled it "Murieta"; Scanland and others, "Murrieta"; the postal officials of San Diego county, "Murietta"; a history of San Luis Obispo County, "Muriatta"; and "Muriete" by what purports to be a book about him. Jackson says there was no Murieta, or "at any rate not much of a Murieta." A half-breed named John Ridge, in 1854, in San Francisco, published a thin paper-backed book that was "reprinted in several languages, was poetized, dramatized, remade three-quarters of a century later into a 'biography' and even brought to the screen." Ridge's Murieta, probably nothing more than invention, was the latest thing in manly perfection up to that time, endowed with all the virtues, who by the most

The Nahl painting of Murieta

vile and unspeakable persecutions was driven to take an oath that still echoes in Hollywood; and "Fearfully did he keep his oath, as the following pages will show." The Ridge book could have been taken seriously only by persons with less sense of the realities than Alice had when bugging her eyes in Wonderland. Doctors of philosophy, even full professors, have accepted him with the most side-splitting seriousness.

Ridge hoped to make a pile of money but his publishers "after selling 7,000 copies and putting the money in their pockets, fled, bursted up, *tee totally* smashed, and left me, with a hundred others, to whistle for our money." Poor betrayed author. The *California Police Gazette* had a writer revamp it for a serial, and from there it took off into the wild blue yonder, as well as into Spain and Mexico, where our Joaquin is a legend bigger than life; and from there to Chile and France and other lands, to lodge by way of Walter Noble Burns in the bosom of Hollywood. The name of Joaquin's wife or woman goes from Rosita to Carmen to Rosita Carmen, to Carmela to Clarina to Belloro, to end up as a descendant of Mexico's royal Montezuma. By the time she found a home in the tender mind of Marcus Stewart and his deathless poem she had become Marie. And Joaquin did not have his head cut off after all but went back to Mexico, "a land far south of California's strand," to live the life of a retired Robin Hood.

Jackson traces in detail the hysterical mythmaking and reminds us that all the legend needed to make it fact was solemn acceptance by recognized historians. "Two men stepped forward to perform this office. They were Hubert Howe Bancroft and Theodore Hittell, whose histories of California have been reference works ever since they were published. Both of these men took their Murieta from Ridge's fiction" and Bancroft even "embroidered his account with further fanciful dialogue." Some California magazines thereupon began to publish the recollections of Murieta by old-timers. From this point on Jackson's account is either too sad or too hilarious to be borne, depending on how you look at such matters. Ridge himself said pathetically in 1871 that a spurious edition was out which added fiction to his "true" story. In essence the legend is that of a manly and admirable man who was shamefully and grievously persecuted by the white miners, who forced him to see his wife raped and his brother hanged, and who then was bound to a tree and flogged within an inch of his life. No wonder (say Burns and Hollywood) that he went forth to redress old wrongs, to take from the evil rich and give to the righteous poor, and to become a scourge on wicked men. "In every particular he was the typical folk-hero of all countries and ages." He was incredibly handsome, of course, with eyes flashing fire and tenderness, and a superb body ever ready to bow at the sight of a lady, or draw a knife and make incisions along the rib-cage of wicked men. After a while there was hardly a town or camp along the Mother Lode without its cavern, cellar, or tunnel in which Joaquin had been hidden, or where his frightful henchman, Three-Fingered Jack, a gorilla-monster who killed with laughing joy, gulped his whiskey and snored it off.

Jackson thinks there may have been five Joaquin bandits, one of whom was named Murieta. It is little wonder that Joaquin seemed to be all over the place, or that the governor approved an act passed by the legislature that authorized form-

WILL BE EXHIBITED

FOR ONE DAY ONLY!

AT THE STOCKTON HOUSE!

THIS DAY, AUG. 12, FROM 9 A. M., UNTIL 6, P. M.

THE HEAD

Of the renowned Bandit!

JOAQUIN!

AND THE

HAND OF THREE FINGERED JACK!

THE NOTORIOUS ROBBER AND MURDERER.

"JOAQUIN" and "THREE-FINGERED JACK" were captured by the *State Rangers*, under the command of Capt. Harry Love, at the Arroyo Cantina, July 24th. No reasonable doubt can be entertained in regard to the identification of the head now on exhibition, as being that of the notorious robber, *Joaquin Murietta*, as it has been recognized by hundreds of persons who have formerly seen him.

er Texan Harry Love to raise a company of twenty rangers and bring Joaquin in—almost any Joaquin would do. Which Joaquin was captured and had his head cut off by Love and his rangers, no man knows. Several Mexicans were killed, wounded, or captured, one of whom is thought to have been Jack; but his head was so badly shot up that Love, who was after the reward of a thousand dollars, is said to have been afraid he couldn't prove it was Joaquin's. So they cut off the hand minus three fingers and put it in alcohol; and because one of those captured had said he was the leader they cut his head off, put it in a huge jar of alcohol, and went in triumph back to Sacramento. Jackson thinks the head in the jug was probably not that of any of the five Joaquins or of any Joaquin. But it was "shown in different museums for many years," with furious argument around and about it.

The editor of the San Francisco *Alta* had had enough: "The humbug is so transparent that it is surprising any sensible person can be imposed upon by the statements of the affair which have appeared in the prints." He concluded that Joaquin was a "fabulous" character and that the head in the jug was certainly not that of a man named Murieta. Many writers since then have not been so skeptical. Jackson says Murieta is "considered historical truth by most Californians." Though one of the state's "soundest historians" (whose name he does not give) has said that Murieta was "never anything but a vicious and abandoned character, low, brutal, and cruel, intrinsically and at heart a thief and a cut-throat," Hunt* and Sanchez have written: "He has been called cavalier as well as outlaw; but even after rejecting innumerable stories as apocryphal, he will doubtless remain the superbandit of California history. Harry S. Love, an experienced and intrepid mountaineer and rider, finally captured him, his head being later exhibited in alcohol as a hint to others."

Carleton Beals knew all about him: "Daring eighteen-year-old outlaw Joaquin Murieta (alias Carrillo), a circus horse-trainer, and his wife Rosa Felix, who rode with him in male attire, hair cropped close, performed exploits soon celebrated in Spanish, even English, ballads. They were the terror of all 'white' land thieves, and Murieta con-

sidered himself a true patriot, called to free his 'native' countrymen. He was the futile Aguinaldo, the Sandino, of the raped province of the West. He was finally betrayed for a few hundred dollars, by his friend, gambler William Burns."

Major Edwin A. Sherman, veteran of the war with Mexico, knew all about it:

As we were coming out of the foothills we halted to cook breakfast, and were nearly ready to eat it, when about a half dozen Mexicans rode up. I recognized the leader as Joaquin Murieta, a dreaded bandit. I walked up to him, and, speaking in Spanish, invited him and his party to take breakfast as our guests. They alighted and most

* Rockwell D. Hunt was at that time Dean of the Graduate School, University of Southern California.

courteously partook of our hospitality, thanked us, mounted their horses, and rode away. . . . When we had traveled about four miles, we came upon the bodies of three men, who had been shot and their throats cut. Continuing on about three miles farther, there were two more bodies in like condition, who undoubtedly had been murdered by Joaquin Murieta and his band. Our fearlessness and hospitality, no doubt, saved our lives; . . .

A curious item appeared in the Santa Clara *Mercury* and was reprinted in the Idaho *World* in 1867. Of one John R. Ridge, it "is generally known" that he was "the son of the distinguished Cherokee Chief who was deposed by John Ross, and was murdered in the following war. Thirty-six ruffians, siding with the Ross party, surrounded Major Ridge's house, in the night, and having called him to the door, shot and stabbed him to death. . . . The subject of this article, then a mere child, escaped from the house, hid in the weeds and finally eluded their pursuit. When about seventeen years of age, the boy, true to the instincts of his race, which had not been subdued by civilization, went upon the warpath as the avenger of his father's murder." If this is the same John R. Ridge it may be that his story of Murieta was another expression of a half-breed's hostility to the "whites."

Early in his career the robber known as Black Bart was thoughtful enough to leave a clue inside an empty express box:

> here I lay me down to sleep to wait the
>   coming morrow
> perhaps success perhaps defeat *and*
>   *everlasting Sorrow*
> I've labored long and hard for bread
>   for honor and for riches
> but on *my corns too* long you've tred
>   *you fine haired* sons of *Bitches*
> let come what will Ill try it on My
>   condition cant be worse
> and if theres money in that Box Tis
>   munny in my purse
>   BLACK BART
>   THE PO8

After studying *Robber's Record,* a compilation by Wells, Fargo's detectives of more than two hundred stage robbers sent to jail or hanged in California, 1875-1883, Jackson thought Black Bart "the greatest—and mildest" of them all. He certainly was the most notorious. He committed

twenty-seven successful robberies before he was taken in the twenty-eighth, and the twenty-seven Jackson lists in Appendix II of *Bad Company.* For years about all the stage drivers or guards knew about him was that he always appeared alone in a long linen duster with a flour sack over his head and a double-barrel shotgun in his grasp; and that after stopping the stage on a sharp bend or steep hill in a sepulchral voice he told the driver to throw down the box.

Then one day he left behind the "poem" given above, on which has been spent a lot of sentiment and speculation. Though some writers have been fairly breathless with admiration when writing about him for us he is very plainly a psychopath —a chronic liar and thief and loafer who lived by his wits. The governor was so disturbed by Bart's insolence that he posted a reward of $300; Wells, Fargo added $200 to that, and the Post Office, $200. A woman gave a description of him—brown hair turning gray, eyes of a brilliant paralyzing blue, two front teeth missing, and slender hands that had never known toil. He did not smoke or drink, she said, but loved coffee, for at breakfast he asked for two cups. With the notion that they might be looking for a blue-eyed poet who loved coffee and was going bald the police chased after him for five years; and in that time Bart robbed twenty-seven stages and never fired a shot.

Wells, Fargo got out a handbill on him, in which those pursuing him were told that his habit was to appear suddenly in front of the lead team, in a stooped position; that he always brought with him an ax, with which to open the box; that he also stole the United States mail and opened the sacks with a sharp knife; that he was polite, especially to women; and that leaving the shattered box and the ax behind he always vanished on his own feet, and so was thought to be a "good walker." And when he was brought in at last those who had been chasing him for five years couldn't take much credit for it. A driver named McConnell had taken his stage out of Sonora before daylight; he had, of treasure, 228 ounces of amalgam, $550 in gold coin, and some gold dust. There were no passengers until a James Rolleri came aboard, saying he was going out to shoot a deer or a bear. After a few miles he got off, and was watched by Bart, who was hiding and waiting. Bart asked the driver who the youngster was and what he was doing out there but possibly the chief worry on his psychopathic mind was knowledge of the fact that the company, sick and tired of his pilfering ways, had nailed the express box to the floor of the coach.

When Bart ordered the driver down he said the brakes were bad and he had to keep his foot on them. As the story has been told neither man showed much gumption. The driver said the stage would roll down the hill, Bart told him to put a rock behind a wheel, and McConnell told Bart to do it. The story says McConnell was not armed! It says Bart told him to unhitch the teams and helped him do it! One writer tries to make all this plausible by saying Bart was flustered. Flustered or not, his career as a stage looter would be interrupted for some time. Jackson admits that there is some "wild absurdity" in all this, for McConnell was now told to take a walk, and did, so that Bart (says the sheriff who had the bandit act out the scene) could use a "sledgehammer, a big crowbar, a wedge for splitting wood, a couple of picks and an axe"—and, according to a later account, a "miner's gad, drill and chisel" to smash the box. At this point Jackson uses exclamation points—and well he might, for it would have taken a Paul Bunyan to carry in such a load of tools, not to mention the fact that only a complete idiot would have brought two picks.

Finding his sense of the plausible outraged Jackson hastens to tell the reader how large Bart was—a little fellow, in number six shoes. It takes an act of the imagination to see a man in number six shoes wielding crowbar, sledgehammer, ax and two picks on an express box inside a small coach. In addition to all these tools Bart came to the scene with a "varied assortment of baggage," including field glasses, a round derby hat, *two* pairs of detachable cuffs, and no doubt a fresh volume of poetry. All this nonsense could easily have been blown away by any historian who realized that he was dealing with a pathological liar, and with a sheriff who had never heard of such liars.

Jimmy appeared, was hailed by McConnell, and the two men took turns shooting at Bart as he backed out of the coach and made off. Practically all the deadshots seem to be under contract to Hollywood; the best these two men could do was to scare Bart so thoroughly that he dropped a packet of papers but still clutched the gold as he vanished into the thickets. Behind a large rock

Black Bart

not far from where the coach was stopped Bart had made himself a robber's roost, and there an "amazing profusion of clues" was found—uneaten sugar and crackers, a leather case for field glasses, a magnifying glass, a belt, a razor, linen cuffs, flour sacks—and on some of them the names of merchants which for smart sheriffs were as footprints to bloodhounds. When captured and asked why he had adopted such a ferocious name he said he had read "The Case of Summerfield" in which there was a Black Bart. He was sentenced to six years but would be free in four and a half. Meanwhile the San Francisco newspapers "were busy remaking the Black Bart legend." Various writers have been remaking it ever since. For years Bart had been a "towering predatory terror" and when brought in was a little man in size six shoes and a round derby hat. That would never do: one writer found that Bart had all the flourishes of a fancy bookkeeper, and another said that as he robbed his victims he "harrowed their souls with rhyme."

The psychopath in him came out in full view, as it always does in such characters for those who can read the evidence. He had of course deserted a wife and children and now in a letter to them he wrote from a broken heart: "Oh my dear family, how little you know of the terrible ordeal I have passed through, and how few of what the world calls good men are worth the giant powder it would take to blow them into eternity." He swore that he was a changed man and would commit no more crimes, that all his thoughts were now for his dear family. Jackson says he wrote letters to his family "that were unfailingly affectionate." Ambrose Bierce was looking deeper:

> "What's that?—you ne'er again will rob a
>   stage?
> What! did you so? Faith, I didn't know
>   it.
> Was that what threw poor Themis in a
>   rage?
> I thought you were convicted as a poet!"

Oh, my constant loving Mary and my children, I did hope and had good reason for hoping to be able to come to you and end all this terrible uncertainty but it seems it will end only with my life. Although I am "Free" and in fair health, I am most miserable. My dear family, I wish you would give me up forever and be happy, for I feel I shall be a burthen to you as I live no matter where I am. My loving family, I would willingly sacrifice my life to enjoy your loving company for a single week. . . . I love you but I fear you will not believe me & I know the world will scoff at the idea.

Not Joseph Henry Jackson: "The sensible reader will not scoff. Bart was sincere enough. He held his wife and his family in the highest esteem." Jackson forgot to tell us why if Bart would give his life for one week with "my constant loving Mary and my children" he didn't go back to Missouri and do it; he was free and in fair health and nobody had a rope on him. Two pages later Jackson's sympathy for this worthless parasite must have given him a twinge of nausea, for he bursts forth with the question, "Why, however, didn't Bart join his wife?" For the reason that all psychopathic wife-deserters do not—and probably because he was laying his plans to resume a life of crime.

Though there "is no good proof that anybody anywhere" ever saw him again there was a rumor that Wells, Fargo gave him an annual income to leave their stages alone. Jackson thought that was no more than rumor but Beebe points out that another writer accepted it as fact. There was, says Beebe, "a good deal in Wells Fargo's conduct of business in those years which never got on the ledgers." After he was freed there was a robbery now and then that seemed to have his mark on it. He was not fool enough to write more doggerel for sheriffs to admire or to peer through holes in a flour sack, but a valise sent to a Wells, Fargo agent from a hotel had two pairs of cuffs with the laundry mark F.X.O.7, crackers and sugar; and, says Jackson, "perhaps the most touching memento Bart left behind him in all his career, a glass of currant jelly."

Leaving Mr. Jackson to that sob-sistering with which he charged various writers, and the Black Barts to the sentimentalists, we pass on to Joseph A. Slade, another kind of criminal.

"From Fort Kearney, west, he was feared a great deal more than the Almighty," wrote Mark Twain.

We had gradually come to have a realizing sense of the fact that Slade was a man whose heart and hand and soul were steeped in the blood of offenders against his dignity; a man who awfully avenged all injuries, affronts, insults or slights of whatever kind—on the spot if he could, years afterward if lack of earlier opportunity compelled it; a man whose hate tortured him day and night till vengeance appeased it—and not an ordinary vengeance either, but his enemy's absolute death —nothing less; a man whose face would light up with a terrible joy when he surprised a foe and had him at a disadvantage. A high and efficient servant of the Overland, an outlaw among outlaws and yet their relentless scourge, Slade was at once the most bloody, the most dangerous, and the most valuable citizen that inhabited the savage fastnesses of the mountains. . . .

Twain seems to have based his impressions on Dimsdale, on a view of the man, and on what he heard along the Overland route; for he says most of the talk of drivers was about this man. "In order that Eastern readers may have a clear conception of what a Rocky Mountain desperado is, in his highest state of development," Twain said he would tell the story. At twenty-six, in Illinois,

From a Denver newspaper; it is not known that this is Slade
but it is thought to be.

Slade killed a man and fled; Mark has him cutting the ears off Indians and sending them to their chief, until he became division agent of the Overland stages at Julesburg. He thereupon made short work of all the enemies of the stage company, his chief feud being with an ex-agent named Jules, who got the jump on Slade and "poured the contents of his gun into him"—and it was a double-barrel shotgun! Jules was also shot and both men spent a long time nursing their wounds and planning vengeance.

Murders were done in the open day, and with sparkling frequency, and nobody thought of inquiring into them. It was considered that the parties who did the killing had their private reasons for it; for other people to meddle would have been looked upon as indelicate. After a murder, all that Rocky Mountain etiquette required of a spectator was that he should help the gentleman bury his game—otherwise his churlishness would surely be remembered against him the first time he killed a man himself and needed a neighborly turn in interring him.

As for Slade he was such a dead shot that if a man were unfortunate enough to offend him he went in haste to make his will. Slade's vengeance on Jules has been variously told: as Mark Twain tells it, Slade's face lit up with "something fearful to contemplate" on being told by his assistants that Jules had been captured; and he bound Jules and had a good sleep "before enjoying the luxury of killing him." He then "practised on him with his revolver, nipping the flesh here and there, and occasionally clipping off a finger, while Jules begged him to kill him outright and put him out of his misery." After Jules was dead "he lay there half a day" before Slade cut off both ears; some writers say that he carried both in his pockets for so long that they became as soft as kid leather; others say that he nailed one to the corral and made a charm of the other.

Journeying on the Overland Twain came to a station and saw there, at the head of the table, a quiet affable officer who seemed to be all gentleman. "Never youth stared and shivered as I did when I heard him called Slade!" When the coffee pot was about empty Slade told Mark to fill his cup but he was still shivering, so Slade filled it for him. "I thanked him and drank it, but it gave me no comfort, for I could not feel sure that he would not be sorry, . . . and proceed to kill me to distract his thoughts from the loss."

The *Dictionary of American Biography* says that in the spring of 1860 Slade was made division agent of the Overland company and told to destroy all the bandits who were preying on the company's property. This Slade seems to have done with remarkable competence and joy. Two years later he was transferred to a station north of Denver and became such a drunkard that he was fired. He had "a dashing and attractive frontier woman" and with her he set out for the Montana diggings; there he did a bit of ranching but still drank heavily and became an intolerable nuisance, defying the vigilantes and threatening to shoot their judge. Root, another agent on the line, says that when a party of men captured

Slade and locked him up, intending to hang him, he begged them to send for his "wife"; and that, being promptly sent for, she mounted a fast horse and rode in from the ranch. The guards let her in without searching her, and a few moments later Slade came out with a gun in each hand.

The final capture of Slade also has several versions—Dimsdale's, Toponce's, and Tom Rivington's, to name only three. Rivington says Slade rode up to a saloon and was met there by a group of men, one of whom knocked him down with a club, and that Slade would have died there but at once they took him away and hanged him. Toponce says Slade in Virginia City was "the ringleader of a hard bunch and he still carried poor old Jules' ears in his vest pocket. His favorite stunt was to . . . get drunk and stand in the middle of the crowded street until there was no one in sight. Then he would laugh and go into the saloon for more drinks." The Toponce version says that Williams, head of the vigilantes, asked Toponce to persuade Slade to leave town, after being found at his usual pastime of shooting up a saloon. Go-

ing to the saloon Toponce saw that Slade had chased the barkeeper under the bar "and with both elbows resting on the bar he was amusing himself shooting off the necks of the bottles. . . ." Slade said he would go home after one more drink but "he got meaner and more defiant and said he was not afraid of the 'Stranglers.'" Toponce reported to Williams, who sent X. Beidler to bring Slade to vigilante headquarters, knowing that Beidler was the man who would do it. Slade, after being brought in, was asked if he wished to write anything; and on demanding why he should, Williams said, "You will be hung in just thirty minutes." Slade then begged Toponce to save him but Toponce said he had done all he could. Slade was hanged back of a corral, and his woman, hearing of it, rode in with two guns at her belt. Williams told her to go back to the ranch or they might hang her also.

Probably the most trustworthy account by far is Dimsdale's. He says that Slade when free of alcohol was a kindhearted intelligent man who had hundreds of friends, most of whom lamented

Slade defying the court

his death, insisted that he was a perfect gentleman, and called his execution a murder. "Those who saw him in his natural state only would pronounce him to be a kind husband, a most hospitable host, and a courteous gentleman. . . . those who met him when maddened with liquor and surrounded by a gang of armed roughs would pronounce him a fiend incarnate." When he came in to "take the town" he and one of his henchmen "might often be seen on one horse, galloping through the streets, shouting and yelling, firing revolvers etc. On many occasions he would ride his horse into stores, break up bars, toss the scales out of doors, and use most insulting language to parties present. Just previous to the day of his arrest he had given a fearful beating to one of his followers; but such was his influence over them that the man wept bitterly at the gallows, and begged for his life with all his power. It had become quite common when Slade was on a spree for the shopkeepers and citizens to close the stores and put out all the lights, . . ." But one store, kept by the Lott Brothers, he never molested, for he had been told that he would be killed if he did.

After cutting up all night and making the town "a perfect hell," he was arrested and taken to the miners' court. When the warrant was being read to him he "became uncontrollably furious, and seizing the writ, he tore it up, threw it upon the ground, and stamped upon it." His henchmen around him cocked their guns. The Committee had become "well aware that they must submit to his rule without murmur" or hang him. But he was urged to go home. He mounted his horse and started but turned back, became even louder in his abuse, and actually pressed a cocked sixshooter to the head of the judge.

Thereupon the "miners turned out almost en masse, leaving their work and forming in solid column, about six hundred strong, armed to the teeth, . . . the executive officer of the Committee stepped forward and arrested Slade, who was at once informed of his doom. . . . He never ceased his entreaties for his life, and to see his dear wife." By the time he reached the gallows he "had so exhausted himself by tears, prayers, and lamentations, that he had scarcely strength left to stand under the fatal beam. He repeatedly exclaimed, 'My God! my God! must I die? Oh, my dear wife!'" His dear wife had been notified

and on a fast horse was riding full speed toward him. When allowed to see his body in the Virginia Hotel her "grief and heart-piercing cries were terrible evidences of the depth of her attachment for her lost husband, . . ."

Since Slade left no will and was, Dimsdale says, in debt to many persons when he died, one George Parker became administrator. Among the effects was a pair of ivory-handled, silver-mounted Colt revolvers, which Mrs. Slade claimed and Parker refused to give up. A typescript in the Montana Historical Society has a note attached which says that it is, "perhaps," from the reminiscences of Granville Stuart. It says: "She walked out of the store without comment but returned soon after and walking up to Mr. Parker, said, 'You say you cannot give me my revolvers, perhaps I can persuade you to change your mind.' At this she presented the muzzle of a .45 Colt in his face. George took no time to reason with the lady but promptly handed her the revolvers which she coolly took possession of and left the store."

Toponce says that when he last heard of her the Slade woman was living in Salt Lake City.

Jefferson Randolph (Soapy) Smith was admired by Wilson Mizner, king of the con men, who said that in a Dawson gambling house he "got fifty per cent of everything I stole, and did pretty well." Perhaps Soapy did no better but he had a large payroll to meet, including bribes to policemen, judges, and wages to a gang of ruthless killers. Like Black Bart he was a psychopath; he had abandoned a wife and six children back in St. Louis, and if he had been hanged, as Slade was, no doubt he would have wanted nothing on earth so much as to be with his dear wife and little ones.

After watching a man with a shell game he decided that taking money from suckers was what he was born for but he wanted a better trick than three walnuts and a pea. With their unwashed itching scalps soap was what the miners needed. So he would set up a table, scatter over it small bars of soap, colorful paper for wrapping, and a number of bills from one to one hundred dollars. "How are you fixed for soap, boys, how are you fixed for soap? Move on up to the table, boys, how are you fixed for soap?" He was an artist at teasing and confusing and hypnotizing dullards and he thought Barnum's statement was a gem of purest wisdom, "I have never lost a dime underrating the intelligence of people." The bug-eyed miners, staring at his hands, thought they saw him wrap a hundred-dollar bill with one bar of soap, a fifty with another, and fives and tens

Soapy Smith

with still others; and for five dollars a chance they were eager to pick out the one with the hundred-dollar bill. After he became a practiced scoundrel Soapy recruited a crew of shills and cappers, all dressed for the part; and when one who looked like a grizzled old miner allowed as how he knowed which bar had the hundred and slapped down his five for it, the others, crowding around, allowed as how they knowed which bars had the fifties and the twenties.

Soapy did all right with his shell game but he was ambitious. The ease with which he could take care of his competitors gave him an idea. There was Kid Barnett, also an accomplished swindler, who moved into Soapy's territory and was told by a henchman to choose another area or make arrangements for his funeral. The Kid marched into Soapy's saloon with shooting irons and daggers hanging all over him; the bouncers

took him from behind, stripped him clean, and tossed him into the street. The Kid bought more guns and knives and came back, firing, and one shot took both thumbs off one of Soapy's strong men. But the Kid and other zealous competitors stood no chance; Soapy had already announced that he was the law in Creede, one of Colorado's most hell-roaring camps, and in fact he was. But while fleecing that town, and Denver, he had an eye out for opportunities worthy of his genius.

He decided that Dictator Diaz down in Mexico needed a foreign legion organized by Jefferson Randolph Smith; and that murderous looter was so impressed that he gave Soapy a pile of money and told him to go ahead. Soapy returned to Denver, where he calculated most of the more adventurous criminals were then gathered, and was busy getting up a small army, of which he was to be the commander, when Diaz called the whole thing off. Soapy next tried to muscle into Seattle's underworld and for a little while was an emperor without an empire. Then he heard of the gold rush in Alaska. The completeness with which he and his thugs took over Skagway indicates that after Diaz put spies on him and learned that all his credentials were phoney, Soapy had been doing some thinking. Like any good politician he had awakened to the fact that to be really successful in a community he had to take up collections for the widows and children of the men whom his thugs robbed and killed; establish a home for homeless dogs; give liberally to all churches and charities; parade his patriotism on public occasions; and say the right things about law and order and human decency.

All these things he learned to do with superb effrontery but like all psychopaths he was vain, he was infatuated with himself, and he was contemptuous of the boobs he exploited. And his thugs were careless. Mizner thought that Soapy's gang killed at least fifty persons; a historian of Alaska has said that eight bodies were picked up on White Pass in one day, all slain for the gold dust they were bringing out. If Soapy had had assistants who were not all greed and brutality he might have lasted a long time, for he was taking care of his side of it with considerable dash. Within a few days after he heard of the war with Spain he had made himself captain of Company A, First Regiment, National Guard of Alaska, and was ready for business. He opened an office, recruited men, and drilled them. Freedom for Cuba! Remember the *Maine!* said his badges of white, blue, and gold, with the compliments of the Company, Jeff R. Smith, Captain. It was hard for even a vigilance committee to get around a guy like that.

News had leaked south that he was head of a gang of cutthroats. The San Francisco *Examiner* sent a reporter up, who was soon so groggy with Smith's hospitality that he wrote the paper that Soapy was "not a desperado. He is not a scoundrel. He is not a criminal. . . . he bitterly resents the imputation that he is a thief and a vagrant." This reporter even got off an ode to Soapy's patriotism. The editor of the *Alaskan* was on his payroll. When the New York *World* sent a man out he was soon penniless, but Smith put him on the *Alaskan* at a handsome salary and the bewitched reporter was soon writing back home that Soapy was "the most gracious, kind-hearted man I've ever met." Berton surmises that by this time in his career Soapy thought of himself as a great public benefactor.

Half the people in Skagway thought he was the Devil up from hell and half of them thought he was a great and noble man trying to bring order out of chaos. Early in 1898 a vigilance committee was formed, and though Soapy did not suspect it he had less than six months to live. But his arrogance never faltered. When the Committee of 101 posted a notice that a word to the wise should be sufficient and that all "con-men, bunco and sure-thing men and all other objectionable characters" should leave Skagway at once, Soapy called a public meeting by his Committee of Law and Order, to put a stop to lawless acts and punish all persons who were disgracing the city. In a harangue to his audience he notified the Committee of 101 that any acts committed by them would be promptly punished "by the law-abiding citizens of Skagway . . . and the Law and Order Society, consisting of 317 citizens." Not even the Hounds in San Francisco, or Plummer and Updyke, or the leader of any other mining-camp gang, exhibited such incredible arrogance.

Soapy might have bluffed the thing through if he could have controlled his thugs. On the Fourth of July, following the organization of his regiment, it was roses, roses everywhere, and flags and patriotism like mad; and it was the Governor

Harry Orchard

of Alaska beside him on the platform. What a great fellow Jefferson R. Smith was! What was Frank Reid, city engineer and a moving spirit in the vigilance committee, thinking as he saw Soapy leading the parade and sitting with the governor? What was a man named Stewart thinking, as he took his weary way up over White Pass, clutching a poke with $2,800 in gold dust, his savings for many a long month? Of his people, possibly, for he was on his way home. He made the mistake that many had made: he was lured by the thugs into the back room of Soapy's saloon, and when he came to his senses he didn't have his poke or a single cent. He went to the deputy U.S. marshal, but that gentleman was on Soapy's payroll. He went to businessmen in the town, and a dormant vigilance committee came suddenly and powerfully to life.

A meeting was called, but so many of Smith's henchmen were present that it was disrupted. It met next on a long pier that ran out into the bay; at the town end of the pier four guards were posted. Frank Reid was one. One source says that when the news was carried to Smith he said he would chase the sons of bitches into the bay. That is plausible, for with two pistols and a rifle across his arm he set off for the pier, with some of his bolder scoundrels at his heels. There is a difference of opinion on what happened at the pier; as Berton tells it, Smith marched up to Reid and thrust his rifle at him. Reid knocked it aside and tried to fire but his gun missed. The two men then fired at about the same moment; Smith received a bullet in his chest and fell and died almost at once; Reid was struck in his lower belly and lingered in agony for many days.

Most of Stewart's gold was found in Smith's trunk. Eleven members of his gang were sent to Sitka for trial, and were given light sentences. Some others, including the editor of the *Alaskan* and the author of the ode, were exported to Seattle. Reid and Smith were buried not far apart; it need surprise no one to learn that for many years a Mrs. Harriet Pullen sent money for the maintenance not of Reid's but of Smith's grave. Today, Reid has a marble slab, "like his memory, half-forgotten," while thousands of tourists still go out to stare at the earth that covers Jefferson

Randolph Smith, public benefactor, friend of widows and orphans and homeless dogs, and captain of Company A.

It may be appropriate to conclude this depressing chapter with a badman who became a good man. Arrested as Thomas Hogan New Year's Day, 1906, for the murder of ex-governor Steunenberg of Idaho, he was tried as Harry Orchard, but his real name seems to have been Alfred Horsley. According to MacLane, who knew him and was at his trial, he confessed to these crimes:

1. Blowing up a mill of the Bunker Hill and Sullivan mine and killing two men
2. Secreting a bomb in a Cripple Creek mine and killing two men
3. Planning but failing in the assassination of Governor Peabody of Colorado
4. Shooting and killing a deputy named Gregory in Denver
5. Planning and helping to blow up a railway station in Colorado, killing fourteen
6. Attempting to murder a mine manager in the Coeur d'Alenes by putting poison in milk left on the front doorstep, and failing in this, arranging a bomb that blew the man

Waiting for the undertaker: Bill Powers, Bob Dalton, Grat Dalton, Dick Broadwell "from his doorway into the street," severely wounding but not killing him

7. Setting a bomb for a justice of the Colorado Supreme Court and killing another man instead

8. Setting a bomb for Judge Goddard of the Colorado Supreme Court, which failed to explode

9. Lying in wait to kill Sheriff Bell of Colorado

10. Planning and abandoning the murder of a man named Mayberry

11. Topping it all off by blowing up Steunenberg.

It is altogether possible that he forgot or chose not to tell about some of his crimes. In his confession he implicated Big Bill Haywood and some other members of the top council of the Western Federation of Miners.

James H. Hawley, who would be an Idaho governor, and William E. Borah, its most distinguished senator, went with accomplices to Colorado and virtually kidnapped Haywood and two others and brought them to Idaho for trial. Haywood and one of the others were acquitted and the third dismissed. That Orchard told the truth about them no one who reads the evidence can doubt.

A Pinkerton agent assigned to Orchard decided that since the man had grown up devoutly religious he could be broken. He was right. Orchard, when working for the Council, was a dead ringer for some of the inquisitors in the Middle Ages; he had no doubt at all that he was right, that the end justified the means. He was convicted on his confession and sentenced to be hanged, but Hawley, who had taken a liking to him, worked like a beaver for years, getting one postponement after another, and finally commutation to life. Orchard became nationally famous as a redeemed criminal; it certainly is true that if ever a man expiated his sins this man did. He lived to be an old man and during the later part of his life was a trusty who lived in a little cottage outside the jail's walls. He had many friends, including Hawley's son Jess.

The Vigilantes

One of the most controversial aspects of life in the mining camps is the work of the vigilance committees. Some writers who seem to have been unfamiliar with the facts have severely condemned them, as some are condemned today whose sense of justice does not embrace an endless indulgence of the chronic criminal. We'll present some of the facts that made the formation of the committees inevitable; some of the typical criticism of the committees; and some account of the work of the committees in a few of the major camps.

Bancroft set at the head of Chapter 1 of his *Popular Tribunals* Edmund Burke's words, "When bad men combine the good must associate; else they will fall, one by one, an unpitied sacrifice in a contemptible struggle." Law-abiding men fell one by one, and by the hundreds, before "lynch law" destroyed the worst of the criminals. It seems not to be generally known that vigilance committees were not a phenomenon in the mining camps but were fairly commonplace in some of the states before the first big gold rush of 1848-49. When, for instance, it was learned in Tennessee that a certain desperado had been hanged in Montana the *Appeal* of Memphis said on November 24, 1865, that if this man's "career of desperation and crime" before he went west were told it "would cause the bloodthirsty tales of the yellow-covered trash to pale for their very puerility and tameness." For the fact is that professional criminals from the eastern states and from many lands of the earth headed for the American West as soon as they heard of the discovery of gold.

Said Delano, after inspecting conditions in San Francisco:

The gold of California had attracted to its shores the dissolute and dishonest from all countries of the civilized globe. Situated within reach of the penal colonies of Great Britain, as well as being in proximity with the semi-barbarous hordes of Spanish America, whose whole history is that of revolution and disorder, it was soon flooded by great numbers from those countries, who were accomplished in crime, . . . their only aim to obtain gold by any means, . . . and owing to the weakness of the constituted authorities, joined to the vicious among our own people, they succeeded in their frauds and crimes to an amazing extent, and rendered the security of life and property a paradox on legislation, hitherto unprecedented in the annals of modern history.

Bancroft examined the criminal records in San Francisco for 1855, the year before the committee of 1856 went to work, and discovered that there and in the areas roundabout hundreds of persons were dying violently every year. Trimble does not exaggerate when he says that "emboldened by the unorganized and unprotected condition of society, these villains banded themselves together for most atrocious rapine and murder." In San Francisco

I saw, [says Kelly] at the Eagle Saloon, in Montgomery street, a *monte* dealer deliberately draw a pistol from beneath the cloth, and shoot a young lad, who was, I believe, honestly scuffling for his stake; and then, with the most perfect *sang froid*, call the "coroner," whom he recognized amongst the bystanders, to hold an inquest, which actually took place on the spot where the bloody deed was committed, in presence of the murderer, a volunteer jury of "pals" returning a verdict of "accidental death". . . .

Armed robbery and murder, says Helper, were everyday occurrences and the "most daring outrages were every where committed with impunity. Unoffending men were shot down and pillaged in broad daylight; shops were broken open; . . . in short, the country was in a state of siege, . . ." He had no patience with the fact that vigilantes were being "harshly judged and unjustly condemned," and his Chapter Seventeen is the most spirited defense of them that we have found, together with a vivid description of the hanging of two brutal killers. In a barbershop in Montana Richardson saw guns drawn and a man killed and another wounded; "but such affairs were so common that the barber did not even stop lathering his patron's face, nor did the patron leave his chair."

After the murder of Dillingham, conditions so "rapidly deteriorated," says Dimsdale, in Virginia City and Nevada City that men didn't dare walk abroad after dark. "Wounded men lay almost unnoticed about the city, . . ." Parts of Sir Henry Huntley's two volumes are catalogs of crimes and punishments: "Last Friday, three men were shot in the vicinity of Stockton; on Saturday evening, an attempt was made to shoot two men, without provocation; and the same night, a Mexican was most inhumanly butchered. . . ." He tells of a physician who "claimed the head, as he had made a bargain with the man for it before he was shot, and quietly pulling out a bowie-knife, whacked off the head, and placing it in a sack, marched off with it."

Little wonder signs were posted, such as those Otero gives from Las Vegas, New Mexico:

A TIMELY WARNING

To Murderers, Confidence Men, Thieves:
    The Citizens of Las Vegas have tired of robbery, murder, and other crimes that have made this town a byword in every civilized community. They have resolved to put a stop to crime even if in attaining that end they have to forget the law, and resort to a speedier justice than it will afford. All such characters are, therefore, notified that they must either leave this town or conform themselves to the requirement of law, or they will be summarily dealt with. The flow of blood MUST and SHALL be stopped in this community, and the good citizens of both the old and the new towns have determined to stop it, if they have to HANG by the strong arm of FORCE every violator of the law in this country.

VIGILANTES

Otero gives another poster in which a number of the criminals were called by name and told they would be hanged if they did not leave the area by ten o'clock that night. They left—for other camps! It was largely because camps exported their unwanted citizens, instead of hanging them, that there were so many vigilance committees in

so many places, over so long a period of time. The organized criminals did not stop at armed robbery and murder; they actually set fire to the camps, so that in the panic and confusion they could loot the saloons and shops. San Francisco was almost destroyed six times in a year and a half. Frank Soulé, who was there, wrote that there was "no doubt" that the fire of June 22, 1851, was the work of arsonists. After one of the fires the *Herald* said that it did not "advocate the rash and vengeful infliction of punishment on any person against whom the proof is not positive. . . . but we believe that some startling and extraordinary correction is necessary . . . to arrest the alarming increase of crimes against property and life, and to save the remainder of the city from destruction."

As for the laws that were supposed to govern, the reader can find in another chapter a characterization of the courts, judges, juries, and lawyers. Marryat found himself puzzled: "It has been very difficult to get a jury to convict a murderer in this country; I am puzzled to say why, for self-interest would dictate an unusual degree of severity—still the fact stands, that in twelve hundred murders, but two men have been publicly executed." For the most part, says Bancroft, only the petty and the penniless were punished.

Able counsel was secured by money, false witnesses were suborned, and judges and jailers made lenient. I do not mean to say that all officials, nor the half of them, were open to bribery. There were some as pure judges on the bench then as now. Yet money, if not directly, then indirectly, would buy acquittal or pardon. . . . Money would impanel a jury favorable to the accused; if not at the first, then the case would be postponed from time to time, . . . All the technicalities of court procedure were employed for the acquittal of the accused. Criminal cases were held in abeyance until witnesses could be spirited away.*

Men, like nations, are likely to get into fights if they are heavily armed. The human male is an aggressive animal, and the restrictions and disciplines which laws impose on him often develop buried hostilities. Among the reasons for the vigilantes is the fact that thousands of men walked around not only armed but often grotesquely armed. Many of them liked to think of themselves as gunmen, many liked to show off. Otero

_____
* All of which is commonplace today.

tells of one who, showing off, killed a man standing nearby. That seemed not to dash him at all, for he continued to handle his guns, and shot a woman. When he was arrested he argued that both killings were accidents. The vigilantes hung him from an old windmill in the plaza, and put on him a sign which said:

This was *no* accident

Borthwick tells how theatergoers gave up their weapons at the door:

... each man as he entered delivered up his knife or his pistol, receiving a check for it, just as one does for his cane or umbrella at the door of a picture-gallery. Most men drew a pistol from behind their back, and very often a knife along with it; some carried their bowie-knife down the back of their neck, or in their breast; demure, pious-looking men in white neckcloths, lifted up the bottom of their waistcoat, and revealed the butt of a revolver; others, after having already disgorged a pistol, pulled up the leg of their trousers, and abstracted a huge bowie-knife from their boot; and there were men, terrible fellows, no doubt, but who were more likely to frighten themselves than any one else, who produced a revolver from each trouser-pocket, and a bowie-knife from their belt. If any man declared that he had no weapon, the statement was so incredible that he had to submit to be searched; ...

According to Beebe, Bernard Baruch said in an interview after he had visited a mining camp that he had met George Wingfield:

He was a man with a gun. ... He was carrying five revolvers—two here, two here, and one here. He also had four Pinkerton detectives with him. ... he'd been having some labor trouble. ... One reason was a company rule he had, to the effect that the men working in the high-grade mines had to strip as they started home, and then they had to jump over a bar, so that any nuggets of gold they had hidden under their arms or between their legs would fall on the floor. ... the best shot I ever saw.

Was vigilance law effective? All the writers on the subject agree that it was. In Idaho the *Avalanche* told its readers:

Blackstone says. "the main strength and force of law consist in the penalty attached to it," and we will add that its strength and force are greatly increased by certain execution of the penalty. Locke remarks, in his *Essay concerning Human Understanding*, that "the dread of evil is a much more forcible principle of human actions than the prospect of good." From these principles of law and ethics, we may fairly assert the truth of converse propositions. The weakness of a law consists in the mildness or absence of its penalties, and doubts of the execution of such as are annexed; and the hope of impunity is a more forcible inducement to commit crime than the prospect of punishment. ... The power to award clemency or pardon, or both, to criminals is a great one, and as dangerous as great in the hands of yielding officials. The temptation to use it is enormous.

Borthwick never saw a vigilante hanging but undoubtedly talked to members of the committee of 1851. It was his opinion that

when Lynch law prevails, it strikes terror to the heart of the evil-doer. He had no hazy and undefined view of his ultimate fate in the distant future, but a vivid picture is before him of the sure and speedy consequences of crime. ... the criminal knows that, instead of being tried by the elaborate and intricate process of law, his very ignorance of which leads him to over-estimate his chance of escape, he will have to stand before a tribunal of men who will try him, not by law, but by hard, straightforward common-sense, and from whom he can hope for no other verdict than that which his own conscience awards him; ...

Why, then, was there such vigorous and often virulent opposition to vigilante justice, after legal justice had failed? One reason was that here and there the vigilantes themselves became a mob. Dame Shirley saw a man hanged for stealing—"a piece of cruel butchery, though *that* was not intentional. ... life was only crushed out of him, by hauling the writhing body up and down several times. ... many of the drunkards, ... laughed and shouted, as if it were a spectacle got up for their particular amusement." As Carl I. Wheat says in his comment on her letter, both Josiah Royce and Shinn (*Mining Camps*) took note of the "occasional fury and brutality" of vigilante justice. Her long "Letter Nineteenth" describes a mob bent on the indiscriminate punishment of Spaniards because of the villainy of a few of them. "The rowdies have formed themselves into a company called the 'Moguls,' and they parade the streets all night, shouting, breaking into houses, taking wearied miners out of their beds and throwing them into the river, ..." But what the Dame saw was far from the cool disciplined devotion to justice of the principal committees.

Shinn in his study of frontier government says on one page that "on the whole, lynch law is manifestly selfish, cowardly, passionate, un-American," a harsh judgment that is simply untrue; but on another page he says,

When, after a reign of terror almost unexampled in American frontier history, the tried and true merchants of Montana organized during the winter of 1863-4 and in a few weeks hung twenty-four desperadoes and murderers, they performed a solemn duty laid upon them as American citizens. The present peace (1884), order, and prosperity . . . are the result of this acceptance of weighty responsibilities. Not until nearly a hundred persons had been waylaid, robbed, and slain . . . did society accept the challenge. . . . It is a matter of history that this organization, like that of San Francisco, never hung an innocent man, and that, when its work was done, it quietly disbanded.

One wonders, then, why he said that "on the whole" the committees were selfish, cowardly, and un-American.

The fact is that there was a lot of ignorant writing against the committees, and perhaps no man fostered so much misunderstanding of the essentials of the matter as General William T. Sherman in the *Overland Monthly* and in his *Memoirs.* As Bancroft said, "Sherman may be a great soldier, but he made a great mistake when he undertook to write a book. . . . I never saw so great a man with so little common-sense." Bancroft exposes some of Sherman's more flagrant errors, such as his statement that Hetherington and Brace were hanged without a trial.*

But the severest critics in some areas were the newspapers. In San Francisco they were divided. The *Alta* said, "How many murders have been committed in this city within a year? And who has been hung or shot or punished for the crime? Nobody." According to Bancroft there were 1,200 murders in the four years beginning with 1850, and only one murderer, a Mexican, was hanged. Historian Hittell says in his history of California that "There probably had never been in the United States a deeper depth of political degradation reached than in San Francisco in 1854 and 1855." The moral degradation was just as deep. It had been almost as deep in 1851 when the first vigilance committee was established, whose pur-

pose, according to William T. Coleman, its great and fearless leader, was "to vigilantly watch and pursue the outlaws who were infesting the city, and bring them to justice, through the regularly constituted courts, if that could be, through more summary and direct process, if must be. Each member pledged his word of honor, his life and fortune if need be, for the protection of his fellow members, for the protection of the life and property of the citizens of the community, and for purging the city of bad characters who were making themselves odious in it."

But the leaders of this committee were of such high character, and were so reluctant to take into their own hands the duties of the courts and police, that they hesitated and wavered. In one instance they were backing away from punishing a brutal killer when one of them, named Howard, spoke up, saying, "Gentlemen, as I understand it, we came here to hang somebody." Years later in Montana a man named Beidler was to speak up, when a committee there wavered. But though Howard momentarily stiffened their spines the Committee of 1851 did not do the job it should have done. Possibly its leaders never fully realized how completely the town was in the grasp of ruthless criminals; or possibly—and this seems more likely—they hoped that punishment of a few would make law-abiding citizens of the others. They were completely mistaken.

They might have done the job then and there if persons in responsible positions had not been among their severest critics. A Judge Campbell, for instance, was calling them murderers. The *California Courier* and the *Herald* both attacked him, the latter saying that when a man with his infant in his arms and his wife at his side was murdered "the murderers were never stigmatized by the presiding judge with half the rigor exhibited in this charge by Judge Campbell against the Vigilance Committee for the public execution, after a fair trial, of a desperado whose life from childhood had been a long series of bold and successful crimes."

But the Committee continued, as Coblentz says, "to exercise moderation and restraint" when what was needed was the firmness of the Montana committee, over ten years later, or of the San Francisco committee of 1856; for the moderation and

---

* The reader interested in the quality of Sherman's judgments can consult Bancroft's *Popular Tribunals.*

restraint of 1851 made the committee of 1856 inevitable. After hanging two of the worst criminals the Committee, smarting under criticism and unsure of itself, abandoned its efforts; and the thieves, muggers, and murderers who had gone into hiding now emerged, twice as arrogant as before and ten times as brutal. There came to life with them James King of William,* without whom it is hard to say whether the second committee would ever have got the job done. King had been a miner; then a banker, who had failed in the financial crisis of 1850; and at last the crusading editor of a small four-page newspaper called the *Daily Evening Bulletin*. King, though a crusader, was still relying on the organized processes of law and justice when an odious and perhaps feeble-minded coward named Cora shot United States Marshal Richardson and again brought things to a head. Apparently the only reason for the murder was a trivial quarrel in a theater between Richardson's wife and Belle Cora, a whore, who was Cora's woman. When the murder became known (Richardson was unarmed at the time), anger consumed the lawful citizens of the town, and the vote of an assembled multitude was overwhelmingly for the lynching of Cora.

Cora's trial in a court of law was the kind of farce all the mining camps were used to. Belle seems to have had substantial savings, and to save her man, whom no woman without perverted tastes could possibly have wanted, bribed various persons to swear to lies. The jury disagreed, and Cora was returned to the jail. The *Alta* summed it up January 17, 1856:

It has been understood for some time past that criminals having money or friends, could not be punished in this community. It has been a subject of common remarks that witnesses could be procured to testify to almost anything; that juries could be made away with at pleasure. The feeling was expressed very generally, as soon as it was understood that Cora's friends had money, that he would never be punished. . . . Indeed, so confident were persons in this respect, that they even pointed out men on the jury who would never consent to a conviction; and bets were offered that such and such men would go for acquittal. . . . the public has been grossly abused. . . . It is not very agreeable to state that the conviction is almost universal, that crime cannot be punished in San Francisco. . . . It is well for every man

James King

to understand that life here is to be protected at the muzzle of the pistol.

Cora was still in jail but living in the style his madam had made him accustomed to when James King tossed discretion out his newspaper window and excoriated the criminals with all the vocabulary at his command. Bancroft, who knew him, thought King "approached the morbid where his cause was concerned. He seemed to regard all lawyers, judges, and officials alike as evil-minded enemies of the public." Whether without a humorless and dedicated man the Committee could have been reanimated and stiffened to do its job is something Bancroft did not tell

* Because in Washington, D.C., where he grew up, there were several persons named James King, he added to his own the Christian name of his father.

us. Whether King approached the morbid and was too intemperate every person must decide in his own book of values. On November 22, 1855, King was writing in the *Bulletin*, "Bets are now offered, we have been told, that the editor of the *Bulletin* will not be in existence twenty days longer. . . . War, then, is the cry, is it? War between the prostitutes and gamblers on one side, and the virtuous and respectable on the other! War to the knife, and the knife to the hilt! Be it so, then! Gamblers of San Francisco, you have made your election, and we are ready on our side for the issue!"

These words inspired Bancroft to write, "Mr. King was, in my opinion, unjustly severe upon gamblers. In politics and morals he was a pessimist. He would have no man in office if he ever had been a gambler, without the most convincing proofs of his repentance."

After saying that he had never gambled and had never associated with gamblers and had no sympathy for them, Bancroft adds, "why not recognize . . . morality in a gambler, if such qualities be there!" Coblentz, who did not know King, saw him as a man

slightly under middle height, but nevertheless of an imposing appearance. Glance at him closely —dark of complexion, with big piercing eyes, aquiline nose, and black full-beard and side whiskers, he boasts a large well-formed head with a wide, moderately high forehead, but walks with a slight stoop. . . . Yet is there not in his aspect the suggestion of a strength that is more than merely physical? His lips, beneath the mustache, are firm and broad; his jaws are resolutely set; his eyes are straightforward and determined under the beetling brows; his whole manner bespeaks a decisiveness, a flinty, uncompromising quality of will that would make him a formidable foe.

A formidable foe he certainly was, and among those who smarted under his lashings was James Casey, of whom Bancroft says that after being discharged from Sing Sing, where he had served a term for grand larceny, he went to San Francisco and soon held a high position

in one of the corrupt cliques which largely con-

Casey and Cora

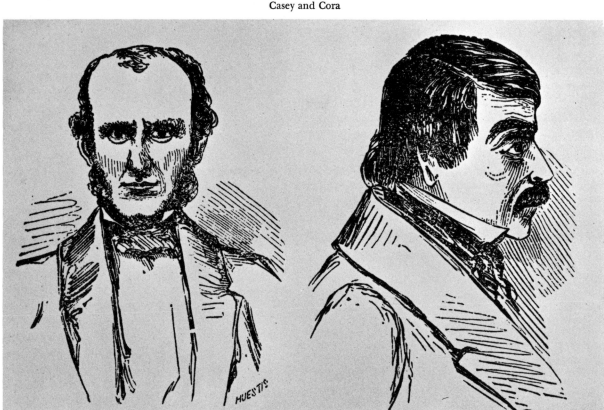

trolled both city and state governments. Like . . . Broderick,* he was a leader among the roughs. . . . Short of stature, slightly built, with delicate features, bright intelligent blue eyes, and very large brain, he possessed altogether an intellectual cast of countenance, in marked contrast to that of his brother assassin Cora. . . . Thief, fireman, ballot-box stuffer, supervisor, editor, murderer. And this man had a host of friends.

Including some who were powerful in state and city government.

In one of his withering editorials King said,

It does not matter how bad a man Casey has been, or how much benefit it might be to the public to have him out of the way, we cannot accord to any one citizen the right to kill him or even beat him, without justifiable provocation. The fact that Casey has been an inmate of Sing Sing prison in New York is no offense against the laws of this State; nor is the fact that having stuffed himself through the ballot box, as elected to the Board of Supervisors from a district where it is said he was not even a candidate, any justification for Mr. Bagley to shoot Casey, however richly the latter may deserve to have his neck stretched for such a fraud on the people.

Those words brought the matter to a head, as King should have known they would, if he had any sense of the wrath and violence mounting against him. Casey in a fury rushed to King's office, and as reported by two men who listened in an adjoining office the substance of the words between them was about as follows: "What do you mean by that article?—that which says I was in Sing Sing?" "Isn't it true?" "That's not the question. I don't want my past raked up. I'm sensitive." It was said that King then leapt from his chair, shouting, "Go, and never show your face here again!" At the door Casey turned and said he would defend himself. Choked with rage, King shouted, "Go!" It has been said that in a saloon among his crooked cronies Casey boasted that he had scared the daylights out of King, and that a certain Judge McGowan, a drunkard and fugitive from justice, slipped to Casey his heavy navy revolver and told him to shoot King. Whether that is true or not, late in the afternoon Casey hid in street shadows, and when King came along, unarmed and unsuspecting, shot him at close range. Soon afterwards the engine house bell that

had summoned the vigilantes of 1851 began to toll, the streets filled with armed men, and the *Herald* the next day dismissed the murder as a mere "affray" and said there was no reason for mobs to gather. But in that newspaper, and in others that morning, was a small advertisement which said:

The Vigilance Committee
The members of the Vigilance Committee in good standing will please meet at number 105½ Sacramento Street, this day, Thursday, fifteenth instant, at nine o'clock A.M.
By order of the COMMITTEE OF THIRTEEN

The crooked bankers, the corrupt office holders and newspaper editors, and the whole criminal world of California had had enough of James King, and with his assassination the law-abiding people had had enough of the courts and judges and juries of San Francisco. The uprising of an outraged and long-suffering people is extremely dramatic and impressive. The efforts of a pusillanimous if not corrupt governor to abash and disband the vigilantes; the various attempts to bring military forces to the scene; the wildly impulsive behavior (he even stabbed one of the vigilantes) of Judge David S. Terry of the supreme bench; the extraordinary discipline and courage of the thousands of vigilantes and the cool firm leadership of the committee at the top—all these and more make a fascinating story for which we have no space here.

Did the vigilantes find support in other California camps? San Jose, forty miles south, offered to furnish a thousand men. Coloma sent word, "If you need help, let the sea speak to the mountains." Nevada, Grass Valley, and Auburn draped themselves in mourning on hearing that King had been slain. Sacramento, Stockton, and other towns "performed the solemn obsequies simultaneously with those of San Francisco," and funds were collected in many places for the editor's wife and children. Sacramento said they were thinking of forming a committee of vigilance there and asked for instructions on how to proceed; and in Sonora five thousand men gathered to demand a committee there. And groups of women in San Francisco said that if the men didn't hang Casey, the women would.

---

* Whom see, in the chapter entitled "Duels."

But the men intended to hang him. Vigilantes by the thousands in massed columns marched against and surrounded the jail, while twenty thousand persons watched from the hilltops. Two carriages were slipped through to the jail door and a note was sent in to the sheriff: Sir: You are hereby required to surrender forthwith the possession of the county jail now under your charge to the citizens who present this demand, and prevent the effusion of blood by instant compliance. By order of the Committee of Vigilance. Cora and Casey were taken. They were tried in turn before the Committee; Cora gave a long list of names of persons whom he wanted summoned as witnesses. Not a one of them could be found, possibly for the reason that hundreds of criminals were going at top speed to other areas. Cora asked for a priest, who refused to obsolve him "unless he first married his *inamorata*." This he was allowed to do: that fact that a brutal murderer was allowed to marry a whore certainly shows that the Committee was not without indulgence. The mar-

riage moved Bancroft to one of his emotional flights: "What may this thing be, oh thou whose name is Love . . . that the presence of a prostitute for a few short hours should seem to take the sting from death itself?" The good women of the town did not share Bancroft's raptures but asked that Belle Cora be driven out of the city: "The truly virtuous of our sex will not feel that the Vigilance Committee have done their whole duty till they comply with this request." This want of pity for a fallen sister touched Bancroft off to typical moralizing: "Had Belle Cora been born, reared, and circumstanced as you were, a thousand to one she would be as pure and proud of her purity to-day as you are. Had you been born, reared, and circumstanced as was Belle Cora, a thousand to one you would be Belle Cora." That seems very neatly to leave heredity out of it.

Casey made a terrible fuss. His last night on earth he paced the floor, crying, "Oh, my God! Has it come to this? Must I be hung like a dog?" Because he wanted to speak to the multitude

Taking the jail

Hanging of the Ruggles Brothers

he was permitted to appear upon his platform as he desired, with head bared and white handkerchief in his hand. With pale face and bloodshot eyes he stammered through an incoherent speech, made mostly of ejaculations, denials of guilt, expressions of concern lest his name should be immortalized as that of a murderer, and of confidence of having made his peace with his maker.

Over and over he cried that he was no murderer and had always done what he thought was right. Over and over he cried, "Oh, my poor mother, my poor mother!" He had already written a note to the Committee in which he asked them to keep news of his hanging from New York and his mother. He said he forgave the Committee this persecution of him and hoped that God would pardon them.

Casey's body had an escort of eighty-four carriages and eighty horsemen, and hundreds of followers on foot, chiefly the criminals who had not yet fled. Cora had only "six hacks." After the hanging of these two murderers the Committee made a blacklist and sent to those chosen for banishment this note:

The Committee of Vigilance, after full investigation and deliberation, have declared you guilty of being a notoriously bad character and dangerous person, a disturber of the peace, a violator of the purity and integrity of the ballot-box, and have accordingly adjudged the following sentence: That you, James Cusick, leave the State of Cali-

Execution of Casey and Cora by the San Francisco Vigilance Committee May 22, 1856.

fornia on or before the twentieth day of June 1856, never to return, under the severest penalties. By order of the Committee. No. 33, *Secretary.*

Like the dumping of unwanted cats, the chasing of criminals off to the neighbors to prey on them is, of course, a fine old Christian custom. If the Committee had steadily resisted the mindless clamor and had hanged or jailed all the more brutal offenders it may be that there would have been no need of a committee in Montana and Idaho and other areas. For it is an established fact that some of those hanged in Montana and Idaho (to mention only two areas) were fugitives from California. As it was, the Committee, says White, hanged only four and banished only about thirty. "But the effect was the same as though four hundred had been executed. It is significant that not less than eight hundred went into voluntary exile." The effect was felt only in California; scores of criminals who fled were soon busy preying on their fellows in other areas, and the bulk of them moved, like Soapy Smith and Plummer, from gold rush to gold rush.

Undoubtedly the chief reason the Committee merely put to flight many whom it should have hanged or put into jail for long terms was the tremendous outcry against it. A group that called itself the Party of Law and Order, led by a slippery governor and a violent judge with strong Southern sympathies, never rested in its efforts to harass and intimidate. Though the *Union* said that if Federal troops were to fire on the vigilantes "such a storm of revolution would sweep over the state as never has been witnessed in these United States," other newspapers took another view of it. The *Sun* said, "The brand of treason is upon you, and you can no more escape it than could Benedict Arnold in other days." It excoriated the *"Bulletins* that cry for more blood; the namby-pamby *Altas* that have wriggled and wriggled out, until they have wriggled into your patronage; the so-called *Californias* that have fungus-like sprung from the putrefactions of the day, to fill the air with their noisome stench of treason, and the lesser fry of traitors of various vileness and calibre, . . ." "Never had there arisen in any state," says Bancroft, "an issue which aroused such bitter feelings in the breasts of the minority against the majority." Apparently the chief reason for the bitter censure of the vigilantes, there, as in Idaho

later, was the fear in proslavery and pro-South people that after the Committee was disbanded its leaders would be the future leaders in their areas. Bancroft says, "It was to make of the Pacific States a slave empire that the chiefs of that party had proposed dismemberment from the federal union, even while begging the aid of the federal authority against the only true federal supporters." When, for instance, Judge Terry, a rabid proslavery man, was brought to trial by the vigilantes for driving a dagger into one of them, the *Herald* and the *Sun* "editors raved like men in delirium" and the Governor "threatened to pour the forces of the federal government into San Francisco bay inside of ninety days." Bancroft says the Committee never wavered but it did waver: against its honest judgment it brought in a verdict of guilty but discharged, and Terry was set free. This verdict so infuriated the thousands of vigilantes who had not sat as jury that it looked for a while as if they might resort to violence. They did not understand that, as a member of the Committee put it, the judge "was bigger game than we calculated to bag." Editors now heaped the "most scurrilous abuse upon men who but yesterday were gods." The editors, like the vigilantes on the outside, did not know that so many powerful persons and influences came to Terry's aid that the Committee was glad to connive at getting him out of town and on a United States ship, before he could be lynched by a mob.

It is a major irony that during the long trial of Terry, who was eventually to go free for an assault that he did not deny, the Committee arrested and tried and hanged Philander Brace and Joseph Hetherington. Both men were brutal murderers but most of the people could not understand why two men should be swiftly tried and hanged, while one who did his best to kill, but had powerful friends, was allowed to escape.

In Montana camps seven years later the Committee that identified itself as 3-7-77 was formed to hang some of the criminals who had been driven out of California. There is no better introduction to this committee than the murder of Lloyd Magruder. He was a man with many friends, this packer and freighter who had plans to be a candidate for Congress. There are different versions of his murder—how much gold dust he had with him, how far the murderers journeyed with him, and so on—but these are the es-

Hill Beachy, who ran down the killers of Magruder

sential facts. He was packing between Virginia City and Lewiston, at the junction of the Snake and Clearwater rivers, which at that time was a hangout for Henry Plummer and his road bandits. On learning that Magruder had a small fortune in his bags Plummer chose four of his bandits—Lowry, Howard, Romain, and Page—to trail him and rob him. He was an ingenuous man, and apparently they ingratiated themselves into his confidence by helping with his packing and unpacking and other chores, until in a mountainous area they chose a spot for their crime. What happened we know only from Page. He and Lowry and Magruder were standing night watch; when Magruder bent over to tend a fire Lowry crushed his skull with an ax. Magruder's sleeping men were then killed in their sleep. After choosing five mules to ride and some equipment, the four bandits burned the remainder of the camp equipment, drove the other beasts into a canyon and slaughtered them, and headed for Lewiston. There Hill Beechy, a deputy marshal and Ma-

gruder friend, recognized some of the equipment and took the trail of the four murderers, who by that time had fled. His tracking them to California and bringing them back to stand trial was a superlative feat. According to the Boise *News* it was one of the "most deliberate, foul, and cold-blooded murders that has ever been recorded in the annals of crime." It was all of that, and it was typical of Henry Plummer's methods. Page, who turned state's evidence, was shot dead in Lewiston soon afterward by a fellow criminal. The other three were hanged.

Of Henry Plummer and his gang Bancroft says they killed "over one hundred and twenty citizens, and plundered stages, express shipments, and private individuals," until no one dared to leave the Virginia City and Bannack areas with gold dust on him. Most writers estimate the number of persons murdered at around a hundred, but "many atrocities were committed which are unrecorded and unknown." Plummer was in Nevada City, California, in the early fifties and was made marshal of that camp in 1856. While marshal he was convicted of murdering a German whose wife he had seduced, and though sentenced to ten years he pled ill health and was pardoned. Soon afterward he robbed a stage, murdered another man, was jailed, escaped, and headed for Idaho Territory after stops along the way. Drury says he paused in Carson, Nevada, and was there hidden by another criminal, named Mayfield. When a sheriff was foiled in an attempt to arrest Plummer and then "rushed the gambler with fury" he was killed with a knife. Mayfield escaped and was "shot down in Idaho a couple of years later after a dispute over a card game." Plummer headed north and for a while made Lewiston his headquarters. No doubt he was looking for a camp that would elect him sheriff, for this was a common trick with the leaders of gangs, including Frank Stilwell in the Tombstone area, Updyke in the Boise area, and Jack Williams, a deputy marshal, of whom Mark Twain says that he "had the common reputation of being a burglar, a highwayman, and a desperado. It was said that he had several times drawn his revolver and levied money contributions on citizens at dead of night in the public streets. . . ." In Bannack in the spring of 1863 Plummer managed to get himself elected as sheriff, and was all set for the big time. In the fall of the previous year he had got rid of a pal-in-crime named Jack

Not authenticated but thought to be Henry Plummer

John X. Beidler

Cleveland, whom, it is said, he no longer dared trust with his secrets. In a saloon in Bannack he fired at least four times at close range: the first shot missed Cleveland by a mile and lodged in the ceiling; the second touched his ribs and the third hit his head, but the fourth barely missed a group of men sitting by a wall. Cleveland was unable to get his guns up, probably because he was too drunk to know where they were.

Plummer was almost lynched, then given a mock trial and acquitted; and a few months later in the same camp was elected sheriff! As deputy sheriffs he chose some of his principal gunmen, Gallagher, Stinson, and Ray, though the chief deputy, strangely enough, was an honest man named Dillingham, who soon knew the names of the road agents and their plans. When this fact was learned by the gang, three gunmen, including two of the "deputy sheriffs," were given the task of murdering Dillingham, though it was not this murder that brought the vigilantes into being. But that story, like the one of the San Francisco committees, is much too long to be given here.

When Plummer was discovered for what he was, and his three deputies, and when it was realized that scores of robberies and murders, heretofore mysterious and unexplained, had been the work of this gang, the fury and outrage of the law-abiding citizens must have been something to behold. An instance of a public conscience goaded beyond control is the destruction of one of Plummer's bandits named Pizanthia. When the vigilantes were hanging the worst of the gang a small party, including men named Ball and Copley, two prominent citizens, went to a shack to arrest Pizanthia. On calling to him and receiving no reply they imprudently pushed the door open to enter. Copley was shot through the chest and died soon afterward; Ball was shot in the hip. This so infuriated the vigilantes that they brought up a howitzer and almost demolished the shack. Some of them then went in and found Pizanthia wounded, and as he was dragged out Ball was allowed to empty his revolver into him. Pizanthia was then strung up and over a hundred bullets were fired into him. Because this was not enough to appease the furious men they set fire to the shack and tossed Pizanthia's bullet-riddled body into the flames. They brought other fuel and piled it on and so completely burned him that one who examined the ashes said there wasn't even a scrap of bone to be found. A delightful part of this execution was the work of some harlots: convinced that Pizanthia had gold dust in the shack they panned the ashes.

On receiving a report of vigilante action in Montana the Boise *News* said: "This is what we call doing the thing with a vengeance. The San Francisco Vigilance Committee is entirely eclipsed and overshadowed by the Beaver Head institution." That was not an exaggeration. No other vigilance committee in the West equaled this Montana one in zeal, singlemindedness of purpose, and the ridding of camps of the worst criminals. This was due in generous measure to the man who was known as X. Beidler. After a Dutchman was found brutally slain and dragged off into the sagebrush, and the killer, George Ives, arrested and brought before the vigilantes, begged for more time to write his mother and get in touch with friends, Beidler, sitting on a cabin roof, called out to the chairman, "Ask him how much time he gave the Dutchman!" That settled it, and from that moment the Committee wavered only once or twice, as when after the brutal Stinson was found guilty and women pushed for-

Col. Sanders, 1869

Gallows built by Plummer while sheriff; he was hanged on it.

Plummer was buried and covered with stones

ward weeping and begging for mercy, his sentence was changed from hanging to banishment and he was given an hour to leave. He said he could do it in fifteen minutes if the horse didn't buck.

Plummer, like most remorseless killers, was a coward. As Dimsdale tells the story—and he was there—Plummer "begged for his life . . . exhausted every argument and plea that his imagination could suggest. . . . He begged to be chained down in the meanest cabin; offered to leave the country forever; wanted a jury trial; implored time to settle his affairs; asked to see his sister-in-law; and, falling on his knees, with tears and sighs declared to God that he was too wicked to die." He said they could cut off his feet and hands if only they would spare his life. He told them he had a fortune in gold dust buried and would give it to them, and ever since then treasure-hunters have dug in the area of a certain corral site in an effort to find what Plummer is supposed to have hidden. Some of them did not beg for their lives but cursed everything under the sky; and a few with silent contempt accepted their fate. Jean Davis somewhere found the story of Plummer's arm. According to this tale a Dr. Glick had, under threat of death, treated Plummer for a gunshot wound but had not been able to find the bullet. After Plummer was buried he stole out in the dead of night, dug under the stones to the corpse, and cutting off the arm hid it in a snowbank. That night a dog entered a dance hall dragging the arm and laid it at his owner's feet. Even if true, that was nothing to upset people in a mining camp.

Vigilance committees were formed in Idaho, as they were elsewhere, because of corrupt officials

Vigilantes' cabin, Bannack, Montana

Old store in Virginia City in which Helm, Gallagher, Parrish, Lyons and Clubfoot George were hanged.

As it looks today

and brutal crime. Typical reporting is that in the *Statesman* of a "most horrible murder and robbery"—two men lying in a road, one dead, the other breathing but unable to speak; and at a little distance away a dead miner who had been seen in camp a few hours earlier with a poke of dust. There was a rumor, said the editor, that a vigilance committee was to be established. He still did not know that Sheriff Updyke was the leader of a gang of criminals. The next month "Our citizens were intensely shocked to hear the intelligence of the horrid murders and daring robbery. . . . these horrible, inhuman massacres— more cowardly than assassination—because wanting the element of malice or revenge. . . ." A week later: "report that Mr. Pinkham was foully murdered. . . . This most atrocious affair is only one of a number. . . . It is about time for that kind of rule to come to an end—a rope's end." When a grand jury up at Idaho City, where Marshal Pinkham was murdered, stood eleven jurors for and four against indicting the murderer,* the editor cried, "This is the most alarming event that has ever transpired in Idaho Territory. . . . a most appalling and unprovoked murder has been committed in open day. . . . How long is this state of things to last?" The editor still did not know that Updyke was leader of the road bandits.

Three weeks later a long editorial said that up in Idaho City men were threatening to storm the jail and hang the criminals; for up there the record was "some sixty deaths by violence . . . yet not a single conviction for murder." But he hoped there would be no lynch law. "If men in authority did their duty there would be none." Murderer Updyke was still sheriff; and of the territorial governor, who loved to call himself Caleb Lyon of Lyonsdale, the *Statesman* (June 8, 1867) reprinted from a Washington newspaper: "Yesterday Caleb Lyon, of Lyonsdale, the governor of Idaho, arrived in this city, and immediately went to police headquarters and said that on the previous night—on his way from New York here he had a belt on his person containing forty-seven thousand dollars, which he had safely carried all through the Rocky Mountains, California and Idaho. . . . he . . . put it under his pillow in the sleeping car, and on waking in the morning found the belt at his feet without a dollar. . . ." The

editor of the *Statesman* commented: "This is the way he settles his accounts. Did any one ever hear such a fools story to cover up a clear defalcation— (steal.) " Later he added: "A private letter from New York concerning Old Kale's story of being robbed, that 'people there are ungenerous enough to believe it to be an entire fabrication.' "

The *Avalanche* was to say a little later, "In every way he could devise he sought, begged, piteously pleaded for popular applause; and, with a meanness exhibited by few men, denounced the journals that would not blindly indorse his entire course and excuse his vanity, idleness, incapacity, and roguery. . . . Of all the political mountebanks that ever attained position, Caleb stands at the head of the gang. . . . basest of ingrates. . . . In the history of official conduct, the two flights of Lyon from Idaho stand unequaled in disgrace. . . . His few friends now claim that he was actually a lunatic." With an insane governor, and a sheriff directing the criminals, the law-abiding citizens didn't stand much of a chance. Pages could be filled with details of the most wanton crimes and the outraged protests of the Boise editor, but the most interesting aspect of the vigilantes in Idaho is the abusive language which the newspaper publishers heaped upon one another.

The tirades against vigilantes by H. C. Street, publisher of the *World* in Idaho City, were an expression of the bitter North-South feud in the area. Says Harlow, "There was even a newspaper, the *Idaho World,* which fought the vigilantes and was in effect an organ of the stage-robbing clan, though not acknowledging them as such. . . . called vociferously on the Democratic party to incorporate a plank in its platform against the vigilantes." Of the *World's* editor the *Avalanche* offered a scathing appraisal typical of the times: "Finding the world—which most people accept as pretty good—not adapted to his wants he created a World of his own and sees himself as its chief luminary. The effulgent beams of wisdom shed over it have no parallel in the world 'God made.' As a writer he never makes mistakes—hence his World is governed with a degree of loveliness and serenity unknown to other spheres. His most remarkable trait is power of perception and even

---

* The account of the murder of Pinkham will be found in the chapter, "Brave Men and Cowards."

Headboards of some of Plummer's assassins. Because original headboards were carried away splinter by splinter new ones have been set up.

prophecy. He divines and tells his readers the main-springs and motives of others in perfect confidence." And so on. When the *World* mentioned vigilantes anywhere in the nation it was to denounce them and call them assassins. When a vigilance committee was formed in the Boise area the *World* wrote that "Matters are being carried with a high hand at Boise City. Lawless recklessness has found a vent, and broken out in the shape of murder, committed by the shadowy, veiled form of a Vigilance Committee,—an assassin stealing in the darkness to his crimes." The *Statesman,* said the editor, "is running with the Vigilantes, now. They suit its style. Hang every one who differs from you, and do it before the moon is up. . . . In encouraging a Vigilance Committee to disturb the peace of the community, to go prowling about at midnight, murdering people, and stirring up bad blood in its neighborhood, the 'Statesman' is consistent; and we judge the work agreeable to the editor's natural inclinations."

By the following Saturday the editor had got his blood up:

The Vigilance organ accused us of using "intemperate" language about that illustrious body of patriots. Intemperate! *Intemperate!* Well Sir, what sort of language would you have us use, when a whole community is put in terror by a band of midnight murderers straggling about the country, waylaying defenseless men, committing deeds outraging the laws of both God and man? What sort of language is fit for men who in times of peace, array themselves in open rebellion against the laws of the country,* and by brute force pale the faces of a whole community? "Intemperate language." It may be all very "temperate" for the editor of the Statesman, ensconced in his sanctum, to cooly applaud the wholesale massacre of his fellow townsmen, but to us it looks more like the gluttony of a malignant and cowardly revenge. "Intemperate language." Does the Statesman expect temperate language applied to brutal murder repeated nearly every day of the week? . . . What evidence is there to show that the members of the Vigilance organization who sent Updyke into another world, are any better than their victims?

In the next issue he calls the whole thing a "political movement. . . . hanging political opponents, and driving them from the country."

The sole purpose was "the destruction of political foes." He surmises that they hanged Updyke for his money, and that a "new victim will be needed soon to replenish the exhausted treasury." The *Statesman* said, "The editor of the World evidently knows nothing about the condition of affairs or what has transpired in Boise City, and if he did would not publish the truth if he could help it." Street retorted, " 'The editor of the World' knows more about the 'condition of affairs' there than is creditable to you, and has 'published' more of the 'truth' than you like to see." He wants to know what they did with Updyke's property and advises a search of the office of the *Statesman's* editor, who had been "able to hold his own when any dead man's trunks were to be gone through, and for lack of a trunk he may have 'kept his hand in' a dead man's pocket."

The Montana *Post* got into the act; and Street, eager to take on all comers, reprinted its editorial:

The Idaho "World" comes out in denunciation of the Vigilance Committee, now happily inaugurated there, and as a quasi champion of the "roughs." This is the inevitable tendency of its logic; for all the bosh about trial by jury, civil courts, constitutional law, etc., when it is all a notoriously inefficient mockery, is too flimsy to stop the vision of the most purblind old lady, who ever ejaculated, "What, hang a man on a tree? How dreadful! In my time they used to hang them on the gallows." Now we take it that the question whether a murderer is hanged by the Sheriff or by the Sheriff's master—the power that made him Sheriff, is of little consequence in an exigency like that which now exists in Idaho. The Portneuf Canon massacre, and the slaughter of 102 innocent human beings for money, is what the want of a Vigilance Committee in Idaho (from whence all our road agents come) cost us of Montana, and we no more regard the shameless advocacy of practical immunity to wrong doing, in the "World," than we do the curses of a desperado on his way to the gallows. Over sixty victims lie buried in one county of that ill starred Territory, and no conviction of murder has been had. Over 300 individuals have perished in Idaho, and the hangings, how many are they? Let the "World" count. Turn we to Montana. The man who says—otherwise as mere hearsay—that an innocent man or a good citizen has died by the hands or at the instance of the committee, states

---

* Street was a rabid Secessionist; he heaped shockingly abusive language upon Lincoln, even after he was dead. It seems to be a historical fact that most of the chief criminals were from the South and that some of them had served for a while in the war. For a fuller account of the bitter controversy between the publishers of the *World* and *Statesman* see the chapter on "Feuds."

an unqualified falsehood, and all men know it who live here. There is no jury as immovably fair, impartial and unassailable, as the cold, stern, lynx-eyed, iron-willed and even-handed Executive Committee. Gold could not buy a conviction, nor the treasures of earth a reprieve. Men know it; the good rejoice, and the bad tremble. When Idaho shall be able to boast of its thousands of good men, banded, armed and disciplined, to maintain law and order, to support and assist her now derided judges, and to enforce the decrees of her powerless courts, then will the "gem of the mountains" blaze forth in true radiance, and the writer of the article in the "World" shall call on the rocks to hide him, and seek to conceal his shame in the *Lethe* of a people's contempt. Let him look at the two Territories. In Montana there is peace, order and security. In Idaho, there is insecurity, bloodshed and rampant lawlessness. We have made our choice.

To which the *World's* editor hotly retorted that road agents did not go from Idaho into Montana.* Then with heavy sarcasm he says that perhaps the people in Montana had "ascended into a higher circle of humanity than it has been the privilege of the rest of frail mortality to know. . . . The Post knows that these unauthorized hangings are in fact murders, and those who participate in them are nothing more than outlaws, and for whom the gallows yearns." He sees "no difference" between the road bandits and the vigilantes, and thinks the *Post* "an organ of one gang in its depredations upon the other." He persists in his opinion that the officials and courts were adequate to punish the criminals; and in spite of the persons in its cemeteries who died a violent death he says the Idaho Territory had been almost "crimeless" since its beginning. Then he comes to the matter that so incensed him, his belief that the Vigilance Committee in Boise City was a "political organization, to be used as an engine to carry the Summer elections. If there had been no election to come off next August, there would have been no Vigilance Committee in Idaho Territory today." At the same time he approves the President's veto of a rights bill.

Then he goes after the editor of the *Avalanche* down in Silver City and wonders how much he is paid annually to play second fiddle to the *Statesman*. By July 14, 1866, he is again enraged by "the outrages of these lawless scoundrels. . . .

A hanging on Helena's famous Hangman's Tree

this band of midnight assassins. . . . If the time has come when the lives, liberty and property of the best citizens of the county are to be sacrificed by a gang of masked desperadoes, herding in stone cellars, and murdering or banishing every honest man who will not participate in their damnable deeds, it is time we understood it." By July 21 he was rejoicing: "The vigilante desperadoes are becoming very sick in this county, and are getting more and more desperate as the certainty of their defeat becomes every day more apparent. . . . The day of these Loyal League

---
* Plummer and many of his worst agents did.

stranglers is past." In another column on the same page he burst forth: "The negro-worshiping stranglers, desperate with the certainty of an overwhelming defeat, have taken a new trail, and are attempting to clear the way to the polls by an indiscriminate use of the most villainous falsification, abuse and blackguardism. Democrats are anathematised as dough-faces, sons of bitches, damned Micks, California Know Nothings, Missouri Pukes, dirt eaters." From the *Idaho Union* he reprints a paragraph: "Query—Does Street and his political 'chums' admire the stars and stripes, or do they prefer the stars and bars? Speak out my elongated boy, like a little man, and define your position. We'll wager a dark lantern . . . that the small-headed, big-eyed editor remains mum. . . ." Street opened his retort with the words, "The above precious specimen of Abolition intellect. . . ." A week later: "In Boise Valley and Boise City, they have created a Reign of Terror, strangled, robbed, and driven out of the county every Democrat they feared. They have renewed the horrible and brutal scenes of perse-

cution against Democrats, which has everywhere prevailed where the Loyal League has been dominant during the past six years. They are now desperately attempting to extend their power over this county. . . . Democrats strangled in the darkness of midnight. . . ."

It was an election year just after the Civil War and on both sides the language was the kind that in some of the camps led straight to duels. By August 11 Street was saying: "We are told that next Monday Democratic blood shall flow; that 'Democratic heads may prepare to suffer.' Stand by your colors! If blood must flow, let it flow—and 'may God have mercy on their souls.'" In the same issue was a long article under the headline, HOW JEFF DAVIS WAS PUT IN IRONS. A Brave, Magnificent and Glorious Achievement. The Democrats won in Boise County, as usual, and August 18 Street was saying, "'We have met the enemy and they are ours'. . . . The Republican party has fought its last fight in Idaho, and gone home to its final rest. We would say, 'Peace to its ashes,' but if we write the words, we

Two hangings on Helena's famous Hangman's Tree

remember the millions of human lives for which it is responsible." On September 29 Street left the paper, but James O'Meara, the new publisher, carried on:

we wish to call attention to this morbid, bad, actually criminal desire for the organization of Vigilance Committees, and disposition to mob law, which is peculiar to the Radical party. . . . It is founded in a spirit which would set at defiance all laws not exactly in accordance with the wishes, greed, or bad passions of the individuals concerned; it is but the low and brutal and ruffanly phase of the "higher law" proclaimed by the Radicals in former years.

The political aspect of the matter he revealed October 13:

Many of our people do not fully appreciate the infernal spirit of proscription which actuates the leading Radicals against the whole Southern people. Their real scheme is to absolutely deprive that people of all constitutional rights, to disfranchise them, . . . to place them completely at the mercy of the few Radicals in the South and under the blighting rule of Puritanism, and to forever cut off from them all means or hope of release from, or amelioration of their prostrate, cruel condition.

In the same issue he defends Patterson, the killer of Marshal Pinkham.

The different owners of the *Avalanche* took no strong position for or against the vigilance committees in Idaho, until that newspaper was bought by John M'Gonigle, whose blood seems to have boiled at the very sight of the words. On reading in the Cincinnati *Times* that four notorious road bandits had been taken from jail and lynched he let himself go:

On the first page of this paper we publish some of the performances of one of the above cowardly, cut-throat organizations. . . . If there is a dark pit in the darkest regions of the damned, ten times hotter and more sulphurous than all the rest combined, it is certainly reserved by a just, exacting God for these libels on humanity—these infamous, dastardly spawns of iniquity, who, despite the tears, prayers and expostulations of loving wives and sisters, tore their helpless and unresisting victims* from their cells, and regardless of their entreaties for one short moment, in which to pray to their Creator for mercy for a life of sin, hurried them into eternity in a manner so cowardly and

brutal as would excite feelings of repugnance and loathing in the breast of the most barbarous brute of the Piute nation. And yet those miserable, midnight assassins, pretend to be christians . . . appear in church pews on each Sunday, wearing a sanctimonious leer on their pusillanimous countenances, with their hands red and dripping with the life-blood of their fellows . . . pounce upon their bound and shackled victims in the dead hour of night. . . . neither of these selfsame midnight stranglers would have the courage to confront even the shadow of a brave man in daylight. . . . these law-defying, God-cursed creatures, yclept vigilantes . . . resort to this dirty work for the purpose of wreaking a personal revenge or filling their pockets with plunder.

The reader is not surprised to find M'Gonigle filling his newspaper with stories of horrible murders (horrible is his favorite word), suicides, knifings, violent lunatics and premature burials. In one of his editorials on "lynch law" he makes a statement that has been made by many who share his views: "Better by far ninety-nine guilty persons go unscathed, than that one innocent being should suffer." The statement takes no account of all the innocent beings who suffer if the ninety-nine guilty ones go free.

As late as February 6, 1868, the *Statesman-World* feud over the vigilantes was still in full cry. On that day the *Statesman* replied to an attack from the *World*:

And as to Updyke, he was guilty of almost every crime against the statute, and you defended him and are still defending him. . . . four-fifths of the people of Ada county rejoiced when it was out of his power to do further harm. . . . You are still subject to fitful howls and unavailing regrets that he is not yet alive to assist murderers to escape and to harbor highway robbers. . . . The course of the Statesman has been from its first number to denounce crime in all its most hideous exhibitions in this Territory. The World has just as invariably apologized for crime and defended criminals.

When murder, stage robbery, and general ruffianism prevailed; when public officers broke open safes to steal the people's money, and legislatures conspired together to plunder the public funds; when counterfeiters had it all their own way in the community, and when systematic frauds upon the ballot box were openly engaged in, and when it was impossible to conduct even a civil suit in a justices' court in Boise and Ada county without danger of assassination, the Statesman was con-

---

* The only information he had was the *Times* story he printed, which said that one of the bandits knocked down three men and all fought desperately.

stantly exposing these villainies and their perpetrators and pointing them out by name. It is superfluous to state what was the position of the World. Its pages are but a continued defense of these same enormities. It has with singular industry defended them each and all. No assassination has ever been committed in the Territory but the World found some extenuation or a positive excuse for it. No public plunder has ever been made it did not defend, often beguiling the people by false statements into assurances of security, that the felons might escape or gain time to cover up their wickedness.

Space forbids or we would enumerate so as to refresh your memory, but will leave you now by saying further, Jim O'Meara, since you choose to be unprofessionally familiar, are mourning in your depraved heart because the times are changed and we rejoice because they are.

The first vigilance committee in southern Idaho was formed by W. J. McConnell, who was to become a territorial governor. As McConnell tells the story, practically everything he raised as a rancher in the Payette Valley was stolen from him over a period of two years. After five of his horses and four of his mules disappeared he and a neighbor set off in pursuit of the thieves and overtook them in eastern Oregon. "How we ever won that battle has always seemed a miracle!" When he returned to his ranch his neighbors insisted on sharing the expenses of the pursuit and making joint cause with him; and so a vigilance committee was born, to the number of fifteen, and sworn to catch and punish all criminals who came against them. So many other men clamored to join that soon the organization reached all the way to Walla Walla in Washington, and over the mountains into Montana. The reticent McConnell says that "it must suffice to say that we accomplished our purpose." The Boise committee sprang, according to Donaldson, out of the murder of Pinkham and the escape and flight of the killer. At that time, according to him, Boise contained "a splendid assortment of murderers, robbers, and tinhorn gamblers. They were the offscourings of all the abandoned and worn-out mining camps in the territory. . . . In my time some criminals and felons were pardoned in gross cases, and the public marveled much."

He says the Updyke and Dixon gangs were in full control of Boise in 1864, when a man named Parks was robbed on a stage and in trying to defend himself accidentally shot himself and died a few days later. It was believed that the gangs committed the robbery. By March of the next year gangsters had persuaded the legislature to make Updyke sheriff of the newly created Ada County, and he thereupon became Idaho's Henry Plummer. In April, 1866, a youth of nineteen named Reuben Raymond had courageously testified in court against members of the gangs, and about nine o'clock Tuesday morning April 3, according to the *Statesman's* story (April 5), Raymond and another man were disputing over the testimony Raymond had given when they were joined by a member of the gangs named Clark. Either Clark called Raymond a liar or Raymond called him a liar; Clark then rushed Raymond and struck him, and Raymond, dodging back, drew his gun. Clark then drew, and a fourth man tried to wrest the gun from Clark and failed. Raymond is reported to have said, "Don't be afraid, I'm not going to shoot." It is said that Clark then told Raymond to go ahead and fire, and that Raymond gave this incredible reply, "I don't want to shoot, I'll give you the first shot." That seems to have been good enough for Clark. Aiming his gun at Raymond he twice pulled the trigger and twice it snapped. "The third time, taking deliberate aim, he shot Raymond through the abdomen. Raymond did not attempt to shoot at all. He lingered until about dark of the same day and died." That night the vigilance committee took Clark from the Fort Boise guardhouse and hanged him. Nine days later Updyke and Dixon fled, with fifteen vigilantes in pursuit. They overtook the two men in the mountains northeast of Boise and hanged them on the spot. When news of the hangings swept Boise criminals on fast horses took off north and west, the exodus being so great according to one writer that the place looked deserted. In Updyke's stable on Main Street those who did not flee could read the following:

Dave Updyke

Accessory after the fact to the Portneuf stage robbery (and murders)

Accessory and accomplice to the robbery of the stage near Boise City

Chief conspirator in burning property on the Overland stage route

Guilty of aiding and assisting West Jenkins, the murderer, and other criminals to escape, while you were sheriff of Ada County

Steve Young, hanged at Laramie October 19, 1868

Threatening the lives and property of an already
  outraged and suffering community
Justice has overtaken you

X.X.X.

Jake Dixon
Horse thief, counterfeiter and road agent generally
  A dupe and tool of Dave Updyke

X.X.X.

All the living accomplices in the above crimes are
  known through Updyke's confession, and will
  surely be attended to.
The roll is being called.

X.X.X.

The accomplices saw those words before they
fled. Bancroft says, "if we may believe the district attorney, within the limits of Boise County
there had occurred some sixty deaths by violence
without a single conviction of murder in the first
degree." This was Idaho City's county, where
Street had the *World*. He says that in one camp
of cattle rustlers the Payette committee found "all
the paraphernalia for Indian imitation, bows, arrows, tomahawks, skins, scalps, and the like."

There were vigilance committees in other areas.
For instance, the firmness and precision with
which San Francisco was made safe after dark had
not gone unnoticed by men who were to become
civic leaders in Virginia City, Nevada. When
they perceived that criminals were taking over
their town they decided to hang a few. Arthur P.
Heffernan didn't know of that decision when at
the bar of the camp's finest hotel he killed a man.
There had been more brutal and unregenerate
killers but he would do for a start and he would
not be allowed to stride arrogantly from a courtroom to drink at a bar with the jurors who had
just freed him, as some murderers had done before
him. So masked men were stationed around the
jail, other masked men entered it after dark, and
a cannon boomed, though in San Francisco it was
the ringing of a fire bell that proclaimed a hanging. At daylight Arthur was seen hanging from
the framework over an old shaft, with the numbers 601 on his chest. That hanging had about as
much effect on crime there as the strangling of
Julia Bulette had on the profession of harlotry,
and so soon afterward a gentleman by the name
of Kirk was seen swinging in the wind, with 601
flashing in the sunlight as the body turned. Kirk
had been told to get out and stay out; he went
away for a few days but slipped back, and was
taken when resting easily in the arms of his soiled
dove in a white cottage. Davis, in the second volume of his history of Nevada, says one of the
most sensational lynchings was the hanging of a
man named Perkins from the rafters of Piper's
Opera House. Perkins, like all bullies and cow-

Big Ned, Con Wagner, Ace Moore at Laramie, October 18, 1868

ards,* shot a man because of some trivial remark he made and by the authorities was taken to Carson City. When he was returned for trial to Virginia City the vigilantes took care of him, and with remarkable thoroughness, for, Davis says, "No sooner did the body swing clear than a dozen or twenty shots were fired into it."

By April, 1859, the law-abiding citizens of Denver reached the conclusion that they had been bilked and terrorized long enough. They organized a committee of a hundred but on learning that some of the worst criminals had slipped into it they reduced the number to about a dozen, who came to be known as The Stranglers. This committee was so eager to get the job done that when some of the Montana vigilantes, on a visit to look things over, said they had seen two men there who should be hanged, The Stranglers promptly hanged them, without investigation or trial. After a committee in Leadville hanged a gang leader and brutal killer about four hundred criminals, it is said, left that area within twenty-four hours. Until then, says Gard, "Finding one to three or four dead men in the alleys back of the saloons and dance halls in the morning was nothing unusual. Thugs took over mining claims by intimidation. Others set up toll stations on Ten Mile Road to levy tribute from travelers. So many highwaymen operated along the stage roads that the express companies refused to haul bullion or coin."

In some of the areas justice moved swiftly. Gard tells of a man who was observed while committing a robbery and who was immediately seized and hanged from the nearest tree. It was often swift but was it always impartial and fair? Not always. Toponce, a tough freighter, says, "In regard to the vigilantes, in the early days in Montana. I don't think they made any mistakes in hanging anybody. The only mistake they made was that about fifty per cent of those whom they merely banished should have been hung instead, . . ." Was there any sadism? There must have been a good deal of it. It is said that after a notorious criminal known as Big Nose George was lynched in Wyoming, a doctor named Osborne, who afterwards became a governor of Wyoming, skinned enough hide off George's

Sam Dugan hanged by vigilantes at Denver, December 1, 1868.

hairy chest to make a pair of boots. Gard tells the story; it sounds like a legend.

Berton tells of a flogging. The one who volunteered to whip a thief "seized the knotted thong and swung it with his full weight across the miscreant's naked back, so that two purple stripes, showing every twist of the rope, followed

---

* See, for instance, Zamonski and Keller on the fantastic killer James Gordon, in Denver.

each blow. The miserable Hansen writhed and leaped in the air as the whip descended, but this only seemed to increase the passion of the executioner, who shrieked and howled crazily as the vigor of his blows increased. By this time the crowd, too, was caught up in a frenzy, some shouting 'More!' while others cried 'Enough!' " There certainly was no lack of opportunity for sadists to vent their brutal lusts. Two thieves were captured on the Washoe, named Ruspas and Reise. Big Jim Sturtevant was chosen to inflict the punishment. He brought forth a big knife, honed it across a stone, and took a firm hold on the upper half of Reise's left ear. With one grandiose swipe he sliced the ear off and tossed it to the delighted jury. Ruspas had no left ear, having lost it for an earlier theft. Since the verdict said the left ear was to be cut off the jury went into solemn deliberation and decided that the right ear would do as well. After the right ear of Ruspas was sliced off and someone pointed out that they now had a pair of ears, instead of two odd ones, some of the onlookers began to sing. "Anyone need a pair of ears besides Ruspas?" one of the men asked, and the jury headed, as was the custom, for the nearest saloon.

Did the vigilantes sometimes go too far? *Idaho Lore* has a lot of stories of mining camp violence. In Warren, for instance, there were so many Chinese that the white miners, with petty persecutions, tried to drive them out, and went so far as to hang three of them from the rafters of an old mine in Slaughter Gulch. Their only offense had been the theft of several pairs of boots. Another story concerns Pierce City. When a white man was murdered there seven Chinese, for reasons unknown, were suspected, and after being chained behind wagons were led through the camp. They were then hanged one by one. The story goes on to say that many years later an old miner confessed that he was the murderer, and that he had been the leader of the mob that lynched the Chinese. An almost identical story appears in Colorado. The most famous Idaho lynching is known as that of Yellow Dog Creek. This tale says that because Plummer's road bandits were using Chinese for target practice a half-dozen of them fled from Virginia City westward and while crossing the mountains saw two prospectors panning gold. They are said to have crept up on the two men and murdered them by striking them with clubs from behind, and then to have put on some of the dead men's clothing, including their shoes. A group of miners coming in met the Chinese on their way out and observing that they were wearing shoes that belonged to their fellow miners they forced them to return with them, and, on finding the two miners slain, shot the Chinese down one by one "for the yellow dogs that they were."

General Meagher, secretary of Montana Territory, pardoned one James Daniels, who had been given a prison sentence for murder. The vigilantes took him out and hanged him, and pinned on his coat this message to Meagher: Do this again and you'll meet the same fate. Howard cites a Montana historian who says Daniels was "indicted for a crime he did not commit, tried by a court without jurisdiction, reprieved by a governor by mistake, and lynched by a mob." There is no reason to doubt that Daniels committed the crime for which he was first imprisoned and then hanged.

What seemed to be an excess of zeal sometimes aroused citizens who were neither vigilantes nor criminals. The Montana *Post* of March 2, 1867, said the following notice had been posted in Red Mountain City: We, now, as a sworn band of law-abiding citizens, do hereby solemnly swear that the first man that is hung by the Vigilantes of this place we will retaliate five for one, unless it be done in broad daylight, so that all may know what it is for. We are well satisfied that, in times past, you did some glorious work, but the time has come when the law should be enforced. Old fellow-members, the time is not like it was. We had good men with us, but now there is a great change. There is not a thief comes to this country but what "rings" himself into the present Committee. We know you all. You must not think you can do as you please. We are American citizens, and you shall not drive and hang whom you please.

Five for One

The notice, said the editor of the *Post*, was not from law-abiding citizens but from disgruntled former vigilantes.

Once in a while an innocent man was hanged, or in any case a man not guilty of the crime brought against him. In the Grass Valley area a man was lynched for stealing horses; when it was learned that he was not the thief a chastened committee erected over his grave a monument, on which was said,

LYNCHED BY MISTAKE
The joke's on us

The joke, of course, was on the poor devil who was hanged. And sometimes the committee was not right. In the Black Hills two horse thieves took into their camp a wandering youngster known as Kid Hall. When vigilantes captured the three of them the two thieves tried to convince their captors that the Kid was innocent but all three were hanged. The *Avalanche* reported July 16, 1870, that the vigilantes took an obnoxious person and ducked him in cold water. The chairman is reported to have said, "We took the thief down to the river, made a hole in the ice and proceeded to duck him—but he slipped out of our hands and hid under the ice; and as he has been there over eight hours now it is supposed he is drowned."

In Nevada, according to the historians, a man named Snow was hanged for a crime he did not commit. The story says they put a rope around his neck and pulled him up; let him down to see if he would confess; pulled him up a second time and let him down; and pulling him up a third time left him hanging. And Caughey cites some instances, for what they are worth.

Was a woman ever hanged by vigilantes? At least two of them. Cattle Kate, queen of the female rustlers, is thought to be the only woman ever hanged for stealing cattle. The vigilantes did a good job of it by hanging her man with her. The other case, that of Juanita, is still a matter of controversy. Statements on the affair range all the way across from extreme to extreme. Carleton Beals says in *Brass-Knuckle Crusade*: "At Downieville, the Anglos made a carnival of lynching a Mexican girl for stabbing an American miner who broke into her cabin to rape her." Caughey says: "At Downieville a year earlier a miner who had indulged too freely broke down a door and intruded on a certain Juanita. The next day he went to her place, assertedly to apologize, but gave further insult, and she stabbed him."

Major Downie, for whom the camp was named, left us a pen-portrait of this dance-hall girl, "of the Spanish-Mexican mixture, and the blood of her fathers flowed fast and warmly in her veins. She was proud, and self-possessed, and her bearing was graceful, almost majestic. She was, in the miners' parlance, 'well put up.' Her figure was richly developed . . . her features delicate, and her olive complexion lent them a pleasing softness. . . . her temper, which was the only thing not well balanced about her."

Because the various statements made about this dramatic episode in the history of the West are a fine instance of irresponsible writing we shall present the matter here in some detail. Our first version of it is Joseph Henry Jackson's. According to him, not even her name is known; some say she was called Josefa. She was Mexican, Spanish, Chilean, Peruvian, or a mixture. She was in Downieville in 1851 in an adobe hut, with a Mexican named José. There was a Fourth-of-July celebration and in the multitude were two friends named Cannon and Lawson. On their way home sometime during the night one or both of them, deliberately or accidentally, knocked in the door of the girl's shack. One version of the story says that Cannon in former times had tried to woo her. The next morning, as Jackson tells it, Cannon, with Lawson in tow, went past the house, and José appeared to demand payment for the door. There was an argument, and Lawson warned José not to draw a weapon. "Juanita snatched a knife from her dress and buried it in Cannon's breast." A crowd gathered. A doctor told the multitude that Juanita was pregnant and a few men made statements in her defense; but she was hanged. One version says her hands were tied; the other says they were not and that she helped to place the noose around her neck.

Stellman in his book *Mother Lode* says that when the two men kicked in the door the woman spat abuse at them, and that Cannon, a gay laughing lad, "called Juanita a name that only a prostitute of the lowest type would brook." With a "fiery glance at her traducer" she ran into the house and returning "came up to him slowly, one hand behind her back, but he seems to have suspected nothing. Suddenly the hidden hand came forth and in it flashed a knife. Someone cried a warning, but it was too late. She sank the steel in Cannon's breast and he fell, dying, at her feet. A roar of horrified indignation came from the crowd. Juanita fled."

Bancroft told the story this way.

Joe Cannon killed! Cut to the heart, and by a woman! . . . Joe was the favorite of the camp, the finest fellow that ever swung a pick or dislodged a bowlder. He was over six feet high, straight as a poplar, with limbs as clean as those of a newly barked madroño. In weight he fell not far short of two hundred and forty, and it was all muscle. . . . And yet he would not harm a fly; his heart was as tender as his sinews were tough. Joe gone! . . . He was the soul of honor, was Joe Cannon. He knew not the meaning of the words cheating and chicanery. He was not very learned; single and simple were his thoughts, and double-dealing found no place in his accounts. . . . There was no poison in his heart that the most fiery liquid could bring to the surface. In nobleness only was he a giant; in guile he was a child. Joe Cannon dead! Stabbed in the breast, and by a woman!

And so

All along the muddy Yuba, and up its muddy tributaries, the accursed tidings sped like an electric message. . . . round Downieville were heard the cries of "Murder!" "Joe Cannon killed!" "Cut to the heart by a woman!" Then dropped pick, pan and shovel as from palsied hands; . . . and from up and down the muddy Yuba, . . . streams set in, streams of angry miners, silently flowing, though hot with melted emotions. Five thousand men and more gathered in Downieville that day. . . . At ten o'clock the deed was done. . . . the slaughtered miner ceased to breathe; then from the deep stillness there rose a murmur. . . . Pregnant enough with purpose were now the miners. You could see it in their eyes, in their step, in the movement of the hand; their pipes smoked of sulphur, and with their tobacco-juice they spat fire. . . . It was a little woman; young, too—only twenty-four. Scarcely five feet in height, with a slender symmetrical figure, agile and extremely graceful in her movements, with soft skin of olive hue, long black hair, and dark, deep, lustrous eyes, opening like a window to the fagot-flames which, kindled with love or hate, shone brightly from within. . . . The man she killed with one hand could have picked her up and tossed her into the Yuba River.

Joe Cannon had been celebrating the Fourth of July, and he was "very drunk, and consequently very happy. From store to store, from house to house, up and down the streets and through all the streets they went, rapping up the inmates, compelling the master of the house to treat and then to join them. It was rare fun." They came to "the premises of a Mexican monte-dealer. Joe Cannon kicked at the door. As he was not in

condition to stand steadily on one foot . . . he may have given it a little harder blow than he intended. . . . the boys told him next morning—that he had kicked in the Mexican's door. That was all right, Joe said. . . . he would go around after breakfast, pay it, apologize. True, there was the wife, or she whom the man lived with as with wife . . . but she was a bashful, retiring little thing. . . ."

So Joe went to the shack, alone, and

found the door still down. The Mexican was within; and placing a hand on either door-post to steady himself . . . he began talking to the man in broken Spanish as best he could. Suddenly from a corner where she had lain concealed, quick as a flash the little woman sprang up, threw herself upon the strong man's breast, and buried her knife in his bosom. . . . he, the picture of physical perfection, the pride of the camp, lay as dead. Why did she do it? Did this man visit her house to insult her? Had they ever met or had intercourse at any time; and was there ill-will existing on either side? No one knew. Who shall fathom a woman's heart? . . . And now, when the enraged miners with a blow of the fist burst her door and stood before her, Juanita manifested not the slightest fear; and yet she knew that she must die.

Bancroft has her tidy up the shack and put on her best garments. Then off she went with her guard "of two thousand." A jury was chosen.

Probably in the history of mobs there never was a form of trial more farcical than this. . . . Never have I met an instance where so many men, or a tenth of them, were so thoroughly ravenous in their revenge. It was wholly unlike them. It seemed that on the instant they had not only thrown aside their usual chivalrous adoration of sex, but that now they would wreak their relentless disaffection on the object of their abhorrence in proportion to their strength and her weakness.

A physician, Bancroft says, told the mob the woman was in no condition to be hanged and the "good doctor was driven from the stand, driven from the town." A lawyer spoke in her defense and was "kicked out of town, being glad to get across the river with his hat and mule behind him." After that "No one dared to say a good word for her; no one was allowed to defend her. . . . In the four hours allowed her before her execution, Juanita made her will verbally, arranged her affairs. . . . At a time when men tremble and pray she was her natural self" and she went to the bridge where she was to be hanged

with a light elastic step, surrounded by her friends, chatting with them quietly on the way. . . . She shook hands with them all, not a tear, not a tremor, was visible. By means of a step-ladder she mounted to a scantling which had been tied for her to stand on between the uprights underneath the beam, took from her head a man's hat which had been kindly placed there by a friend, shied it with unerring accuracy to its own-er, meanwhile smiling her thanks, then with quick dexterity she twisted up her long black tresses, smoothed her dress, placed the noose over her head . . . and finally, lifting her hands, which she refused to have tied, exclaimed, *Adios, senores!* and the fatal signal was given.

The Sacramento *Times and Transcript* told it this way:

The act for which the victim suffered was one entirely justifiable under the provocation. She had stabbed a man who had persisted in making a disturbance in her house and had greatly out-raged her rights. The violent proceedings of an indignant and excited mob, led on by the enemies of the unfortunate woman, were a blot upon the history of the state. Had she committed a crime of really heinous character, a real American would have revolted at such a course as was pursued to-ward this friendless and unprotected foreigner. We had hoped the story was fabricated. As it is, the perpetrators of the deed have shown them-selves and their race.

Of that paragraph Bancroft said,

This editor goes far out of his way both to dis-tort the facts and then to draw from them false conclusions. The woman was not a friendless for-eigner, nor was her act justifiable. The man she murdered offered her no violence, and she had no right to kill him. The people were right to hang her, but they were wrong to do it madly and in the heat of passion.*

The Sonora *Union Democrat* reported it as fol-lows:

At Downieville was perpetrated the greatest possible crime against humanity, Christianity and civilization. A drunken gambler, with criminal audacity, had thrust himself into the presence of a Spanish woman of doubtful character. She was *enciente;* enraged at his demeanor she seized a

knife in a moment of passion, thrust it into his breast and killed him. The gambler was popular, as gamblers may be and as many are. The mob, in angry excitement, seized the poor, trembling, wretched woman, tried, convicted and hanged her; hanged her in spite of her situation, her entreaties, and the fact that she killed her assail-ant in defense of her person.

The San Francisco *Alta's* reporter, who appar-ently was there, had this to say:

It would be difficult to find in the annals of human wickedness an instance of a murder so un-provoked and so unjustifiable. . . . Scarcely a dozen persons of the thousands who witnessed it could see the slightest injustice. It was true that, for the defence, it was stated by "her physician" that she was pregnant; but at the consultation that followed, it was declared by competent pro-fessional men that "there was no reason to be-lieve that pregnancy existed". . . . The victim in this instance was not the first nor the second who had been stabbed by the same female. He was a man who by his good conduct and peaceable de-meanor made hosts of friends, by whom his un-timely fate is deplored. . . . I have lived long enough in Downieville to know that its inhabi-tants are not the bloodthirsty, diabolical mon-sters they have been represented; on the contrary they have heretofore been too mild in their pun-ishment of offenders; and in the case before us, nothing induced them to pursue the course they did but retributive justice.

Borthwick, an educated man, an artist, and a good observer, was in Downieville soon after she was hanged. He says,

A Mexican woman one forenoon had, without provocation, stabbed a miner to the heart, killing him on the spot. The news of the murder spread rapidly up and down the river, . . . The woman, an hour or two after she committed the murder, was formally tried by a jury of twelve, found guil-ty, . . . The case was so clear that it admitted of no doubt, several men having been witnesses of the whole occurrence.

It may be appropriate to close this chapter with a sampling of how those lynched met their ends, bearing in mind that the password in Plummer's gang was, "I am innocent." The *Avalanche* re-ported that John Hoag, hanged in Canada, said on the gallows, "All young boys, take my advice

---

* Since Bancroft cites no sources and does not say that he was on the scene or talked to any persons who were, we must won-der on what grounds he can be so positive about a matter when there is so little trustworthy evidence to support either view.

and be obedient to your parents, before you bring disgrace upon them as I have done. Especially shun and keep away from whisky, fast young women and the United States." The British prig, Sir Henry Huntley, told of the execution of a horse thief, who, after being led out, took a deep drink of raw whisky and told the assembled throng that he was of good family and this was his first and only offense. He refused to have his eyes bandaged and asked that the six execution-ers with rifles stand close so that they would not botch the job. Three huge balls were fired into his heart, and the moment he toppled over a doc-tor in the crowd stepped forward and cut off the head, explaining to anyone who was curious that the thief had given it to him. He seized the gory thing by its hair, dropped it into a sack, and marched away, no doubt with a professional air.

But the hardened criminals never died as gal-lantly as those two. Beidler and Dimsdale, two of the Montana vigilantes, left records of how some of Plummer's most vicious killers met their ends. Beidler says,

On the arrival of the wagon bearing the coffin with Keene seated indifferently upon it reaching Hangsman Tree in Dry Gulch at about half past eleven o'clock in the morning, Keene remarked, "He was d----- dry," and inquired if they were going to hang a thirsty man. A messenger was sent at once to the nearest saloon, returning short-ly with the desired liquid. Keene took his last drink and savagely remarked, "I killed the son of a b---- and would do the same thing over again and if you don't like that you can drive your d---- old cart off as fast as you d---- please."

This execution, it is true, was in the Helena area, and Keene may never have been a Plummer bandit; but the six given next not only were but were some of the most dehumanized and brutal.

The cannibal Boone Helm "was the most hard-ened, cool, and deliberate scoundrel of the whole band." When arrested he said, "I am as inno-cent as the babe unborn; . . . I am willing to swear it on the Bible." Says Dimsdale, "Anxious to see if he was so abandoned a villain as to swear this, the book was handed to him, and he, with the utmost solemnity, repeated an oath to that effect, invoking most terrific penalties on his soul, in case he was swearing falsely. He kissed the book most impressively." Asked to give information about the gang he told them to go to Jack Gal-lagher, at that moment separated from him by a partition; and Gallagher, hearing him, "burst out into a volley of execrations, saying that it was just such cowardly sons of ----- and traitors that had brought him into that scrape." On being told that he would be hanged Gallagher "sat down crying; after which he jumped up, cursing in the most ferocious manner. . . . His general conduct and profanity were awful." Boone meanwhile was telling him not to make a damned fool of

John Heith (also Heath) in Tombstone

himself, that "there was no use in being afraid to die." When Gallagher was put on the box he "cried all the time, using the most profane and dreadful language." After the box was kicked away and he dropped, Boone Helm "looking cooly at his quivering form, said, 'Kick away, old fellow, I'll be in Hell with you in a minute.'"

In sharp contrast to Gallagher's behavior was that of George Shears. When arrested he said, "I knew I should have to go up, sometime, but I thought I could run another season." "On being taken into a barn, where a rope was thrown over a beam, he was asked to walk up a ladder, to save trouble about procuring a drop. He at once complied, addressing his captors in the following unique phraseology, 'Gentlemen, I am not used to this business, never having been hung before. Shall I jump off or slide off?' Being told to jump off, he said, 'All right; good-by,' and leaped into the air with as much *sang-froid* as if bathing." Dutch John "begged hard for life, praying them to cut off his arms and legs, and then to let him go. He said, 'You know I could do nothing then.'" When told that he must hang he asked to write a letter to his mother.

Unusual in another way was the behavior of Johnny Cooper and Bill Hunter. When they moved to put the noose around his neck Cooper kept throwing his head from side to side, dodging it, but when commanded to hold his head still and submit, he did so, and "died without a struggle." Hunter also seems to have died quickly, or to have lost consciousness, but "strange to say, he reached as if for his pistol, and went through the pantomime of cocking and discharging his revolver six times. This is no effort of fancy. Everyone saw it, . . . It was a singular instance of 'the ruling passion, strong in death.'"

Found hanged and thoroughly baked by the Arizona sun

PART IV

Special Characters and Situations

CHAPTER 21

Rags to Riches and Riches to Rags

Possibly in no other area of human endeavor have so many huge fortunes been possessed (one can hardly say earned) so quickly and easily, often with no relationship to intelligence, knowledge, and the application of talent; or, having been won largely by luck and chance, been lost so quickly by persons for whom money was indeed a curse. From one point of view the history of the western mining camps is a story of the fantastic rise from poverty to wealth, and the fall from wealth to the most abject need. It is a story that has been written only in fragments, the full scope of which would fill many books. We have space here only to suggest the nature and range of it.

One who reads widely in this field must be impressed by the thing called luck—if by luck is meant the accumulation of riches almost wholly by chance. The story in Leadville of the Little Jonny mine could stand as the perfect instance of all such tales. A greenhorn, poking around with pick and shovel, and with only the vaguest notion of what prospecting meant, asked a geologist, busy near by, where to dig. Annoyed, the geologist said, perhaps a bit sharply, "Oh, anywhere around here. Under that tree over there." The greenhorn dug under the tree and found ore that made him a millionaire. Another story is told of a Negro who wandered into one of the California diggings and asked how he could find some of that-there gold in the hills. Pranksters sent him to a high hill in which everyone knew there was no gold and would never be any gold; and for a week they laughed their heads off over their rum and beer as they told the story, and evoked pictures of a poor black man up there, digging holes all over the place like a man after water. But they didn't laugh when the Negro came down the hill one afternoon with a small

fortune in flakes and nuggets. His mine was named Nigger Hill.

Mildretta Adams, for whom Idaho's Silver City is a hobby, says that W. H. Dewey, convinced that there was a rich lode in a certain mountain, spent a whole summer literally crawling on hands and knees all over the mountain, and he found a lode that made him a very wealthy man, who built a railroad, and a hotel that until recent years was a landmark in the Boise Valley. That was more effort than luck. When Bummer Dan came to Alder Gulch he was not known as Bummer; but he was such a lazy parasite that the miners soon despised him; and when they caught him in a small theft they hustled him to an unworked area of the gulch below the camp and told him to dig. Hours afterward they went to the spot to see if he had obeyed orders, and found him in a feverish sweat, with gold dust and small nuggets all around him. When, in 1847, Colonel Frémont decided to speculate in land he gave three thousand dollars to the American consul at Monterey and told him to buy a tract near San Jose. Misunderstanding the instructions the consul bought the 45,000-acre Mariposa grant. Frémont was furious, for he thought the tract was worthless, until Kit Carson found on it the first quartz vein of California's famous Mother Lode.

During the stampede to Cripple Creek two brothers spotted a man who owed them thirty-seven dollars. They demanded their money then and there; having none, the debtor gave them a half interest in a claim he had just staked. That half made the two brothers millionaires. Charles Higgins, later mining editor of the *Gazette* in Reno, had prospected for years and found no colors. When his wife told him to try once more it "made a new man of me." He said he would never come back unless his pockets were full of gold. He staked a claim for which he was offered

$17,000; refused it and was offered $20,000; and accepted when the offer rose to $22,500. Getting his money in bills and gold pieces, he stuffed every pocket full and rode back to his wife.

Most of those who grubstaked the ever-hopeful prospectors never got paid, much less became rich, but there were a few exceptions. A man named Parker grubstaked a man named Lockhart for twenty years, and at last received half of $28,000 when Lockhart sold a claim to Charles M. Schwab, whose agents were everywhere. Lockhart then bought control of the Florence mine for $6,000, and Parker shared in its millions because he had staked a lot of grub and a lot of faith in a man.

An old-timer told Glasscock a story of what a gamble with five thousand dollars did for him.

Within the thirty days we cashed in the first shipment of $419 a ton ore at the Salt Lake smelter, and paid off the $2,500 due. Within the year we paid the remaining $70,000 from ore taken out of the shaft, and in addition to that put $80,000 aside for a mill, invested $40,000 in a pipe-line to Alkali Springs, paid $5,000 for the springs, used $35,000 for mine equipment and declared a $16,000 dividend. In two more years we paid $1,350,000 in dividends and then sold for $4,000,000.

A man named Patrick, one of the first arrivals in Tonopah, also took a chance on five thousand dollars, and found that "business was so good that the first manager we hired stole ten thousand dollars in the first month without our even suspecting it. After that, it got better." It was also in the Tonopah area that Jim Butler came in from the hills with some pieces of rock, which he showed to T. L. Oddie, an attorney. Jim said he was broke: if Oddie would get the rock assayed he would give him a part of the claim. Oddie was broke too but remembered that an assayer named Gayhart was teaching in Austin; so he sent the samples to him and promised him half of his part, if the rock panned out. Gayhart could hardly believe his senses; the ore ran as high as $575 to the ton. The good news was rushed to Butler but at the moment he was harvesting his wild hay and could not be bothered. Hearing of the discovery, men rushed all over the country, trying to find the spot. Butler's luck held. In those days a man's word was his bond, or Jim's was: he granted more than a hundred oral leases to men to mine his claim. The first of them re-

ceived a check for $574,958.39 for just one shipment of 48 tons. According to F. C. Lincoln's "Mining Districts and Mining Resources of Nevada" the leasers, as they were called, took four millions' worth of ore the first year, without drawing up a document of any kind. Butler turned out to be the discoverer of the greatest bonanza of its time.

The story of a Swede named Anderson has been told by various writers. Our version here is Rickard's. Ignorant, naïve, and not very bright, Anderson came to a Klondike camp with six hundred dollars. After he was drunk two old prospectors persuaded him to buy Number 29 on Eldorado Creek, a hole that they thought was completely worked out. The next morning the Swede tried to find the two men, for he wanted his money returned to him, but they were gone; so, unable to think of anything else to do, he went to No. 29 and began to dig. He had been told by other miners that he had the poorest hole on the creek but he kept on digging, and in a day or two he was eighteen feet down. There he struck bedrock and he must have rubbed hard at his eyes, unable to believe what he saw. "The layer of sediment, four inches thick, was more than half gold." The two who thought to swindle him came along and asked how he was doing. "Ay tank ay got gold here," said the Swede, and showed them a pan with $1,400 in it. Number 29 on Eldorado gave up $1,250,000 but no one knows how much of it the Swede got.

In Idaho City a householder hired a well driller to find water. The driller struck gold-bearing bedrock within eighteen feet, and "within a half-hour a string of claims a half-mile long was staked and all of them proved to be fabulously wealthy." The Goldfield *News* told the story of "An old Irishman, bent with labor, his hands trembling," who "shuffled into a local bank a few days ago holding between his fingers a slip of paper. 'Is this good?' he asked timidly, presenting a check for seventeen thousand five hundred dollars. It was signed by a well-known firm of brokers." Told that it was as good as gold the old fellow asked to have it all in cash. This he counted and shoved back to the cashier, telling him it was the only easy money he had made in a long life. For three hundred dollars he had a year ago bought some shares in a mine and this check was the price of it a year later.

Sometimes it was not luck but knowledge. Standing in a Wyoming town one day a man named Conger saw a carload of silver ore on its way from a mine. He studied the nature of the stone, telling himself meanwhile that on a certain mountain he had seen stone much like it. And so, in 1869, there came into being the silver mine that was to be known as the Caribou.

Bob Womack, discoverer of Cripple Creek, was a man who had luck but not much gumption. He didn't become one of the twenty or more millionaires in that area. He never became much of anything except a name and a face in a few books many decades later. Fame overtook him once, briefly; in 1904 he returned to have a look at Battle Mountain, and said, "My, you have a big city now." He caught a cold, had a paralytic stroke, and taking to his bed lay there for years. There were stories about his "career" and on its front page a newspaper reported the progress of efforts to raise a fund of five thousand dollars for the man who started one of the largest gold rushes in the West. The sum of $812 was the best they could do, but Winfield Scott Stratton was dead by that time. For years Bob lay on a bed, and died destitute at the age of sixty-six. Stratton appears in the chapter, "Display of Wealth"; he is mentioned here only because of a character named Popejoy, who was typical of the many who squandered everything they could lay their hands on. Once, during his many years of prospecting, Popejoy grubstaked Stratton. Stratton searched the mountains around Pikes Peak for weeks and returned, hungry and luckless. When Popejoy said he had had enough of him, Stratton promised to repay him, with interest, in three months. Stratton then went out and made his rich strike. When he was a millionaire many times over, Popejoy, with his attorneys, appeared on the scene and threatened to sue for half of it. Though he had paid the grubstake with interest, Stratton settled out of court for $40,000, and Popejoy went away with his part of it and drank it up and came again. According to Frank Waters, he "obtained two shyster lawyers, new in town," who thought they could get a million or two. By this time Stratton was dead, and the administrators had not only Popejoy threatening suit but a horde of other people. The estate's attorney suggested that Popejoy should be paid off again. Though the heirs were furious, Popejoy was paid $43,000,

and with his $40,000 of that he went off for another drunk, wondering, no doubt, how he could get hold of half of the eight or ten millions.

What kind of men were they who went from rags to riches? All kinds. Of the Big Four on the Comstock, two had been saloonkeepers, the other two were ordinary miners. Of twenty-eight in Colorado who became millionaires, Sprague gives the background: four had been in real estate, one was a schoolteacher, two were grocers, one was a butcher, three were druggists, two were lawyers, one was a promoter, one a lumberman, one a lather, one an engineer, a millman, a plumber, a handyman, a shopkeeper, a coal dealer, a cigar store proprietor, a department store owner, two roustabouts, and two prospectors. Not a broker, banker, or industrialist in the whole lot of them.

How many of them went from riches back to rags? A lot of them, including a few who were worth millions. Presented here are some who are among the most impressive in their ignorance, or in their combination of ignorance and greed; and in combination of ignorance and greed Eilley Orrum Bowers had few peers. She and her equally illiterate husband, Lemuel (Sandy) Bowers, took at least half a million dollars, worth about three million in 1967 dollars, out of their twenty feet of mine, and squandered most of it, believing, as Tabor over in Colorado was to believe, that the mine would never run out. Sandy died early at thirty-five a pathetic incompetent who had no sense of fortune or any right to it; and Eilley, after spending or being bilked of what she had left, "suffered dire privations." For thirty-five years she fought against poverty, as Baby Doe would in Colorado, believing to the end, as Baby believed, that her mine was still a treasure and would pay her more millions. Eilley, who in a few weeks in Europe had spent more than half a million (in today's dollars) and who cultivated the notion that she was a seeress and the queen of the Comstock, died in a poorhouse, a woman too stupid and greedy to arouse pity.

The names are not even known today of those mentioned in Dame Shirley's third letter, "who worked in these mines during the fall of 1850." They were

extremely fortunate; but, alas! the Monte fiend ruined hundreds! Shall I tell you the fate of two of the most successful of these gold hunters? From poor men, they found themselves at the end of a few weeks, absolutely rich. Elated with their good fortune, seized with a mania for Monte, in less than a year, these unfortunates,—so lately respectable and intelligent,—became a pair of drunken gamblers. One of them at this present writing, works for five dollars a day and boards himself out of that; the other actually suffers for the necessaries of life,—a too common result of scenes in the mines.

Thousands gambled and lost all they had in things besides monte. In the summer of 1875 the Comstock mining stocks plunged sixty million dollars in a week, and the Bank of California was forced to close its doors. Some have thought that the Big Four—Mackay, Fair, Flood, and O'Brien—outmaneuvered their financial rivals and brought on the collapse. They bought stocks at bargain prices. William C. Ralston was found to owe the bank over four million, and when his resignation was demanded he went to San Francisco's North Beach, a favorite swimming spot with him, and died of a stroke or drowned or (some have thought this most likely) committed suicide. His death and the closing of the bank forced the closing of the Pacific stock exchange and the financial panic.

Of the fabulous twelve hundred feet of the Gould and Curry, on the Comstock, Mark Twain says: "Curry owned two-thirds of it—and he said that he sold it out for twenty-five hundred dollars in cash, and an old plug horse that ate up his market value in barley and hay in seventeen days. . . . Gould sold out for a pair of second-hand government blankets and a bottle of whisky that killed nine men in three hours and that an unoffending stranger that smelt the cork was disabled for life." Stephen J. Field, one of the founders of Marysville, California, whom Lincoln later appointed to the Supreme Court, has told us that of his sixty-five lots in the new town "Within ninety days I sold twenty-five thousand dollars' worth and still had most of my lots left. My frame and zinc houses (shipped up after a trip to San Francisco) rented for one thousand dollars a month. The emoluments of my office as *alcalde* were large. At one time I had fourteen thousand dollars in gold in my safe." A year later he was broke and in debt.

How did those fare who discovered the big bonanzas? It's a sad story. Of them all, James Wilson Marshall, who burst in on Sutter with his hands full of gold, has been given the most sympathy. He seems to have been the complaining self-martyred kind who found it easy to succumb to indolence, alcohol, and notions of persecution. Bancroft, who knew him, says he "probably rendered himself exceedingly obnoxious." His last years were spent in bitter complaints and poverty and drunkenness. There were those who thought he deserved a state pension, and as early as 1860 the proposal was made in the legislature but a bill to provide one was killed with amendments. In 1870 a statement was sent to the press: "J. W. Marshall, the discoverer of gold in California . . . is old and poor, and so feeble that he is compelled to work for his board and clothes." The Pioneers of Sacramento sent him a hundred dollars. In 1877 a bill was introduced to give him a pension of two hundred dollars a month for two years. It was passed. By 1883 he "was still walking straight and upright, and apparently promising to outlive many a younger man." As for Sutter, one writer has said that when gold was found on his property he was the second wealthiest man in the United States. He lost just about all he had as the stampedes swept over him, and spent his last years in Pennsylvania, poor and forgotten.

Old Pancake Comstock, discoverer of the big bonanza in Nevada, sold his few feet of a fabulous mine for $10,000, according to Drury; for $11,000, according to other writers. He then bought a farm, went broke, wandered up to Montana and there "committed suicide or was murdered." Of those originally in the claim with him, Finney sold his part for a song, drank it up, and fell off a horse and killed himself. Alvah Gould, of what was to become the famous Gould and Curry, sold his half for $500 and boasted to his friends that he had out-foxed the smart boys in California. While he peddled peanuts in Reno the out-foxed boys took out millions. O'Riley, another of the men in a claim with Pancake, managed to get $40,000 for his part, but then went insane and died penniless. Bill Fairweather was one of six men who on May 26, 1863, found the riches in a gully that they named Alder Gulch, and Bill and the five had their choice of claims. Bill scattered small nuggets as if they were no more than kernels of wheat, and bought drinks for all the bums that swarmed around him. He became for a while a familiar sot in Virginia City. Hearing of strikes in Canada and Alaska he took off but returned, broke and sick, to die at Robbers' Roost at the age of thirty-nine.

Henderson of the Klondike, Manuel of the Homestake, Feenan of the Pelican—the story is much the same with them and many more. Manuel, it is true, was given $70,000 for his strike, or $105,000, according to others. Whatever the price it was a piddling sum for a mine that would build the Hearst empire; by 1900 nearly sixty millions in gold had been taken out of it and that was only the beginning. Of the twenty-eight millionaires or near-millionaires produced by Cripple Creek Sam Strong was a tough, crude, heavy drinker who decided to abandon the soiled doves and settle down. When thirty-eight, he married a girl half his age, and at once Luella Vance slapped a breach of promise suit on him for $250,000, and, to make it really good, Nellie Lewis charged him with having dynamited a mine and sued him for $200,000. Sam fled to Paris with his bride but after his return it took half a million dollars and a full year to heal the hearts he had broken. Having learned by that time that a rich man had a lot of enemies he was ready for some heavy drinking, and one day a gambler named Crumley put a shotgun to Sam's drunken head and pulled the trigger.

There is much pathos in the riches to rags stories but the individuals, no matter how tragic, seem trivial when compared to the mass movements and ruin, such as speculation in stocks. On the Comstock in the spring of 1871 Crown Point stock, which Jones and Hayward had been secretly buying for two dollars reached more than a thousand; in another year it topped $1,825, making the mine worth, on paper, twenty-two millions that a year and a half earlier had been worth $24,000. Widows, waiters, farmers, clerks, and shopkeepers hastened to buy. The Belcher, close to it, controlled by Sharon (whom see), jumped from $1.50 to $1,525. The Comstock had become the biggest mining bonanza in the history of the world. One group of mines would climb in paper value from forty thousand to one hundred and sixty millions. "It was to establish fortunes and power which are still a notable influence in the world. And eventually it was to bring banks, business houses, and speculators throughout the Pacific Slope crashing down in chaos with a loss of 386 million dollars in three months."

Mackay and Fair were deep underground, sleuthing around, peering and touching; and their two partners, Flood and O'Brien, were at the mining exchange in San Francisco. They were soon paying themselves one million dollars a month in dividends, but when they knew that the rich ore was reaching its end they began to unload at boom prices, and ignorant speculators begged, borrowed, and stole in their frenzy to buy stock that was about to collapse. Said a writer in San Francisco: "Bankers, retired capitalists, manufacturers, merchants, shopkeepers, clerks, farmers, mechanics, hod-carriers, servant men and servant women, clergymen, lawyers, doctors, wives and widows poured in their orders for purchase of bonanza stocks." The excitement spread to European cities. While manipulating stocks for their own selfish ends the four Irishmen asked Dan DeQuille of the *Enterprise* to go into their mines and inspect them for all he was worth. What the beguiled DeQuille found was "the finest chloride ore filled with streaks and bunches of the richest black sulphurets" that would assay thousands of dollars per ton. He estimated the whole of it, cut his figures in two to be on the safe side, and told the world that there was $116,748,000 in sight. The Irishmen then called in Philip Deidesheimer, a mining expert known to mining men everywhere, and after a week in the mines he told the world that he could see a billion and a half dollars in sight in just one mine, and that the Consolidated and Virginia alone ought to pay five thousand dollars a share! It is said that he invested every cent he could beg or borrow, but DeQuille apparently bought none.

In the mid-seventies the Big Four increased the C and V's monthly dividend from $324,000 to $1,080,000 a month, and at once wild rumors swept the West that it would soon be doubled and then trebled. Stocks climbed steeply as thousands of people mortgaged everything they had and bought. Many of them were soon wealthy, *on paper*, and as with bated breath they watched their stocks climb they began to buy every extravagant luxury they could find—"servants and gardeners were imported from England, chefs from Paris, blooded horses direct from Arabia, rugs from the Orient, objects of art from Italy, furnishings from the world's centers of fine craftsmanship, food and drink from the world's greatest caterers." They were all millionaires, weren't they? On paper for a few months or a year or two. San Francisco, in 1874, was outstanding in the distribution of its wealth; by 1877 it was bankrupt. The thousands who owned property in the city had mortgaged it to buy stocks, and there was no capital left. In January of 1877 the Consolidated and Virginia, most fabulous of all the Comstock mines, passed its dividend. The market crashed. It literally plunged downward, and people who had imported chefs and blooded horses were reduced overnight to begging in the streets. "Men who had owned and directed large enterprises, employed scores of wage-earners, maintained luxurious homes . . . were begging handouts at back doors or lunch counters along with those who had been in their employ."

Shinn saw it:

In Pauper Alley one can walk any time in business hours and see creatures that once were millionaires and leading operators. Now they live by free lunches in beer cellars and on stray dimes tossed to them "for luck." Women too form a part of the wretched crowd that haunt the end of the alley . . . and beg speculators to give them a pointer or to carry a share of stock for them. . . . Old friends you thought were prosperous . . . shove themselves out of the huddle and beg for the price of a glass of whiskey. . . . The ghost of many a murdered happiness walks among these half-insane paupers as they chatter like apes of lost fortunes and of the prospects of their favorite stocks.

After thousands were reduced to beggary the Big Four wondered why they were not loved. The San Francisco newspapers were giving them hell. The *Mail* said, "The magnificent Mackay, who is indignant at the public want of appreciation of the disinterested course pursued by his firm in kindly looking after Consolidated Virginia for the trifling remuneration of about ninety cents of every dollar it produced. . . ." The *Chronicle*: "The whole history of this bonanza deal is a history of duplicity, fraud and cunning venality without precedent or excuse of any kind. They have won the memorable distinction of having preferred to be millionaires by tricky stock jobbing, when they might have been millionaires by honest mining. So they must expect the natural reward—the hatred and contempt of mankind." If the public was in a condition of "financial hysteria" at least three of the Big Four were laying

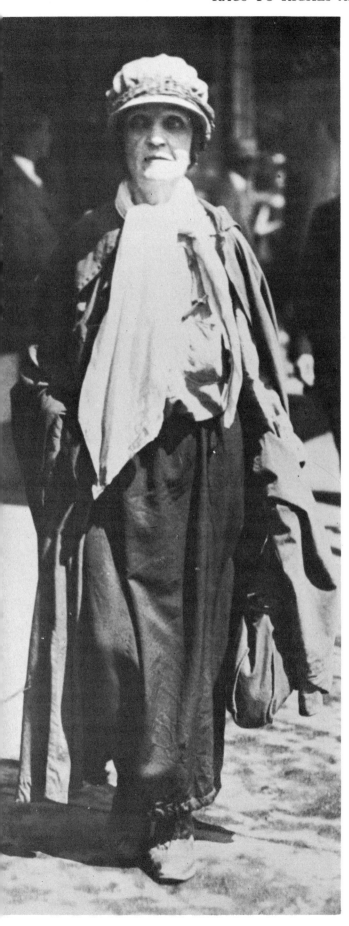

Baby Doe: riches to rags

plans for mansions and social climbing.* Banks throughout California and Nevada went to the wall, and thousands were financially ruined by the cunning manipulations of four Irishmen; but this has been the way of stock manipulation the world over, and the fruits of such stock jobbing have been known to put at least one man in the White House.

Of all those who rose to enormous riches and sank back to poverty possibly the one who can stand as the archetype of all, in incompetence and ignorant mismanagement and abuse of wealth, is H. A. W. Tabor, of Colorado, son of a Vermont tenant farmer and a mother obsessed with religion. An ineffectual and unprepossessing young man, he managed to marry Augusta Pierce, a spare, homely, no-nonsense woman, and together they went west to find a fortune. That he found it and became Colorado's first bonanza king is merely a part of the fabulous American story. He set up a small store in a mining area and staked one prospector after another, taking as pay, as was the custom, a percentage of any riches that might be found. Legends, says Frank Waters, have obscured the facts: instead of a check for $50,000 to build the Grand Hotel and the same amount to furnish his own suite "the hotel cost only $12,000 and the cost of furnishing his private suite did not exceed $1,000." The yield from the Matchless mine was "seldom over $25,000 a month instead of the fanciful figure of $2,000 daily" and the total from all his mines probably did not exceed five million. Well, that was probably twenty-five million in 1968 dollars; and he and his Baby blew it all.

For a while, everything that Horace Tabor touched "seemed to turn to gold. Even those who tried to cheat him lived to rue their deeds." Such as Chicken Bill who salted an old prospect hole. Without even looking into it Tabor wrote out a check for $40,000 and hired some men to dig. On learning that he had bought a salted hole he was not at all dashed; he told the men to keep on digging, and out of this mine, once thought worthless, he is said to have taken three millions. He had so much money he didn't know what to do with it, and so for $117,000 bought a claim called the Matchless, for which no other man in Colorado would have given more than a fraction of

_____
* See the chapter, "Display of Wealth."

that sum. This also turned out to be a bonanza.

But he was such an utterly incompetent man that he steadily went broke. He thought he had huge and valuable holdings in Honduras and Mexico but in these he had been swindled; and spending, as apparently all simple persons do who come into great riches, as if there could be no end of his luck and his income, he lost one by one the Tabor block, the Opera House, his mansion on Capitol Hill, and just about everything else, except his Matchless mine, which was full of water and worthless. The collapse of silver under Cleveland in 1892 would have been, in itself, almost enough to clean him out. When he became old and destitute he went to see Stratton. Stratton had known him when he had nothing and Tabor had millions. Stratton was very kind to him; he addressed him as Senator Tabor, though Tabor had served as senator only a few weeks and had been a national scandal and a disgrace. He wanted a loan, so that he could recover his great fortune and his position in life. Like Eilley

Bowers and so many others he didn't know that he was through. He had a mine, he said, in which there was plenty of gold, and all he needed was money to hire men. Stratton wrote him a check for $15,000, and Tabor gave him as security a deed to a worked-out mine and a promise to pay, and signed it with his customary flourishes by The Tabor Mines and Mill Company, by H. A. W. Tabor, President.

For a little while he lived on a small salary as a postmaster but he was to die in utter poverty, this man who had been Colorado's first multimillionaire, a United States senator with huge diamonds on his fingers, a lieutenant-governor who dreamed of being governor, and General Tabor of the Tabor Light Cavalry, for like Soapy Smith he loved to see himself rigged out in a glittering uniform. But his last years are so intimately bound in with those of his second wife that they will be considered with Baby Doe in the chapter on adventuresses.

CHAPTER 22

Display of Wealth

A chairman of the San Francisco stock and exchange board said that men worth a hundred thousand were thought to be well off; men with five hundred thousand were addressed as "Mister." He did not say by what title the millionaire was addressed. Donaldson states that if a man aspired to political office and was not addressed as "Judge" or "Colonel" he was thought to have a screw loose somewhere. Though such titles "signified prior service neither at bench nor bar" they apparently were freely bestowed over the whole country on men who rose in wealth or position above their fellows. The custom is still with us and is of course a sign of envy and prestige-seeking in those who still look across to the titles in the old kingdoms and monarchies. There were actually social levels in the mining camps; in Virginia City, Nevada, they were actually tiered, with the soiled doves in a row of whitewashed cabins next below C Street, and below them the unsightly shacks and hovels of the Chinese, and still farther down the Indians, who climbed the slope to pick over the garbage.

It seems not to have mattered at all how a man got his money. Charles L. Brace, a fairly enlightened person, was as enraptured by the flowers and trees of California as he was shocked by the way some men became millionaires. In Virginia City he found that

cunning directors are occupied in "freezing out" unfortunate stockholders thousands of miles away, or are forcing up stocks, whose worthlessness they know, to incredible values, or are preparing new reports and statements to beguile the unhappy public. A single day will alter the apparent value of property here by millions of dollars. . . . No such sharpers exist in the world as deal in mining stock speculations in California and Nevada. Beside them Wall Street itself is rural and moral.*

---

* See "Rags to Riches."

We need not be surprised to learn that such persons after they became wealthy revealed themselves as vulgarians. We shall look at a few of them here, keeping in mind the fact that vulgar display was nationwide and respected, as it still is. When it became known in April, 1873, that U. S. Grant was coming to Central City, the enterprising town managers hustled over to the Caribou mine for a fortune in silver bricks, and laid in front of the Teller House a sidewalk of solid silver. That was not unusual in these times. A typical man of wealth was James E. Birch, who in a few years in transportation in California accumulated such riches that he hardly knew what to do with them. "Besides his new colonial mansion, replete from basement to cupola with such rare treasures as his taste could conceive, there were huge and elaborate stables, filled with blooded horses, various types of coaches and sleighs, together with the finest carriages for Mrs. Birch." That was only ordinary spending—Eilley Orrum Bowers did better than that—Eilley Orrum Bowers, whom we met in the previous chapter, tried to make more display than that—with her illiterate and simple-witted man with his sloping shoulders, tubercular countenance, ski nose, and the look of one born to be dominated by a woman and to die young. Eilley's mansion was hardly more than a square box compared to Flood's, but just the same, according to Drury, it had doorknobs and hinges of silver, windows of French plate glass, a bronze piano with mother-of-pearl keys, and a fortune in statuary, paintings, and books, not a one of which her husband could read or she could understand. Eilley tried to ransack Europe for treasures and she did spend a substantial fortune over there, but she was a pitiable piker compared to Mrs. Mackay, to whom

Eilley Orrum Bowers

we are coming. Up in Utah the Silver King Thomas Kearns, built his mansion on fashionable Brigham Street in Salt Lake City, now called South Temple; it had thirty-two rooms, "including an all-marble kitchen, six marble bathrooms, bowling alley, ballroom, billiard room, two parlors, two dining rooms and three vaults for silver, jewels, and wine." He imported artisans from Europe to carve the magnificent staircase of French oak, as well as such allegorical scenes as The Rape of the Sabines and Botticelli's Graces. This mansion was given over to Utah governors for twenty years but since 1957 it has housed the Utah State Historical Society, though the state dining room has been preserved in "its original beauty."

William C. Ralston, San Francisco banker, who made his pile speculating in Washoe stocks and became boss of the Bank of California, "entered upon such an orgy of promotional spending," according to Beebe, "as a Solomon or a Caesar might

Eilley's famous mansion as it looks today. It has been put to many uses since her death—at one time it had a public bar. Today it is a kind of museum but the visitor should put a skeptical eye on some of the things said to be authentic. Her children were buried on the hillside above.

have envied." One expensive plaything which he set upon the earth was the Palace Hotel, at a cost of six million, a fantastic price for a hotel in those times. It was for Beebe "the noblest hotel of the Western world," to accommodate which a brick factory and a furniture factory were built, since Ralston was unable to find furnishings that looked costly enough. To decorate it a New York firm set up a branch office in San Francisco, and skilled workers labored for three years to build into the 755 rooms "luxury generally believed to have been unparalleled since the time of Nero." When in 1875 it welcomed the world with a banquet "the whole world marveled"—at the first electric call buttons ever seen, the electric clocks, the intercoms, the five miles of fire-control standpipes and four miles of fire hose—and at the elevators. Of these a newspaper said there were thirty-four (there were five)—four for passengers, ten for luggage, and twenty for mixed drinks. "Each elevator contains a piano and a bowling alley." The hotel had "thousands upon thousands" of plates, cups, saucers, from Haviland in France; and today, says Beebe, when the hotel lays out a state dinner for 24-carat princes and premiers it sends an armed guard across the street to the Wells, Fargo vaults for the "priceless gold dinner service." When the hotel first opened it had nine thousand spitoons for the tobacco-chewers, carloads of "fringed portieres, marquetry tables, cloisonné, ormolu and Turkey carpets," as well as "literally acres of 'hand-painted' oil paintings, mostly depicting the Golden Gate at sunset or the Sierra at sunrise." The "glory of the hotel's cuisine" fetched gourmets from many lands, "to be deposited in a thicket of potted palms under a vast glass dome" after being helped out of their carriages. When in later years the elder J. P.

Morgan arrived with his own chef, personal steward and cellarman "and other members of an imperial entourage in a six-car private train," San Francisco was "pleasantly outraged." And Beebe is authority for the statement that the Grand Duke Boris found it practically impossible to get enough of the oyster omelet and dry champagne at the breakfast grill.

Ralston also built a huge ornate palace for himself, and other bonanza multimillionaires followed his example, until a part of the Pacific coast was ablaze with self-indulgence. The financial situation in the West at the top of the Comstock boom was much like that in the country in 1929. We have pointed out in other chapters that Ralston helped to bring the whole phoney flamboyant edifice down in ruins, out of which he crawled to slip away to the North Beach and drown himself. His unsecured liabilities were $4,655,973.

The four Irishmen sat on top of the ruins with millions piled around them, and so it is natural that we should look to them for the most extravagant spending. The court record of Flood's testimony in a certain case reads like this:

Who are the owners of the Pacific Mill and Mining Company?
Mackay, Fair, Flood and O'Brien are the principal stockholders.
Who are the owners of the Pacific Wood, Lumber and Flume Company?
Mackay, Fair, Flood and O'Brien are the principal stockholders.
Who are the owners of the Virginia and Gold Hill Water Company?
Mackay, Fair, Flood and O'Brien . . . and others.
Is there any other corporation from which the company draws supplies of any character of which Mackay, Fair, Flood, and O'Brien are not the trustees and principal owners?
I don't know of any.

These four Irish pillagers who did exactly what Brace said they did, bringing thousands to financial ruin and many to suicide, are treated with gentleness by many writers, possibly because John W. Mackay became a philanthropist; but as Professor Paul says, "The Comstock Lode produced some of the world's great exhibitionists . . . who built houses that looked like grotesque wedding cakes and sent their wives to Europe with unlimited bank accounts to purchase titled sons-in-law. What was the relationship between these indecently happy few and the great majority of the community, including unknown simple folk . . .?" To answer the question whether the four Irishmen had any interest in the unknown simple folk we have to take a brief look at the kind of men they were.

O'Brien seems to have been the simplest of the four, the least vain and pretentious, the most gregarious and the most devoted to the bottle. He and Flood had failed in business in San Francisco and had opened a saloon in 1857. From that they graduated to speculation and stock brokerage. Apparently with no wish to show off his wealth he spent much of his time drinking and playing cards with cronies in a saloon which stood on the ground where the San Francisco *Chronicle* now stands. It is said that he kept stacks of silver dollars piled around him on the table, from which his friends were free to help themselves.

James G. Fair, known as Slippery Jim, and James C. Flood were undoubtedly the most ruthless and single-minded. Fair had spent a year or two in Shaw's Flat, bent over pick, shovel and

William S. O'Brien

sluice box, the sweat running down his naked back and belly while his eyes, intelligent, quick, and cold, scanned the country roundabout. He trailed a party of Chileans who had struck it rich, and staking a claim close by soon had a pile of gold worth two hundred thousand. O'Brien said that among ten thousand men who used their hands it was enough to have one who used his brains. Fair was determined to be that one. A newspaper of the time called him "gross, dull, greedy, grasping, mean, and malignant" and said he "sacrificed family, friends, and honor to gratify his money-making instinct. . . ." His wife sued him on "habitual adultery," a charge which he did not deny; the settlement cost him the custody of his two daughters (he kept the two sons) plus nearly five million to his wife and $200,000 to her attorney. Whether the judge who assessed that fantastic fee participated in it is not known. Fair then applied all his talents to the accumulation of more wealth, and at the height of his fortune owned sixty acres of the choicest parts of San Francisco, on which his buildings returned him an annual rental of around three million. As disease and age overtook him he drank more and more, and even ate more, who had always been a voracious eater—steak, four eggs, and a dozen slices of toast for breakfast. But if he was a monster of selfishness and greed he was to pay for it. One son, while still young, became a drunkard and at twenty-seven died in an alcoholic stupor or killed himself. The other son was also a wastrel, often drunk, always in debt. Defying his father's wishes he married the madam of a "questionable resort." Fair then sent him on a long journey, and told the woman she could have $25,000 to annul the marriage or she would get nothing. Being a sharp madam she took it, fled to Europe, met the son there and married him again. The son then took to racing cars and he and his wife were killed on a track in Europe.

As with so many rich men the troubles did not end when Fair died. The two daughters had gone east with their mother, where one married Herman Oelrichs, the other, William K. Vanderbilt, Jr. Fair's will had made a few small cash bequests and left the bulk of the estate to his two daughters and to the son in Europe. Greedy and selfish like their father, they hired lawyers to break the trust, and the litigation, which dragged on for seven years and was a national sensation,

James G. Fair

is surely the most fantastic of its kind in American history. The will vanished from the safe in the office of the probate judge, and while the heirs were viewing that development with satisfaction a Mrs. Nettie Craven, principal of a grammar school, came forth with what she said was Fair's last will, scrawled in pencil, and made three days after the estate was put in trust. It gave the estate outright to the children. They had no trouble at all believing that the scrawl was their father's writing, until the amazing Nettie came forth with another document, also in pencil scrawl, which gave her two pieces of property worth a million and a half. The children then said that both were forgeries. But Nettie was not through with them. Her next move after a few weeks was to dig up another document, which floored everyone: over what purported to be the signature of James G. Fair were the words, "I take Nettie R. Craven to be my lawful wife." Over her signature were the words, "I take for my lawful husband James G. Fair." Sarah Hill, the Rose of Sharon, was not as clumsy as that.* After a year and a half there were enough attorneys representing interests, real and alleged, to fill an auditorium. Five and a half more years would pass before the courts would be done with trust officials, greedy heirs, Nettie, and a swarm of men, women, and girls who swore that they were ex-wives, illegitimate waifs, abandoned sons, and betrayed mistresses. In perjury and vulgarity this litigation should be compared with that following the deaths of Sharon and Stratton.

O'Brien's vanity did make one display—he built for himself a mausoleum that was long unrivaled, Lewis says, "for sheer ugliness in the field of cemetery architecture." But of the four Irishmen Flood was the most accomplished master of vulgarity. A biographer says he was "proud of the fact that his taste for luxury had developed early." A carpenter and then a saloonkeeper, he had married the daughter of a fisherman. He could have passed for a gymnast or weight-lifter—short, with thick neck and shoulders, heavy obtuse features, a florid complexion, he was known as a jolly fellow who kept the hot-tempered Mackay and Fair from one another's throats when the four met in conference to plot their next moves against the unknown simple folk. Four years after he and O'Brien left their saloon, known as the Auction Lunch, they were "jointly enjoying an income in excess of half a million dollars a month." Great wealth changed Flood, as it changes most people. The "oldest and least forceful of the bonanza group, but always the most popular," and with more friends, a San Francisco newspaper said, "than falls to the lot of most rich men," Flood began to get ideas, and they ran chiefly to huge ugly mansions. He indulged "his liking for luxurious living on a scale that set new standards in the West." One house he bought from an ex-mayor, and the press reported that he intended to spend a million remodeling and furnishing it. That was only the first of a whole series of mansions. His eye was on Nob Hill.

Each house cost more than its predecessor, and the whole lot of them had San Franciscans gawk-

James C. Flood

ing for nearly half a century. The ex-mayor's house was a palace to awe most citizens, but Flood didn't stay in it long. He went down the peninsula, as Ralston had done, and on thirty-five acres at Menlo Park he built, at fantastic cost, the wedding-cake thing that he called Linden Towers, with its incredible mass of turrets and gables, topped by a tower rising 150 feet above them. Flood and his wife, like all the *nouveau riche*, faced problems when they moved into an area of snobs: among them was Gertrude Atherton, whose sharp eyes saw a great deal. The Floods managed to get accepted, after some of the stiffest spines "sipped tea from Dresden cups and admired the handsome service of Comstock silver" and took a tour through the mansion. The drawing room, as big as a football stadium, had walls of embossed velvet and a music room in rose and satinwood; and outside there was a house for the carriages "as large as a hotel ballroom." It is said that U. S. Grant, a simple soul, who was flabbergasted by Mrs. Mackay's shack in Paris,

* See "Adventuresses."

looked on Flood's expanse with "unmeasured admiration." The nation, or that part of it which measures everything in money, held its breath over the announcement that a romance was budding between Grant's son Buck and the Flood heiress known as Jennie; but however much he may have been impressed by Flood's millions Buck took a dim view of Jennie after he returned to the East and thought it over. But with her Pa's money she managed to corral a Vanderbilt.

The Wedding Cake aroused Flood to his next venture, which had to be, naturally, on top of Nob Hill. His wish to be surrounded by *visible* evidence of his riches grew stronger as he grew older. He had, for instance, dozens of carriages of all kinds, their wheels "polished to the brilliance of mirrors," and "burnished silver harness fittings" on perfectly matched horses. The *Chronicle* said that when Flood drove down Montgomery Street a hush settled over it that was like that over London when the Queen arrived to open Parliament. Flood intended his mansion on Nob Hill to put a hush over the whole city, and it did, with its eighty-foot tower and a thirty-thousand-dollar brass rail that surrounded it. A man was employed full time to keep the rail shining, which

led a wag to remark that it was only an extension of Flood's saloon days.

Did fabulous wealth bring happiness to any of the four bonanza kings? A little to O'Brien possibly, but not to the other three. Fair had planned to build the most costly mansion in the West but never took time off from his obsessed wish to add to his fortune. Flood became old and ill at an early age, and after going to Germany for relief died in a hotel room there. A woman popped up after his death and claimed that he was her father; whether he was or not we don't know but the settlement cost the heirs a million dollars.

Of the four Mackay was the only one with any sense of the responsibilities that inhere in great wealth. He gave much of his wealth away, as did his son after him, which led Drury to say that " 'good deeds stand beside him like Angels of Light.' " Paul says that he had a "quiet, modest, truthful nature, together with a notable generosity," but he didn't allow such traits to get in his way when he was bilking the unknown simple folk. He accumulated so much money with his stock rigging that he could no longer find pleasure in winning at cards: Drury says that after raking in a huge pot "all of a sudden came the chilling

Flood's wedding cake Linden Towers, razed in the thirties

His Nob Hill mansion

John W. Mackay

thought, What of it? Even if I should win every cent in sight it would not make the slightest difference to me!" He walked away, leaving his pile of gold on the table, an act that would have been angrily censured by his wife over in Europe, if she had seen it; for though wealth meant little to Mackay, a shy and lonely man, it meant everything under heaven to his wife, whose vulgar display of wealth in Europe made Flood look like a piker.

Marie Louise Antoinette Hungerford was distinguished by neither training nor background when, at the age of sixteen, on January 1, 1860, she married Dr. Edmund Bryant in Downieville, where Juanita had been hanged. Her father, who had cut a few capers in the Mexican War, had been reduced to cutting hair in a barbershop but by his son-in-law was now elevated to work more socially acceptable. Two daughters were born, one of whom soon died. For reasons unknown the marriage did not last. The doctor became an alcoholic and did not long survive his daughter. Not long afterward Mrs. Bryant found herself in Virginia City, where, to support herself and child, she advertised for sewing and gave

The Virginia City house he built for his wife

lessons in French, for her mother had been born in France.

Mackay took to her, according to Manter, a contribution from himself and others and "his shyness appealed to the strong-minded girl." We suspect that what appealed to her was his millions. As Lewis tells it, Jim Fair and his. wife Theresa, an enthusiastic matchmaker, had Mackay to dinner, with the grass widow who looked as if she needed a good meal. What Mackay saw in Marie the world may never know. What Marie saw when she looked at Mackay was a man so rich that his money bored him. He took her to the opera in a five-dollar private box when dress circle and orchestra seats were selling for only a dollar, and she "was already dreaming of being the leader of America's social world." They were married, and Mackay, whose tastes to the end of his life would be simple and modest, built a two-story cottage for Marie and her daughter. But the seamstress wasn't settling for anything like that; at heart she was only a high-class Eilley

Bowers and so she talked her husband into going abroad. Along with her went her daughter, an infant son born to them, father-in-law, mother-in-law, and a nurse. Brooklyn-born Marie was then twenty-eight. We can be sure that in Europe her black eyes looked hard at the possibilities for a woman with millions to spend.

They returned in the fall of 1873, Mackay homesick for Virginia City, his wife shuddering at the thought of it; and so before long she was buying a thirty-thousand dollar residence (about $150,000 in our shrunken dollars) and furnishing it in the "ornate French style fashionable at the time." That was in San Francisco, and there she laid siege to the top snobs such as the Crockers, Huntingtons, and Stanfords, while up in Virginia City, according to the *Chronicle,* Mackay "dresses no better than an ordinary gent in Virginia, eats no better food than a conscientious editor ought to have," though he was so rich that he could have given a two-million endowment to a college in every state in the Union and opened

Newspaper sketch of Meissonier's "suppressed" portrait of Mrs. Mackay. From the *San Francisco Examiner*, February 22, 1891.

Mrs. John W. Mackay. Taken during the period when her social triumphs in Paris had made her an international figure.

up the Darien Canal. After three years Marie felt that she had panned out San Francisco and was ready for Europe, while Mackay, who must have wondered a thousand times why he hadn't remained single, lived in Virginia City as simply as he had lived before he married.

Only once in the next twenty-five years would that ruthless woman return to her native land and her husband, and then only for a brief stay. In Paris with her children and nursemaids she laid her plans. She intended to crash society in a big way, and the first thing, her limited knowledge told her, was to get a big mansion. She bought a huge French Renaissance palace at 9, Rue Tilsit that filled a whole square close by the Arc de Triomphe. Her furnishing and their cost must be left to the imagination. Like many other American women whose trashy ambition was to be accepted by royal parasites she then tried to figure out how to lure guests to her house. U. S. Grant was sent from heaven or somewhere to help her out. Since he and Hungerford had known one another in the Mexican War, why not, asked the father, invite the general to stay with them? Marie must have nearly fainted at the thought. She cabled the general and he accepted.

Well, now! It looked as if the girl whose father had barbered in Downieville was on her way. The scope and cost of her preparations to entertain the general must also be left to imagination. A whole army of artisans and caterers moved to the Rue Tilsit and set to work. Of course the whole enormous lower floor had to be redecorated in red, white, and blue, for Marie's taste was about on a par with Eilley's. All the furniture was reupholstered in red, white, and blue satin. What John Mackay thought when the canceled checks came through his bank will never be known. He was living alone in a small room while his wife had an entire wall yanked out of the drawing room so that it would open on a hastily built pavilion covered with costly tapestries. Paris was so fascinated that mobs came daily to gawk at the way things were being torn down and rebuilt. Marie had to hire a gang of policemen to keep order.

Her compulsion to spend reached such insane extremes that she was determined to decorate the Arc de Triomphe, probably in red, white, and blue. The story says that when her proposal was haughtily rejected she slapped the jewels that hung round her neck and cried, "All right, I'll buy it and decorate it myself!" It seems to be a blessing that General de Gaulle was not around then. The French were staggered and outraged by the gall of a woman who seemed to be compounded of gold and bad taste. A biographer of Grant says that to the tables groaning with the finest liquors and food three hundred guests came, while John Mackay, six thousand miles away, was eating a modest lunch in a small cafe. There was a sprinkling of distinguished Frenchmen with American sympathies but most of the guests were the cream off the top of the American colony in Paris.

For the next quarter of a century "her social triumphs were staple reading for millions all over America." John Mackay, in his hotel room, read that his missus was entertaining the exiled queen of Spain and the president of the Third French Republic. Those were only two among hundreds. Newspapers filled columns with descriptions of her jewels, "the finest collection outside the royal treasuries."* How Eilley, who had spent twenty thousand in Paris in one day, must have seethed with resentment when, dead broke, she read about the necklace of sapphires "with a pendant gem the size of a pigeon's egg, set in large diamonds." It had cost the barber's daughter only a million and a half in present dollars, this trinket that was shown at the Paris Exposition in 1878. Did Eilley, green with envy, read about the diamond bracelet that "encircles the arm above the elbow five times; it is formed of a single row of very large diamonds, three hundred in number." Her collection of jewels was estimated to be worth five million 1968 dollars, and it is said that she wore practically the whole thing when she appeared at dinners and balls. An old man who slipped up on Mackay in an alley and shot him in the back had a note in his pocket which said that Mackay had spent enough on jewels for his wife "to assure a comfortable old age for five hundred of his indigent countrymen." This man's story was that Mackay's stock market rigging had cleaned him out.

Marie's daughter Eva had, of course, to marry into royalty, so Marie found for her an Italian count. The marriage didn't last and it cost Mackay a pretty sum to get the girl out of it. And Marie—why, yes—had to be painted. For this su-

Winfield Scott Stratton

premely important job she commissioned M. Jean Meissonier, most famous of French portrait painters; but she didn't like what she saw and she wouldn't pay for it. God only knows what she expected to see. Her haughty rejection of the work choked so many patriotic Frenchmen with fury that it's a wonder it didn't lead to war; again we must give thanks that le grand Charley was not on the scene. There was such an awful uproar in the press that Marie paid eighty thousand francs, invited some friends to her home, and, with a gesture befitting a barber's daughter whose husband was supporting all his in-laws, hurled the portrait into the flames.

In a great tiff the American queen closed her French mansion and moved on London. There her pad at No. 6 Carlton House Terrace excited such awe that descriptions of its "colossal dimensions" and of its immense rooms were cabled across the Atlantic. The marble staircase itself had cost the

_____

* She had a silver service made of Comstock silver by Tiffany, each piece embossed with the head of William Cullen Bryant, a distant relative of her doctor-husband!

original owner, the Duke of Leinster, three hundred thousand dollars. He had had to get rid of it, for England at that time was not on the American relief rolls. Marie made the most of it and her vulgar countrymen, or women anyway, hung breathless on every move she made. The Magnificent Mrs. Mackay, the San Francisco *Call* told its readers, was the "hostess beyond compare." She was an "honor to her sex and to her countrymen abroad." She was, of course, a colossal bore from any point of view, but a nation pining for royalty took her to its heart. Her husband went over briefly once in a while but he preferred to sit in Virginia City in his twenty-five-dollar suit, with a whiskey and a cigar, and the fellowship of men who had worked with him when he was a common miner at four dollars a day. While his wife squandered millions he lent millions to friends and parasites—it is said that at one time he tore up over a million dollars' worth of notes, saying, "This one can never pay, this one is dead, it all doesn't matter."

When one of his two sons died, abroad (the one who was always called Willie), he was "buried with the pomp of a prince"—by his mother, of course. She reopened the Tilsit house and covered it with crepe and draperies; spent a fortune draping the nave and aisles of the church where High Mass was said; and bought "incomparably the handsomest mausoleum in the world" at a cost of a million and a half of present dollars. Her husband meanwhile was doing his best to add to his fortune; after he was dead and sealed in the handsomest mausoleum in the world it was estimated that he was still worth from thirty to sixty millions.

Were all those who took vast treasures from the earth as depressing in their vulgarity as the Jim Floods, Mrs. Mackays, Tom Walshes, Haw Tabors? Not quite all. There was now and then an exception and this seems an appropriate place to give one, for there is a world between the two, the barber's daughter who knew nothing of either manners or taste, and Winfield Scott Stratton, who had an innate sense of both.

Stratton was a tall, slender man with light blue-gray or gray-blue eyes and a mustache; by the age of forty-three, just before his "stroke of inconceivable luck," he was thin and frail, his hair "white and shiny as bleached silk." He was an excellent carpenter and supported himself at this trade when raising grubstakes for his prospecting, to which he gave seventeen years. Sprague says his mother had twelve children, eight of them girls, and that of the four boys only Winfield reached adulthood. Sprague thinks that early in his life he learned to hate the female sex and swore that he would never have anything to do with it. If he made such a promise he almost kept it, except in his relations with the soiled doves. He did, however, have a brief marriage with one Zeurah Stewart, but he was a man of violent temper and furious jealousies, and night and day the neighbors could hear the bride weeping and the husband shouting at her. She became pregnant months before the marriage and Stratton doubted that the child was his; Frank Waters seems to think that it was. Anyway, he sent her back to her people and never saw her again.

Mabel Barbee Lee, who spent her childhood in the Cripple Creek area and whose father was one of Stratton's friends, tells many an anecdote, which may or may not be true, and some tall tales, such as the story that Stratton went to Denver's Brown Palace Hotel and was not allowed to register because he had with him Lola Livingston, madam of Cripple Creek's Mikado House. The legend says that next day Stratton bought the hotel so that he could fire the clerk. Frank Waters talked to many old-timers who had known Stratton, and they told many fascinating stories, such as those about his dreadful tantrums and his suffering but faithful housekeeper Eliza. If Stratton was served oysters or turkey but happened to want sausages he would fling out of his small modest cottage in an awful rage but would return after a while, chastened and sorry. It is said that he told the devoted Eliza that he wanted fried chicken always to be in the icebox. And so it was. Further, she would set a plate of it by his bed at night, eager to please him, until at last in a fury he shrieked at her, "What are you trying to do, starve me? Don't I give you enough money for food? For God's sake get rid of that fried chicken! It's in the icebox, my bedroom, all over the damned house!"

A stickler for cleanliness and neatness, he played mean little tricks on Eliza. He would put a spot of cigar ashes out of sight in a spittoon and the next day he would look in to see if she had found it; and on hands and knees he would crawl about the house looking for dust. These stories and others easily give the impression that he was a brutal and insufferable tyrant, when as a matter of fact his heart was too big for his frail body. What we need to know is from whom and when he got the notion that all women were whores. Parasites and spongers dogged him day and night. If a woman came to his door for a handout he as likely as not would shout at Eliza, "Women! Women! They'll do anything for money, any damn thing at all! Give her some and get her out of here!"

His rich strike seems to have been born of a combination of dream, fantasy, superstition, and fool's luck, though he was quite a shrewd businessman when he had to be. He got control of some of the richest parts of the lode and was soon a multimillionaire. Then his troubles started and he learned that money became, as Waters puts it, "an insupportable weight he could not shake off." We surmise that he was haunted by a very deep sense of guilt that was related to his father, at whom, when only about ten years old, he fired a rifle, according to Sprague, or a shotgun, according to Lee. There probably are things in his relationship to parents and his eight sisters, and to women after he had fled home and come West, that we know nothing about. For it is as plain as day that a deep bitterness entered his soul, and that all his adult life he was a suspicious, lonely, unloved, and very unhappy man.

He was an extremely modest and unassuming man; he never put on the dog, the only luxury he allowed himself being handmade boots. Other Cripple Creek millionaires set up in mansions on Millionaire Row in Colorado Springs. The Denver *Post* thought Stratton a fool because he didn't buy yachts and mansions and blooded horses. The people generally expected him to buy a senatorship, as Tabor had done, as bonanza millionaires all over the West had done or tried to do—along with a countess, Arabian horses, Russian necklaces, and the Arc de Triomphe. While other Cripple Creek millionaires were making a vulgar display of their riches Stratton went on living in the small house he had built for himself many years before his strike. If the snobs called they found that they were not welcome. If Stratton was invited to tea he declined—and he even refused to go to a Mining Exchange banquet because he was expected to wear a tux. He told a friend that only once in his life had he enjoyed

dining out: that was in Mexico when his host couldn't speak English and he couldn't speak Spanish.

The epitaph of this proud lonely man was his giving; after selling his Independence mine for ten million to a London group he felt more strongly than ever that a person who had taken vast wealth from natural resources "should use it to develop the region from which it had come." That was drawing it a bit fine, but even so, his feeling is in staggering contrast to that of most of the men who ransacked this nation's resources to build enormous wedding cakes, buy titled parasites for their daughters, and let their wives make vulgar fools of themselves all over Europe.

Though Frank Waters in his biography of him calls him "The Midas of the Rockies," Stratton's generosity was phenomenal compared to that of all but Mackay among the bonanza kings. He was a bit eccentric in it—for instance, on what seems to have been impulse he would at Christmas time send a $50,000 check to each of his key employes, and build houses for still others; and have loads of fuel taken to both strangers and friends. His gifts included $85,000 to the Salvation Army, $20,000 to Colorado College, $25,000 to the School of Mines, and other bequests; but, as Mrs. Lee says, "it was the city of Colorado Springs that really blossomed under his philanthrophy." He put up buildings for civic centers, and a park, where free concerts were offered, and all together spent, Sprague says, four and a half millions in a few months. But possibly the best of the memorials to his generosity is the Myron Stratton Home (named for his father), to be a refuge for destitute children and old people of El Paso County. His will said:

It is my special desire and command that the inmates of said home shall not be clothed and fed as paupers usually are at public expense, but that they shall be decently and comfortably clothed and amply provided with good wholesome food and necessary medical attendance, care and nursing to protect their health and insure their comfort.

Which suggests that the paternal ran deep in the man.

When still young in years but wasted away by disease and alcohol and his inner torments, he died, Lee says almost no one in Cripple Creek and Colorado Springs felt any loss and that he was "hardly in his grave" when people were speculating about his will. When its contents became known "the rumble of disappointment and indignation shook the foundations of homes and institutions all over Colorado." A half million was given to relatives, $50,000 to the young man who was or was not his son, and the remainder, about six million, was to found and maintain the home for the children and the old. Says Waters: "City, State, newspapers, citizens and the Court began a battle for Stratton's millions." Only two weeks after his death "his properties looked like armed camps." Colorado's attorney-general announced, "We are going to get that $7,000,000 for Colorado." Colorado Springs felt the same way; indeed, so many persons and groups entered suit that it was "like a riot." Of the man who had been so generous it was now alleged that he had been a drunkard and a drug addict, a monomaniac, a lunatic, a mental wreck, a man wholly without principle—there seems to be nothing that some lawyers won't say and do, and the marvel is not that they are beneath contempt but that judges tolerate them. The incompetent son appeared, lawyers took him in hand, and the estate settled for $350,000; the lawyers took half and the son soon lost his half in speculations.

Before his death Stratton had said, "I wish I were worth ten million and I would spend every cent of it to put every lawyer in Colorado in jail." The fifteen years of litigation were to be one of the hardest fought and most bitter, vindictive, and shameful legal battles in American history. The total of the suits was over thirty-five million dollars. Scores of persons swore that they had lent him money, or were bastard daughters or sons, or common-law wives; and on their oath at least twelve women said they were his widows. The most amazing of them was Sophia Gertrude Poor Kennedy Chellew, and the reader can find her story in the chapter, "Tall Tales."

The editor of a weekly called *Facts* wrote:

Stratton pays the penalty of his great possessions in a thousand annoyances, in threats, blackmailing schemes, . . . However, if a kind Providence burdens me that way I hope He will give me the same kindly heart, the same reckless extravagance in doing what I please with my money that He has given Winfield S. Stratton. This is not a prayer, but O Lord! keep me from being like some of these Cripple Creek millionaires, afraid of being bored by old friends, insolent in feeling toward those who have not had my luck, and scared to death lest a dollar go by and I not get it.

In spite of the fantastic perjury and greed and complete lack of decency in those who tried to make a travesty of a sick and lonely man's last wish the Myron Stratton Home is still there and it still serves.

The mansion of a Utah mining king now houses the Utah State Historical Society.

CHAPTER 23

Four Adventuresses

The bonanza camps with their fabulous riches, bachelor millionaires, and spending of money as if it were going out of style were an irresistible lure to the parasitic and predatory human female. We shall briefly characterize four of them who have had a great deal written about them. Though they were only high-class trollops the ending of three of them was tragic, and of two of them so pathetic as to touch even the stoniest heart.

Lola Montez In May, 1853, a ship came into the San Francisco port, bringing, according to the papers, William Gwin, who hadn't yet had his battle with Broderick; Sam Brannan, the Mormon who was to breathe fire into the first vigilance committee; and a woman under the name of Lola Montez, with "Countess of Landsfelt" in parentheses after her name, "easily the most widely known and thoroughly discussed woman of the decade."* The London *Times* helped to take care of that: it had given her so much publicity that when the woman arrived in New York hundreds of persons filed through her hotel suite. She herself was to say that she had been notorious but never famous. But she played it that way. She would be called "the international bad girl of the mid-Victorians" but even that gives a bit the best of it to an Irish exhibitionist who parlayed her small talent into an international sensation.

Jackson says she was plain Eliza Gilbert but Lewis presents her as Maria Dolores Eliza Gilbert. When hardly more than a child she let a British officer take her to his post in India, and there she began to raise the unholy hell that was to characterize her. She was pretty—some have said she had black eyes and hair but Rourke says "her hair, which curled almost childishly back from her face, was bronze with dark shadows, her eyes were blue." She was not the raving beauty that some writers have made her out to be, and like her kind she was more than normally vain, selfish, ruthless, and unmoral. Deserting her husband and hastening back to London, she next appeared as Lola Montez, a danseuse. That was in 1843. Newspapers then, as later, skipped over her dancing and acting, which would never rise above the mediocre, and capitulated to her "Spanish beauty." From London she rushed to Paris, and only four years after her elopement with the officer she was mistress of King Ludwig of Bavaria and had the simple-minded king on his knees before her. A year later she was Countess of Landsfelt. Ludwig had never been more than a stepping-stone to higher things, and so she left him at about the time that his kingdom collapsed in revolution and he fled for his life. Because, like all psychopaths, she cared nothing about such legal formalities as divorce she was soon up for bigamy for having married a "snub-nosed young officer." The judge, like so many then and now, was bewitched by her, and in a fatherly way explained that she was married not to this officer but to the one in India. He told her to be a good girl. Being a good girl was the last thing in her mind, so she took off for New York, her reputation going ahead of her. In a city that has always welcomed foreign phonies she had a splendid time, before taking off for Boston and New Orleans.

This high-class Irish whore would not be in this book if she had not heard of gold in California. By boat she headed for the diggin's, and on her arrival in San Francisco a writer for the *Herald* pulled all the stops—"This distinguished wonder, this world-bewildering puzzle, Marie de

---

* These pages about her are based on Jackson, Lewis, and Rourke and all quotations are from them, unless noted otherwise.

Lola Montez

Landsfelt-Heald has actually come to San Francisco, and her coming has acted like the application of fire to the combustible matter that creates public curiosity, excitement, or *furore*." She was discovered to be the "very *comet* of her sex; and we watch her course with the same emotions we follow the brilliant movements of that erratic body flying through space, alone, unguided, reckless, and undestined." The silly fellow was probably very proud of that sentence. She was even compared to Mirabeau. The *Herald's* completely undone writer went on to say, "Whether she comes as danseuse, authoress, politician, beauty, bluestocking or noble lady, she is welcomed—so she permit herself to be seen, admired, sung, courted and gone mad over here as elsewhere." Sounds for all the world like Hollywood!

Well, that's the reason Eliza was there—that and the gold. By eleven in the morning half the town was standing in line before a theater for her opening performance. Seats were put to bidding, the first going for $65. In her spider dance she hopped around in a frenzy flinging rubber spiders off her. Helmets were flung to the stage, bouquets tossed to her, speeches demanded; and she gave a curtain talk, remembering, as Rourke says, that she was Spanish and must speak with a Spanish accent. In a fury she forgot to do that. After her spider dance some verses appeared:

> Joaquin to the mountains was advancing
> When he saw Lola Montez a-dancing;
> When she danced the spider dance he was
>   bound to run her off,
> And he'd feed her eggs and chickens, to
>   make her cackle, crow and cough.

A wretched thing which she called "Lola Montez in Bavaria" so excited the scorn of the Chapmans (local celebrities of the theater) that Miss Chapman put on an extravaganza with the same title; and though many were outraged by the fun made of their heroine the theater was crowded as Chapman "obstinately stamped, coquetted, and credibly impersonated the Countess." She transformed the spider dance into a fanciful ballet that she called Spy-Bear, which was a "glittering weapon of satire against Lola as an actress who never knew her lines." Both the newspapers and Lola were in an uproar. It is said that in a fit of fury and frustration she challenged one editor to a duel with pistols at five paces or his choice of two pills in a box, one of which would be poisoned. Another editor said that "History pays her a higher compliment than her own play" and that she was not a spoiled court pet but a "dignified and resolute woman." With such manly defenders it was no trouble at all to remain the center of interest, whether on stage or when strolling with her poodle and smoking a cigar. Though at least one newspaper thought that Caroline Chapman's "Mula, Countess of Bohemia" lacked wit, point, common sense, and had bad "rymes" it was enough to send Eliza up to Grass Valley to think things over.

It was time, of course, to get married again, and this time she chose Patrick Purdy Hull, an Irishman who had come in on the boat with her. Eliza said she liked him because he could tell a story better than any other man she had known. The *Shasta Courier* came out swooning: "Lola Montez is married to a man of California! . . . this celebrated actress of unblemished virtue, the charming and peerless Madame Marie Elise Rosanna Dolores, Countess of Landsfelt, Baroness of Rosenthal, was united in the holy bonds of matrimony. . . . It affords us the most exquisite happiness. . . . This is certainly a very great country." The Irishman, said the writer, had "immortalized himself" and that's about the only thing he said that had any truth in it. Sacramento gave her a great hurrah until a few members of the audience had the impudence to laugh. In a furious pout she left the stage. When Marysville didn't fall to its knees she went into another sulk and had a hell-roaring fight with her new "husband." One story, probably pure fiction, says she tossed his baggage out the window and tumbled him downstairs.

As Jackson says, most of the stories about her are inventions. After taking leave of Hull she went to Grass Valley and to the amazement of all who have written about her lived there two years. She is said to have bought a house, put on old clothes and planted a garden, and established a salon in which she kept a collection of parrots, monkeys, and bears. Two of the many stories that

some writers have been happy to take seriously are these, that she horsewhipped an editor who said she was not the world's most beautiful woman and greatest actress; and that she was the nugget at the core of a scheme to set California up as a kingdom and make her the queen. The truth is that she came, put on a few mediocre acts, turned a few editors out of their wits, and now in folklore is, along with Black Bart and the Mexican Robin Hood, half as big as all creation. Lewis says the house she bought was a white cottage. There is a large house today, occupied by an eccentric, with an admission price at the door, which is put forward as Lola's cottage; and in the yard behind it is a smaller house, said by some old-timers there to be the authentic and immortal structure. It is very probable that neither is.

After two years in Grass Valley she gave five farewell performances in San Francisco and sailed for New York. She went to Europe and returned to the States as an expert on beauty, fascination, and love, and went on a lecture tour. Nobody had told her that the public is fickle; it wants its phenomena to make grander and grander appearances, in steadily more awesome designs. Eliza had just about shot her wad. No audiences rose to cheer; no newspaper remembered the days when the London *Times* lavished on her endless pages of fantastic nonsense. Rourke, a woman and her most severe critic, says she arrived in Grass Valley "shorn of glory" and "trailing clouds of wickedness"—such a brazen hussy that "by any usual rule of thumb the whole town should have turned against her." On the contrary, the miners named a peak for her.

But the Irish are tough and she was not through. Some writers ignore or pass briefly over what was probably the most significant period of her life, her journey to Australia with a maid, and a young actor in tow. She was gone about a year. On the return journey the actor drowned, though whether he jumped in or fell overboard is not known. The word usually applied to her after her return is subdued. Certainly she was in some measure changed. She is said to have sold her jewels and settled the money on the two children

Of two houses in Grass Valley claiming to be hers this seems to be the more likely one.

of the actor; this would indicate that she felt herself guilty because of the actor's death, or had become very fond of the children. For she didn't have much to settle on anyone. According to Jackson only five years after she left Grass Valley she died, January 17, 1861, destitute and probably not yet forty, in New York and was quietly buried

Elizabeth McCourt

without fanfare. For what it is worth—and it may be worth only a little—James Reynolds wrote in his Boise *Statesman*, March 16, 1866, that "the best and most attractive" theatrical production yet offered in Boise was "Pas de Fascination, or Lola Montez, Countess for an Hour." It had been, Reynolds said, a success everywhere it had been presented, largely because it was "about the widely celebrated and fascinating Countess of Landsfelt, Lola Montez, and anything said or written of her must be of interest. This singularly brilliant woman" had been "actress, authoress, and lecturess. . . . When she was in the zenith of her glory, and her life but a daily routine of gayety and dissipa-

tion, she is said to have horse-whipped more men than any other woman that ever lived. Finally, seeing the evil of her ways, she retired to private life, and became a repentant woman. . . . she professed religion and died a true christian."

Baby Doe   President Chester A. Arthur bowed over her hand and said, "I have never seen a more beautiful bride." The *Albany Journal* said she was

without doubt the handsomest woman in Colorado. She is young, tall, and well proportioned, with a complexion so clear that it reminds one of the rose blush mingling with the pure white lily; a great wealth of light brown hair . . . large, dreamy blue eyes . . . a Mary Anderson mouth and chin, and a shoulder and bust that no Colorado Venus can compare with; delicate feet and a tiny white hand. . . .

Another newsman found in her "a perfect blonde, with a magnificent suit of golden hair. . . . there never was a more beautiful set of teeth. . . . her eyes are large and full of expression. Her nose is slightly retroussé, but that only adds to the piquancy of a face that would be called beautiful among beauties."

Elizabeth Nellis McCourt, like Lola, was Irish, born in Oshkosh, and called Baby by her brother James; at twenty-two she married Harvey Doe, who was twenty-three, and a part of that American institution known as the mom-boy. His mother, a part of the dragon known as Mom, didn't want him to marry Baby or anyone else; from the beginning his marriage, like that of millions of mom-boys since then, didn't stand a ghost of a chance. The only son, he had an adoring mother, five sisters, and a father who took a dim view of him. He was good-looking, shy, inept, and completely lost when out of sight of Mom. Baby wrote in her scrapbook, "I met my love on the street, April 3, 1876." She would never be that mistaken again. She gazed at William Harvey Doe, Jr., abandoned all thoughts of becoming an actress, and married him, while his mother faked heart attacks, moaned that she was dying, and thought the world was coming to an end. Papa Doe had some mining interests in the Central City area of Colorado, so bride and groom, with the bride leading the groom by the nose, left Mom Doe prostrate and headed west to make their fortune. Their troubles had just begun.

Lizzie McCourt was obsessed with a drive toward riches; it would dominate her life. When Harvey proved to be pathetically ineffectual in digging down to a rich lode she put on the rough garb of a miner and dug with him. The respectable women of the area were scandalized. In those days women, including prostitutes, not only did not appear in public in two G-strings or skintight pants; they didn't even show their ankles. But Lizzie was after riches, not public approval, about which she was never to care a damn; and she went right on digging with her man, forlorn and homesick for Mother, until she was forced to realize that she had married not a man but a child. And at that moment there providentially entered her life a handsome Jew named Jacob Sandelowsky, whom Miss Bancroft disguises as Jacob Smith, and who would soon drop the "elowsky" and be plain Jake Sand. He was a bachelor merchant about thirty years old, and he paid the blonde "extravagant compliments," gave her some expensive jewelry, and before long took her to bed.

Harvey's sufferings during this period were extreme enough to drive him to drink. While yearning across distance for his mother and enduring the ill-concealed contempt of his father he could not hide from himself the rapturous affair his wife was having with another man. When she became pregnant Harvey could not tolerate the thought that it was Jake's child. Whether it was we do not know but a divine providence took it away at birth. One author gravely assures us that Jake was in love with Baby but as we size him up he played the whole thing coolly for what there was in it. Was Baby in love with him? Her scrapbook indicates that she thought she was. She must have heard of his reputation as a lady-killer and a gambler but under pressed gentians she wrote: "Jake gave me these September 25, 1878, the night of the festival in his store when we sat on the school house steps together. He kissed BT three times and oh! how he loved me and he does now." She pasted around his photograph poems with such titles as "I Love Thee." Oh, how she had been thrilled when she danced with the tall dark man, for Harvey's mom had refused to let him learn how to dance. Out of his grief and loneliness he gave his Baby a gift, "From the man whome loves and worships you. HARVEY." His mother also had never let him learn how to spell.

After Baby had left him and divorced him he wrote a pathetic letter to his parents that tells all anyone needs to know about Harvey:

You have of corse herd before this of my sad sad loss in loosing my darling Babe I am heart broken about it I shal go crazey about it I know I shal for my dear Father & Mother it was not my fault that I went into that parlor house. Let me tell you all about it and then I hope you will not blame me there is a man in this town who was trying to sell my mine for me. We had been looking all over town for a man. . . .

Baby had heard that he was going to one of the fancy brothels and with an officer she marched in and caught him there. But the mom-boys are the perfect gentlemen:

. . . just as I was going to go out Babe came there and caught me and she did act like a perfect lady and conducted herself so nicely in such a place as that. Now my dear Father & Mother do not blame me. . . . no no I love my darling wife babe to much ever to disgrace her in that manner. . . . Even my own Father has worked against me and wherever he could hinder me from making money he has done so. I hope and sincerely pray my dear parents you will not blame me and do try and get babe to come back to me for she is all I have got in this world do try and persuade her. . . .

This letter, all of which Mr. Hall publishes in his book, is surely one of the most pathetic ever wrung from a young man whose ogre-mom never gave him a chance.

After she got her divorce his Baby probably never gave him a second thought. She was off for "leadville—and Jake," for Jake had gone into business there. A shrewd guess is that she now expected him to marry her. He didn't, and again Providence stepped in. She was eating in a restaurant when, according to Miss Bancroft's fictionized version, in which Baby tells her story, she says, "Glancing up, my eyes met Mr. Tabor's piercing dark ones . . . I knew in an instant that I was falling in love. Love at first sight!" That is the way she talked about it in old age but it is nonsense. She had heard of Tabor, then forty-nine and famous as Colorado's first millionaire, as the Silver King, and as a man determined to become governor. She lost no time in telling Jake that she loved Tabor, and returned his diamond ring. The cool way in which Jake accepted her statement suggests that those writers who have him in love with her are matchmakers at heart.

H. A. W. Tabor

The wife he left

Very revealing are the words she copied in her scrapbook: "Stolen love, oh sweetest is the stolen love, the apples shaken from the bough unseen, unseen, though eyes are keen, and such a love was ours but now." Under the verses she pasted three words that she had cut from a newspaper: "Lizzie was frightened." Lizzie had no need to be, unless she had heard of a newspaper reporter's story that Haw (Hall calls him Hod) * Tabor kept as many as five mistresses at one time, including an Amazon whose hobby was feats of strength. It all makes one take a second look at the photograph of a man eager to get rid of his wife and cultivate his taste in mistresses, and shirts with diamond buttons as big as beans. The trouble with his marriage was that the Vermont hillbilly when he got his hands on millions wanted to run wild and spend it, whereas his literal-minded and practical wife wanted to live much the kind of life she had always lived. He put her up in a

mansion in Denver that cost him forty thousand, equal to about four or five times that sum now, but Augusta didn't want to be loaded with jewels and rushed off every evening to fancy restaurants and balls. She did flee to Europe in the hope of improving her "social graces" but putting on the dog was simply not for her. Her preference was for a quiet life at home with a quiet husband. Possibly she was shrewd enough to foresee that he would make a fool of himself, which he proceeded to do in a big way; before he was done with it he was an object of ridicule all over the nation, with his decorated velvet caps, nightshirts with pointed lace, huge cuff links with onyx and diamond settings, and a huge diamond on a finger of each hand.

It is a bit difficult to understand but it is a fact that Lizzie McCourt was just the woman for him. She was, of course, hypnotized by his millions and these undoubtedly were the chief lure

---

* H. A. W. Tabor appears on other pages; he is, for instance, almost the archetype of the rags to riches to rags characters.

but she seems to have been genuinely fond of the man. Perhaps she had no reason not to be: he loved money and the things money can buy, and so did she, and so in their strange way they made a perfect team. If he had told Augusta he wanted to build the fanciest opera house in the West she would have thought he was losing his mind. Baby loved such ideas. She must have thought it an act of genius to send architects to Europe to ransack the continent for treasures he could build into his "modified Egyptian Moresque"; to have Marshall Field ship huge loads of carpets and furnishings, most of them red because red was her favorite color; and to have marble lintels, and pilasters shipped from Italy, heavy silk Louis XIV tapestries at fifty dollars a yard from Paris, and above the proscenium to have Hector tell Andromache to be of stout heart. The structure which, Horace maintained, combined the choicest features of Covent Garden and the Academy of Music in Paris, built, as the New York *Herald* observed, out in a wilderness of Indians, grizzlies,

and stopes, at a cost of half a million dollars, became and still is one of the most famous buildings of the early American West.

Erecting fine buildings was all right, but in the way of bonanza millionaires, Horace developed political ambitions. He became lieutenant-governor of Colorado, and United States senator for thirty days. He was constantly ridiculed by Eugene Field and others but he was not stupid and he was not the oaf they tried to make him out to be. The worst charge that can be brought against him is innocence and naïvete. He had had a bitter fight to get a divorce. Before becoming senator he settled a fortune on Lizzie's parents, who had fallen on poor luck; slipped away from Denver to marry Baby secretly; and then from the United States Senate wrote her that he would have a "grand dinner party tomorrow night the best that has been held here for a long time"—Capon à la Godard; sweetbreads, à la Conti; canvasback duck, and many other delicacies. But those in fact were some of the things on the menu after Baby, with

The Tabor Mansion

her parents, two brothers, two sisters and their husbands, had gone to Washington to be with the Senator. He gave the dinner to the "few people who had been polite to him," as a Washington paper put it, and the invitations were engraved in silver with silver margins a quarter of an inch wide. But envy and malice are tough enemies and Horace was a clumsy fighter. It did him no good to marry Baby a second time, in Washington, to pay $7,000 for her wedding gown ($30,000 today) —for in the American way he was trying to intimidate or to win with money. He merely caused a national sensation, and forced upon his homely ex-wife a deeper humiliation. She said she cried herself to sleep that night, and every night for the next ten years. For all his banquets, innocence, and money, Tabor got no more than this savage malice from the New York *Tribune*: "A cold shudder has gone through Washington at the rumor that the Colorado man (Senator Tabor) liked Washington, intended to build there and live."

But in this chapter the story is Baby's, not Horace's, though she remained in the background, happy, it seems, to let him have the limelight and spend the money. He had millions but he was to die almost a pauper, and though the collapse of silver would help to bring him down he would undoubtedly have gone broke anyway. Like Eilley Orrum Bowers and a few other essentially simple persons who struck it rich he actually believed that his mines would never run dry; and when

one by one he had lost everything he had, and was dying, at the age of sixty-nine, his last words were to his Baby, when he told her to hang on to the Matchless. That had been the richest of his mines.

The story was really his as long as he lived but after he died it was Baby's, and even Augusta would have been moved to pity if she could have seen her struggle year after year as she strove to support herself and hang onto the Matchless. Hanging onto it was no problem, for nobody wanted it; what she tried to do was to borrow money, again and again, to hire men to keep digging in a mine that was dug out. Tabor died in 1899, his wife thirty-five years later; and in all those years she remained on Fryer Hill, in that toolhouse shack just out of Leadville, with a worthless mine. Her lovely daughter, whom she called Silver Dollar, went to the dogs in Chicago, and a neighbor brought to Baby the headline in a Denver paper—SILVER DOLLAR SLAIN BY FIENDS OF CHICAGO SLUMS. It was in 1911 that Baby moved into the "cold dark toolhouse cabin of the Matchless." Miss Bancroft, with her engineer father, saw her there in 1927, and again before her death. Did she lose her mind? In the way a person loses it who becomes obsessed. For twenty-four years she lived in that shack, above ten thousand feet, in deep snows and bitter cold; and most of time in her last years she seems to have had little to eat.

As Bancroft tells it, "The last day anyone saw

Baby's wedding dress

Mrs. Tabor alive was February 20, 1935. On that morning, she broke her way through deep snow around the Robert E. Lee mine which adjoined the Matchless, . . . and walked the mile or more into the town of Leadville. Her old black dress was horribly torn and the twine and gunnysack wrappings on her feet were dripping wet." After that nobody saw her for two weeks; she was then found frozen to death on the cabin floor. Bancroft says that "fourteen trunks of her former belongings turned up in Denver and Leadville warehouses. But there was no other estate." The shack was ransacked and almost demolished by loutish treasure hunters who had heard the senseless rumor that she had hidden a fortune in it. Her husband's body had been moved to the Mount Olivet Cemetery in Denver; she was buried at his side, to whom he had written six years before his death: "My darling, darling wife, I long to hold you in my arms and whisper my love and tell you my plans. . . ."

Relevant to Baby Doe's last years, and to many others in the American West whose last years in various ways were just as sad, is an item we found in the *Overland Monthly*. John Milton Hoffman was exploring in the Keya Paha Mountain area when he found a lone woman, the only inhabitant of a ghost town,

a strange creature with flaxen hair and blue eyes. Every feature of the face was marked with sorrow; every movement was that of a person in deep distress—not physical, but mental agony. . . . She seemed to disregard her situation utterly, and went about the town peering in at the broken windows, pulling the weeds from the doors, and propping up the tottering houses with poles and pieces of rock. Sometimes she would be heard to moan and cry piteously, then pray, then burst out in wild frenzy; but, for the most part, she bore her sufferings, whatever they were, mutely. And another thing, she seemed unconscious of the presence of anybody; would answer no questions. . . . The rough people of the mountains all seemed to know her, although she had nothing to say to them; not a week passed that some honest miner did not leave food at her door. . . . I can not tell you how beautiful she seemed, even in her wild sorrow; how white her face, and beseeching and tender; how wonderfully beautiful her hair, how blue her eyes. Above her the rugged mountains and the brazen hills; at her feet the river running by like mad; desolation all around her, rocks, trees, hills, and abandoned mines, and, more than all else, abandoned houses, and deserted, weed-grown streets.

In her last years at the Matchless shack

Nearby was a grave she protected, and in the grave were the bones of her husband.

Rose of Sharon The court fight, one of the longest, most bitter, most expensive, and most amazing in American history, turned on the question of whether the following was a forgery.*

In the City and County of San Francisco, State of California, on the 25th day of August, A.D. 1880, I, Sarah Althea Hill, of the City and County of San Francisco, State of California, age 27 years, do here in the presence of Almighty God, take Senator William Sharon, of the State of Nevada, to be my lawful and wedded husband, and do here acknowledge and declare myself to be the wife of Senator William Sharon of the State of Nevada.

SARAH ALTHEA HILL
August 25th, 1880, San Francisco, Cal.

I agree not to make known the contents of this paper or its existence for two years unless Mr. Sharon himself sees fit to make it known.

In the City and County of San Francisco, State of California, on the 25th day of August, A.D. 1880, I, Senator William Sharon, of the State of Nevada, age 60 years, do here, in the presence of Almighty God, take Sarah Althea Hill, of the City of San Francisco, Cal., to be my lawful and wedded wife, and do here acknowledge myself to be the husband of Sarah Althea Hill.

WM. SHARON, Nevada
Aug. 25, 1880

Sarah swore that Sharon dictated and she copied, on a piece of ordinary note paper, ruled on one side, blank on the other. Sharon said she asked for his autograph and he gave it to her, well toward the top of the paper, on the unruled side. Leaving out the attorneys on both sides the fact is that some persons have thought the document genuine, others have thought it a palpable forgery. When much later Sarah went insane she was found to be schizophrenic. The schizoid type, with its child fantasies, can be very cunning, unscrupulous, and dangerous. Sarah came from a good Missouri family but the evidence said that she was a born flirt and was betrothed to several men at the same time. After she came west she had her eye on men with money, and William Sharon had an incredible pile of it.

Lewis calls him a "pale little man with a large head, ladylike hands and feet, and cold blue eyes."

_____
* A photograph of the document can be found in Lewis and in Kroninger; see Notes.

Loot-hunters ransacked the shack after her death

He was the Bank of California's agent on the Comstock when wild stock speculation closed the mines and brought on financial collapse. Sharon saw his opportunity and in a few years became a millionaire twenty times over, and by 1875 was called King of the Comstock. The King then cultivated his taste in poetry, wines, and mistresses, the latter of which, all young and pretty, he bought in batches. He made no bones about it. In 1875 the legislature made him a senator and in that year his wife died; he then by his own admission put attractive young women on a monthly salary, "requiring only that they make themselves readily available and that they be discreet." Sarah said he gave her a thousand a month, but on second thought reduced it to $500. Sharon said it was $500. He could well afford to have a whole harem, for in one period his income was $500,000 a month. He didn't care for mansions and display, or even for food, but the frail little man, who looked like anything but a lover, did have an appetite for lovely women and freely indulged it. He had fathered a child by a Miss Dietz; to get her out of his life he gave her some money and sent her back East. But to be sued for adultery? In Chicago he said to a group of newsmen, "I'm a pretty old fish and that kind of bait won't catch me."

It almost caught him but he was never to know how close it came: the trial was dragged out for so many years that he died before the last judgment was in. If he had been a judge of women he would have run from Sarah the moment he saw her. In the next chapter we shall see Attorney Stewart, a huge man, call the bluff of notorious killer Sam Brown. Sarah boasted in court that she could hit a four-bit piece nine times out of ten and she pulled a gun from her reticule and pointed it at Stewart, who was examining her. When he asked if she wanted to shoot him she said, "I am not going to shoot you just now unless you would like to be shot and think you deserve it." To say that she had a temper and a sharp tongue would be an understatement; she flabbergasted judge and attorneys again and again. One time she slapped a lawyer's face and told the press he had made love to her. But her most spectacular display of temper in a courtroom occurred after the trial had dragged on for years and her chief attorney was David S. Terry—the same furious giant who, when on the Supreme Court, had

rushed from Sacramento to San Francisco to plunge his knife into a vigilante, and who, in 1859, killed Senator Broderick in a duel.

The court asked her to hand over what she alleged to be a marriage contract. Sarah jumped to her feet and cried in his face, "Judge, are you going to take the responsibility of ordering me to deliver up that marriage contract?" His honor said, "Madam, sit down." He was no judge of women either. Almost screaming the words she said, "You have been paid for this decision! How much did Newlands pay you?" Newlands was a Sharon son-in-law. His honor said, "Mr. Marshal, remove that woman from the courtroom!" Judge Terry, who by that time had married her, was on his feet, sixty-five years old but six foot three and tough and always ready for a fight. He shouted, "No God damn man shall touch my wife!" It took a half-dozen men to push him down to a chair. Scratching and screaming and kicking, Sarah was taken out, with Terry following; and when at the jail door he was told that he could not join his wife he drew a knife and said he guessed he would. Valor moved aside for prudence and he was allowed to pass.

The trial (eighty-three days) was so long and vulgar, so obviously packed with lying witnesses on both sides, and so tiresome that even the principals did not always appear. The Rose (the press was calling her that) was violent and abusive; Sharon was cool and contemptuous: "I offered her five hundred dollars a month to live with me, and she accepted it." He had tried to settle with her for five thousand and be done with her, but she had demanded ten thousand. If Sharon had given it to her it would have been the bargain of his life. To cronies in Chicago he was reported by a newspaper to have said that Sarah had been an adventuress from her twentieth year. "That's enough to say. Oh, I've been ready for her at every turn. What is it Pope says in 'The Wife of Bath'? The mouse that trusts to one poor hole can never be a mouse of any soul." In the next he said he was quoting Thompson: "He saw her charming, but he saw not half the charms her downcast modesty concealed." It's a good guess that there was more than half of Sarah that he didn't see when he was admiring her downcast modesty.

The first judge—there would be a lot of judges —gave the decision to Sarah, and she must have

been elated at the prospect of seven or eight millions, even though her attorneys would get half. The California Supreme Court sustained it, but when it was again appealed there was a new set of judges and they reversed it. Then the long fight was on. An appeal went at last to the United States Supreme Court. At that time appeals were handled by sending single justices over the land. Sent to California was Justice Field, a Californian, between whom and Terry there had been ill will for a long time. It was Field who told her to sit down, had her carried to the jail, sentenced her to three months, and Terry to six. One can imagine what bitterness and what thoughts of vengeance seethed in Terry after that. United States Marshal David Neagle was appointed as Field's bodyguard.

Terry and his wife happened to embark for San Francisco on a train on which Field and his guard were passengers. At Lathrop the passengers left the train to eat breakfast in the station restaurant. Terry saw Field sitting at a table and marched over to him. There are different views of what happened. The evidence seems to support the view that Terry was not armed and that he went over only to slap Field rather lightly across his cheek, probably hoping for a challenge. Neagle was eager to earn his pay. He shot Terry, and as Terry fell he fired a second time, and that was the end of the man who in his most impassioned voice had thundered in the courtroom: "She goes from this courtroom either vindicated as an honest and virtuous wife, or branded as an adventuress, a blackmailer, a perjurer, and a harlot!" Sarah ran to him and flinging herself down poured forth hysterical expressions of love and grief. There can be little doubt that this violent man with his ardent sympathies for the South and his hatred of the North was the only love of her life.

The rest of the story is Sarah's and it is a very sad story. We don't know if she saw in the press the sensational accounts of Terry's death. Back East the *Nation* said, "Somebody ought to have killed Terry a quarter of a century ago." The Stockton *Mail* turned to satire: "Terry had a bowie knife all the way from a foot to eighteen inches long, with the blood of his last victim still upon the blade. He stood picking his teeth with it when the Rev. Mr. Nagle [sic], a distinguished prelate from Arizona entered the room upon the arm of Stephen J. Field, a sacred personage descended from Heaven to execute the will of God upon earth." But she had no need of such stimuli to put her out of her mind, for her childhood, whatever it was, had shaped her to be a tragic one. Long interested in fortune-tellers and spiritualists she turned again to them, and at times believed that she was in communication with her dead husband. She reported the theft of diamonds and lace which she had never owned. And her behavior as she walked in the rain of San Francisco streets, a forlorn figure in filthy dripping garments, or tried in one way or another to pay for a room where she could sleep, became so clearly insane that a newspaper foretold that she "will end her days in an asylum" and a doctor who examined her said she was insane, and imagined herself attended by a host of spirits. Shabby, unkempt, hair streaked with gray, she was a pathetic caricature of the woman who had briefly bewitched the King of the Comstock and aroused the intense devotion of David S. Terry.

Reporters for a while trailed the woman who

stood unprotected against the pouring rain. . . . dripping and soggy with wet. . . . covered with mud . . . Mrs. Sarah Althea Hill Terry"; and the acidulous Ambrose Bierce wrote: "The male Californian—idolater of sex and proud of abasement at the feet of his own female—has now a fine example of the results entailed by his unnatural worship. Mrs. Terry, traipsing the streets, uncommonly civic, problematically harmless but indubitably daft, is all his own work, and he ought to be proud of her."

It was not that simple but the satirist is always a sentimentalist under it. At last she had to be locked up. At a court appearance, said the *Chronicle*, Mammy Pleasant, a Negress madam who, in the opinion of some, gave Sarah the idea for the marriage contract, was the only person present who "showed any sympathy for the demented creature." She still had enough wit to cross-examine a doctor who declared her insane, and to retire him "in some confusion." She was committed to the asylum at Stockton and was there for nearly forty-five years, outliving every other principal in the sensational case.

In 1936, when she was eighty-six, Oscar Lewis talked to her. "The neat, white-haired little figure sat in a rocking-chair . . . regarding her visitors with bright, shrewd eyes." When Terry was mentioned she chuckled and said, "He was one of my husbands. He was a big man. He could hardly get through that door. Do you know him?" What about Senator Sharon? "He was a rich man. Is he dead?" Did the Senator recite poetry to her? Indeed, he did—and she recalled six words from "Maid of Athens." She said wistfully that she would like to go to San Francisco. Seven months later she was dead.

Klondike Kate   During the Klondike rush many women in the eastern states hoped to find a rich husband and moved toward their goal in a number of ways. One woman sent her photograph to a Dawson paper, with an advertisement, and a note saying that the first man who called was to pay for the advertisement. The newspaper was swamped by men eager to pay for the advertisement and get the woman's address, but not all women who headed for Alaska were as successful as Irish Kate Mulrooney, who is said to have snared both a fortune and a French count. Lucia says that one shipload of husband-hunters, mostly widows, were wrecked rounding Cape Horn, and that of the five hundred only one ever reached Alaska. Another boat had fifty prostitutes. Most of the harlots strove to disguise themselves with such names as Nellie the Pig, Diamond Lil, Oregon Mare, and Gumboat Sue, but Klondike Kate was not a name that was intended to separate a girl from her past and her people. Kathleen Rockwell loved publicity too much to hide any part of herself from anyone.

Mr. Lucia frankly calls his biography the life and the legend of this woman, and it is probably more legend than life. On top of that is his tendency to glamorize her. On page 18 she is "willowy"; on 19, as a dancer, she "floated about the stage, as gracefully as a white-breasted gull riding the coastal winds"; on 30 she moves with the "swiftness and agility of an antelope"; on 34 she is "spinning and leaping gracefully over the lawns"; on 91 she is a "slip of a girl in pink tights"; on 112 "Pantages was rankled by seeing this slip of a girl earning fantastic sums"; and on 246 she is, at age 53 "still slender, graceful." The reader will look at her photograph and decide for himself. For her biographer "She turned men from suicide, inspired them to 'mush on and smile,' reminded them to write to their folks, and sent them home to sweethearts and

wives. . . . She was all things to all men: sweet-heart, wife, mother, sister, mistress, good friend. Through misery, unhappiness and disappointment shone the image of this sprightly girl." He says that among the titles lavished on her were "Queen of the Klondike," the "Belle of Dawson," "Flower of the North," "Toast of the Yukon."

As a character she seems to have been much like Calamity Jane, though not so masculine oriented: both were extroverted, forthright, compassionate, hell-raising, and during a part of their lives alcoholic; and neither when presenting her image to the public bothered to distinguish between fact and fiction. Kate surely knew her age and she must have known where she was born but it was not her way to tell the same story twice. She "put flowers in her hair, flirted with each Dapper Dan that came her way, and became the talk of the town. She didn't mind at all, for Kitty gloried in being the center of attention." So did all the adventuresses; they worked hard to find the center and stay there. And they worked hard to shake the men loose from their money, even while playing mother, wife, and sister to them, with mistress thrown in for good measure.

Mr. Lucia presents her in a typical shakedown. A young man was broken-hearted because his girl had given him the gate, and when he said to Kitty, "She looked like you," Kitty knew she had him hooked. In from the diggin's with a poke of dust, the man ordered "bottle after bottle of champagne" at fifteen dollars a bottle. We are told that Kitty "restricted her drinking"—she had to, if she was going to clean him out; and then, gravely, that her "heart went out to him." The reader can judge how her heart went out to him: she took him for five hundred dollars that night (or that is the story), as her percentage of it; we are not told how much the house got. But she was decent enough at daylight to call a man and ask him to "take care of the boy"; and after looking at the strain in her face she "bawled herself to sleep." Possibly that is the story she told, but if it is, it is only another of her nimble lies. Her full character came out, as we shall see in a moment, in her affair with Johnny Matson.

In Dawson Kate became enamored of a scoundrel named Alexander Pantages, and he preyed on her, as all the more generous free-loves were preyed on by their lovers. She set him up in business and lavished on him the money she took from all the lonely men that "her heart went out to." When Alex became restless and wanted to go south she agreed to leave the north, little dreaming what she was heading for. She is another case of the ruthless woman who stripped gullible men, and in turn was stripped by the man she loved. When she left Dawson she is said to have had $100,000 in gold coins and $50,000 in jewelry (this may be only her story), which was a substantial fortune in those years. Of whatever she had she seems to have given a large part of it to Pantages, the Greek, who was eager to establish a vaudeville chain, which he eventually did. The woman to whom the house boss had said, "How much does this gay blade have on him? Is he loaded for bear, this sourdough from Eldorado? Play your cards well, Katie girl, so we can get it all tonight"—this slip in pink tights found herself, when she came to her senses, outside with the door locked. She had bought him silk shirts, expensive cigars, had paid his board bills, had been his night-love—and when she found that he was married and was through with her she took it hard.

All her life she cultivated the image of herself as a pure woman: "Everybody knew that Alex Pantages and I lived together, but it was the custom of the country and nobody thought anything of it. . . . Almost every girl in the north had her man. It was a rough, hard country for women at best, and if a girl could get a good, comfortable cabin for the winter, she was lucky." Especially if a lover were tossed into the bargain. Now and then a woman did offer to share a man's bed if he would give her shelter for the winter, but Kate's words are typical of her lifelong rationalizing: after all, she furnished the cabin and the victuals, and Alex, professing undying love, simply moved in. After he moved out she sued him for breach of promise; he refused to answer most of the questions, saying over and over again, "I was never engaged to her, I never intended to marry her." Her biographer says, "Kitty just couldn't comprehend this. It was beyond belief that Alex would do this to her. A door had been slammed in her face, leaving her out in the cold and not understanding why."

That is probably what she said, for again it is typical of her dishonest attempts to cover her tracks. A woman who could strip hard-working

miners, in from loneliness and the cold, of their last cent certainly understood why. When the Seattle *Daily Times* headlined, USES HER MONEY, THEN JILTS THE GIRL, she got a little more than was coming to her. When Robert Marshall has a character say in his *Arctic Village* that Pantages treated her "like I wouldn't treat a bitch dog," she got a lot more. Strange it is, as Ambrose Bierce pointed out, that men who are eager themselves to debauch women rise in righteous wrath against the other human bulls. Kate, Lucia tells us, "was swept by waves of tears, followed by numbness and pain, and then by acid bitterness," and began to drink a lot and didn't give a damn what happened to her. It must indeed have been a blow to her vanity to read in the *Times* the Pantages statement that he didn't even know her.

But in spite of drink and tears she would live a long time yet and she would get her vengeance. In 1929, after the Greek had become the "multimillionaire king of vaudeville," he was hauled into court on a charge of raping a seventeen-year-old dancer. Lucia says the "scandal ruined him" but it didn't take from him his millions or his health. Kate was there in the courtroom, probably staring at him every minute, and, for the reporters, "dabbing at her tearless eyes." She told the whole wide world that she still loved him but she didn't mean a word of it. As she looked at him she must have been thinking, You cheap son of a bitch, I gave you most of what I had when you had nothing up in Alaska, but now that you have millions you don't know me, do you? Though she stared at him he did not look at her, and she must have known at last that he would not have cared if she had died the day he left Dawson. But she had her vengeance. Reporters, eager to elevate some persons to godhood, are just as eager to strike others down, and they struck hard at Pantages, without knowing whether he was guilty or not.

Many years before the Pantages trial Kate had fallen in love with the Bend area of Oregon and had settled there, an eccentric with a fading reputation, who at thirty-nine married a farmhand half her age. It probably was a desperate effort to seize and hold her vanishing youth. She pretended for a while to farm with her giddy young husband but on one occasion he surprised her coming down a hotel stairway on the arm of a man, and on another occasion she dressed in her silks and ruffles, red high heels, huge hat, and a lot of jewelry, and went on the town, where she picked a man clean. Though she was to become known as Aunt Kate of Bend she was not forgetting her old habits or letting the world forget her, for she made a pest of herself at the newspaper office trying to keep her name in print.

As Mr. Lucia tells the story, when Kate was a toast in the Klondike there was a small Swede named John Matson, who came in one day from his claim, far out in the mountains, and watched Kate dance, and forced himself at last to tell her in his broken English that she was the most beautiful woman he had ever seen. Kate forgot him but he did not forget her, and when more than thirty years later he saw her name and address in a newspaper he wrote to her and for two years they corresponded. Mr. Lucia is very probably right in thinking that there rose before Kate a vision of all the gold Johnny must have accumulated during that long time, for she became impatient if his replies did not come promptly, and, having got rid of her youngster-husband, she was eager to marry John. And at last down he came out of the far north, and again the Belle of the Klondike made "international news." Johnny was small, his English was poor, he had dentures, he had been shrunk and crippled by the many long, cold winters, and some of the reporters treated him as if he were a Simple Simon; but he had the woman he had wanted to have and he took her to Dawson, where they received a great welcome, and Kate shed a few tears in the old theater where she had dazzled the miners. If Johnny had the notion that his bride would mush along with him forty or fifty miles back into the mountains, to his cabin, which must have been a crude one-room thing with an earth floor, he was to get the first of several hard lessons in the nature of women. Kate was soon back in Seattle, and Johnny was back in his cabin.

She told reporters that her husband would build a home for her while she settled her own affairs, but it was the end of a marriage that had been no marriage, though for a while there would be infrequent letters between them. For a while Kate would go to Dawson each summer and pick up a poke of gold. But the legend would go on growing: "As long as the Arctic winds howl across the tundra and yarns about the old days are spun, . . . the story of how Johnny Matson loved and

Kate and her Yonny

won his queen will be told and retold." Kate didn't care a hoot about the reticent little hermit who for more than fifty years would live in his shack and dig for gold, but she did care with all her being for the headlines and the tales about her in the Sunday supplements. "A man who will wait that long for a woman," she said, "well, you don't find them every day." She was right that time: most men are not that simple, and most women are not that brazen.

In her last years Kate would have her story told and retold; she would appear in a Louella Parsons column; she would teach Jinx Falkenburg how to roll a cigarette; and she would be summoned to Hollywood to act as technical adviser on the story of her life. The time would come when there would be no more letters from Johnny. He was found dead one day on the claim where he had spent his whole adult life, and he was buried there. Kate asked one of his friends if there were things of value in the cabin or roundabout, and was told that "A man in the woods can get along with a tin can." Two years later she was married again, giving her age as sixty-eight, though she was seventy-one. The legend kept her in the news almost to that February 1, 1957, when she died in her sleep at the age of eighty. *Time* and *Newsweek* took note of her passing.

"What a gal!" an old-timer cried, on hearing of her death. We all have to concede her that much.

CHAPTER 24

Brave Men and Cowards

Shaw tells it this way in *El Dorado*:

A solid ounce ball from an Indian rifle struck him in the ankle, and tore through the flesh and bone. It was a terrible wound, even had there been a surgeon to amputate and dress it. What then must it have been when no medical or surgical assistance was to be had! But the leg must be amputated or the man would die. It was done. Taking a beaver knife, the edge was hacked into a saw while another was sharpened to its keenest edge, and with these rude implements Sublette amputated his own leg. The plates of beaver traps were heated redhot and applied to the raw and bleeding stump, charring the veins and arteries and stopping the flow of blood.

This famous mountain man's experience was not a rare one. Richardson tells of a trapper whose leg was shattered by gunfire.

Amputation was necessary; but no surgeon within hundreds of miles. He whetted one edge of his hunting knife to its utmost sharpness; filed the other into a saw; and with his own hand, cut the flesh, sawed the bone, and seared the veins with a red-hot iron. He still lives in California, walking upon a wooden leg.

In her fourth letter Dame Shirley tells her sister about a young man whom she called W, who with hard work had accumulated four thousand dollars and was planning to go home; but while working on a hillside he had a leg "crushed . . . in the most shocking manner." While a doctor was striving to save the leg W was attacked by typhoid, and then by what she thought was erysipelas; and "the latter disease settling in the fractured leg, rendered a cure utterly hopeless. His sufferings have been of the most intense description." Her husband, Dr. Clappe, over the objections of others said an amputation was necessary or W would die. Others doctors said he would die anyway and that it would be senseless cruelty to saw the leg off. W was now being fed milk with a teaspoon;

his four thousand dollars was almost spent. "Poor fellow! the philosophy and cheerful resignation, with which he has endured his terrible martyrdom, is beautiful to behold."

There were many as cowardly as these were brave. Says Richardson, "Frequently one drew his revolver upon some peaceful citizen, compelling him to fall upon his knees, submit to every vile epithet and beg piteously for his life." Haskins tells of a bully whose sport was to accost men with drawn knife or pistol and demand to know if they had said their prayers. He would force them to kneel and pray. When one time he drew his knife on a man at a bar and put the question the man said he had prayed enough to last him a lifetime, and drawing his pistol shot the bully through the head. "No inquest, as the coroner did not think it was necessary." Typical of such bullies was one in Austin named Vance, in 1868, who told an Irishman named Carberry to get his gun. When Carberry came down the street with his gun, Vance, in a sheltered position, began to fire and kept firing. Carberry kept coming, and when he thought he was close enough to hit his enemy he coolly rested his gun across his left arm and fired. Vance, like all cowards, lost his nerve when he perceived that the one he had challenged would not.

Because she galloped over the hills with her skirts up and her red flannel drawers flapping in the wind she was called Madame Featherlegs, whose house of joy was in the Black Hills. One of her male friends was Dangerous Dick Davis, who killed her one night for the money she had saved. He is said to have confessed the murder years later when the vigilantes had a rope around his neck. Much like him was a Sam Porter in a Colorado camp; well filled with whiskey he proclaimed to the world his eagerness to kill the first man he met. He did, and was lynched that day. Quiett tells of a Tom Ryan who, with knife in one hand and a stone in the other, terrorized a camp and chased the other bully out. Tom had things all his way until an enraged miner hurled a red-hot stove lid against his face. Tom's humiliation was so deep that he fled. Marryat tells of a similar case—of a "fierce brute of a man" known as Cut-throat Jack, who "was more feared in the mines than I should have supposed any man to have been." One evening when some miners were bringing a pile of quartz to a white heat Jack strolled up and began to boast. A miner knocked him down. The bully "unfortunately fell on the red-hot quartz, and the sensation was so new to him that . . . he drew neither pistol nor knife, but was instantly lost to sight in the surrounding gloom, and never swaggered into our camp again from that night forth."

Though the boastful Wyatt Earp has taken most of the cheers, among the frontier marshals and sheriffs there were some remarkably brave men.* T. G. Smith, in Nevada, was only one of many sheriffs shot by criminals. "At midnight he became conscious that he must die and calmly made his preparations. His business affairs were arranged; he made requests as to his burial, and then called his weeping family and friends around his bedside, and bade them an eternal farewell, and with quiet fortitude, such as distinguished him in life, awaited death." Tom, John, Bill, and Jim were four young bullies who liked to terrorize the camps. In one of them Joe Carson was marshal. When he asked them to check their guns while in camp they poured upon him such profane and insulting language that he took on all four. In the duel that followed, Otero says about forty shots were fired, nine of which hit the marshal. Bill was mortally wounded, and Jim too severely to be able to drag himself away; Tom and John, both wounded, escaped but were overtaken by vigilantes and hanged. With nine bullet wounds the marshal crawled on hands and knees toward his house, hoping, but in vain, to say farewell to his wife and children before he died.

Dr. Knower, a Forty-niner, tells of a Texas ranger named Jack Hays, a veteran of the Mexican War. A "Mexican officer, splendidly equipped, came forward on horseback, and challenged any American to meet him in single com-

bat between the two forces. Jack Hays volunteered to go, and he killed him." The Mexican's horse, gold watch, and other personal articles he took but afterward sent them to the widow. Hays was elected sheriff of San Francisco County, California, 1850-53.

McConnell tells of an Irishman who, when half-sotted, was lured by gamblers to their table. The Irishman had a full poke of dust; on realizing that he was steadily losing it he placed his hand over the pot. One of the gamblers shot him. The Irishman was hit so hard that he was knocked from his chair but he was tough: drawing his gun he rose to an elbow and killed the three gamblers before they could get away from the table and out of the room. He then fell back and died.

An oft-told story is that of the meeting between two men named Fairfax and Lee. During an exchange of sharp words Lee drew his sword cane and drove the blade clear through Fairfax; and though Fairfax then drew a derringer and thrust it almost against Lee's head, he changed his mind: "You have a wife and children, so I will spare your life."

Somewhat similar is the story of one named Dibble, in Nevada City, California. When he called a man named Lundy a liar he was cussed out for it and challenged to a duel. At the signal Lundy fired. Dibble opened his shirt to show that he had been hit, and in a calm voice said, "You fired too soon." The story says that he refused all aid and walked away to die.

Of cowards there is no end of the tales. Colton, a fine observer, left for us the story of two miners, whom he knew, who on their way to Stockton took a nap under a tree. A Hessian and an Irishman crept up in the dark and killed them for the gold they had. While hastening south they met three deserters from the Pacific Squadron, who were glad to join them. The party of five came to the house of a prosperous rancher named Reade, who, like all ranchers of that time, offered them the best he had in food and lodging. The next morning the five men paused in their journey and decided to go back and rob Reade. They not only robbed their generous host; in cold blood they killed him, and his wife and their three children, a female relative and her four children, and

---

* See also the chapter, "The Badmen."

two Indian servants. After ransacking the premises they headed south again and had reached a cove in the Santa Barbara area when overtaken by a posse of outraged citizens. The posse were few in number and poorly armed but they fought it out. One of the killers was shot down; a second leapt into the ocean and drowned. The other three were brought before a jury of twelve men and before the sun set were in their graves. Today, pickets would march in their behalf, hysterics would clamor for their freedom, experts would declare them sick, lawyers would drag their case out for six years, and an author with a repressed sense of guilt would write a book about them.

McConnell and others have pointed to the fact that there is a breed of subhuman creatures who want to be chiefs—all the way from the Hitlers to the Eldorado Johnnies. They want power over men, and killings and notches on their guns. Such chiefs were in all the camps and practically all of them were brutal cowards. McConnell tells of a packtrain that entered Placerville in the Boise Basin in the summer of 1864. The owner, after passing the Magnolia saloon, paused at a well to draw a pail of water. Three gamblers were loafing at the saloon door, waiting for a victim. One of them is alleged to have said to the others, "Watch me fix that feller," and to have strolled over to the well. Taking the pail of water he hurled the contents into the packer's face, and while the astonished and confused man was trying to see, the gambler drew his gun and shot him dead. He ran no risk, for he had two companions to swear that the packer drew on him and he then shot in self-defense. Another coward in Placerville was an ex-colonel in the Confederate armies, who, eager to put another notch on his gun, chose a butcher named Brown. Going up behind Brown as the man bent over his books the bully shot him three times; but Brown was still able to seize his gun and pursue his assailant and shoot him in the neck. Both men survived their wounds. Brown was tried for simple assault and given ninety days. It was such a mockery of justice that led to the vigilance committees.

Stoll, an attorney in the Eagle City camp of northern Idaho, tells this story. When the tent city was snowbound he and some friends went one evening to a gambling hall to watch a game of stud poker. "Now gamblers are, as a rule,

superstitious, believing in mysterious and secret influences at work behind their cards. One callous-hearted gamester, after a consistent run of heavy losses, began looking for his hoodoo. He quickly found it in the person of Oregon John, a worthless underworld follower and camp hanger-on." John had one foot on a rung of the gambler's chair. "The gambler, his hoodoo discovered, rose deliberately from his seat and shot Oregon John through the heart. There was no demonstration. Someone carted the body out without ceremony and buried it." As editor for a while in Candelaria, Drury saw Blue Dick, a gunfighter, and Shagnasty Joe, the camp bully. Blue Dick one day was persuaded to play dead on a billiard table, with a white sheet over him; and when Joe came in he was told in hushed voices that Dick was dead. Joe at once began to give his opinion of Dick—an ornery coyote, a quitter, a braggart, and a contemptible coward. Blue Dick then leapt from the table, gun in hand, but he got entangled in the sheet, and by the time he was free of it Shagnasty Joe was vanishing through the doorway, his eyes on the road to Columbus. Dick was after him, "shooting and shouting like a demon possessed. Fear gave Shagnasty wings and he was soon out of sight."

Not all the killers were cowards. There was Spillman, the first man hanged by the Montana vigilantes, who for Granville Stuart was "a large, fine looking man." He "made no defense" and "walked to his death with a step as firm and countenance as unchanged as if he had been the merest spectator. . . . It was the firmness of a brave man." Borthwick tells of a Mexican thief sentenced to be hanged: "On being told of the decision of the jury, and that he was to be hung the next day, he received the information as a piece of news which no way concerned him, merely shrugging his shoulders and saying, 'sta bueno,' in the tone of utter indifference in which the Mexicans generally use the expression." When taken to be hanged "he walked along with as much nonchalance as any of the crowd, and when told at the place of execution that he might say whatever he had to say, he gracefully took off his hat, and blowing a farewell whiff of smoke through his nostrils he threw away the cigarita he had been smoking, and, addressing the crowd, he asked forgiveness for the numerous acts of villainy to which he had already confessed, and politely took leave of the world with 'Adios, ca-balleros.' "

Some of the bravest men of the early West were newspaper editors, among whom seems to have been James S. Reynolds. In his *Statesman* he wrote, August 15, 1865: "A malicious individual by the name of A. Heed made a most cowardly assault upon the editor of this paper last Sunday evening, under the following circumstances: It had been reported to us that he had . . . read on the stump garbled extracts out of the *Statesman* that were published there from other papers, and that he represented them to be the language of the *Statesman*, and added that the *Statesman* favored Negro suffrage. In Saturday's issue we made the statement that if any person had done so he was guilty of the meanest kind of lying. Heed came to us on the street as before stated, and in a short conversation admitted that he had done as reported, and demanded that we retract what we had said about it; to which we replied we never would. He then aimed a heavy blow with a sort of bludgeon cane, which failing to take effect he immediately presented a cocked derringer and still demanded the retraction, and threatened to 'send us to hell in a minute.' He didn't get the retraction, and he didn't 'send us to hell,' though he had a good chance to shoot a dozen times without danger to himself, as we were entirely unarmed." Donaldson, who knew Reynolds well, says that Judge Heed thrust a gun against Reynolds, and Reynolds shouted, "Shoot, you coward, shoot!" and that a moment later Reynolds seized the gun and "beat the judge over the head until he fainted, kicked him into the street." Donaldson adds this strange comment: "His nerve was so superb that not a bystander interfered."

Sometimes it is not easy to tell whether a man was brave or was a coward trying to be. There is John Dennis, known in the books as Austin Johnny, because he came from that camp, or as Eldorado Johnny. As Gillis tells the story, the bully in the Virginia City area was Farner (Langford) Peel, a Harvard man, handsome and polite, with six notches on his gun, all from killings in self-defense. His reputation as a gunfighter "had spread over the whole country," at last reaching the unbelieving ears of Johnny in eastern Nevada. Johnny seems to have had the notion that the world was not big enough for two gunmen of his stature, and so he strapped on his arsenal and

headed west. On the Comstock he was told that if he loved life he ought to go back home, but Johnny said with a fine swagger that one of them must die. He was then told that if he was tired of life he should get his boots polished. This Johnny did, and he had his whiskers trimmed and his hair curled. Some books say it was in Lynch's saloon but Gillis says it was in Ben Irwin's that Johnny found Peel and told him it wasn't fitting and proper for Nevada to have two killers with such formidable reputations. Peel, every inch a Harvard man, said softly that he had no quarrel with Johnny and would rather buy him a drink than shoot him. It is nice to think that Johnny said that fancy college talk cut no ice with him and that Harvard manners were no defense against his kind of weapons. Peel said at last that if Johnny insisted on a little target practice they ought to go outside, where innocent bystanders would not be hurt. Imagine Johnny in that moment! He had expected to settle it right where he stood: he must have stared at Peel a few moments, when, with the most gallant courage or the most abysmal halfwittedness, he backed toward the door, with Peel following. Johnny, as Gillis tells it, backed all the way across C Street, and there, very unlike a deadshot, took a kneeling or squatting position, and was aiming across his left arm when Peel appeared in the doorway and shot Johnny "squarely between the eyes." Peel is alleged to have said afterward that Johnny seemed not to know how to handle a gun and that shooting him was like shooting a child.

Virginia City must have agreed with him, for lo and behold, Johnny was covered with glory in death. The camp put him in an elegantly upholstered casket and the finest burial robes, and as he lay in state from Friday till Sunday seven bartenders were kept busy refreshing the multitude who came to look upon the simple-witted face from Austin. He was escorted to his grave by the largest funeral procession known in Nevada up to that day, and probably since.

The doors of every saloon in Virginia City, draped in mourning, were closed, and an air of sadness seemed to pervade the whole community. . . . A stranger viewing this great demonstration would have thought that some great statesman or benefactor of his race was being honored and mourned by a grief-stricken community. After Johnny's body was laid in the grave, the people marched back to the city to the tune of "When Johnny Comes Marching Home Again" . . .

Whether the one known in books as Jack Mc-Call, murderer of Wild Bill Hickok, was another Johnny with an itch to be a killer-chief; or was hired to kill the great civilizer by criminals who didn't want to be civilized; or killed in spite because he had lost a poke of gold dust, is not known. Nor is it known whether he was a mentally subnormal runt. It is known that he was stupid—so stupid that he actually used an old-model six-shooter in which five of the cartridges

Bob Ford, killer of Jesse James

were defective. If some kind of evil genius had not put under the hammer the only cartridge that would fire Jack would have been carted off to the graveyard that day. He certainly was a coward, for he slipped up behind Hickok as he sat at a card game, shouted, "Take that, damn you!" or similar words, and shot him in the back of the head. It is sometimes said that he was firm and composed when, after a fair trial, he faced hanging, but that is not unusual in criminals of low intelligence and inordinate vanity, with thousands of spectators before them. This creature and Johnny and Billy the Kid belong to that breed of killers who are small, unattractive, and stupid. Much like them, but handsomer, is Bob Ford, who slipped up behind Jesse James and shot him in the back of the head. Ford then went to Creede where he opened a saloon and proclaimed himself the camp's boss, in the manner of Soapy Smith. He was such a repulsive criminal that the camp was soon sick of him, and a committee of vigilantes gave him an hour to get out of the area. When he asked for time to settle his affairs he was given forty-eight hours, which was time enough for another bully named Ed O. Kelly (most writers spell it "O'Kelly") to slip into the saloon and shoot Ford in the back. The drinking party at the boothill, celebrating the end of Ford, is said to have lasted three days and nights.

Some of the killings in the early West are still the subject of acrimonious controversy, and in Idaho Territory none more than the murder of Sheriff Pinkham by Ferd Patterson. Bancroft says Patterson was "a gambler, a secessionist and democrat. . . . Pinkham a union man and republican." There are several versions of Patterson's earlier life. The one that seems to be most widely accepted says that when he fancied himself insulted by the captain of a ship he caught him unawares in a hotel lobby and killed him; and that when he became enraged at the woman he traveled with he seized her hair and slashed at it and cut away a piece of her scalp. In both cases he was arrested but soon freed. McConnell says few men were willing to stand up against Patterson, who is reported to have told his cronies that he would kill Pinkham if they would swear that he did it in self-defense.

The killing occurred at a warm springs resort a mile south of Idaho City, but what happened there depends on what source you read. After the

Sumner Pinkham

Ferd Patterson

killing Patterson mounted a horse and fled, but Pinkham's deputy, Rube Robbins, overtook him and brought him back. The remainder of the story is delightfully ironic.

Because hundreds of men were ready to lynch Patterson, a vigilance committee decided to take him from the jail at two in the morning and hang him. The acting sheriff, forewarned, manned the jail "with practically all the thugs and tinhorn gamblers in the city, and was prepared to defend his prisoner. Thus a comical side was presented." It was, indeed: the sheriff, supported chiefly by criminals, faced McConnell and his vigilantes, and if the sheriff had not backed down there would have been appalling slaughter. The heart of the thing was actually not the murder of Pinkham but the bad blood between those for the Union and those for the South. As Donaldson tells it, "Patterson was indicted and had his trial, but capable counsel, aided by powerful influences, brought about an acquittal. The feeling against the murderer was very bitter." So bitter that Patterson fled the country. One

version of the story says he went to Walla Walla and was there killed in a hotel corridor by a man named Donohue. Another version says he was shot in a barbershop in Portland by an officer who had seen his brutal manhandling of his woman. There are still other versions, and in Idaho today there are persons whose blood pressure rises when they talk about this killing.

Who the bravest man was in the early West, who the most arrant coward, no one knows; it would be foolish to speculate on it. We are presenting here instances of both, and will conclude with a couple of officers who were among the bravest, and with a coward whom some writers have chosen as the most odious and contemptible of his tribe.

Thomas J. Smith, of Irish parents, stood an inch under six feet and weighed about 170. He did not drink, smoke, or gamble, and when he came to Abilene, Kansas, and applied for the job of town marshal the city fathers did not think he looked the part. Their own boys had failed as fast as they had been appointed; they then asked the chief of police in St. Louis to send over a couple of tough ones. These were soon on a train, headed back where they came from. Desperate, they then turned to Smith. Did he really think he could handle the bullies? He thought he could. One who knew him has said, "Silently he moved off, and I watched him with misgivings disappear down town."

Almost at once Smith encountered Big Hank, a desperado who had sworn that no man would ever disarm him. When Smith asked him to hand his gun over the response was "profanity and abuse. Instantly he sprang forward and landed a terrible blow," which knocked the bully out. Smith had similar experiences with other bullies; in a saloon after he had pistol-whipped one of them and told him to get out of town, the saloon-keeper said it was the most amazing piece of nerve he had ever seen. But Smith lasted as marshal only from May to November, 1870.

His last fight we base on Drago. Two neighbors on Chapman Creek named McConnell and Shea had a quarrel, in which Shea was killed. When a warrant for McConnell was taken out by the sheriff, McConnell obtained refuge in a dugout and drove the sheriff off. When the sheriff then found himself very busy with other matters, his deputy, named McDonald, asked Smith to go with him. A man named Miles was chopping wood; McConnell, in the dugout, warned Smith not to approach, but he went in. Miles then tried to shoot McDonald but his gun missed, and as McDonald retreated, Miles kept snapping his gun at him. Inside the dugout there was a shot. McDonald fled. When a posse arrived from town it found Smith's body lying in the yard with its head cut off, except the skin on the back of the neck. Miles and McConnell had fled. A man named Stambaugh, who reached the scene ahead of the posse, says there had been a dreadful fight between Smith, shot in the chest, and McConnell, shot in one hand. Smith had been trying to handcuff McConnell when Miles came up from behind and swung against Smith's neck with an ax. "Frontier Abilene never again witnessed such an outpouring of grief." The brave marshal was practically forgotten until the twentieth century, when a bronze plaque was placed above him.

Gard tells the story of a sheriff named Owens, who with his Winchester went to the house of a criminal named Blevens to arrest him. Glancing

Thomas Smith

through a window he saw three other men, but called to Blevens to come out. A moment later he was covered by guns from two doors. The shooting then started. Blevens missed, and in the same instant the sheriff gave him a "mortal wound." Another Blevens fired from the other door but missed the sheriff and hit a horse tied out front. The sheriff shot him in the shoulder and he staggered back inside. Then a man named Roberts dashed out with a six-shooter but "Owens beat him to the draw and sent him running back with a fatal shot." The fourth male was another Blevens, a lad of sixteen, who, breaking away from his mother's grasp, seized a gun from a dying hand and rushed outside, "just in time to get a bullet in his heart and stagger back into his mother's arms." The coroner's jury said, "Too much credit cannot be given to sheriff Owens. It required more than ordinary courage for a man to go single-handed to a house where it was known that there were four desperate men inside and to demand the surrender of one of them." For courage and cool shooting there is nothing in the record of all the Wild Bills and Earps put together to match it.

The coward, Lyman says, came from California to Nevada with sixteen notches on his gun; he had killed fifteen men, says Drury, who knew him, and was "a great swaggerer"; and DeQuille, who also knew him, says he entered a saloon "one side at a time." Sam Brown, thickset, bull-voiced, sandy-whiskered, has been cited by some authors as the archetype of the bully—"simply a brutal monster," says Emrich, "with every instinct brutish, utterly mean, treacherous, and an arrant coward." He was, says Angel in his history of Nevada, "an arrant coward, and did his killing mainly when he had been stimulated to courage by strong drink." So far as is known he never picked a fight with a known gunman but chose inoffensive and usually unarmed men, such as a miner named McKenzie, in Virginia City. This murder has been described by many writers. According to Emrich, Brown "ran a knife into his victim, and then turned it around, completely cutting the heart out"; he then "wiped his bloody knife and lay down on a billiard table and went to sleep." At that he was no more cowardly than the men who watched him kill and then allowed him to lie in peace. Haskins knew him when he was terrorizing California camps, "a noted desperado, a fine looking man, with long curly hair of a sandy color; he was rather of a good natured disposition when perfectly sober, but a demon when drinking." In Tuolumne a man called Tex had said that the area would be a better place if Sam Brown were out of it. Hearing of this, Brown accosted Tex and demanded to know if he had made the statement. Tex said, "I did and I meant it too and I am willing to make it to your face, if 'tis any accommodation to you; and now what are you going to do about it?" As Haskins tells it, Sam took his hand off his pistol butt, muttered something unintelligible, and walked away. That's about par for the bullies.

Jackson says that in Coloma Brown was a braggart and a thief, and that after he "wantonly killed a bystander in a saloon" he was saved from lynching but was given three years. From there

he went to the Comstock, and in Jackson's version celebrated his arrival by shooting a defenseless drunk in a saloon. The owner of the place then brought forth a short-barrel shotgun "and blew Mr. Brown out of Washoe County for good." That sounds fine but Jackson was mixed in his chronology. Angel, on whom is based the murder of McKenzie, says that as early as February, 1859, Brown killed a man named Bilboa in Carson City.*

How Sam came to his end is a tale with many versions. Basing his account on the Davis and Angel histories of Nevada, on Stewart's *Reminiscences,* on the Sacramento *Union,* on Van Sickle's statement, and on still other sources, Lyman tells the story in his *Comstock Lode;* and his may be as nearly correct as any other.

One of Sam's gang had killed an inoffensive man and was on trial; and William Stewart, huge and loud and able to blow almost any man down, who would be not only a leading mining engineer but a United States senator, had been engaged to give the prosecuting attorney a hand. When Stewart announced that the killer would be hanged, something was set off in the dark murderous mind of Sam. "So he braided his sidewhiskers, tied them under his chin to protect his windpipe," hung around his waist his finest man-killing equipment, and headed for the courtroom. Some writers have been tempted beyond their strength when they described the scene that followed. It is possible that when Sam entered the courtroom he sent chills down the spines of judge and jury, but it may not be true that some of the jurors were so terrified that they leapt through windows. In this more romantic version Stewart drew his guns, and we can imagine his stentorian voice when he covered Sam and roared, "Get your hands up!" If we follow this version he then said to the marshal, "Disarm him!" and then to Sam, "Take the witness stand!" Sam, with his hands up, and the bravado rapidly running out of him, took the stand; and possibly in some such words as these Stewart addressed him: "I hear you've been bragging all over Sun Mountain, and the hills and the desert roundabout, that you would march in here and set a murderer free; and you would make the judge like it, and the jury like it, and me. ME!" Stewart thundered. Then

---

* Angel's history gives many pages of murders committed in Nevada camps.

stepping close to Sam he fixed on him his most ferocious scowl and in low tones of pure menace said, "Now get this firmly fixed in that cabbage you use for a head—if you make one wrong move in this courtroom, or tell one lie, I'll blow your small sample of brains all the way from here to Gold Hill!"

This version says Stewart kept his guns on Sam, to fix his small mind on the truth; and that Sam admitted that his pal had a villain's reputation. When a defense lawyer rose in legal heat and shouted to the judge that his client was being intimidated, Stewart blandly asked Sam if he was being bluffed, coerced, threatened, or intimidated in any way whatsoever, and Sam's stupid head, thatched with sandy hair, gave a vigorous no. How absurd it would be for the bully boy of the Lode to admit that he was being intimidated by anyone! Then out of his wolf brain came a surprising statement: he had merely dropped in to ask Stewart to be his attorney, in a case of assault, and had five hundred dollars to put down on it; and allowing as how judge, jurors, and lawyers might be in need of refreshment he proposed to set up the drinks. Stewart said many years later that in the bar "I continued to hold my pistols in front of him."

Just what took place in the courtroom we'll never know but it seems likely that the bully rode away with his *amour-propre* in ruins. It was his thirtieth birthday, and his idea of a proper memorial of it was the slaughter of another inoffensive man. But this time his judgment was bad. He rode to the inn of a Dutchman named Henry Van Sickle or Sickles, who was not used to being pushed around. Again we cannot be sure of exactly what happened. Stewart says the waiter didn't serve Brown's dinner in the style he was accustomed to and that he knocked the man down. Van, who ought to be about the best authority on what happened, told it, much later, this way:

The killing of the notorious Sam Brown was as follows: Brown was a heavy man about 200 lbs weight was noted as a lawless desperado whose name was terror to all who knew him or had heard of him. he made his brags on the day of his death which was his 30th birthday that he had killed eleven men & was going to have the 12th one for his supper Brown had frequently stopped at my station in previous times always acting like any other civil traveller paing his bills and behaving himself and while he bore a bad reputation I had never had occasion to feel that he would in any manner interfere with me or my business, but on the day of his death he in company with a young man who subsequently state that Brown compelled him to come with him, rode up to the door of my then public house and while in the act dismounting as I supposed to stop for the night, I stepped out with the remark shall I put your horse up Mr. Brown, just as I would to any traveller who seemed to be desirous of staying with me. He in a very gruff manner said No you Son of a ----- I have come to kill you and at once drew his gun. being entirely unarmed and knowing the character of the man I at once left the scene, for he following with a gun & cocked I passed in through the dining room where there were some 20 men seated eating supper, they of course were alarmed by the sudden appearance of my hasty entrance, followed by so formidable a character as Brown was known to be, Brown exclaimed in a loud tone where is the Son of a -----?

Seeing so many men he dropped his gun from its position and put out and got on his horse and rode away in company with his young traveling companion. I secured my own gun, got on another horse and at once went in pursuit of him overhauling him in ¾ of a mile from the house, and when as I supposed within shooting distance I called to his companion to look out, and as he pulled away from Brown, I shot at him, but being at too long a range failed to hit him. Brown turned & returned my fire, I again shot, this time relieving him of his hat and burning his face with my fire, but again failing to bring what I was after, namely his head. He now pulls out and putting spurs to his horse drove away at a furious rate, I reloaded my gun and took after him again. in 3 miles I again got within what I supposed was gunshot range and turned loose again, he returning the compliment dark coming on and he losing his way I headed him off and thus reached the point where he intended to go, ahead of him, & awaited his arrival, I knowing he would surely come abided my time, well knowing that it was a matter of self protection with me, as if I failed to kill him he would without a doubt kill me the first time he got a chance. I therefore waited till he rode up within short range, when stepping out I said as he had previously said to me you son of a b---h I have got you now, and at once ended his career putting seven buckshots right through the center of his body, death being instantaneous, he falling from his horse without muttering a word that could be distinguished from a groan, thus ended the life of a man that had few if any redeeming traits to relieve a life of blackness and infamy.

That quiet telling of it is surprisingly different, in some of the details, from other versions of it.

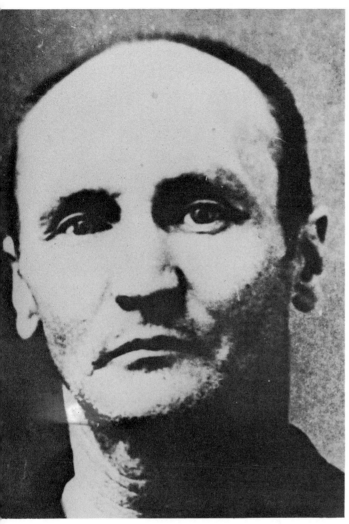

The man-eater of the San Juans

Van says that at his own expense he saw Brown "well buried." A jury said that Sam was a bad one and got what he deserved, and that it was the verdict of an all-wise Providence. According to Stewart, Van was elected sheriff of Douglas County after it was organized, and so deserves a place on the roster of brave officers of the early West.

In Colorado the choice for most notorious coward of the early West would probably not be Sam but Al Packer. With five companions he was lost and winterbound while trying to reach the area of a new gold discovery, and when at last he came out of the mountains alone some of the men who saw him were suspicious, and particu-larly after Packer, on being pressed, admitted that all his companions were dead. He gave more than one version of what had happened but confessed in all of them that he had eaten parts of the other five. He insisted to the end that on returning to camp one day he saw that four of them had been slain, and that the fifth, still alive, a man named Bell, was coming toward him with a hatchet. Packer said that he then shot Bell. Arrested and jailed, Packer escaped and nine years passed before he was found; he was then convicted on five charges of manslaughter and given forty years. His attorneys continued to fight the case for him, and three separate times it reached the Colorado Supreme Court. As with many notorious criminals, well-meaning but deluded persons in many walks of life signed petitions in Packer's behalf and sought parole and even pardon for him. In this they were at last to be successful. Scores of persons who had never seen him actually believed his story that he had killed only one man and that man in self-defense.

But what was the opinion of Cyrus Wells Shores, an outstanding Western sheriff? Doc Shores was the man who first jailed him; he got to know him well. He had this to say:

For the three years that I was associated with Packer, I learned a lot about him and none of it was good. Of all the prisoners that I held in custody during my eight years as sheriff Packer was the only one in whom I failed to find at least a few good qualities. He was slow-witted, cowardly, vicious, and a natural bully. From his crudely written letters and my conversations with him, I learned that he had committed other serious crimes for which he was never arrested or prosecuted. Shortly after his escape from the Saguache County jail, for example, he murdered two young men east of Colorado Springs and stole their team and wagon. Later on near Tombstone in Arizona Territory he killed a prospector and took possession of his horse and pack mule. . . . Packer wrote some of the most depraved letters that I have ever read to his sister in Pennsylvania. Among other things he accused her of neglecting him, and threatened to kill her when he was released from prison.*

He is known today in various books as the man-eater of the San Juans. The Governor paroled

---

* The reader familiar with the Capote book may be reminded of the two subhuman creatures who deliberately planned the slaughter of the Clutter family, and their guests, if they happened to have any; and who, after they were caught and condemned to death, expressed the regret that the sister of one and the first wife of the other had not been present in the Clutter house, so that they could have had the pleasure of killing them.

him only because he could or would no longer endure the furious abuse of one Polly Pry, a Denver *Post* professional sob sister, and of the *Post's* owners, Bonfils and Tammen, whom she had convinced that Packer was a pathetic and tragic victim of hatred and persecution, and that the governor was a man with neither heart nor soul. This newspaper's virulent crusade in behalf of a subhuman brute is one of the most disgusting chapters in American history. For the established facts leave no doubt that Packer killed his five exhausted companions while they slept, and as a coward was possibly even more infamous than Sam Brown.

BOOTHILL

ITEMS

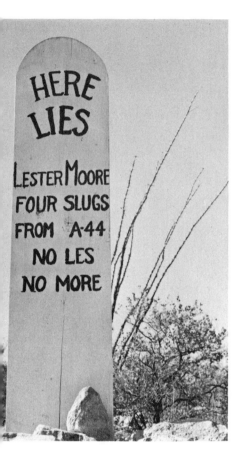

Time has obliterated nearly all the genuine boothills, though tourists are here and there directed to what professes to be one, as in Tombstone. Margarita and Lester Moore are in it but the boards probably are not above actual boothill graves. Repeated search in Idaho City's cemetery failed to discover the authentic boothill of either gunmen or harlots; the one here with the tree growing out of a grave is typical of graves that are old but not boothill, so far is is known. The Helm and two other boards, and the three weathered boards in Hangman's Gulch, all with the names of Plummer bandits, at Virginia City, Montana, may or may not stand above the actual graves.

CHAPTER 25

Duels

Duels, Bancroft says, "were in order; of the one hundred fought in California about one-third were fatal to one of the combatants." Though for a short time in San Francisco a blight was put on this form of childlike behavior in boys who refuse to grow up, by the murder of James King, a search of the newspapers reveals the fact that there was an astonishing number of duels in the mining camps. In Nevada, for instance, dueling "suddenly became a fashion," Mark Twain says, "and by 1864 everybody was anxious to have a chance in the new sport, mainly for the reason that he was not able to thoroughly respect himself so long as he had not killed or crippled somebody in a duel or been killed or crippled in one himself." Mark "had no desire to fight a duel. I had no intention of provoking one. I did not feel respectable, but I got a certain amount of satisfaction out of feeling safe. I was ashamed of myself, the rest of the staff were ashamed of me—but I got along well enough." He did at last provoke a duel, as we shall see.

Usually the newspapers reported them as if they were no more than morning strolls. The Boise *News* told its readers only that "A duel between Thomas Fitch and Joseph T. Goodman came off at 6 o'clock this morning. Fitch was shot in the knee at the first fire, and the firing was then stopped, Fitch being unable to stand. The bone was shattered. The distance was ten yards, weapons, Colt's five shooters; fire and advance." At Mokelumne Hill a German and a Frenchman, neither of whom could speak English or the other's language, quarreled over their claims; and when their talk in gestures failed to settle the difference the German said in signs that they would get their guns and fight it out. The German was killed. On learning that there had been no real reason for argument between them the Frenchman "cried like a child and begged them to hang him."

Marryat found that dueling had become quite the rage. Taking up the newspaper one day, I observed a conspicuous advertisement, in which one gentleman gave notice to the public that another gentleman "was a scoundrel, liar, villain, and poltroon," . . . The next day it was understood that the gentleman with the unenviable titles intended to shoot his traducer "on sight." . . . When I reached the Plaza, I found a large concourse of people already assembled to see the sport; and it was such a novel and delicious excitement to stand in a circle and see two men inside of you exchange six shots a-piece. . . . they commenced walking about the square as if they did not know each other, and when within shot, one said to the other, "Draw and defend yourself!" which the latter did by sending a bullet through the assailant's arm. The fire then became warm; six shots were exchanged in rapid succession, and both combatants were taken wounded from the field. . . .

Up in Idaho a few years later

One Col. O. H. Hall, Esq., of Idaho City, grossly insulted Judge Rosborough last week, for which he instantly got a black eye, and would have got a severe pummelling but for the interference of friends. The Col.'s Southern blood was fired with indignation and he sent a challenge to the Judge, demanding satisfaction for his wounded honor. The challenge was accepted, weapons, ground, and seconds chosen. The affair was to come off early next morning on the hill near the Gambrinus Mill, with rifles, at forty paces. The Judge was cool, kept the matter quiet and was on the ground ready for duty. The Col. blowed the whole affair around town, and tried to get himself arrested. The officers knew his weakness and let him run. In the morning the Col. began early, and after bidding his wife good bye three or four times he succeeded in finding an officer that consented to relieve him of the unforeseen difficulty he had got himself into. So the matter ended,

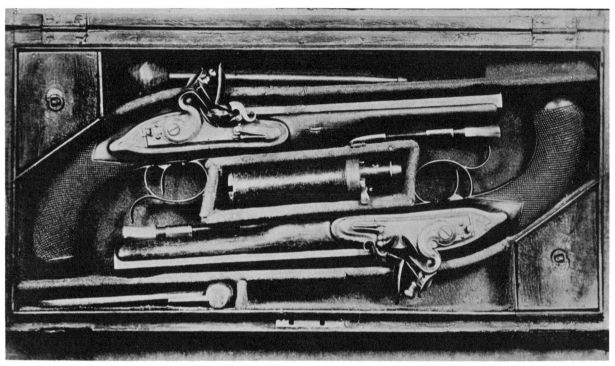

Flint lock dueling pistols owned by Aaron Burr, and used in the famous Hamilton-Burr duel

(as the Idahoans say) without much credit to the Colonel's blood.

Thorp cites a writer who found that it was

thought a virtue in American duels that they always mean real business, and are not those caricatures of a barbarous custom which, in nine cases out of ten, serve to appease wounded honor in England. . . . The U. S. Secretary of the Navy told a young officer who protested against duelling restrictions that he didn't care how much the men duelled just so they didn't kill citizens.

Possibly no form of dueling was more ferocious than that with knives. After the invention of the bowie, schools were established to teach the art of "cutting and slicing" and Thorp quotes from a book left by an expert: "I had taken regular lessons in the knife school in St. Louis, where once I saw two Frenchmen fight for half an hour with Bowie-knives—cut and parry—and all the harm done was, that one of them lost his little finger by a clean slash, and the other bit the first man's thumb off, after first missing gouging him." The knife was of many sizes. Thorp quotes a New Orleans paper which described a "most terrific slaughter with a Bowie-Knife, 23 inches in length, and weighing 5 pounds." A man named

Brown made remarks about a woman that were resented; when men attacked him he stabbed the the first, cut open the second from shoulder almost to navel, and sliced the arm of a third man so dreadfully that it had to be amputated. Thorp tells of a Colonel Crain who had the habit of not paying his debts and of challenging to a duel any creditor who dunned him. This led to a feud with General Cuny. Both men were in the notorious fight with knives at Vidalia Sandbar.

In a gambling house in Hornitos in 1857 a Mexican and a Chilean exchanged insults and went outside, where each wrapped his shirt around his left arm and faced his opponent with a long sharp knife. One who saw the fight has this to say:

A few feints and strokes were made on each side, when the Mexican made a great lunge at the throat of his adversary. The blade slashed the Chilean from chin to eyebrow, diagonally across the cheek and nose. The wounded man struck out desperately, burying his knife in the Mexican's shoulder. The gladiators, smarting with pain, now fell furiously upon each other, slashing wildly. At length, covered with wounds and faint from loss of blood, they closed in a death wrestle, each grasping the other's knife-arm at the wrist. Suddenly the Chilean broke his hold and sprang backward, and his adversary, slipping on the gory turf, fell forward upon the point of his weapon, which pierced his breast. With a profane shout of triumph the Chilean reeled away to a barroom, where he gulped a tumblerful of brandy and sat down to resume play at the table, covered with wounds and gore as he was. His wounds, thus neglected, proved fatal shortly afterward, and the two gladiators were buried side by side in Dead Man's Gulch.

That gory duel is only a foretaste of what is now to come: perhaps it would be well to remind the reader that there were humorous moments in this bloody chapter of the old West and to give him a couple of instances. There is that of two young men hopelessly in love with a girl who found it impossible to choose between them. They thereupon decided to fight a duel, the winner to have her; but when they faced the ordeal, back to back, and were told to walk ten paces each and turn and fire until one of them was dead, they both kept on walking. Stellman says he was told of an early duel that "I believed without question." Two elderly friends of Yuba County fell out over a trivial matter and each swore to shoot the other on sight. After a few weeks one of them, smarting under taunts, cleaned his shooting iron and went gunning. At the door of his enemy he shouted a challenge. When there was no answer he peered in. He then entered and found on a cot his longtime friend more dead than alive and out of his mind. Being an old "bach" he lived alone. From this point on the story is exactly as O. Henry might have written it. The man on the warpath put his gun down and went to work. He slopped the pigs and fed the chickens and he nursed his friend-enemy back to health; but at the door when ready to leave he turned with a fierce scowl and growled, "Look out fer yerself ye son of a bitch, fer if I ever ketch ye over my way I'll fill ye so full of lead ye'll set down on yore heels and stay there."

Quite frequently the duel didn't come off. In Denver one Riley sounded big and brave until Harrison, his opponent, imposed terms so savage and murderous that Riley's friends persuaded him to withdraw. Sometimes there was unexpected gallantry. In Arizona a candidate for Congress and a newspaper editor developed a feud; and one morning they met with rifles and "four

rounds were fired without effect." Then, when the candidate's rifle refused to fire it was decided that he was entitled to a shot. "The editor stood, unarmed, to receive it" but the candidate fired into the air.

How many spots were salted for greenhorns no man knows. A Major Duffield, known as a crack pistol shot, and a crook named Abbott sold a salted claim and then quarreled over the booty. Feeling insulted, Abbott challenged the major and was told by friends that his death was certain. At sunrise they met, and stood back to back, a few feet apart. At the call of "One!" they turned; at "Two!" they raised their weapons. "Excuse me, sir," Abbott said, in a voice as bland as honey. "May I have a word with you, Major?"

"Now what in hell is it?" cried the impatient major, and lowered his weapon.

"Why," said Abbott, "you had your gun pointed at me. It's all right now, let's go ahead, sir," and in that moment Abbott fired, his bullet striking the major's pistol hand and shattering it. Duffield lived to become a United States marshal in Arizona and was killed when he tried to drive off some claim jumpers.

As Horace Bell tells it a Frenchman challenged Roy Bean* to fire with him at a target and Bean said he would accept the challenge only if they fired at one another. The news went out and people poured in from the country roundabout to watch the duel. An obliging sheriff said it would be all right for them to shoot from horses if they did not kill any of the spectators. The duelists had to maneuver so that when they fired at one another there would be no people in the line of fire. Bean is said to have accomplished this, and to have killed his opponent. Among the spectators were high-spirited señoritas, who now made Roy their hero: when he was arrested and lodged in jail they stormed it "with baskets and shawls filled with flowers, cold chicken, tamales, enchiladas, dulces, wines, and cigars and crowded for positions at the gratings to hand their gifts through to their Adonis."

In Tombstone in 1881 Luke Short and Charlie Storms had reputations as gunmen. Myers gravely tells us that "Short was in the upper brackets —firmly enough entrenched to have earned a

_____
* See the chapter, "Frontier Judges."

permanent niche in the records of a good few of the West's liveliest towns—and Storms had, if anything, a higher rating." Short was dealing faro when Storms, well armed, entered the Oriental Saloon and tossed a few insults and told Short that if he would come out to the street he would be happy to finish him off. We are told that in the street, Storms, with the higher rating, "shot often enough but not straight enough." Short put him down with two bullets and returned to the faro game. Charlie, hauled off to the boothill, has today a marker above the spot where he fell.

In Downieville Bob Tevis wanted to be the Fourth-of-July orator, and after announcing his plan to run for Congress gave such a long tedious harangue that some of the miners fired above his head, and a state senator excoriated him in the *Sierra Citizen*. Bob called the senator a slanderer and a liar and was challenged to a duel. Friends tried to bring peace but Bob said his honor was horribly outraged and demanded a double-barrel shotgun loaded with one-ounce balls. Both men fired on the signal. A lock of hair was knocked off the senator's head, who lived to become a brigadier-general and Secretary of State of Illinois. The one running for Congress tumbled over backwards with enough lead in him to kill half a dozen men.

A man of foul tongue named Priest had a grudge against a shift boss named Sullivan, and spread the rumor that if Sullivan's wife did not stop running after him he would have to shoot Sullivan. Both men then went armed and wary. One day Priest approached the Sullivan home, ready for trouble, and Sullivan came out, gun in hand. They fired at about the same moment and both were mortally wounded. The chief concern of each was not that he was dying but whether his enemy would die, and so they demanded frequent news of the other's condition, nourishing their strength with the "violence of their hatred." One of them is said to have muttered, "I won't die till that son of a bitch is dead." Priest died at midnight, Sullivan about an hour later.

A gunfight in a Florence, Arizona, saloon is said by Cronin to have had "perfection of technic of combat." An Arizona judge has said that Sheriff Pete Gabriel was the most fearless man he had known. His deputy was Joe Phy, who hoped to become sheriff when Pete retired. But one day Phy arrested and flogged a man so severely that he almost died, and Pete thereupon fired him. Nursing a grudge, Phy stalked him and for a while Gabriel kept out of his way, but the hour came when Phy loaded his shotgun and said he was going to get the son of a bitch. He was talked out of it that time but the people of Florence knew that it would be settled some day. According to Judge Thomas, Pete was doing some drinking with the boys when Joe again went on the warpath. Thomas, playing cards in a hotel, heard two shots "so close together as to appear as almost one." He rushed to a saloon and saw Pete standing "midway between the door and the edge of the sidewalk, with his feet spread,

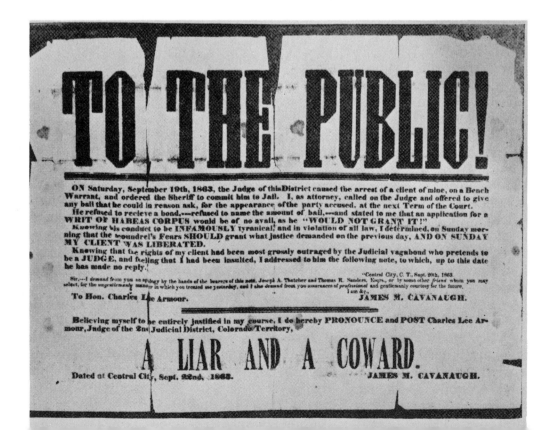

arms hanging down, his gun in his right hand. Just as I got to him he began to sag and sink, like a half-filled sack of grain. . . . Phy was in the street, but a few feet away, and had raised upon one elbow as Dave Gibson approached him. I heard Dave ask the question, 'Are you hurt much, Joe?' Phy replied, 'Go away from me, you murdering son of a bitch!' and made a slash at Dave, cutting him to the bone in the leg above the knee."

When the moment of shooting came Pete was standing at the bar drinking with a friend. The swinging doors were kicked in and there stood Phy, gun in one hand, knife in the other. His first shot struck Pete in his left chest; before Joe could fire a second shot Pete drew and shot Joe in his stomach. As Pete advanced on the enemy Joe shot him three times—low in the right side, through his body, and through a wrist. "As Gabriel neared the door and reached out for Phy, Joe turned, crashed through the swinging doors and pitched across the sidewalk, where he fell." Gabriel said he did not fire a second time because he was too weak. Phy died the next morning. We are told that with his last breath he asked if Gabriel would die, and died happily on being told that he would. But the sheriff recovered. In a corral Joe's horse was found, saddled and ready for instant flight. He neither drank nor smoked, and so was stone sober and had it all figured out, but luck was not with him that day.

Dr. J. C. Tucker tells this story in *Six Horses*. Two Frenchmen and two Frenchwomen boarded a stagecoach, on which was a tall Texan. The younger of the two men was haughty and supercilious, and seemed to be jealous of the younger of the two women, an attractive girl. The Texan, handsome in boots and buckskin, turned his gaze on the girl. From the driver it was learned that the Frenchmen were gamblers—bold, arrogant, and good with their weapons. Such knowledge did not dash the Texan at all; he continued to flirt with the girl, at one stop gathering flowers for her, at another assisting her as she entered the coach. All this put the young Frenchman in a towering rage. When the Texan through a window shot the head off a hawk, and then offered his gun to the girl for target practice, the Frenchman could stand no more. He swore an oath in English and ordered the Texan to address no more remarks to the lady. The Texan replied,

"Frenchy, you do speak English well enough to apologize at the next station."

The girl burst into sobs as the two Frenchmen carried on a violent discussion, in French, and cast flashing murderous glances at the tall man in buckskin. When the stage pulled up at a supper station, Texas was the first man to spring out. Dr. Tucker followed him and was soon addressed by the older Frenchman who said the Texan must "give my friend the satisfaction—here—now!" Tucker said he thought that was a part of the Texan's plans.

A duel with pistols was quickly arranged. Back of the station was a large empty corral, with two gates opposite to each other. Texas asked Tucker to tell the Frenchman to come in by the north gate, ready to defend himself, and to tell the other Frenchman to keep out of it, "while I fix the frog-eater." The older Frenchman wanted the duel fought with his two beautiful pistols but the Texan, being the challenged party, had the choice of weapons, and chose his. This made the Frenchman hiss through his teeth and cry, "Damn him, he shall be shot in the corral like the beast he is! I will bring my friend."

Tucker told the two Frenchmen that only the two duelists would enter the corral, and that they would fight it out there, unmolested by anyone. Texas gave the doctor his address and that of a friend, to whom he was to write if he were killed.

The two gates were about two hundred feet apart, the ground between them level and without obstructions. Tucker tossed his handkerchief into the air as a signal, the two duelists entered from separate gates, their pistols cocked, the Frenchman's pointed toward the sky, the Texan's pointed down at his side. The Frenchman, eyes "gleaming malignant hatred," advanced in a sideways movement; the Texan came broadside toward him. Suddenly the Frenchman lowered his gun a little and fired two shots, one of which struck the Texan and half spun him and made him stagger. The Texan then rushed forward a few paces and fired but missed. The Frenchman fired again and missed. "The blood was pouring from Texas' wounded arm" as he again advanced; and again they both fired. The Frenchman's bullet knocked off the Texan's hat. Texas then sank quickly to one knee, leveled his pistol barrel across his wounded arm, and fired. The bullet hit a mortal spot. The Frenchman threw his arms "wildly in the air" and "fell dead." The two women now rushed into the corral and made loud lamentations over their fallen lord; the driver, who had eaten his supper, waited for neither man nor woman. Dr. Tucker grabbed a towel and some splints and he and the Texan tumbled into the stage as it was moving off; of the fallen man and the three around him the records have no more to tell. The driver said the two Frenchmen were notorious gamblers and killers; the older woman was a faro dealer; the younger one was possibly a new recruit, for hurdy-gurdy and bordello.

The Idaho *World* gave an account of a duel, as reported by one Dr. V. C. Lawrence, of Colorado. Two brothers, Joe and Charley Bigger, with men named Norton and Jackson, had sold their livestock, and flush with money had chosen a cozy spot for a game of cards. There rode up three professional gamblers, one Watrous, known as Cockeyed Wat, Dick Bradford, and a third named Allison. Watrous, Bradford, Jackson, and Joe Bigger decided to play, and before long Watrous was charged with cheating. It was decided that Bigger and Watrous would fight it out on horseback with bowie knives. "They were divested of their coats and shirts, and their knives were bound to their right hands. They were then placed sixty yards apart, with orders to ride at each other with full speed, passing on the left. Both were splendid horsemen. Bigger was mounted on a clean-limbed, fiery pony, a little over fourteen hands high. while Watrous rode a large 'watch-eyed' vicious roan." At the word go the two rode furiously toward each other, and passed, without injuries. On the third run the roan received a cut in the flank. "On the fourth round, Bigger, as he passed Watrous, threw himself on the offside of his pony, so as to expose no portion of his person, and drove his knife deep into the neck of his adversary's steed. Watrous, divining the manoeuvre, wheeled as the blow was struck, and attempted to hamstring Bigger's pony, but succeeded only in inflicting a severe wound. This style of fighting was then abandoned, and both men and horses appeared to become infuriated at the sight of blood."

As they neared each other on the fifth run Bigger struck Watrous with his left fist in the face, at the same moment cutting a fearful gash in his thigh; but before he could get away, Watrous succeeded in driving his knife into Bigger's shoulder. The combatants and horses were becoming weak from loss of blood, when Watrous determined, if possible, to end the combat, by riding down his adversary, which he thought the superior weight of his horse would enable him to do. Accordingly, on the sixth round he made directly upon Bigger's pony, and Bigger, in attempting to avoid the collision, was severely cut in the arm and face. The pony, however, was game, and although very lame, seized the roan by the cheek, lacerating it in a fearful manner.

At the seventh encounter the horses came together with a fearful shock, the pony being thrown, falling upon his rider, but both immediately regained themselves. Watrous' horse was fast bleeding to death from the stab in the neck, and Watrous himself could scarce keep his seat from the wound in the thigh. Bigger succeeded in again striking Watrous in the thigh, and was struck in return in the side. Several blows were interchanged or evaded, or fell only upon the horses. The fight had now lasted more than half an hour, when Dr. Lawrence rode up in time to witness the final round. As they came together Watrous endeavored to rise in his stirrups and to throw himself upon Bigger, but neither horse could stand the encounter, and both fell. Bigger was streaming with gore from the cuts in the face, back, and arms, but was able to extricate himself, and rushed upon Watrous with the fury of a fiend, and almost in a moment his knife had reached the unfortunate gambler's heart, and Bradford, seeing the fate of his friend, raised his pistol, fired, and Bigger fell dead across the corpse of Watrous. A free fight at once ensued, Charley Bigger, Norton, and Jackson firing upon Bradford and Allison. Bradford was killed in the melee, and Charley Bigger and Jackson severely wounded. The wounded were taken to a cabin about half a mile from the battle field, and their wounds dressed by Dr. Lawrence, who pronounces them in a fair way of recovery.

Perhaps the most famous political duel in the American West was that between David C. Broderick and David S. Terry. It is well to know the characters of the men. Broderick, says Bancroft, was a "professional politician of the New York type, rough and self-reliant. . . . Possessed of many objectionable qualities, he was not without redeeming traits." For Hittell he was "one of the worst samples of a very bad class,—the American political boss" under whom San Francisco "reached a most disgracefully corrupt condition in the spring of 1856, and against whose dominion the labors of the Vigilance Committee of that year were directed with signal and admirable suc-

David C. Broderick

David S. Terry

cess." In the matter of gall, says Wagstaff, he had no peer in California. A big brutal man he had been a power in Tammany in New York before coming to California in 1849, where in no time he was in the state senate and began "to connive for a seat in the United States Senate." Failing to get Frémont's, he fixed his eye on that of a man named Gwin and "schemed and struggled for it by day and night, ceaselessly, without scruple as to means, and with an energy that would have exhausted a less robust and less indomitable nature." The campaign ran for more than two years and was so savage that it alarmed the nation. It brought on Broderick several challenges and eventually the duel with Terry, a justice of the State Supreme Court.

Terry the reader can also meet in the chapters on vigilantes and adventuresses. One of the most amazing characters of the early West he stood six feet three and weighed 220, and he was practically all impulses and emotions—an ex-Confederate officer, a man of ungovernable temper and prejudices, who would fight with fists, knives, or guns at the drop of a hat. For some time the two men had been enemies and during the bitter

campaign Terry had to have his say: "Who have we opposed to us? . . . A miserable remnant of a faction, sailing under false colors, trying to obtain votes under false pretenses. . . . They belong heart and soul, body and breeches, to David C. Broderick. They are ashamed to acknowledge their master." Broderick made scathing remarks in return. Terry, says Wagstaff, who knew him for years, had "outbursts of irrepressible anger caused by breaches of his self-constituted ideas of dignity and respect," and probably shook with outrage and impatience. Friends tried to prevent the duel but Broderick said it had to come and "this is the best time for it." Police got wind of it and broke up the first early-morning meeting; but the next morning they met again, with several doctors present, one of whom came with no instruments and threw himself down in his overcoat as if for a nap; another of whom "dragged this horrid-looking sack, with its rattle of instruments, its ugly protruding saw, and its plethora of linen rags for bandages." The *Morning Call* had said: "It is generally understood that Judge Terry is a first-rate shot; but it is doubtful whether he is as unerring with the pistol as Senator

Broderick. This gentleman, recently, in practicing in a gallery, fired two hundred shots at the usual distance, and plumped the mark every time. As he is also a man of firmer nerve than his opponent, we may look this morning for unpleasant news from the field." The odds were against the judge, but on the field, it is said, he carefully studied his adversary, while Broderick pulled his hat low and refused to look at Terry.

When the principals took their stations, Broderick as the challenged had his back to the sun, and though it was barely above the hills, Terry had to face it. It has been said that while pistols were being examined and loaded Terry told a friend that Broderick was in no condition to fight and that he intended only to "lame him." That sounds apocryphal. Wagstaff thought that Broderick did not want to fight Terry but did want to fight Gwin and intended to; and says, further, that when a friend handed his pistol to Terry he told him this was no monkey business, that out with Broderick were eighty men "here to see you killed and to rejoice over it." All of which sounds like Monday morning quarterbacking. Wagstaff says Terry received his pistol, held it behind him a few moments, and then rested it across his left arm. Broderick's demands, as the challenged, had astonished Terry, for the pistols were to be held with their muzzles pointed *down*, and instead of the words "Ready, aim, fire," they were to be "Fire, one, two," with no firing allowed before the word "fire," or after the word "two." Much has been written about the reasons for this, and the conclusion has been that Broderick, the better shot of the two and with his back toward the sun, had demanded for himself a cowardly advantage: a faster man in raising a heavy pistol and aiming, he expected to hit Terry on the word fire, and to be able to fire again before the word two, if he should wish to. Terry's seconds objected to these changes in the rules and were told that the words were not unusual or unfair, though contrary to the accepted code.

The pistols provided had been used many times in affairs of honor between boys who refused to grow up and were said to be ususually fine ones, both with hair triggers so "evenly and equally adjusted" that there was no perceptible difference between them. It is said that Terry in practice with one of them made a couple of line shots, and finding that he was shooting under the target

"tried them no more." Whether he figured that he would come to his end under the gun of Deadshot Broderick, nobody knows. It seems likely that he did not, for if we are to believe the biographies about him he had been a hotheaded fighter since childhood and had been in a number of duels with knives and pistols.

Whether Broderick was in fighting condition that morning is not known; some have argued that the "prodigious mental and physical strain of the campaign" had left him "shaken and exhausted" and that he had allowed himself to be prevailed on by his "rash and reckless advisers." On the other hand it may be that he was unnerved by the way Terry fixed on him an intent and implacable gaze. He knew of Terry's impetuous and ferocious assaults on people in times past; in San Francisco he knew that Terry had buried a knife in the neck of a vigilante, and indeed in his exchange of insults with Terry had pointed out that while Terry was in jail awaiting trial by the vigilantes "I paid $200 a week to support a newspaper" in Terry's defense. "I have also stated heretofore that I considered him the only honest man on the Supreme Bench; but I take it all back. You are the best judge as to whether this language affords good ground of offense." It afforded whole acres of ground for a man like Terry.

Whatever the reasons it is said that though Broderick's "facial expression was of imperturbable composure" his face was pale and his attitude was rigid. Further, he held his pistol not with the muzzle down, "as the articles required, but pointed outward in obtuse angle"; and when an effort was made by his second to correct the position "his rigor of frame was so intense that, in the effort to adjust his pistol to the required position, he was obliged to use his left hand to bring his right arm into proper form; and in the effort he also so swerved his whole body that his right leg was pressed out of place. . . . He held his pistol in vise-like grip; and his wrist, instead of being in condition for ease of motion, was as an iron bolt, to move only with and as rigidly as the arm." Terry, on the contrary, "stood erect and firm, but in an easy attitude."

At the word "Fire" Broderick fired, but the ball "entered the ground just nine feet from where he stood." Terry fired an instant before the word Two and his bullet struck his adversary with such force that a "visible shuddering of the body was instantly perceptible, then a violent contraction of the right arm, from which the pistol dropped to the ground. A heavy convulsion shook his quivering form, he turned toward the left, his head dropped, his body sunk, his left knee first gave way, then the right, and in a moment he was half prostrate on the sod. . . ." The doctors thought he was not mortally wounded but three days later he was dead. Broderick's friends created such an uproar that Terry fled the state; took up a mining claim in Virginia City, where he knocked down a couple of Irish claim jumpers; and returned to California to live out with increasing melodrama his tempestuous life.*

There were other duels with a political base, the most notorious in Idaho Territory being that between Holbrook, a member of Congress, and a gambler named Douglass or Douglas. The Idaho *World* says the two men met about eight one evening at the corner of Main and Wall streets in Idaho City, and after an exchange of angry words both men drew their guns. Eleven shots were fired before an officer arrived; Holbrook was then helped into his law office a few feet away and a doctor was summoned. The congressman had been shot low in the abdomen and died the next morning; Douglas, according to Donaldson, "left the country very suddenly in the night." Some of the witnesses said Holbrook fired the first shot, and some, that Douglas did, but they all agreed that the firing continued until both guns were empty. In the same issue is a eulogy of Holbrook, twice a delegate to Congress, whom year after year the *World* had defended against the *Statesman*: "He made a bitter and unrelenting war against the Radical Federal office-holders in the Territory . . . untiring efforts to secure the removal of Radical office-holders secured for him the hatred and fear of his political opponents." But the *Statesman* had said earlier:

It is well known to the outside world that Holbrook sits on the radical side of the house in Congress, and plays himself off for a good Union man. . . . Wouldn't he be a good free soil, free suffrage, nigger-loving democrat if elected for two years more. . . . the great defender of crime and criminals. In Idaho City soon after the fire,

when the merchants had lost thousands of dollars worth of stolen goods and talked of getting out search warrants to search the premises of the thieves . . . this great criminal apostle publicly denounced the proceedings and advocated armed resistance to such a vigilant policy.

Two months later the *World* said it had refused to publish rumor and gossip about the killing: "It has been heralded far and wide, through mercenary motives and for political effect, that Douglas assassinated Holbrook. We pronounce the statement as unauthorized by the testimony, and as being a willful, deliberate, and premeditated falsehood, conceived in sin and iniquity, and promulgated for the most base, bloodthirsty . . . designs." After a column and a half of bitter denunciation the editor told his readers that the *Statesman* was at last praising a man about whom only recently it had said: "Some men seem born with a special genius for evil. . . . They toil incessantly in the cause of 'pure cussedness'. . . . of shallow mental capacity. . . . The shallowest brain, intent upon mischief, by adopting the crooked ways of a reptile. . . . Such is the *genius* of a man who swaggers daily through the streets of our city and boasts that he 'is going to crush his enemies' . . ." Though Holbrook's name did not appear in that editorial, inasmuch as the *World* had the habit of calling the man a genius it seems likely that the *Statesman* had him in mind. The *World* goes on to say that the weight of evidence supported the belief that Holbrook had fired first and had fired six times to his adversary's five. It concluded with a lament for the fact that Douglas had been captured and was in jail, "maligned and hounded down . . . by a malicious, mercenary, and unprincipled press, with none to refute the vile slanders and falsehoods circulated by his bitterest enemies . . . lies so industriously circulated by a few unprincipled, malicious and venomous curs. . . ." If North-South bitterness had not been such a strong factor in the case we doubt that much would have been written about it on either side.

After so much billingsgate it is a pleasure to come to the most amusing duel in American history if we are to believe the story as told by Steve Gillis, Mark Twain's friend.* When Joe Good-

_____
* See the chapter, "Adventuresses."

* Twain's biographer, Paine, says his "memory had become capricious and his vivid imagination did not always supply his story with details of crystal accuracy."

man was called away Mark found himself in charge of the *Enterprise,* but at first he could think of no better subject for an editorial than Shakespeare. The next day there were no more Shakespeares, so he "woke up Mr. Laird," a rival editor, "with some courtesies of the kind that were fashionable among newspaper editors in that region, and he came back at me the next day in a most vitriolic way." His courtesies were a charge that in Carson City funds raised for sick and wounded soldiers were being diverted by the good ladies there to other uses. Mark did not know that he had aroused some of the husbands to fighting pitch. He did know that one of the furious wives sent a letter to the newspaper in which she called Mark's attempt at fun a fabric of malice and lies and demanded to know the name of the author. The rival paper's editor called the anonymous writer a "liar, a poltroon, and a puppy," said he had a groveling disregard for truth, decency, and courtesy and apparently sought to be known only as a vulgar liar. Those were fighting words, and of course Steve Gillis and his pals, all of them hoaxers, were tickled silly. Gillis told Mark that such insults could only be wiped out with blood and reminded him that he himself had fought several duels, that a year earlier Goodman had crippled an editor for life, and that it was about time that Mark had his baptism of blood. "He was not eager for it; . . ." That was quite an understatement. Mark had been having his jokes and enraging a lot of humorless people with his hoaxes but when the long Gillis face gravely explained that this was not a matter for jest but for bloodletting, he didn't feel at all like a humorist. But with the help of Gillis he wrote a challenge and Gillis delivered it in person. They then had to wait, and while waiting Gillis and his cronies tried to build up in Mark a fund of self-confidence. He needed it: a small slender man who hardly knew one end of a gun from the other, or what a bowie knife was for, he was about as eager to fight a duel as he was to go off into the hills and shoot himself.

When no reply came at once, Mark wrote *forty years* after the event that "I began to feel quite comfortable. . . . it seemed to me that I was accumulating a great and valuable reputation at no expense, and my delight in this grew and grew as challenge after challenge was declined." According to Gillis it wasn't that way at all. A re-

ply from a mere reporter said he wrote the article, was responsible for it, and had nothing to retract. "A mere reporter!" Gillis cried—and explained to Mark that an inflexible rule of the dueling code was that duels were fought only by equals. So that evening Mark got off another letter, this time to Laird, the editor. It said that he had demanded an apology, and a reply had come from an underling with whom he had no business. If Laird was not a "cowardly sneak" he would give the "satisfaction due a gentleman, without alternative." Again Gillis delivered the letter. Laird was as frightened as Mark; he actually went to a wealthy acquaintance and offered to sell him a part of his newspaper if he would take over the duel with Mark! The man said he would be happy to buy into the paper but not to shorten the life of such a promising young man as Mark Twain.

Laird was driven at last to write a reply. It was a long one, full of self-righteousness, and peace on earth, good will toward men. The reporter, he said, was a former army man and fully qualified to accept the challenge; and if Mark was not in fact a liar, poltroon and puppy he would go out with the army man at daylight and get it over with. It has been surmised that this letter was so craven that Mark began to feel like a gladiator after all. In his next letter to Laird, in which he said that the man still tried to hide his "craven carcass" behind another, he concluded: "I have twice challenged you. . . . if you do not wish yourself posted as a coward, you will at once accept my peremptory challenge, which I now reiterate." Laird replied again, saying that he had had no part in the matter and begging to receive no more "long epistles from you." Mark's attack on the dear ladies was on May 17. Laird's expression of weariness with the whole thing was dated May 23. If Gillis had not been goading Mark and beating into his addled brain the notion that he must avenge insult to his honor the whole ridiculous thing could easily have been settled at this point. As it was, Mark posted Laird as a coward and he made it strong: "I denounce Laird as an unmitigated liar. . . . he is a liar on general principles, and from natural instinct. I denounce him as an abject coward. . . . he insults me . . . yet refuses to fight me. Finally, he is a fool, because he cannot understand that a publisher is bound to stand responsible for any and all articles printed by him, whether he wants to do it or not."

That left Laird no choice but to accept the challenge or leave that part of the country. The day passed and the evening, and about midnight the challenge was accepted. Whether Mark's heart sank to his navel we don't know; he had acted like a lion and now he had to make some show of being one. He might have been thinking with horror of facing the black muzzle of a Colt when he received a challenge from a husband named Cutler, whose wife was still in shock from Mark's story about her, as well as word from other husbands, whose wives had suffered grave damage to their minds and health, that they were coming over to horsewhip him. From any point of view he had his hands full. Nearly all of that night, Gillis says, he talked to Mark about his own duels, the men he had killed, and the fine points in shooting your adversary before he could shoot you. He begged Mark to make a will, saying it was always best to have one, in case the Almighty took a kindly view of your opponent. Mark got no sleep and he must have felt pretty "ganted up" when he staggered out into the dark and over to a ravine. On the way Gillis swiped a door, saying that Mark would have a little time for target practice; and somewhere he picked up a squash to represent a man's head. About daylight Mark began to blaze away with his big six-shooter but he couldn't hit the squash or the door and it is possible that he even missed the hillside. As Gillis says, "Of course he didn't hit anything."

At that moment Providence intervened to save a humorist for a nation in need of one. They heard shooting over in the next ravine. "What was that?" asked Mark anxiously.

"That's Laird. His seconds are practising him over there." It didn't make my principal any more cheerful to hear that pistol go off every few seconds. . . .

A mud hen then came down to sit on a bush about thirty yards away, and with Mark's gun Gillis shot its head off. At that moment Laird's seconds came in sight; they saw the head fall, and with awe written all over their morning faces they advanced to Gillis and one of them asked, "Who did that?"

"Sam," I said.

"How far off was it?"

"Oh, about thirty yards."

"Can he do it again?"

Of course, Gillis said, every time. "He could do it twice that far."

The second turned to Laird and said, "You don't want to fight that man. It's just like suicide. You'd better settle this thing now."

And so, says Gillis, "there was a settlement. Laird took back all he had said; Mark said he really had nothing against Laird—the discussion had been purely journalistic and did not need to be settled in blood." Gillis says he recalled one thing Laird had said, when told by his second that he had better not fight: "Fight! Hell, no! I am not going to be murdered by that damned desperado!" According to Gillis, every time Mark aimed the gun at the squash and fired he shut his eyes.

Disgraced, Laird left the area. Whether Cutler challenged the damned desperado or Mark challenged him is not certain; Gillis says Mark sent an insulting challenge after receiving an insulting letter. But Mark had made enemies in the area, including a judge, whom he had called Professor Personal Pronoun, and after the Governor sent him a warning his friends thought it was time for him to be gone. So the ferocious duelist one morning slipped into an outgoing stage and fled the country.

If Mark's is not the most amusing duel in American history it could be only because the records divulge one between two bordello madams named Mattie Silks and Katie Fulton.* It was not the only duel between women in the West. There is on record the fight of two Mexican girls from a fandango hall, at five on a cold morning, their left arms wrapped with shawls, their right hands flashing daggers. They made so many holes and incisions in one another that they were both dead by sunup. Parkhill says that two Denver madams who were known as Mollie May and Sallie Purple "were deadly rivals" who carried on a bitter feud for years that at last burst into gunfire, as the two shady ladies thrust guns from the windows of their whorehouses and blazed away at one another. They were worse shots than Mattie and Katie, for they didn't even hit a pimp.

In the reminiscences of a Dr. Palmer, Thorp

_____

* The reader can also find out about Mattie in the chapter, "Girls of the Line."

found the story of a duel between two women with bowie knives. He quotes the doctor: "At San Francisco, a Creole spitfire, or 'ballroom heroine' from New Orleans was stabbed by a jealous Chilean. Between stitches of her wound she cursed alternately her assailant and her Adams revolver that had missed fire." She was barely free of the doctor's hands when "she struck her Chilean enemy a fatal blow with a Bowie-knife."

If Parkhill is right the "only known formal pistol duel ever fought between women" was that of Mattie and Katie over a foot-racer, gambler, pimp and parasite, who for a while was Mattie's man. It has been said that she had owned a nest of soiled doves in Abilene when Wild Bill was there and that he taught her to shoot. It is likely that she taught him a few things, for his favorite women seemed to have been harlots. Somewhere along the way she met Cortez D. Thomson, braggart and showoff and ladies' man. Parkhill gravely tells us that "Stripped for racing, he was a sight to set any woman's pulse aflutter. He wore pink tights and star-spangled blue running trunks. Across the breast of his quarter-sleeve striped racing jersey were pinned rows of gold medals, some won legitimately." It may be prudent to doubt that he could raise *any* woman's pulse but he did seem to trouble the blood pressure of Katie and Mattie, for it is assumed by all writers on the subject that the cause of the duel was jealously over Thomson (sometimes spelled Thompson).

It is easy to imagine him there on Denver's long street of brothels, walking with a stick and a swagger, arms akimbo, according to one who knew him, his sandy mustache "in tight, twin bartender upcurls," his garments of the finest, because Mattie paid for them, as for years she paid his liquor bills and gambling debts. As a matter of investment she had a right to claim him. It may be that Katie made a pitch for him, with money and flattery: Cortez was from the South, where, he said, "nobody works but niggers," and he was a perfumed dandy with the kind of brag and contempt that hugely impressed the women in the houses of joy. He must have thrilled them when he boasted that he had been one of Quantrill's raiders. The mind fails when trying to imagine what stories he told, when full of whiskey and conceit, or the way the eyes of the madams shone upon him. The time had to come when Mattie and Katie would fight for the prize, but how much

of the story is fact and how much is fiction no one knows. There are several versions. It seems probable that Mattie, who fancied herself as a dead shot, challenged Katie to a duel.

She chose Cortez as her second, and Katie chose a gambler named Sam Thatcher. The field of honor was a cottonwood grove on a bank of the river, just beyond Denver's limits, in what was known as the Park or Garden. The time was August 24, 1877. There was an outing in the park of (Miller says) "Denver's sporting world: gambling, proprietors, thimbleriggers, the *maquereaux* (or pimps), faro dealers, and madames, along with some of their most showy girls"—the girls in "flounced gowns, sparkling jewelry, and ostrich plumes," the dandies in top hats and sword canes inlaid with mother-of-pearl and silver or gold. As O'Connor tells the story, under the spell of hallowed dawn-gray of daylight the two madams with pistols cocked stood back to back, stepped forward a few paces, and turned and fired. Parkhill says thirty paces were stepped off; if a man did the stepping the distance could have been anywhere from sixty to a hundred feet. Whatever it was, this version says that at the count of three they fired, that there was a dreadful shriek of pain, and that a dapper little fellow with waxed mustaches fell flat, with blood spouting from the back of his neck. It's such a pretty legend that one hates to disturb it; it is so pleasant to speculate on which madam fired at Cortez, or, if neither did, which madam was such a lousy shot that she hit the pimp instead of her opponent. The weight of opinion in the legend has been with Katie.

The Denver *News* gave this account: "A most disgraceful row occurred late on Friday night at 'Denver Park,' in which two notorious women of the town, Mattie Silks and Katie Fulton, were principals, two men, Thatcher and Thompson, seconds, and five or six participants on each side. There were some disfigured faces and broken noses, and Thompson received a pistol wound in the neck, not regarded as serious."

The *Times* called it "A Free Fight" and gave this version: "According to the statement of Anderson, the manager of the garden, the affair originated in a quarrel between two women, Mattie Silks and Katie Fulton. While this was in progress a man named Thompson stepped up, and with the remark that he would fight Mattie Silks' battles, struck Kate Fulton a blow in the face, knocking her down. Sam Thatcher then came up to restore peace, and was also knocked down. A number of Thompson's friends, among them was Al Mitchell, the hack driver, here came up to attack Thatcher and the woman. Fulton interposed her person to protect him, and while in that position received a kick which broke her nose. Thompson was about this time knocked down, and a pistol which he had drawn fell out of his hand and was picked up by another party. The fight continued some little time and then the parties separated. After it was all over, and when Thompson was returning to the city in a hack, some one came up to the carriage and fired a shot at him, the ball taking effect in the back of his neck and inflicting a slight wound."

The story goes on to say that the next morning Kate took off for Kansas City. She was back by September 6 because on the seventh the *News* had this to say: "Kate Fulton felt terribly aggrieved yesterday. After being knocked down and kicked, and having her nose broken, she was yesterday brought before Justice Lutho on a charge of having on the occasion referred to, threatened

the life of Madame Silks. Some tall swearing was done to make the charge stick, but as a number of parties not connected with the crowd in any manner, denied the utterance of the threats or the drawing of any weapon, as alleged, the defendant was discharged, and went on her way rejoicing."

In its first report the *News* said the park was "owned by the Denver Brewing company whose own reputation is at stake in not taking some measures to prevent a repetition of such affairs as disgraced it on Friday night. The county officials should take the place in hand at once." That is pretty heavy moralizing from a newspaper that must have known that Denver was a wide-open town, with brawls every week or so more disgraceful than this one. After all, most of the shootings were over the sordid matter of money, whereas in this affair two famous madams strove to fight it out for the favors of a dapper pimp, who survived the fracas to become with Mattie the principals in a "long tempestuous love affair that put to shame the fabled amours of Frankie and Johnny."

CHAPTER 26

Feuds

Feuds between brothel madams, mineowners, gamblers, politicians, newspaper editors, and camps were as inevitable and as common as hoaxes and tall tales. There were feuds, often extremely bitter, between camps that were within shouting and shooting distance of one another—Oro Fino and Pierce City in Idaho, Black Hawk and Central City, Kokomo and Carbonateville in Colorado, Virginia City and Gold Hill in Nevada, to name only a few of many. The verbal war between Silverton and Ouray in Colorado was typical of them all. When the editor of the *Solid Muldoon* in Ouray triumphantly announced that Ouray's mines were producing "like mad" the Silverton editor retorted that the mines were not as mad as the men writing about them. Ouray's editor then said that it was natural for Silverton to throw mud, for they had a huge supply of it, all their roads being from six to seventeen feet deep in the stuff. When Ouray's editor wondered when the camp called Ophir would begin to live up to its biblical name, Ophir's miners, with no newspaper in which to hurl insults, drove into Ouray a herd of burros, the rump of each bearing the name of a prominent Ouray miner. When the Ouray judge saw his name on the hind end of one of the beasts, he is said to have been so furious that he threatened to arm himself with six-shooters and shotguns and march alone to make war on the enemy.

But the most picturesque feuds, it seems to us, were those between newspaper editors. Anyone who reads at length in the camp newspapers must be impressed by the fact that gold and silver, lawlessness and hell-raising, and the vigor of the newspapers were all related, and that the decline of any one of them marked the decline of all. When the editor put his billingsgate and his guns away the camp was as good as dead.

As Emrich says, "The early newspapers of the west were essentially personal in character. The owners and reporters were lords of all they surveyed and gave not one tinker's damn for the opinions of the outside world or of their rivals. They potshot at each other editorially and, in literal fact, often resorted to guns to settle their disputes. They had more fun than a barrel of monkeys and kept themselves and their subscribers wide awake." Their newspapers' names are enough to chill the effete world that has followed them—*Reese River Reveille, Fairplay Flume, Missing Link, Boomerang, Daily Rustler, Owyhee Avalanche, Gringo and Greaser, Cripple Creek Crusher, Arizona Kicker.* Some of the editors specialized in practical jokes and hoaxes; some, like the editor of the *Carson Appeal,* invented an opponent, the *Wabuska Mangler,* so that they could vent their hostilities, sharpen their malice, and amuse their readers. The *Appeal* quoted from editorials in the *Mangler* that were vicious attacks on the good people of Carson, and replied in kind. The editor of the *Mono Index* fetched his readers to attention with a graphic account of the domestic troubles of the mayor and his wife and members of the city council, none of whom existed except in the editor's files, and filled his columns with accounts of murder trials, accidents, suits, and robberies that were as mythical as the mayor. Such nonsense was probably inspired by the outrageous inventions of Mark Twain and Dan De Quille.

The spirit of newspaper editors in the camps is set forth in the words of James King of William, when, on becoming editor of the *Bulletin* in San Francisco, he said to his critics: "We make it a rule to keep out of a scrape as long as possible; but if forced into one, we are 'thar'. *Entiende?*" C. S. Bagg of a Tombstone paper wrote: "Last Saturday night, soon after eleven o'clock, some gentleman whose identity is unknown to us, fired a charge of buckshot through the side window of our editorial room directly at the spot where our cot is usually placed. Had the cot been there we should have been inquested and buried ere this. But the cot wasn't there." On the jacket of Drury's *An Editor on the Comstock Lode* the publisher says that when Drury went to Gold Hill as an editor on the *Evening News* "The very first day . . . he covered one murder, two fatal mine accidents, a run-away stagecoach affair and gained a reputation as a gun-fighter by accidentally discharging the office pistol in the general direction of a bully who had decided to horsewhip the editor." After writing about a gunfight Drury's life was threatened by one of the gunmen, whom he accosted before the International Hotel: " 'I understand you're going to shoot me on sight,' said I, as evenly as I could. He noticed my motion, and protested hurriedly, 'I haven't got a gun on me.' 'Well, get your gun and come shooting,' I said. The crowd waited around with me, expecting some excitement, but the blusterer didn't show up."

It delighted Drury, of course, to quote Ambrose Bierce from *The Argonaut:*

One day last week a journalist of this city was severely beaten for something—I do not know what—that had appeared in a newspaper with which he is concerned. . . . I beg leave to state, in the character of an expert who has a practical experience with both methods of redress, that it is more agreeable to a journalist to be shot than beaten. . . . There is no recorded instance of punishment for shooting a newspaper man. . . . I am quite serious in the statement that nobody in the United States has ever been hanged for killing a journalist; public opinion will not permit it.

Dick d'Easum dug out of some old newspaper the story of a hotel owner in Idaho Territory who put five bullets into the editor of the *Wood River Times,* and asked petulantly when arrested, "Why doesn't the old deadbeat pay his bills?" As d'-Easum says, "Shooting an editor ranked as a misdemeanor"; the hotel man was fined one hundred dollars.

When the *Enterprise's* drama critic heard a man say a certain judge's mind was a "howling wilderness," the critic said, "Worse than that, his mind is a regular Death Valley. If an idea ever got in there it would perish of loneliness." As Miller says, "The billingsgate fired back and forth

by early journalists was a unique phenomenon. Such vituperation has become a lost art. . . . Contemporary editorials would have appeared pallid and anemic to both newspapermen and readers of those more robust times." Editors nowadays apparently would rather be dull but safe than immortal but dead. Many of the camp editors were men of furious temper and scathing wit. One named Jernegan, a founder of the *Enterprise,* felt that his partner had wronged him, and thereafter did his best to denigrate and destroy him. After his death it was learned that his bitterness had been so extreme that in his diary he had written: "I call on God to curse the *Enterprise* and all, dead or alive, who robbed me, and I also call down the curses of the Great Jehovah on Nevada by quarter-sections and subdivisions."

One of Drury's most amusing stories concerns Editor Arthur McEwen who was having a nip in a saloon when a "little dried-up looking fellow" sidled up to him and said, "Hello, Mac, old boy." When Mac paid no attention the little fellow said, "Gimme a cocktail too, barkeep. I guess I'll drink with Mac." In sudden rage Mac snorted at him and said, "When I want you to drink with me I'll ask you." To the barkeep he said, "What makes you let a bum like this lie around your place to insult gentlemen?" He then told the bum to get out or he would sock him in the jaw and throw him out. A spectator named Brown told Drury that he almost fainted. A collection was then taken, and McEwen was presented with a tin pail full of dollars. What was it for? asked the astonished editor. Why, for standing up to one of the most notorious gunmen and killers in all the mining camps of the West. It was McEwen's turn almost to faint. He was ready to rush forth and beg the bum's pardon when Brown seized his arm and said, "Don't! If you want to be elected constable you can get it as slick as a whistle, and two years from now you will be solid for sheriff or congressman." Trembling with gratitude McEwen said he had not fired a gun more than a dozen times in his life and reckoned he ought to buy a pistol and do some target shooting. But most of McEwen's shooting was words in a verbal duel with Ambrose Bierce that has become famous. "There was never anything fiercer in western journalism." That is a lot to say. McEwen, said Bierce, was one "whom Providence has for some mysterious purpose permitted to wear a clean shirt and know how to spell." Bierce, McEwen retorted, was "most at home when breaking butterflies on the wheel, when torturing poor poetasters and feeble scribblers of prose, who but for him would remain unheard of." As for the notorious gunman, he confided in friends that he didn't want to fool around with any man who hadn't sense enough to be afraid of the camp's boss.

Like many of the editors, Marion of the Prescott *Miner* kept himself in a continuous and comfortable fury. As northern Idaho editors were enraged when the territorial seal and papers were smuggled out of Lewiston and taken to Boise, so was Marion when the capital of Arizona was removed from Prescott to Tucson. He called John Wasson, editor of the Tucson *Citizen*, "liar, affidavit man, scavenger, scullion. . . . We dare this abominable beggar. . . . thieves and blackguards. . . . Wasson has fouled the pages of his own paper." To which Wasson replied, in part: "He feels badly. Something hurts him; his bones always ache. . . . We wonder what is the matter with the poor sufferer? Is it because a generous people, that have too long stood his abuse, are about to start another paper in Prescott?" To which Marion retorted: "The whiffet of the Carpet Bagger's organ at Tucson appears to be spoiling for a newspaper fight with the Miner, but the Miner cannot lower its dignity and character, or neglect the interest of its supporters, by noticing so contemptible a blockhead, dog-robber, liar, and slanderer. . . . The Miner deals with principals; never with paid hirelings, cowardly calumniators or unmannerly understrappers who live by doing dirty work for third-rate politicians and their tools. . . . Back, dog, to your kennel!" In the same issue he was touched off again: " 'Dare not talk,' indeed! Lying wretch, we dare speak at all times and upon all occasions; not like you, however, from behind a mask, but openly and aboveboard. Furthermore, we wish to inform you that you are a cowardly political bummer, sneaking liar, and tell-tale eaves-dropper, without force, character, or standing. So much for this toady, who 'bends the supple hinges of his knees, that thrift may follow fawning.' . . . This adder; this loafer; this Gubernatorial bootblack and scullion, who bummed his living from the military. . . ."

The most famous Arizona feud between editors was between this same Marion of the *Miner* and the "equally belligerent but more poetically gifted Judge William J. Berry" of the Yuma *Sentinel*, though in the beginning they were friends. On one occasion the Judge said he would sue for libel, but for the adage, Sue a beggar and get a louse. He then continued: "It is unnecessary for us to say that the tirade of that miserable nincompoop against us is a tissue of lies, concocted by the addled brain of its author. . . . We are sorry to have to dirty our columns with any reference to this abandoned creature." A little later they both wrote valedictories and left the newspaper field.

But we are going to concern ourselves here with an Idaho feud, not because it is typical of the best and the worst of them but because practically nothing about it has been published. Probably no other newspaper feud could bring out more fully the savage bitterness that existed between the pro-Union and anti-Union factions. H. P. Langford, a Federal man and vigorously pro-Union, said that in the Virginia City area in Montana "I was in a Territory more disloyal as a whole, than Tennessee or Kentucky ever were. Four-fifths of our citizens were *openly declared secessionists.*" They had their Jefferson Davis Gulch, their Confederate Gulch, and in "local matters we were completely under the rebel rule." It was largely rebel rule in the Idaho City-Boise area that sparked the feud between the *World* and the *Statesman,* as it also was in that between the *Mountaineer* and the *News* in Denver.

When Richardson visited the Idaho Territory, society there

was not attractive. Murders were frequent; for with a majority of industrious, law-abiding settlers, the Territory had also many late rebel soldiers and Missouri runaways; and the worst desperadoes from California, Nevada, Oregon, and Montana. The legislature contained just one Union member; and during the war there was more disloyalty than in any northern community except Utah. Old Parson Strong of Hartford, the fierce political preacher in the days of Federalism, was accused of charging, from the pulpit, that all the democrats were horse-thieves. He replied: "It is a slander; I never asserted anything of the kind. But what I do say, and what I can prove, is, that all the horse-thieves are democrats." So in this community the Disunionists were not all desperadoes, but all the desperadoes were Disunionists.

There seems not to be much exaggeration in what

Richardson says, except in his statement that Idaho had all the worst criminals from the camps around it, though the *Enterprise* down in Nevada nominated Idaho for "the next Copperhead triumph" because "It is overrun with thieves, gamblers, highwaymen, guerrillas and murderers. It is the home of the outlaw, the paradise of crime, the lair of the escaped convicts of every state. Outside of the confines of the lower regions, better material for the formation of a Peace Democratic party cannot be found than in Idaho."

All of that the editor of the *World* dismissed as the usual lies published

in nearly every newspaper in the State of Nevada. California papers also very frequently indulge in a similar strain of ignorance, falsehood and folly. . . . Nevada welters in mortification and furious, unreasonable jealously, because Idaho has surpassed it in general prosperity. While the silver mines of Nevada have proved the sepulcher of thousands of fortunes, and the ruin of crowds of hopeful men, Idaho has sent forth a constant stream of the precious metals. . . . During the past year Nevada has "caved in" . . . its towns have decayed . . . its former population have left the country in huge disgust. . . . Idaho is a better country now than Nevada ever has been or ever will be. In silver ledges it equals, if it does not excel Nevada; in gold ledges it has no equal; in placer diggings it is second only to California in its best days. . . . A California paper calls Idaho "a frozen, desolate country, fit only to emigrate from—a country overrun with outlaws and Vigilance Committees." People who live here will set down the author . . . as an "egregious ass". . . . There has been more robberies, murders, and crimes of every character committed in either of those states during the past year. . . . There is not now, nor has there ever been, a Vigilance Committee . . . nor has there ever been cause for one.

All of which is like the boasting and threats of small boys hurled from a safe distance. Nevada's Comstock bonanza had not reached its summit. As for vigilantes, the Payette committee was even then coming into being, the Updyke and Dixon gangs were in full control of Boise City, and the murder of United States Marshal Pinkham was only seven months away, which would upset Editor Street's placid assumptions. But it was the way of editors to fire verbal shots at editors far removed from them, even clear across the continent. Reynolds of the *Statesman* was not as violent and venomous as some of them but he had a mean verbal arsenal when he took it off the shelf, as we

shall see. For instance, after by deception and stealth the capital of Idaho Territory was moved from Lewiston to Boise and Lewiston was in a fury, Reynolds began an editorial, "The city of bedbugs is in travail," a not very generous way to speak of a fallen opponent.

Because we haven't space to pursue the various elements in the feud between the *World* and *Statesman,* such as corrupt county officials (chiefly Democrats), Fenianism (the *World's* editors were Irish), Negroes, slavery, secession, Lincoln, we shall give only some of the choicest billingsgate. The feud really got underway in early 1865. By

James S. Reynolds

March 25 the *World* was saying that the *Statesman's* editor had "worried himself into a horrible passion" because he had run for political office and taken a licking. He was advised to let his "blood flow agreeably" and take to heart the last verse of the "Iliad Made Easy":

Oh, little children, think of Troy, and never fight at all; or, if you must, select a boy who's very, very small!

Reynolds of the *Statesman* tried to ignore Street of the *World* but by May 13 he mildly observed that the *World* was "so morally oblique that it is sure to take the wrong side of every question"; on May 18 he characterized as fiendish and infamous Street's effort to make a jest of Lincoln's assassination; by May 23 it was "mean-spirited"; and by May 30 two long editorials were given to the *World* but they were pretty mild. May 20 Street had said that Reynolds had been "ignominiously" kicked out of his party because his spirit was one of "natural meanness and malicious mendacity" which added to the "universal derision in which his paper is held." Calling it the *Mulatto Squall* he thanked God that his own was not as "depraved and imbecile" as Boise's, whose editor had never learned "to put his drivel into respectable English." On June 10, under the word DEMORALIZED, Reynolds said, "This is the only term that is applicable to the condition of the democratic press . . . that the government still allows to publish treason. . . . Animated only by hate, they continue to hiss and snap at random, seeking by what means they can do the most harm without endangering themselves. . . ." June 20 he thought the *World* was showing a little improvement: "The last number contains but two paragraphs that show the editor to be a malignant, disappointed secessionist. . . . Keep on, brother Street. . . ." July 11: "The fellow that turns the crank for the *Organ* at Idaho City, for the regular Petticoat Democracy, has discovered an 'isshy.' The democracy of Boise County has nominated a ticket. He indorses it in a leader which reads about as follows: 'Negro, nigger, republicans, issue. Nigger, nigger, chinamen, issue. Issue, Indians, constitution, suffrage, democracy, harmonious nigger, nig-g-e-r-r in Idaho." By July 29 Reynolds had taken down a bigger gun: "That monster viper, the copperhead, is dead. With one slimy fold in each State of the Union, its tail extending over the distant Territories and its head in Richmond, where it received its daily food from the hands of the hermaphrodite Jeff. . . . the tail in Idaho still exhibits a convulsive wiggle. . . ." August 19 he quoted from the *World*, "A more vile, false, and abandoned attack has never been made upon the character of an honest and worthy gentleman." The Reynolds retort begins: "There now, get seven or eight big fellows to hold you up, and keep you from hurting

Henry C. Street

somebody, while we talk to you calmly and coolly like a father."

It becomes clear by September 9 that Street had got under the Reynolds skin:

That puerile concern, the Idaho World, struggles from week to week for a little notoriety, by senseless twaddle thrown at the STATESMAN. If the WORLD had a circulation sufficient to make it an instrument of any considerable harm, or if its dish-water filled columns were written with any considerable degree of ability, it would pay us sometimes to devote a little attention to it, and occasionally show how ridiculous and how mean it is. But since very few persons ever see the paper, and still fewer read its brainless scribbling, it isn't worth while to notice it except as a source of light amusement now and then. We feel kindly enough toward the editor and are willing to notice him occasionally, just to let the public know that he still survives, but he must excuse us from occupying very much space with him.

Reynolds certainly didn't expect to get away with such condescension. A week later he was wondering what was to be done with the "liberated darkey" and said that both politicians and philanthropists were "racking their brains—some of them have a few." A "stroke of the pen" was not enough to make the Negro a responsible citizen. The Negro did not "fight for possession of the ballot box; he does not comprehend it and it never will, of right, belong to him till he does comprehend it."

In the *World* on the same day, by one who signed himself Celt and pretended to be a *World* correspondent in Boise (but who undoubtedly was Street), appeared this blast: "About this man Reynolds I desire to say a few words as I think he is not generally known in Idaho. He is a hatchet-faced creature whose soul and intellect are as narrow as his look is mean and contemptible. He is thoroughly despised, and with just reasons, even by 'Union men' in every community wherever he is known." It is said, the writer continues, that Reynolds formerly was a Federal spy and informer, and that one of his ancestors betrayed two brothers in the Irish rebellion of 1798. He had taken their children to his lap and had "kissed and caressed them" and afterwards "put the rope around the necks of the fathers." He quotes from a book on Irishmen by Madden: " 'He was a man who measured his value by the coffins of his victims'. . . . The Boise City Reynolds, betrays all the mark of having descended from the Irish spy and informer. 'Blood will tell.' And the congeniality with which the Reynolds here takes to the occupation of spy and informer, as well as the natural inate [sic] baseness of the man, betrays his origin. A man who will greenback his brother, and starve his wife to death, who makes his first raise out of the 'secret service money' of Mr. Lincoln's administration, is just fit to be leader of, and to give ideas to, the shoddy party of Idaho Territory."

On September 23 Street outdid himself. After a long paragraph in which he warmed to his subject he wrote:

Too illiterate to be able to find his way about the State of Oregon alone, weak-minded enough to be the object of the derision of even the donkey he rides, his malice is the only noticeable trait of his character. The exhibition which he has made of himself of late can justly be compared to that of a helpless, wounded adder, with malice enough to sting to death without the ability to injure.

Like the writhing serpent he blindly strikes promiscuously at all within reach, including his own flesh, for the worst enemy of the Statesman could not wish it a greater injury than is self-inflicted by the public exhibition of the human depravity reflected from the pen of its editor. "Curses like chickens come home to roost," and the filthy obscenity of that paper can only render the circumambient atmosphere of its own office oderiferous with the perfumes of its vulgarity.

As with most of the mining camp editors correct spelling was not an accomplishment. He continues:

Billingsgate collected from the vilest portions of shunned outskirts, together with all the "honorable points of ignorance appertaining thereunto," are the only portions of the English language with which his breeding and association appear to have made him familiar. The grimances and chattering of a sea-sick monkey are more expressive than his expletives when he ventures beyond his native dialect. His attack upon Mr. Malone is a violent and unsuccessful effort to disgorge his childish wrath into a vernacular sentence,—"a tale told by an idiot full of sound and fury, signifying nothing." It falls into a filthy string of words which might sue out a writ of habeas corpus from any decent court in Christendom.

It is not the first time that language has appeared in that paper which would not be tolerated,—and cannot be found in any decent paper in the United States. To Mr. Reynolds it belongs exclusively. He will be remembered and known by it as long as he retains his name. An indignant community which he has grossly insulted will remember it. The finger of scorn will point its contempt upon him wherever he goes. Shame will cry out and blacken his face with its unforgetting memory. If it be true that "He most lives, who thinks the most, feels the noblest, lives the best," there is but little evidence extant that the editor of the Statesman has known much of life. Nothing but his connection with a profession to which he is a disgrace, preserves him from that oblivion which his own insignificance and the merited contempt of the public have consigned him.

"Here he is—and here he lies,
No one laughs—no one cries;
How long he'll stay or how he fares,
No one knows—no one cares."

Of Jefferson Davis Reynolds had written April 29: "What a picture of unamiable ambition and unwept disgrace does this haughty, scheming villain present!" After Street's consignment of him to oblivion Reynolds stooped to his lowest point in the feud: "H C Street (ed of the world) and Pat Malone—the editor and his correspondent. The former inherited his bestial instincts from the low-flung 'poor white trash' of North Carolina, or else sucked them, during infancy, from the filthy breast of some plantation nigger wench —process of instillation being easy and natural. The latter, too foul to be the conception of a lazar-house maggot, an abortion produced in a 'laystall.' A well-matched pair, truly. Malone is nearly as mean and low as Street—both baser and more nasty than the slime on a dysentery evacuation." Reynolds then left Boise to examine the Owyhee mines and during his absence his paper was quiet. Street kept firing at him and by December 19 Reynolds again took note of his "whining nonsense, mixed with vituperation." By January 30 of the next year he says of Street and another attack, "We will attend to him as soon as we can get time." By February 3 Street was charging Reynolds with the "most dirty and cowardly trick that we have ever seen attempted by any newspaper. . . . fraud, deceit, slander, villification and corruption. Every instrumentality has been seized upon to strangle the truth, to warp the public mind." Caleb Lyon of Lyonsdale, the Territory's insane governor, was now dragged into the feud and both editors used him as an excuse to larrup one another, Reynolds "laying open the rotten and plundering schemes" of Lyon, while Street vigorously defended him. By March 17 Reynolds was a "bald headed liar."

On July 7 Reynolds announced that the *Statesman* would thereafter be managed by a board of directors but there was no change in its policy. July 17 he took on the editor of the San Francisco *Examiner*: "The people know his treason too well. . . . we would as soon think of Satan's cleaning out the cesspools of hell, or planting roses around the mudsill of tophet. The copperhead slime of such sheets will follow the pen, until clogged by its own dirt, it ceases to move." Reynolds looked far afield and saw other editors who in his opinion were unworthy of their trust, yet he lived in Boise month after month when Sheriff Updyke was head of a gang of robbers and murderers, and seems never to have suspected it.

He never realized the extent of the corruption under his nose until after the vigilantes were organized and had hung one of the worst offenders from a tree. Street was violently opposed to the vigilance committees, dismissing them all as midnight assassins, and this now became still another matter for bitter exchanges between the two editors,* as it also had been or would be in other areas. Almost without exception the anti-Union men opposed the vigilantes, and the pro-Union men supported them. This was in part because most of the chief criminals *after* the outbreak of the Civil War and especially in the early postwar years were ex-soldiers of the Confederacy or Confederate sympathizers.

A judge had said in his charge to a grand jury in Boise:

You will visit the County Jail, provided you can find it. The Court has been here two weeks, and after the most diligent enquiry I have been unable to find that there is such a place. . . . You are the servants who are to do this work—cleanse this cess-pool into which the outlaws of Utah, Nevada, California, Kansas and Oregon have emptied themselves. . . . You and I are the Vigilance Committee, not self-constituted but legally appointed, and chosen from the people to be the guardian of their rights. Stand firm in the discharge of that duty. . . .

It was impossible to stand firm where the leaders of crime were public officials, unsuspected even by the judge. Street up in Idaho City believed, or pretended to believe, that all vigilance committees, whether self-constituted or legally appointed, had only one thing in view—to hang all the Democrats. He had, said Reynolds, "vigilante on the brain. . . . he don't like the action of the grand jury." Sheriff Updyke, with a great show of civic zeal, rode away with a posse ostensibly to fight marauding Indians, but after the nature of the man became known it was suspected that he and his bandits had been robbing stages.

Street kept hammering away with his charge that the vigilantes were hanging the Democrats, including all the political leaders, until Reynolds exploded: "Of all the malignant fools that were ever suffered to run loose, the editor of the WORLD seems to be the most untruthful. Under his management the paper has established a

* See the chapter, "The Vigilantes."

character for the most utter disregard of truth or consistency, a moral as well as political leper among newspapers. The only consistent course it has ever taken has been of late, to lose no occasion to proclaim itself the organ of road agents, stage robbers and villains generally." How much Street was aware of we do not know, or how much he concealed of what he knew, but it is a fact that Updyke and his bandits in Idaho, and other gangs and their leaders in other areas, prospered enormously in crime while loudly defended by editors who said they were only law-abiding Democrats persecuted by Unionists. Street withdrew as editor and another Irishman named O'Meara came in but there was no change in policies; the two men were so much alike in beliefs and temperament that the reader would never know that there had been a change if he did not observe the new name on the masthead. After declaring that the *Statesman* had insinuated that he had had a hand in a bogus gold dust swindle O'Meara asked: "What will not the dirty dog resort to in his malicious spleen." Like Street before him he got under the Reynolds skin. January 17 (1867) the *Statesman* said: "It is seldom necessary to contradict falsehoods in the World. They are so apparent and silly as to contradict themselves." January 24: "The editor of the World has been snarling like a bear with a sore head for two weeks past" because a man had been jailed who had been in "the bogus business." February 7: "The editor of the World is a very brave man" who had made "a mean assault upon the Governor of Oregon, who lives five hundred miles away. . . . We expect to hear next of his attacking some widow woman." February 14: "The tadpole of the World says he will make that paper pleasant for us this week. . . . if you succeed in making one number of your paper pleasant or interesting to anybody, it will be the first instance of the kind since your connection with it, . . . Come, to it again, little one."

There was established in Boise City a small newspaper, called the *Democrat,* and the editor of this paper joined O'Meara in venomous attacks on Reynolds, until by February 13, 1868, Reynolds had had enough of J. Walton Browne, the *Democrat's* editor: " 'What can the Statesman say that would lose a vote for a decent man?' " That, said Reynolds,

is the most dignified sentence selected from a half column of vulgar personal slang which usually appears in the *Democrat*. We have purposely avoided any discussion with this thing called the "organ of the democratic party," except in a few instances to point out its affinity with rascals, and its proneness to attack and vilify [sic] the best man in the community. We have even suffered the "organ" to admonish the Statesman that it was conducted in "bad taste," without noticing the character of the beast who conducts the organ, but when he continues to ply the public with his vulgarity, and at the same time talks about "decency" in the Statesman, we think it about time the public should know what kind of man this J. Walton Browne, editor of the Boise *Democrat,* is. Those of our readers who have noticed the manner in which he has conducted the *Democrat* may not be aware, but they will not be surprised to learn that not long before he set up his "decency" shop in this city, he was caught under the dwelling house of a respectable citizen of Silver City, secreted for the purpose of insulting and ruining a young girl; and she gave the alarm, and a gentleman rushed out just in time to collar him before he escaped; that Browne begged for mercy, and excused himself by saying he had gone under the house for a purpose that cannot be mentioned in print. Just think of it! Under a family parlor in the night time! . . .

Browne begged lustily and said, "O, you wouldn't arrest a man now." The gentleman with difficulty restrained himself from punishing him on the spot, and let him go by vigorously applying the toe of his boot to Browne's posterior. We do not propose to tell half the truth about him, but just give the public an idea of what kind of man he is. He has been repeatedly caught peering in at the windows trying to observe through cracks in the curtains respectable ladies while preparing to retire. The only reason why he has not been chastised for this nameless conduct is that people dread the notoriety of contact with such a loathsome being. Over in Owyhee he was a drunken bummer, too lazy to work, subsisting like a lazaroni, and content to wear the cast-off clothes of other men rather than work and earn them. His first appearance in public in this city was at a social party in such a state of beastly intoxication as to be an object of disgust. This is not half of what we might say with truth of this man Browne, but it is enough to indicate to the people of this city what kind of man it is who is preaching "decency" and "taste" to the Statesman, and endeavoring to gain a little notoriety by filling his filthy sheet with vulgar epithets applied to the editor of this paper. We mean no play about this matter, for we can prove the above assertions by gentlemen of respectability now in this city, and we are at liberty to use their names whenever we think proper, only they will if called on state the case worse than we have stated it, and make him appear more disgusting than we can be allowed to do in print. We should have spared our readers anything more than a casual allusion to this nameless thing if he did not sometimes impose upon people who are not acquainted with him. But we do not intend the public shall be beguiled into believing, encouraging, or supporting him long in Boise City, without being fully acquainted with his true character.

Three weeks later the *Statesman* reported that editorial control of the *Democrat* had changed hands. On the same day that Reynolds castigated Browne he was again after O'Meara: " 'The *Statesman's* departed idol—Lincoln.' There be some things that can be more emphatically said by innuendo than by positive affirmative assertion. The above, quoted from the *World,* is a case in point. It reveals the depravity of the writer in a stronger light than a conviction in a court of complicity in the assassination could. The editor of the *World* would no doubt express great regret that

Lincoln was murdered if directly questioned, but here he unwittingly discovers how hypocritical it would be. Lurking in that little sentence, that sneer, are his true sentiments. He might just as well have said of Booth, 'Well done, thou good and faithful servant.' " Two weeks later the *World* said that an editorial in the *Statesman* was "simply one reeking mess of infamous lies" but time was passing and though the bitterness between North and South had hardly softened at all by 1872, in that year Reynolds sold the *Statesman*. He had married a young woman, who had, as Donaldson puts it, "too much hidden past"; she ran away, taking everything Reynolds had that was not nailed down, and after getting rid of the newspaper Reynolds hunted for his wife in California and found her in a place that as newspaper editor he would have thought unfit for mention in print. He "considered himself fortunate in escaping with his life" who had stood up to all enemies during his eight years in Idaho.

CHAPTER 27

Hoaxes

"The literature of scientific hoaxing is rich and varied," says Robert Silverberg,

encompassing the explosive Piltdown Man fraud, which was exposed little more than a decade ago; Dr. Cook's controversial claim to the discovery of the North Pole; the monumental "Etruscan" statues that proved so embarrassing to the Metropolitan Museum of Art in recent years; and many others. Men of seeming rectitude have done their best to confuse, to deceive, and to obfuscate. The psychopathology of hoaxing is a fascinating study in itself, worthy of investigation.

"Of course all right-thinking folk" would agree with Mark Twain, said Drury, "that jokes of the practical kind are vulgarian. It is the more painful to record, therefore, that the camp's Best Minds—lawyers, doctors, bankers, editors, gentlemen gamblers—entered with zest into such gigantic sells, along with the rest of us."

The gold rushes and their camps were perfect setups for the hoaxers, for the reason that avarice and greed and wishful thinking had taken complete possession of thousands of minds. Of the innumerable hoaxes we have space for only a few of the outstanding ones, and it seems appropriate to begin with Henry Vizatelly who, as J. Tyrwhitt Brooks, to make some money and have some fun, wrote (in ten days, it is said) *Four Months Among the Gold-Finders,* published in London in 1849, and put out as a journal of his experiences after a visit to California. Nobody knows how many writers took it with deadly seriousness before Douglas S. Watson exposed it for the fraud it was, almost a century after it appeared! The august London *Times* "with almost un-English enthusiasm" gave it about three columns and said that here was a "gentleman who knows all about it! A live Englishman who has set foot in the diggings, has had his hands in the washings, has scratched for the dust till his back ached, has picked up lumps as big as your fist without any labour at all." But he returned as poor as he went. "Reader, why should you journey over sea and mountain, through difficulty, danger, heat, vexation and trouble to achieve the same foolish and unpofitable purposes?" Because one of Visatelly's invented men had refused to dig on Sunday the *Quarterly Review* thought it would be a smart thing to close the British post office on Sunday; and the *Atheneum,* overgrown with the moss of prestige, gave the hoax six solid columns. The pedantic and bearded *Dictionary of Dates* copied "Dr. Brook's" tale of how Sutter told him exactly how the gold had first been found. But Visatelly's triumph was more than that of a clever deception. Robert E. Cowan, compiler of the well-known bibliography of authentic books about California and the Pacific West said, after reading the Watson exposure: "Despite the fact that as a narrative the . . . book is entirely fictitious, its historical properties are still more nearly accurate than three-fourths of all the similar works of contemporary writers who actually may have visited the gold diggings." That's an extreme statement but if Vizatelly were alive today he certainly would love it.

A number of camp editors besides Mark Twain were skilful hoaxers and possibly the ablest of them all was William Wright who wrote under the name Dan De Quille. Emrich thinks his two best hoaxes were of the traveling stones of Pahranagat Valley and the solar armor. According to the first a prospector discovered a field of round stones about the size of walnuts. "When scattered about on the floor, within two or three feet of each other, they immediately began traveling toward a common center, and then huddled up in a bunch like a lot of eggs in a nest." In the second story a man named Newhouse invented a refrigerated armor as protection against desert heat.

Its secret was a hood and jacket saturated with water from which the sun drew moisture, and the person wearing it could set it at a temperature most comfortable for him. In Death Valley the thing went haywire and the man wearing it froze to death, with an icicle a foot long hanging from his nose.

The amazing thing is not De Quille's inventions but the seriousness with which the scientific world accepted them. Barnum offered him ten thousand dollars to show the stones moving to the center; German scientists charged him with bad faith in withholding details; and by November 11, 1879, the thing had become such an international sensation that De Quille said wearily in his newspaper: "We have stood this thing about fifteen years. . . . We are now growing old and we want peace." After telling of hot springs that tasted like chicken soup and a hill of singing stones that gave off pleasant music, he came up with the Shoo Fly, and an excited entomologist asked for one, saying: "I presume them to be rather a hymenopteron (wasp or bee) than a depteron (two-winged insect)." He promised to take the matter up with the Smithsonian and other learned societies. De Quille published the letter and commented: "What the professor says may be true, but when we view the insect and consider its cuspidated tentacles and the scarabaeus formation of the thoracic pellicle, we are inclined to think it a genuine bug of the genus 'hum.'"

The London *Times* proposed to the Government that it buy a shipload of the refrigerated suits for its soldiers sweltering in India. A journal of engineering was impressed by his cock-and-bull idea for perpetual motion: simply send sand and water into the atmosphere on strong winds, for after the wind died the falling sands would keep windmills turning to pump water. Nothing seems to have been too fantastic, as long as it seemed to be in the fields of science. A different sort of hoax, and one easy for most persons to believe, was Mark Twain's about the bartender who speculated, lost his money, murdered his faithful red-haired wife and nine children, and killed himself. This yarn aroused great consternation and anger, although the bartender was actually a bachelor, and there was no married red-haired woman in the whole area. As Emrich says, De Quille was gentle, Mark Twain was often savage: the bartender not only knocked out the

brains of six of his nine children and scalped his wife, but with his throat cut from ear to ear and the gory scalp in his hand he rode into Carson and fell dying before the Magnolia saloon. The sheriff and a posse took off for the ghastly scene: "The scalpless corpse of Mrs. Hopkins lay across the threshold, with her head split open and her right hand almost severed from the wrist. . . . In one of the bedrooms, six of the children were found." Mark gave names to the children, and gory details to harrow the sensibilities of his readers. He apparently did not foresee what a sensation he would cause with his silly story, for the next day he said in the newspaper, "I take it all back." The uproar so alarmed him that he almost fled the area.

A friend of Charles Russell told of what was done with a piece of Limburger cheese that came in the mail. It was rubbed inside hatbands, on doorknobs, inside drinking cups; and when the pranksters saw a drunk asleep they thoroughly filled his beard and mustache with it. "Next day Charlie Russell saw him out back of the saloon, sitting on a box and looking very tough. He would put his hands over his mouth, breathe into them, drop them and look at them, and shake his head." When Russell asked him how he was feeling he groaned horribly and said he was not so good; this was not like the other times when he had felt bad. He again covered his mouth, breathed against his hands, sniffed the incredible odor, and gave off a despairing groan. Because he now had the awfulest breath any man had ever had he figured that his stay on earth would be short. "O God!" he cried and almost toppled over. Russell was so convinced that the man was going to die right there that he hastened him over to water and got him to wash his beard.

When a hotel proprietor told a gullible public that the waters of the Calistoga Springs carried so much gold that he could take sixteen ounces out of two barrels of it, Mark Twain wrote a letter:

I have just seen your dispatch from San Francisco in Saturday evening's Post. This will surprise many of your readers but it does not surprise me, for I once owned those springs myself. What does surprise me, however, is the falling off of the richness of the water. In my time the yield was a dollar a dipperful. I am not saying this to injure the property in case a sale is contemplated. I am saying it in the interest of history. It may be that the hotel proprietor's process is an inferior one. Yes, that may be the fault. Mine was to take my uncle (I had an extra one at that time on account of his parents dying and leaving him on my hands) and fill him up and let him stand fifteen minutes, to give the water a chance to settle. Well, then I inserted an exhaust receiver, which had the effect of sucking the gold out of his pores. I have taken more than $11,000 out of that old man in less than a day and a half.

Mark goes on for several more paragraphs: one of his faults as a humorist was that sometimes he didn't know when he had said enough or more than enough.

Most of the miners disliked snobs as much as they loved hoaxes. When there came to one of the camps a well-dressed gentleman who preferred to eat his meals alone in his hotel room the miners were so outraged that they went to their accommodating sheriff and had the snob arrested for horse stealing, a capital crime, and thrown into the horrible jail. He was brought out the next day for trial. The prosecutor, demanding the death penalty, painted a picture of one of the slickest horse thieves in the West. The defense attorney pled for the man's life, calling on judge and spectators to see before them a gentleman with such faultless manners that he dressed for dinner, which he took in the quiet of his own room; a man who never bought anyone a drink in the saloons, for the reason that he strictly minded his own business; and a man who couldn't ride a horse without falling off, much less steal one, or tell a horse from a jackass. It was all too much for judge and spectators; laughter got them down, and they all poured across the street to the nearest saloon. The snob who minded his business got the point, that in the frontier camps snubbing his equals was a crime no less serious than horse stealing.

A few miles northeast of Downieville, which is northeast of Sacramento, is a small body of water called Gold Lake. A man named Stoddard appeared in the Downieville saloons with a handful of gold nuggets; and when pressed by greedy and excited men to divulge the spot where he had found them he said he had gathered them from the waters of a lake. That was enough for men already bereft of their senses. With pleading or threats they prevailed on Stoddard to lead them to the spot, and he set off, with a group of eager

men at his heels, who fully expected to be led to a lake where they could pick up solid gold nuggets by the pailful. After a long difficult climb through heavy timber up steep mountains the men became suspicious and then furious, because Stoddard would not or could not take them to the lake. They threatened to lynch him. Bancroft thought it was a hoax and the most memorable of all the hoaxes played in California on greed and greenhorns. Another historian says that "probably it cost the lives of scores of the credulous gold-hunters of '49 and the early '50's." It was a measure of the credulity and greed of many of the miners that even when they were taking from their diggings a pound of gold a day they would rush off on the latest wild-goose chase. Stoddard himself had been one of them. It is possible that he did pick up nuggets in a rich area and was unable to retrace his steps. If that is so, Stoddard's spot is another of the "lost mines."

But by far the two largest hoaxes of which we have found record are the Mount Pisgah and the Diamond Field. Bob Womack, first discoverer of gold in the Cripple Creek area, had been prospecting off and on for years when two men decided to sink a shaft. It is said to have been eighteen feet deep. In their pile of earth they set a sign to tell the world that this was a placer, and the discoverers were S. J. Bradley and D. G. Miller, April 7, 1884. They claimed twenty acres. After reading the sign a rancher in the area named Grose rode forth to tell his neighbors, who told their neighbors, while Bradley and Miller were blandly saying in Canon City that their samples assayed two thousand dollars to the ton. As the glad tidings spread from ranch to ranch and as far away as Salida and Pueblo and Denver, an agent of the Denver and Rio Grande began to send telegrams to his superiors.

The result, if the matter is not exaggerated in such books as Sprague's, was one of the wildest stampedes in mining history; people abandoned their ranches, they abandoned their jobs in the towns, and after frantically getting their hands on food and bedding and tools they took off for the hills, some jolting along in dead-axle wagons and carts, some on horses and mules, and many walking, all of them with "picks, shovels, tents, frying pans, coffee pots, beveled gold pans, whiskey barrels, beer kegs, hams and groceries." Bradley and Miller hadn't even chosen a likely spot; they had dug their hole where digging was easy, on a hillside of luxuriant gramma grass, and it wasn't long before men were digging all around them. Imagine the scene—hundreds of men digging with the energy of a dog after a gopher; pausing again and again to seize a handful of earth and sift it through their fingers and bug their eyes at it; and then furiously going deeper in soil that was fine for orchards and lush pasture but that had hardly a color in a thousand tons of it. Miller and Bradley meanwhile sat on their pile of earth with shotguns across their laps and watched the scene.

And far beyond the scene, which was a flank of Mount McIntyre but which the stampeders had heard was Mount Pisgah, thousands of frenzied gold rushers were moving in. We can imagine them too, for we have seen people just like them in stampedes over the whole West: in every man's mind was the tormenting thought that he would be late, that he would get only a poor claim or none at all; and in every woman's mind, for there were wives in this stampede, was the hope that at last her fumbling and ineffectual man would strike it rich. The Denver and Rio Grande loaded its cars with gold seekers and off the absurd engines went, puffing their black smoke and shrieking their warnings as they slowly climbed the canyons to the camps closest to the Pikes Peak area. Businessmen all over the Colorado country rubbed their hands and discussed the matter and wondered if they should close their shops, grab a pick, and take off. And everywhere the newspapers were busy, spreading the wonderful news.

Bob Womack was busy, too. Though he was not the sharpest prospector who ever peered into a hole or studied a piece of rock, neither was he the dumbest by any means. He knew that you didn't find rich placer beds in lush pastures where there was no sign of ancient rivers. So he slipped away with a sample of earth near the Bradley-Miller hole and a professor of metallurgy at Colorado College got busy with it. It had neither gold nor silver. In another sample on the mountainside there was only three-tenths of an ounce of silver to the ton; and still another had only one-tenth of an ounce of gold. That much gold to the ton could probably have been found anywhere in those mountains.

Why Bradley and Miller set up such a cruel hoax no writer has tried to explain. Possibly they thought that if hundreds of men began to dig

over the mountains gold would be found and they might worm their way into the riches, as Old Pancake had. One of the thousands of stampeders expressed the feelings of many when he said, "We dotta hung the bastards." For Womack, Sprague thinks, it was "a good story. It made people laugh wryly." We don't imagine that men who abandoned jobs and ranches to rush to a gramma grass pasture laughed at all. But we would guess that if they had been given a few years to get over it they could have been hoaxed again.

But the granddaddy of all the hoaxes concerned a fabulous field of emeralds and diamonds. Jackson thinks it could never have happened but for the two fantastic decades of gold fever and gold rushes that immediately preceded it, but others still think that Barnum was right.

In London a banker and real-estate speculator, one Asbury Harpending, formerly of San Francisco, received from his good friend William Ralston, who made his pile on the Comstock, a cable that must have popped his eyes out, for it cost $1,100, or more than $5,000 in present dollars, and it said that somewhere in the United States was a vast diamond field worth at least fifty million. And not the least of the news was the statement that Ralston was practically in control of it. Would Harpending please come to San Francisco to manage this new enterprise? Harpending refused, but it takes more than one refusal to make a prudent man of a banker. When Baron Rothschild showed interest in the glittering enterprise Harpending had second thoughts and took another look. Could it possibly be true, or was Ralston fast slipping into senility? Harpending was losing his grip on the realities when another frantic cable begged him to come, and so, propelled by credulity and greed, he hastened to a ship, and arrived in San Francisco in May, 1872.

Looking as if they had spent their lives being buffeted by wind and weather, two men entered Ralston's Bank of California with a worn parcel and asked to have it put in a safe place. When asked what it contained the weary men, so obviously in need of sleep and rest, unrolled the dirty canvas and revealed to astounded eyes the gleaming wealth of large uncut diamonds. When questions were fired at them they were simply too weary to talk. They took a receipt and left. It didn't take long for their secret to fly from tongue to tongue, but the two men had known that it would not. Their names, they said, were Arnold and Slack, and for a shrewd judge of men like Ralston they had the look and manners of two yokels who had struck it rich. Questioned again, the yokels, still weary, said the diamond field was off yonder somewheres, Arizona-way, to hell and gone. Some men organized scouting groups and took off for Arizona, not even pausing to kiss their wives good-bye. Gold was all right, but a field where you could walk around and pick up diamonds, as you might chestnuts in a forest, was a lot better. Ralston's blood pressure steadily mounted.

Did Arnold and Slack want financial help to develop the field? Well, now, they—that is—why, yes, they might, they would think about it. Fact was, they hadn't quite figured out what to do. For a small—oh, a very nominal—interest in the field would they take a hundred thousand in cash? They rolled their sleepy hillbilly eyes at one another and scratched their unwashed heads. Well, yes, with that much they could get the thing going. Ralston and his pals, by this time pale with avarice and running a fever, felt jubilant, for they believed that they had their camel's nose inside the door. But the two hillbillies turned stubborn. They wanted the money at once. Ralston explained to them that a banker simply could not advance such a sum without a look at the property. Arnold and Slack kallated as how that might be fair enough. They would think it over. So reluctant that assent was obviously dragged out of them they agreed at last to take with them two men of Ralston's choosing, provided that over a part of the journey they were blindfolded. That, Ralston said, was fair enough; and so two blindfolded men went and returned, and had more diamonds and even a few rubies. Their reports glowed with jeweled lights. Ralston's fever went on climbing, and at this happy moment Harpending and another shrewd judge of men arrived from London.

They wanted at once to talk to Arnold and Slack but they had vanished; so Harpending brooded over the reports, and on investigation found that Arnold was a miner with a good background but that Slack was unknown. Ralston, feeling cagily secure, confided to Harpending that he now had in the safe $125,000 worth of diamonds, and that if this was not enough to secure the hundred thousand in cash the men would simply go to their field and pick up a bushel or two. Harpending now felt the onset of fever and a rapid pulse. He allowed as how that diamond field must be quite a sight. While he was congratulating himself on his decision to come from London a telegram arrived from the hillbillies. They were in Reno, on their way back and wanted someone to meet them and help them protect the huge treasure they had with them. Harpending said he would love to do that. What a wonderful country these United States were after all! When he met the men he found Slack exhausted and asleep, and Arnold half-dead with fatigue and anxiety. Did Mr. Harpending understand what an ordeal it was to ride through country thick with bandits, with a king's ransom in diamonds in your poke? Harpending understood it very well. Could he see the jewels? Arnold confessed between yawns that they had found an area in their field where quality diamonds were ten times as thick as in the areas they had formerly explored. Unfortunately when crossing a river suddenly swollen by rains they had lost a good part of their treasure; but they figured that what they had with them, plus the pile Ralston had, ought to be enough security. Oh, by all means, said the enraptured Harpending.

When the party arrived in Oakland Harpending and his fellow bankers were fit to be tied, for they wanted a look at the stones at once; but Arnold said he and Slack were too doggone tired to look at them that night. Still, they did all go to Harpending's quarters and a sheet was spread on a billiard table and handsful of diamonds along with a handful or two of emeralds and sapphires were poured upon it. That did it. It was now made plain to the simple and unsophisticated diamond finders that this was real big business and far beyond their financial and other powers; millions of dollars would have to be spent in development and they ought to be satisfied with a relatively small share, for that would be enough to make them rich. Looking as suspicious and unhappy as only yokels can when in the presence of shrewd and rich and wordly men the two men signed papers that would give them $600,000 for three-fourths of the entire field. A corporation was set up with a capital stock of ten million; the bankers let a few close friends up to the table for very small slices of the pie; and then they thought of Tiffany's.

Well, Tiffany's, at least in those days, must have had a lot yet to learn; for with Horace Greeley and other fascinated bigwigs looking on, Tiffany himself held stones up to the light; and his judgment so entranced Harpending that he left us a record of it: "Gentleman, these are beyond question precious stones of enormous value. But before I give you an exact appraisement, I must submit them to my lapidary." The lapidary was not quite as bewitched but he said it was a fine pile of jewels. Now the smart thing to do was to find a mining expert and take him to the field. Henry Janin was chosen, for he had examined six hundred mines; his fee was stiff but his reputation was almost alone in its eminence. Arnold meanwhile was restless; he had paper, yes, but he had no cash. He seems to have been a superb actor. Conceding that Arnold was not being unreasonable Harpending said they would all return at once to the West and take Mr. Janin and his party to the field. Then the cash would be forthcoming.

In Rawlins Spring, Wyoming, the party rented horses and mules and vanished into the mountains, with Arnold and Slack leading the way. For them this was to be the severest of all the tests. If they were honest men they had nothing to worry about but they were no more honest than the bankers who still sized them up as a couple of lucky hillbillies. After a few days of arduous travel Arnold pointed to a small basin littered with boulders and traversed by a small stream and said that was it. With the supreme confidence of a man who had nothing to hide he told them to go ahead and dig and explore and make up their minds. With picks and other equipment Janin and his assistants set to work and almost at once there was a shout from one of them. The others ran toward him and saw that he had uncovered a very fine diamond. Then they all dug with fresh energy and they turned up diamonds, rubies, sapphires, and emeralds. This was just about the most wonderful act from the hand of God that Harpending and his expert had ever heard of. In his memoirs Harpending confessed that he felt "the intoxication that comes with sudden accession of boundless wealth." What a happy man he was!—and Janin, because wherever the engineer went he dug up more diamonds. Still, he put in a third day at it, to convince himself beyond all doubt; and with six

hundred strikes and no errors he and his party staked off the entire country roundabout. What a field it was! Now to get back to a hotel and a bath and then to their offices, to start the big handsome wheels of finance turning.

It was a great triumph for the two yokels. Incredulous writers have wondered how they could possibly have deceived and swindled such giants of the financial earth. But there is no mystery, if one keeps in mind human avarice and greed. The stones were actually diamonds—inexpensive diamonds that Arnold had bought in the diamond markets and with which he had salted the high mountain basin. He probably used a total of thirty or forty thousand dollars' worth. It is plain enough that Tiffany's lapidary was not worth his board and room, much less his pay; and that Janin was out of his province when he looked at precious stones. The bankers like all good men and true had taken leave of their senses when faced with the prospect of fabulous wealth. Another $300,000 was paid to Arnold, who now had power of attorney for his friend; and like the sensible fellows they were they disappeared. Harpending meanwhile was telling envious friends that Janin was "wildly enthusiastic" when he saw the field, and said it "would certainly control the gem market of the world." News of the wonderful discovery "upset the caution of the wisest heads in the Old World, as well as in the New. There was a wild scramble to get on board, almost at any price." An article in the New York *Sun* located the fields in southeastern Arizona, which was off, says Harpending in *The Great Diamond Hoax,* by seven or eight hundred miles. The same article said a single gem "larger than a pigeon's egg, of matchless purity of color, worth at a low estimate $500,000" had been found, and this glad news started a stampede for Arizona.

The San Francisco *Alta,* August 1, 1872, gave its readers the news of the "American Diamond Fields. One Thousand Diamonds Now in This City. Also Four Pounds of Rubies and Large Sapphires."

We have a wonderful story to tell. We listened to it at first with incredulity. . . . The place of the new mines has not been communicated to us by any of the interested parties, but street rumor says it is New Mexico. About three years ago, they say, an Indian near the diamond deposits gave several diamonds to a white man who brought them to Messrs. Roberts and Harpending in San Francisco. These gentlemen satisfied themselves of the value of the gems and sent men to hunt for more. They met the Indian after a long search, he took them to the place and was subsequently drowned.

In such fashion, using nothing substantial and happy with the wildest rumors, legends are made. But the facts were enough to send any level-headed person in search of medical advice: "Mr. Janin went to the spot, washed a ton and a half of gravel, took out 1,000 diamonds, four pounds of rubies and a dozen sapphires. . . ." A second party went to check on Janin's conclusions and "Mr. Roberts says that if they had been deceived they are the worst deceived and cheated men who ever lived." That was only a slight exaggeration.

The public was keyed up to a point of a speculative craze such as even the Comstock never saw, not alone in San Francisco, but in nearly every financial center of the earth. Millions upon millions would have been invested. The shares would have soared to fabulous figures. Banks would have advanced money on these prime securities, as was the custom in those times. And then the awful crash! There would have been more ruins in financialdom than San Francisco exhibited after the fire. Every day the mails were loaded with letters from eager correspondents making inquiries for stock.

A map had been got up, in the corporation's office, showing the positions of Discovery Claim, Ruby Gulch, Diamond Flat, Sapphire Hollow; and hundreds of envious persons, eager to invest their last dime, stared at the map week after week. "Some fifteen or more bona fide offers were made to purchase a concession for $200,000 cash and a royalty to the parent company of 20 per cent." Then on November 11 came a wire from Clarence King, well-known geologist and engineer, who said the field had been salted and the whole thing was a fraud. The bankers "were simply stunned." A few years later King published a story of his breathtaking ascent of Mount Whitney, in which he risked his life a score of times; and a geologist by the name of Goodyear rode to the top of the mountain on a mule, thereby proving that King had never climbed it at all. "The laugh that followed broke King's heart. He died a few months later."

But he was right about the diamond mine and he headed a party that examined it. The salting had been so clumsy that they found stones that "bore the plain marks of the lapidary's art." The anthills gleaming with diamond dust proved to be, when kicked, "the work of a sinful man, not of the moral insect." They found holes that had been made with a sharp stick, with a diamond at the bottom of them; and on top of a flat rock, rubies and diamonds pressed into crevices to hold them in place. "This was so grotesquely raw that it seems incredible, and led to a story that some of the diamonds were in the forks of trees." Toward the end of November the fraud was revealed, and we can easily believe that the "excitement was intense."

A man named Cooper came forward; he said he had been one of three swindlers but had been cheated of his part of the booty and now wanted vengeance. News came from London that the diamonds were "almost valueless 'niggerheads' from the South African fields" and had been sold to Arnold by the sackful. Looking at the matter with hindsight Harpending wrote: "That diamonds, rubies, emeralds, and sapphires were found associated together—gems found elsewhere in the world under widely different geological conditions—was a fact that ought to have made a goat do some responsible thinking. But it seems to have been entirely overlooked by Tiffany, by Janin, by the House of Rothschild, to say nothing of Ralston, Sam Barlow, General McClellan, General Butler, William M. Lent, General Dodge, and the twenty-five hard-headed business men of San Francisco—" and by Harpending himself, he might have pointed out.

As for Arnold, who had an air of such "simple, rugged honesty" that General Dodge, who thought meanly of human nature, said in a printed interview that he would stake his life on Arnold's integrity," he returned to his native state of Kentucky, apparently with practically all the booty, which he put in a safe. The story has a delightfully ironic conclusion. Some of the men who were bilked put detectives and expensive counsel on Arnold's trail and attached his property. His neighbors were outraged. Arnold said, with the air of a man deeply wounded, that the $550,000 in his safe, his bank account, and his holdings in real estate all represented a long life of toil. To his reply to the charge that he had salted a field with gems he attached the Tiffany appraisement, the

Janin report, and an article in the San Francisco *Chronicle*. He said that if the field was salted it had been done after he last saw it, and that the actions brought against him were diabolical attempts to persecute and destroy him by "California scamps." His friends and neighbors and people all over the nation "gloried in what they were pleased to call his 'spunk'" and Arnold "was the hero of the hour."

He kept the booty. What became of Slack nobody knows. Harpending says it "seems not unlikely that he died somewhere in the Western country, . . ." It may also seem not unlikely that the circumstances of his death were known to Arnold.

CHAPTER 28

Tall Tales and Lost Treasures

We had intended to give a chapter to each of the two subjects in the above title but both tall tales and lost treasure, which have a lot in common, so plunder the mind with their love of fantasy and their unabashed exaggeration that a little of them can go a long way. Still, there are whole books, and many of them, given to lost mines and lost treasures of various kinds; and the tall tales spun in the American West would also fill books, if they were all gathered. It seems best in this book, where space is at a premium, to give only a few representative tall tales and let them serve as an approach to lost treasures, lest the reader take the treasures too seriously. Readers sometimes do; when, many years ago, we included a short chapter on lost mines in a book about Idaho we were bothered for years by readers who eagerly sought more specific directions, maps, guides—anything at almost any price that would lead them to the hidden gold in the City of Rocks or the Craters of the Moon.

Most of the tall tales are as open-faced and obvious as some of Mark Twain's yarns. Glasscock, in one of his books, tells the one about the small cows that could walk between the legs of the giant who milked them, while sitting on a stool and reaching down. Just how small was his smallest cow anyway? "Well, now, she war so small that a gallon of her milk, it wouldn't fill a pint bottle." The Grizzly Bear House in a California bonanza camp was the scene of many a yarn, one of which was about a bear. A hunter fired at the beast twice, without visible effect, and then took to his heels, as Meriwether Lewis had done in a similar situation. Reloading as he ran he turned and fired into the monster's left eye. That only increased its fury, so he reloaded again while running and, turning, fired into the right eye. The bear was now blind but it put its nose down and followed the scent. At this point a big yokel who

had been sitting with mouth agape and Adam's apple running up and down began to laugh his head off. "How now?" demanded the hunter. "Ye think I lie?"

"Naw," said the yokel, choking and wiping at his eyes. "I warn't a-laffin at you. I wuz a-laffin at the bar."

Another story concerns Acme Sulphide who, in from his diggin's on Caribou Creek, took a dead cougar to a taxidermist shop.

"Want him skun out and made in a rug?"

"Not on your life," said Acme. He wouldn't think of walking on Petronius.

"Jist a cougar, ain't he?"

"Not by a damn sight. He's an institution, that's what he is. When he was jist a kitten I ketched him by the mine shaft. Him and Pluto, they was great friends until Petronius growed up. Then one evening when I comes back, why Pluto was gone and Petronius wouldn't eat his supper. I was plum mad but I figgered Pluto was gittin old and wasn't so much account any more. Then, by gum, I missed Mary, the goat. When I missed the last of the chickens and Petronius had feathers in his whiskers I made up my mind to shoot him. But I got to thinking how that goat could butt, and the hens wasn't layun no how. So I let it go but that was a mistake. Liddie—she's my old woman, or she was. Partner of my joys and sorrows nigh onto forty year. One evening she was gone and Petronius agin didn't want no supper. That got me danged mad and I went for my gun; but again I got to thinking. Liddie never was much for looks and she had swore she would leave me. So I put the gun up."

"Then what happened?"

"Last night he jumped me on the trail and I thought I was a goner. So mount him up pretty. He represents my hull damn family."

Our next tall one appeared in the Idaho *World*. While looking one night for a lost cow Len Atkins saw by the light of his lantern a human skull bounding along the path before him. He was no coward but his hair stood on end. The skull was coming toward him, rolling over and over, dancing crazily; and Len, hiding behind a tree, watched the thing as it went on down the path until it came to a colony of prairie dogs. A score of chattering dogs came out of their holes, and from the cavity of the skull the head of another dog emerged, followed in a moment by the body. There was an ancient Indian burial ground in that vicinity. Some persons believed that the animal had somehow become imprisoned in the skull, but those with a keener appetite for wonders insisted that the dog had used it as a vehicle to return home.

The next one we lift from Emrich and rewrite a bit. A group of persons one evening saw a strange figure coming down a white mountainside, breath pouring like steam from nose and mouth, hair and beard looking like wraiths of frost. The figure came up to the astounded group and in a voice like that from a grave vault asked, "Was it cold in Como?" It then braced itself and staring with eyes like two small ice caps it repeated in the same hollow voice, "Was it cold in Como?" They thought an expression like a sneer spread across the forehead and eyes. After the question was asked a second time the people felt a dreadful chill and a sense of horror; and it appeared to them that the figure seemed to be puffing up and growing thinner. Suddenly frost gathered on their beards and over their faces, that miraculously became icicles; their ears felt like pieces of ice and their teeth began to chatter. They understood then that all the people in Como had frozen to death and that this horrible apparition was in a way the ghosts of all of them. With shrieks they turned to flee but were stopped by a sound as of a mountain of ice falling.

A man then came by who looked normal; they asked him if he had just heard a mountain of ice falling, and he said he guessed it was icicles from the eaves and the roof. They looked round them and heard a hollow voice in a deep dark frozen place say, "Was it cold in Como?"

"Are you asking me?" the stranger said.

"No, we thought you were asking us."

"You did, indeed! And what did you think you heard?"

"We thought you asked us if it was cold in Como."

Some people prefer tales that have an affinity with the mysterious, the unknown, the eerie and horrible, the unearthly and spectral and spookish. Some prefer those that rest, like most legends, on a foundation of fact. So we shall conclude with two tall tales related to the Cripple Creek multimillionaire, Winfield Scott Stratton.*

A story that has been told about him again and again by a number of writers, including one of the most popular Colorado journalists, according to Frank Waters, is this: he went to the Brown Palace Hotel with a brothel madam (some say), with a blonde lady (as Waters tells it) and when refused lodging left the place in a fury and bought it for $800,000 cash (various figures are given), "returned to the lobby, fired the clerk, and moved into the bridal suite." Another "journalist-biographer" tells the same old story, except that in his version it was the manager who was fired. Waters gives still another version which has Stratton several stories above the lobby, looking down on it, and dropping bottles of champagne to astound the clerk and guests. When compelled to abandon his fun Stratton bought the hotel for a million dollars—and no doubt went right on dropping bottles. Says Waters: "None of these colorful versions are true." Like so much that passes for history of the West they are merely tall tales.

Another tall tale related to Stratton is that which has as principal character Sophia Gertrude Poor Kennedy Chellew, the thirteenth (Waters says), the twelfth (Lee says), of the women who swore after his death that they were Stratton widows. It was not until December 20, 1913, more than eleven years after his death, that the woman was asked to set up her claims. After "many demurrers, protests and showings, eight different orders for continuance and applications for jury trial, the case came up for trial on May 16, 1916," three and a half years after she was asked to show

---

* See the chapter "Display of Wealth."

her claims. But both sides meanwhile had been busy.

The abandoned and heartbroken Sophia, who brought sobs to the throats of reporters, had first married a brickmason named Poor. After one child he died in 1872, leaving her twelve thousand dollars. In December, 1873, she moved with her daughter and a cousin known as Little Doc Bell into the home of John Henry, close to St. Augustine, Texas. While a Christmas party was in progress a group of men came to the door, including a Russell, a Thompson, and a Bill Scott, who had come to Texas to buy cattle. For Bill and the widow it was love at first sight, and December 31, with Thompson along, they went for a license. She then learned that Scott's name was Stratton. The next day she and Stratton were married, and January 2 she went with her daughter, husband, and the other two men to Fort Worth. Twenty days later, according to this tall tale, Stratton rode into Colorado Springs, where his home was, and told friends he had married down in Texas and his wife had given him most of her money. A year later Stratton went to Fort Worth and learned that he had twins. Sophia waited for him there until the fall of 1875, when with the daughter, twins, and Doc she set out to find him. On her way she heard that three cattle buyers had been killed by Indians; she had no doubt that they were her husband and Russell and Thompson. The twins died there, conveniently, and there another daughter, Mary, was born, November 30, 1876. And Sophia now married a Michael Kennedy.

In the late eighties they moved to Colorado near Leadville; in 1895 Kennedy died and she married a man named Chellew. When she heard two miners talking about Stratton and his millions she went over to Cripple Creek and met him in a restaurant. They agreed to say nothing about their marriage. He gave her money, visited her in Leadville, accepted Mary as his daughter, and promised to make them all independent. Witnesses now began to appear and her lawyer, J. T. Bottom, was busy taking depositions. A Mrs. Nelson who once had a boardinghouse in Colorado Springs told Bottom that Stratton had boarded with her and had told her about his wife and the money she had given him. A man named Shaffer appeared to say that he had prospected with Stratton in the spring of 1878 and that Strat-

ton had often spoken of his wife down in Texas and showed him pictures of her many times. Thompson, or anyway a Thompson, was dug up somewhere, and he told of going with Stratton for the license and of the marriage; in 1878 in Leadville he saw Stratton there and there was a baby in the house and Stratton said he had twins of about that age. Later he said his wife and babies had died. After Thompson had told a detailed story, Russell, or a Russell, was found and he told exactly the kind of story that Thompson had told.

By this time some of the newspapers were full of grief and outrage, one of them saying, "Sorrow leaves impress on the face of the soft voiced Stratton widow." After he became wealthy Stratton had been hated by most people who knew him or read about him, and now, in his grave and unable to speak, he was hated doubly, as a psychopathic liar spun her tall tale. Down in Texas a man named Parker "remembers issuing them the license but the Shelby County records were destroyed by fire." Sophia undoubtedly knew that before she chose a place for the license. A John Henry and a Sam Henley were found who swore that they were present at the marriage. Advertisements down in Texas asked people to come forward who had known the odious fellow named Stratton, and photographs of Stratton were published there; and down in Texas there were stories of the huge estate he had left, and of the sorrowing and disinherited widow who had spent her life marrying one man after another, yet was entitled to half; so come forward, all of you, who have information to give. For a slice of an estate of such proportions there were plenty of persons who had information to give, and they came forth, grave and sympathetic, and ready to do their duty as citizens.

But lawyers for the estate had meanwhile had Burns detectives busy—busy indeed for three full years, gathering facts to prove that everything the woman said was a monstrous lie and that all her witnesses were liars and perjurers. Parker, Thompson, the landlady, and others had been promised a part of what the court would surely award to Sophia Gertrude Poor Kennedy Chellew. Parker actually had a contract with her that would give him a fourth of what he undoubtedly hoped would be half of the trust fund of six mil-

Bringing gold ingots to the surface from the sunken Egypt

lions, that was to establish a home for destitute orphans and old people.

With the lesson of Sophia in mind we can approach the lost treasures with at least a modest amount of skepticism. Though we don't know which tales of lost mines were hoaxes, undoubtedly some of them were; other tales were told by men driven insane by heat or hardship; and still others by old men who, after a long life of disappointments, were senile and deluded. Thousands of persons are looking every year for lost treasures, and some of them make a lifetime profession of it. Some of them struck it rich.* But any reader who intends to desert wife and hearth for one of these wild hunts ought first to consult the Treasure Trove File maintained by the United States Treasury, where up-to-date records can be found of endless possibilities. As Hogg says, "The file is in unceasing demand" and the prospects

"appear to be endless, and their variety almost as great." Coffman, one of the experts in this matter, says:

It has been determined that hidden in the soil of the Western Hemisphere or sunk in its coastal waters are riches to the tune of three hundred billion dollars. To the treasure hunter equipped with the modern devices now available, . . . combined with a general knowledge of the locations of some of these tremendous hoards, let me say that he need search no farther for escapes from the monotony of living, and that his chances of acquiring riches are much greater than the ordinary job or position will ever offer.

Coffman gives the name, date, location, and estimated value of nearly three hundred treasures, whose total worth is, he says, in the hundreds of millions. But that seems trivial when in his Chapter XVIII you find him saying that the

---

* A recent spectacular recovery was reported with photographs by James Atwater in the *Saturday Evening Post* in 1964; these treasure hunters found millions of dollars in old coins in shallow water along an Atlantic beach. Excellent books on this subject include, besides F. L. Coffman's, those by Harold T. Wilkin; R. J. Santchi's *Treasure Trails,* in three volumes; Capt. D. T. Bowen's books on the Great Lakes; Edward R. Snow's ("the world's leading authority" on sunken treasure off the Atlantic and New England coasts); Ken Krippene (a noted treasure hunter himself), *Buried Treasure;* and books by Lieut. Harry E. Rieseberg, the first man to walk the streets of the sunken city of Port Royal "since the disastrous day of June 7, 1692, when it disappeared from sight." "King" of undersea treasure hunters, Rieseberg "has never conducted a salvage expedition that resulted in a loss of a penny to any backer or investor." A popular book is J. Frank Dobie's *Coronado's Children; Tales of Lost Mines and Buried Treasures of the Southwest.* Another spectacular recent find, that of an ancient Spanish ship, is described in *Look,* March 9, 1965.

total of all treasures lost in the oceans is around six hundred billions and that 25 percent of all the gold and silver taken from the earth has been lost at sea. Only 5 percent has been recovered. Over 1,600 documented vessels have gone down in the small English Channel, and "five times that number around the British Isles. Britain leads the world in marine salvage recoveries but has recovered only a little over $400,000,000 since the 16th century." After Spain began to pillage the New World "nearly 1,000,000 vessels have gone to Davy Jones locker with *billions* in valuable cargoes." The United States Navy's Hydrographic Office has records showing that from 1850 to 1950 an average of 2,172 vessels have been lost every year. These do not include thousands of small crafts of all kinds that were not recorded. Known vessels numbering 7,147 have been lost in the Great Lakes, with Erie having over three hundred million in lost cargoes on its bottom, many of them in less than one hundred feet of water. "Treasure salvage," said Coffman, "is fast becoming big business today." The United States Treasury reported in 1950 that over eleven million in value had been found, which was thought to be only a fraction of the total. Coffman, a professional treasure hunter, gives in his books the departments and agencies to be consulted and the equipment to buy. He also lists hundreds of treasures and says that during his eighteen years of research "covering over 42,000 reported locations," he has been able to add information to about 10 percent of them. He says that hundreds of persons have perished in the American West while looking for lost treasure, chiefly the loot of stage robbers and the ore ledges reported to have been found. He advises those who hunt treasure buried in the earth to take the equipment they need—a tent, a gun, food, water (if you are going into the desert), and such tools as hammer, pick, shovel, pan, and a beast or a Jeep to carry them.*

Sometimes a "lost mine" is found. In 1954, James Williams, a prospector out of Spokane, was climbing the hills around Salmon City, Idaho, when he discovered float and then found a mining tunnel so old that in its entrance was a tree that had been growing for over fifty years. The interior was dark and old but he explored it and found in its walls quartz so rich in gold that he

was able to pick out pieces as large as grains of wheat. Some of the ore assayed as high as eleven thousand dollars to the ton. How many hunters had seen that tunnel entrance?—and how many had dared to enter it? Hult wonders which of the lost mines it might be. Probably none of them. Some prospector long ago followed the float lead up that mountain, found the vein, tunneled in, and took a load of ore away by muleback. Something happened to him: he might have perished in a fight with a grizzly, or been captured by Indians, or buried by a snowslide or landslide. There is no reason to doubt that many a prospector found a vein, perished on his way out with ore samples, or was never able to find his way back.

Emrich tells of a mine superintendent in Arizona who, lost in a roaring dust storm, found refuge behind a ledge. He said it was a ledge of quartz rich in free gold. Leaving his coat and revolver to mark the spot, he took samples and headed for home, carefully observing all landmarks on his way out and entering them in his notebook. He was found dead of thirst. His samples assayed twenty-five thousand dollars to the ton but his quartz ledge has never been found. A likely guess is that the sandstorms buried it.

There are many lost mines that have become famous and that are written about year after year in much the same old way—the Gunsight, the Peg-Leg, the Breyfogle, the Dutchman, and others. We shall sketch a few to give the nature of them.

The story says that three Mormon emigrants while trying desperately to get across and out of Death Valley climbed the mountains somewhere along the west side. One of them had knocked the front sight off his gun, and in a ledge he found a soft malleable metal and made a new sight. After he reached safety he discovered that his new sight was almost pure silver. Nobody knows how many men have lost their lives looking for that ledge. Beebe says, "The Gunsight, like the Breyfogle, was constantly being 'discovered' . . . and the news of its locating printed in good faith by optimistic editors. A dispatch to the San Bernardino *Guardian* in April, 1875, announced the discovery by a prospector named Frink, of the authentic and undisputed Gunsight on the edge of Death Valley, halfway between Panamint and the Cerro Gordo. 'The specimens sent us will go

---

* F. L. Coffman, *1001 Lost, Buried or Sunken Treasures.*

up in the thousands of dollars,' said the *Guardian*, 'and we congratulate Mr. Frink on his splendid luck.'" But for all that men are still looking for it.

Long ago Horace Bell said,

The author has little faith in the actual existence of the Peg-Leg Mine because it was reported by that artistic old liar, Peg-Leg Smith, whom he had the honor of knowing in the palmy days of Peg-Leg's lawlessness. Peg-Leg was the biggest horse thief that ever ranged the country between the Missouri River and the Pacific Ocean. . . . a magnificent thief on the wholesale plan and the most superlative liar that ever honored California. . . . In the latter days of the '50's, dilapidated and played-out, he found his way once more to Los Angeles and sat around the old Bella Union bar, telling big lies and drinking free whiskey . . . his alleged mine of fabulous richness . . . a myth, a Peg-Leg lie. But the Goller mine will some day be found, and it will provide plenty of excitement.

Horace Bell is not the most trustworthy reporter of the old West but we'll take a brief look at the Goller, that will some day be found.

John Goller, a robust German of thirty, was a member of a Death Valley party which "struggled for fifty-two days across a blistered volcanic desert. It makes a terrible tale. . . . They covered eight hundred miles of utter wilderness, climbed dead volcanoes never seen by white men before, traversed the hideous Death Valley, dragged along for days without water, ate the hides and the very bones of their starved cattle." This party was in the '49 gold rush but took the southern Old Spanish Trail because of the disaster that overtook the Donner party on the northern trail. "For one stretch of one hundred miles these desperate Illinois people found no water. There were five successive days when they went without a drop. Finally they discovered one small spring, known to this day as Providence Spring." One man they had left by the trail, too weak to stagger, but after finding water some of them went back to bring him on. He was dead. He had crawled after them on hands and knees for four miles. The most pathetic member of the party, in Bell's version, was a Frenchman, "who went crazy one night and wandered off into the desert. Fifteen years later hunters found him a prisoner among the Digger Indians, still wildly insane and for this reason feared and respected. . . ." In the party

was a woman, who made it through and kept her two children alive, ages four and seven. When the survivors at last reached Jose Salazar's house, John Goller had on him some nuggets of gold which he thought he had found somewhere in Death Valley. He "reported that where he found them he could have loaded a pack mule with gold had there remained with the party any such animal uneaten. There must be a mine there, he said, of fabulous richness; and during the whole of his long and prosperous life in Los Angeles John Goller spent his spare time and money fitting out expeditions to search for the lost bonanza. But he never found it."

The Breyfogle, which Beebe calls "the Holy Grail of all true fissures," has, like some others, different versions; it has been written many times and often the authors seem merely to copy one another. In a Glasscock version John Breyfogle *in 1864* packed his burros and headed south from Austin, Nevada, to prospect; his horses strayed and when looking for them he found the ledge, went insane from thirst, was picked up by Indians, but could never again find the ledge. Glasscock says that during the next forty years hundreds of men hunted for it, some of whom perished. J. Ross Browne, who lived at the time, says he was familiar with the area and talked to some members of the party, gives the time as the summer of 1852 and tells a corking tale in his *Adventures in the Apache Country* that is very similar to the Goller tale. He says, "I knew Breyfogle in former years. He was tax-collector of Alameda County, California; and seemed to be a man of good sense, much respected by the community." But this version is so different from another version that is widely accepted that we suspect that the reader is in the realm of legend in all Breyfogle versions, and pass on to the Lost Dutchman, which is almost as bad. For here again the reader can choose among such versions as Miller's, in *Arizona, the Last Frontier,* in which he goes back to 1895 and the Phoenix *Gazette* or that in *Arizona, a State Guide,* produced under the Federal Writers' Project, or a half-dozen others.

About thirty-five miles east of Phoenix are the Superstition Mountains. In one version, that is obviously pure legend, a Mexican in love with a beautiful maid fled north from the wrath of the girl's father; in another version two German prospectors found six Mexicans working a rich lode and killed them. One of the Germans, Jacob Wolz or Walz, then killed his partner. All that may be legend but there was a Jacob who in his old age moved to the vicinity of Phoenix and made infrequent journeys into the mountains, to bring back small sacks filled with rich ore. He is said to have made his last journey in 1884, to have brought back two sacks containing about five hundred dollars' worth of ore, and to have confided his secret to a woman who ministered to him in his last years. The description furnished to her is said to have been about as follows: in a certain gulch in the mountains, identified by certain landmarks, there was a small shack at the mouth of a cave; and across the gulch from it was a tunnel well concealed by bushes. Not far above the tunnel on the hillside was a shaft, down which a man could climb, and there was the ledge of gold. One would think that any man could have found it, with such conspicuous marks to look for; but the woman, taking a miner with her, spent an entire summer, and the next summer, and was unable to find even the ruins of the shack. Many have hunted for it since then, including specialists well equipped. The *Guide* says, in fact, that "literally thousands of prospectors, both professional and amateur" are said to have hunted for this ledge, and that some of them never came back at all and others came back with pieces of human skeleton. It seems that the Dutchman knew of a ledge where he found gold-bearing quartz but just about everything else in this tale is legend. And now, says Miller, the "tragedy and violence connected with the Lost Dutchman have added to the strong convictions of Arizonans that the Superstition Mountains are cursed."

Of tales that seem to have a basis in fact Coffman tells this one in his *1001 Lost, Buried or Sunken Treasure*. Karl Steinheimer, born in Germany in 1793, took to the seaways at the age of twelve. After some years as a pirate and smuggler he turned to mining in Mexico. When Mexicans became hostile after losing Texas, he left the country, with ten mules, some of them carrying all the silver and gold he had accumulated. After crossing the border he sensed the presence of enemies, including both Mexicans and Indians, and looking round him decided to bury **his** treasure and return to it when conditions

were more favorable. The spot he chose was "sixty or seventy" miles north of Austin, the present capital of Texas, "near a place where three streams join into one." Unless the streams have changed their minds since then, as streams have a way of doing, that junction should not be far from the treasure. After journeying into the southeast twelve or fifteen miles he came to what he called a "bunch of knobs on the prairie" from whose summits he saw a valley in the west. He then was attacked by Indians and, though gravely wounded, escaped. In one of those knobs he buried a bar of gold and struggled on until he came to men who tried to help him. But gangrene had filled his wound and he was dying. Knowing this, he wrote directions to his cache, sealed it in a letter, and asked one of the men to mail it to a certain woman in St. Louis. This was done, but because Texas was in turmoil the woman's male relatives delayed for years their journey to find the treasure. They reported that there had been such changes in the land surface that they couldn't determine the spot. Since there could not have been much change in that length of time the presumption is that they lied or didn't know where they were. The story became widely known, and year after year persons have looked for the spot. It is believed that the three rivers must be the Salado, Lampasas, and Leon, which come together not far from the present town of Belton. At their junction they form Little River. Twelve or fifteen miles southeast of this junction there is indeed a group of hills known as the Knobs, that overlook the valley of Little River.

Eberhart tells of a treasure in the Bull Canyon area. It is not a lost mine or buried loot from a stagecoach; it is the spot where a man hid his fortune who distrusted banks. How large the fortune is no one knows but it is "a recorded fact that he added $5,500 to it a few weeks before his death." It seems that an Indian known as Henry Huff had sold a herd of cattle, and this is only one of the facts that offer an advantage to those who seek this treasure. For one thing there is a photograph of Huff's cabin in Bull Canyon, taken by the United States Geological Survey. The man lived there for years and presumably he hid his money at least within easy horseback distance of it. Another advantage is the fact that the state of Colorado not only believed that Huff had hidden a fortune but after his death claimed it and tried to find it.* Huff, who had a friend named Akers, was shot by a long-time enemy named Keski, but when, knowing that he was dying, he dictated a will it was not to Akers but to Mrs. Keski, for reasons apparently unknown. He died before he was able, or chose, to tell where his treasure was. If we were to go looking for that one we'd pitch a tent near the cabin, learn all we could about Henry Huff, work up a lively distrust of all banks, and imagining that we had a fortune to hide we would put ourselves as well as we could inside that Indian, mount a horse, and go look for a suitable spot. One major disadvantage in this instance is the fact that Henry was an active cuss who had livestock and rode the range, and who probably knew every good spot for a cache within a radius of ten or twenty miles.

It is said that the plan originated in the fertile mind of Richard Barter, a Canadian, who made his way south to the California mines and after being twice unjustly charged with crime became a bandit, to be known as Rattlesnake Dick. Among his toughs were the Skinner brothers, George and Cyrus, both accomplished criminals and both headed toward a sad end. From the mines in the Yreka area north of Sacramento the gold did not proceed by stage but was shipped out on muleback. Because the mules wore their owner's brand, Dick and Cyrus undertook to steal a herd, while George and three others, plus an eager Mexican, took the mule train's guards by surprise, tied them to trees, and made off with the booty to a point agreed on. The ambush was close to the basaltic desolation of Trinity Mountain. Because eighty thousand dollars of gold was too heavy a burden for five men to carry any great distance, it is assumed that it was hidden not far from the trail where the train was ambushed. The men with the gold waited two days and nights but Dick and Cyrus did not come. The Mexican's terror became so great that he drew a long knife and demanded his part of the gold then and there. George solved the problem by shooting him.

Afraid that the men bound to trees had by that time given an alarm George and his three accomplices buried about half the gold and staggered off with the remainder of it, and their equipment and weapons. Whether the four men or only

George knew where the gold was buried we do not know. The four men made their way to Folsom and hid half of the gold there, and sallied forth to learn what had happened to Dick and Cyrus. A posse met them. George fell under the first fire, one of the three was wounded, and the other two were captured. The wounded man and one of the other two were given ten years; the third man, hoping for freedom, led officers to the gold that had been brought down the mountain. He said he did not know where George had hidden the other half. Jackson, who investigated this tale, says this man was never known to go within a hundred miles of Trinity Mountain. That may or may not be a fact. Whether after the other two were released from jail they found the gold is not known but Jackson felt that it was still there.

In the early 1860's northern Idaho had felt the surge of several gold rushes. Some years later when Jack Breen came into Coeur d'Alene from the diggings, toting a bag of rich ore, the news spread fast, and in a saloon men plotted to get him drunk. But before that a county commissioner and a businessman had proposed to Breen that they would furnish money for development for a share in the mine. On learning that men intended to get him drunk they hastened to the town marshal and prevailed on him to lock Breen up for the night, on a charge of drunkenness. That probably made Jack Breen awfully mad—for there he was, in jail, the man who had found a rich ledge. No doubt he felt over the inside walls of the wooden structure, seeking a way out; but like most of the mining-camp jails this one had no windows, and its heavy door was barred and bolted. Breen seems to have decided that he could burn a hole through one end of the jail and crawl out. An hour or so later the town's fire alarm system did its best to arouse the citizens, while men chased all over the place trying to find the marshal, who had the only key to the heavy lock. Firemen with axes and crowbars attacked the door but by the time they got it open Jack Breen was dead.

Beyond any doubt a lot of gold was hidden by robbers of coaches and expresses—by lone bandits like Black Bart and by such gangs as the

---

* Colorado seems to have the habit of trying to seize fortunes that belong to others. See under Stratton.

Plummer and Updyke. In the Idaho *Guide* many years ago we gave a few of the tales that Mr. Harrington (see below) thought had substantial fact behind them. One of them we'll give here, which is typical of all. This robbery, especially treacherous and brutal, aroused such fury in the vigilantes that the bandits were immediately run down and hanged. It is thought that justice was so swift that the robbers never had a chance to return to their cache. Frank Williams, a former stage driver, tipped the gang off and it set up an ambush in July, 1865. The scene, since known as Robbers' Roost, is just off the main highway that runs south from Pocatello, a mile or two or three miles north of McCammon. The amount of gold taken has been estimated at $70,000 to $100,000. Four passengers were almost immediately killed inside the coach when the bandits dashed up and opened fire at close range. The moment they appeared, Williams tumbled from the coach and hid. A man named Carpenter was not hit inside the coach but was so covered with blood from those slain that when the bandits looked in to see if they had killed them all they thought he was also dead. At $15 to the ounce $70,000 was about three hundred pounds of gold; it is believed that robbers in flight, knowing that the vigilantes would soon be on their trail, would have paused somewhere in the ambush area to make a cache.

There is a Robbers' Roost in Montana that for a while was the headquarters of Plummer and his gang. The two-story building, which had a corral adjacent, still stands on the main road to Virginia City and is a historic landmark in Madison County. When Plummer was taken by the vigilantes he begged for his life; and the story as told by one of them, William Owsley, is substantially as follows: "A few minutes before the rope was settled around his neck he threw his arms around me and said, 'Owsley, I will take you to a place where I have $300,000 in dust cached.' I pushed him away and he again grabbed me. 'Hanging me will not give back the dust,' the doomed man said. 'I'll take you over to Robbers Roost [was it called that then?] and if the gold is not where I say, you can cut me to pieces. Just give me a horse after I show you the cache and I'll leave the territory forever.' As I started to push him away again, Plummer said, 'Oh, God, I tell you it is within one hundred yards of one corner of the corral!'" If that is true it should have been

no great chore to find it. The corral is no longer standing but in former years a lot of persons dug like prairie dogs all over the area at each of the four corners. One possibility is that Plummer lied, though he had nothing to gain by lying; another is that some member of his gang made off with the gold when the vigilantes began to close in.

J. A. Harrington, an old-timer whom we knew well, made a hobby of collecting tales of lost treasures in Idaho and the surrounding states. He had been a United States marshal, and a prospector; he had been in the Klondike. On May 21, 1939, he published in the Idaho *Statesman* his version of the lost Swim. Isaac T. Swim, of a well-known south-Idaho family, in early September, 1881, headed down the Yankee Fork for home. Somewhere near the junction of this stream with the Salmon River he camped for the night, with a heavy storm and a wild wind blowing over him. The next morning he saw a tree that had been blown down, whose roots, torn from the earth, exposed gold ore. Swim posted a claim September 9 and went down the river to Challis to record it. He took home with him "choice specimens from the ledge and the nearby stream." Though he had intended to confide his secret in only one or two close friends, by the next spring a lot of people had heard about it, and when Swim and a companion set out for the claim they were trailed by a body of men. Because of high spring water the two men camped on the Salmon and waited, and those who had followed them became impatient. One of them, George Blackman, has said that they wanted to push on in spite of the danger in crossing a swollen river, and that at last an agreement was reached with Swim that he would go alone and after staking a claim for his companion would return and lead the others to the site. Those waiting watched him swim the river on his horse. When after a time he did not return they searched up and down the river, and finding no sign of horse tracks emerging on their side they went home, probably feeling that they had been betrayed. "Swim was never seen again. His horse was later found in some driftwood, fifty miles from where it was last seen. Late that summer, an unidentified body, supposed to be Swim's, was found on Salmon River" at a spot since known as Dead Man's Hole. Prospectors and treasure hunters have again and again searched the whole area, including Warm Springs, Rough, Big, and Little Casino creeks.

That is Harrington's version, probably based on what Blackman and others told him. Esther Yarber in her *Land of the Yankee Fork,* who when she wrote her book had never heard of Harrington's story, gives a quite different version, which she based on tales by old-timers, including Arthur McGown, who lives at least part of each year on the Yankee Fork and has a tiny museum there. Says Mrs. Yarber: "On September 9, 1881, a quiet, bearded man named Isaac T. Swim rode into Bonanza City, . . . hitched his horse in front of the Charles Franklin Hotel," and asked where the recorder's office was. Even giving his room number (five) Mrs. Yarber says he came out wearing a clean, checkered flannel shirt, disposed of a deer he had brought in, had his horses fed, and filed on his claim:

SWIM

Notice is hereby given that the undersigned having complied with the requirements of Chapter Six of thirty-two of the Revised Statutes of the United States, and the local custom laws and regulations, has located commencing at this monument and running 750 feet in a northerly direction and 750 feet in a southerly direction and 300 feet on each side. Said location is situated in Custer County (created in January of 1881), Idaho Territory, and the Yankee Fork Mining District, and fifteen miles, *more or less* in a southerly direction from Bonanza City, Custer County. This location is known as the 'Swim Lode'.

Dated Sept. 9th., 1881

Signed: ISAAC T. SWIM

From the county records, Challis, Idaho

"Swim mine"

Notice this Location is to be Known as the "Swim" Lode.

Notice is hereby given that the Undersigned having Complied with the requirements of Chapter Six. of Title thirty two of the revised Statutes of the United States. And the local Customs laws and regulations, has located. Commencing at this monument and running 750 feet in a Northerly direction and 750 in a Southerly direction, and 300 feet on Each Side. Said Location is Situated in Custer County Idaho Territory and Yankee Fork mining District and 15 miles More or less in Southerly direction from Bonanza

Dated Custer County Bonanza Mg district Sept. 9th 1881.

Isaac T. Swim

Territory of Idaho }
County of Custer }

I. T. Swim being duly Sworn deposes and says as follows. That he is a Citizen of the United States and resident of the Territory of Idaho. That the within described Claim and location has not heretofore been located according to law. or if so located, the same has been abandoned or forfeited by Non fulfillment of the laws of the United States and of the Territory of Idaho.

I. T. Swim

Subscribed and Sworn to before me this 10th day of September A.D. 1881 }

E. T. Carr. Deputy Recorder.

Filed for record in this office September 10th, 1881 at 9 oclock A m and duly recorded in Book B. of Quartz Pages. 105 and 106.

E. C. Whitsett. County Recorder

By E. T. Carr. Deputy.

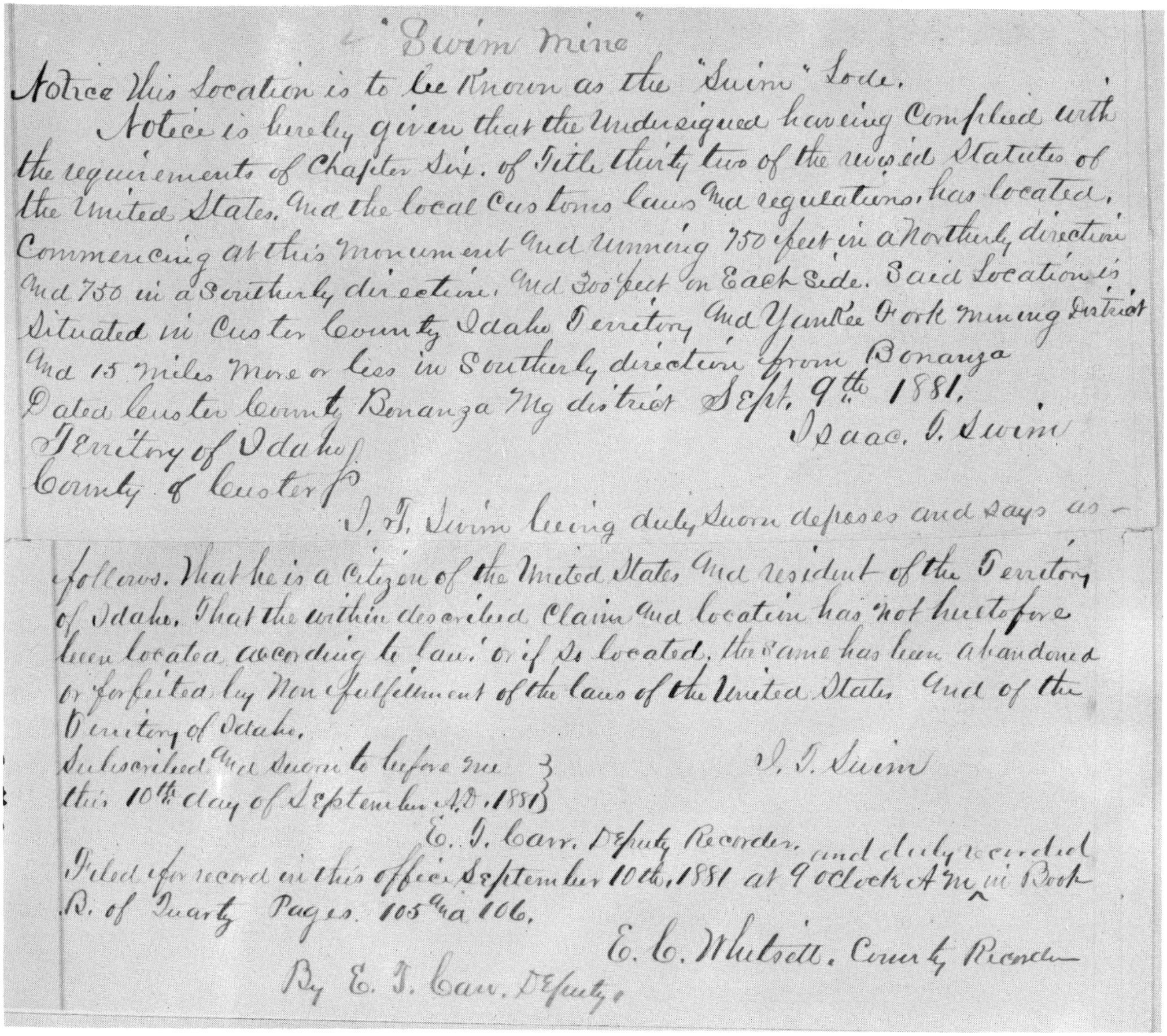

Recorded Sept. 10th., 1881 at 9 o'clock a.m.

E. C. WHITSETT,
Co. Recorder

By E. T. Carr, Dept'y

The ghost town of Bonanza is nine miles up the creek from U.S. 93, which at this point follows the Salmon River. Less than fifteen miles would put the discovery either north or south of the river, but about ten miles would put it south. Mrs. Yarber says Swim *"just disappeared."* She asks if he drowned, was murdered, or died of natural causes. "An unidentified skeleton of a man was found some years later on a flat below Challis, carried there during very high water, and the skeleton of a horse with a pack saddle on it was found on a drift pile on the Salmon. . . ." She says that in July of 1881 Swim found ore samples that assayed "clear out of sight." She says that he was a quiet and reticent man who minded his own business, but on one occasion asked about county boundaries, apparently being uncertain whether his claim would be in Alturas or Custer county. In view of the fact that Alturas County disappeared as other counties were established, a part of it vanishing into Custer, the person who seeks this lost lode would do well to make a thorough study of the old line between Alturas and Custer, in relation to Bonanza and the Salmon River.

Mrs. Yarber's informants told her that by August it was known that Swim had made a strike and that during that month when he helped harvest a crop of hay to pay for a packhorse he was asked questions by the ranch hands. After recording his claim he went to Silver City, and there some persons who saw his ore samples offered to finance him. The next spring with either two or three partners, who had grubstaked him and who therefore were entitled to a portion of whatever he had found, he set out with a wagonload of supplies. Coming to the mountains they found the snow too deep for the wagon, and so Swim went ahead to find the nearest campsite to the mine to which the wagon could be taken. We seem to be getting lost in legend at this point: the mountains that stopped their wagon are fifty or sixty miles south of the junction of Yankee Fork with the Salmon. It is said that he discovered that he was being trailed, though how this is known is not clear, inasmuch as he was never again seen alive. After much of the snow had melted the partners took the wagon north to the Stanley Basin area, not far from Bonanza, and waited for him. It is said that his trail was lost at Lake Creek, which is on the Salmon just over the mountains, and a long way south of Stanley.

After he recorded his claim he willed a part of it to a woman in San Francisco, for reasons unknown, and this time in the description he says his claim was in Calley Gulch in the Yankee Fork district, and that it was about "fifteen miles southerly from Bonanza City," which might or might not put it across the Salmon, depending on how southerly is defined. That Swim found rich ore there is no reason to doubt but anyone who looks for it is face to face with a broad country.

The midnight sun

Notes

Chapter 1.   Waters is quoted from Adams, Snell, from *Western History*, Spring, 1965, 86. Steckmesser and Adams are *passim*.

Chapter 2.   Whitney as quoted by Bowles; Encyclopedias; Hogg, Emmons, Bernewitz, Rickard, Fitzhugh, Quiett, Coll. essays of *London Times*, 1933; Bowles, Sprague, Pefley, Clark, Stoll, Barbee, Waters, Boercke, Patterson, Farish, *Calif. Hist. Soc. Quarterly*, Oct., 1924; Glasscock *Golden Highway*; Hinckley, Colton.

Chapter 3.   Harlow 326; Zollinger 233; on San. Fran. Quiett, *Built West* 190; Stoll 6 f.; Buffum 83, 89, 172, 74, 89, 56; Dana 118, 116; Shinn 125-29, 101-2, 104-5; Morrell *passim*; Bowles 308; Page 272, 286, 207; Delano 242-43, 251-52; Bryant 345; Borthwick 60; Quiett *Pay Dirt* 60, 92; Woods, 170, 60-63; Davis 40; M'Collum 133; Allsop 30-31; Haskins 128, 142; Helper 151; McIlhany 41; Kip 30, 35; Knower 71; Shirley Letter Tenth; Stuart I 232; Caughey 161; Twain 245; Rickard *Hist. of Mining* 23-36, 40.

Chapter 4.   Possibly the best book on this is Hulbert's *Forty-Niners*, based on many diaries and journals and letters by many persons who made the journey, by land or sea. His literary device was to invent a young man who as a member of the overland company writes to his folks back home. *London Times, Gold*, 186; Herald, quoted by Paul; Hafen Journals of 49ers 323-24; Page 105, 167, 257, 116, 185; Quiett 28; Hulbert *passim*; "One came hourly" quoted without source by Hogg; rheumatism, Silver City *Avalanche* Feb. 8, 1868; Andrews 184; Taylor II 39; Twain *Roughing It* 136; Greeley 83-84.

Chapter 5.   Twain 185; Quiett 11; Paul *MF passim*; Drury 16; Bowles 148; Emmons 178; Quiett VI; Richardson 152, 158, 166; Pierce, Quiett 174 ff.; Wells *Idaho Yesterdays* VI No. 1; on other Idaho camps see Wells Bulletin 22 of Idaho Bureau of Mines and Geology; Trimble 89; Stuart I 283, 246; Seattle etc. Quiett Ch. 14; O'Connor 26; Rickard *RM* 312; Hellprin in Hogg 188; Adney 188, 109; dogs, Greever 344, 349; Goddard, Greever 337; Adney and Berton are among the best books on the Klondike.

Chapter 6.   Shinn 225; Harlow 366; Twain in *RI*; "a writer" J. Davis 91; Dimsdale 7; C. Bancroft *Matchless Mine* 9; Shinn 149-50, 139-40; Borthwick 119; Colorado camp Londoner quoted Paul *MF* 117; *Post* Aug. 26, 1865; Richardson 186; Hayne *London Times, Gold* 189; Borthwick 24; Dimsdale 17; Colton 189, 223, 231; Bancroft *Pop T* I 143; Taylor *El Dorado* II 11; Trimble 152; Richardson 443; Twain 195; Page 212; Reid quoted Jackson *BC* xii; Borthwick 146-47; *News* Nov. 14, 1863; Rickard, 261-62; Woods *Sixteen Months*; *Statesman* Aug. 23, 1864; Harlow 192; McConnell *Hist. of Idaho*; Borthwick 229 ff.; Savage

Donaldson 147 ff; Rickard 1-21 *passim* and 70; Stuart quoted J. Davis 29; Schieffelin Meyers 25; Black Hills Peattie 58; Greeley 127-28; Quiett 12, 16; Lewis 126; on Monahan see the Mildretta Adams booklet on Silver City.

Chapter 7.   Visscher 17; Root *Overland Stage passim*; Bowles 52-54; Colton 42; Banning *passim*; newspaper Root and C. 4; Hungerford quoted Beebe 50; Borthwick 107-8; Twain 28; Richardson 159; Huntley II 9; Barnes 70-71; stage line Root and C 11; Bancroft quoted Drury 138; R and C 74; Harlow 181; Drury 141; R and C v, vi; Harlow 179, 182; Ward *Avalanche* Mar. 24, 1866; Richardson 484; Jackson *BC* 119; Barnes 81; M. Adams 13; Huntley I 68; Drury 144; Richardson 384; Cross 185; *Pioneer* May 24, 1879; Drury 139-40; Beebe 95; Haskins 206; Bowles 279-80; Browne *Apache* 312; Farish *GH* 183-85; Jackson in *AG*; Beebe 50; Bowles *Switzerland* 32; Visscher 15; Twain and Pegasus in Banning; Twain 52 ff.; Harlow *passim*; Visscher 26 f.; Borthwick 192-93; Lowden in Harlow 143; Aubrey in Visscher 38 f.

Chapter 8.   Huntley *passim*; Marryat 255; Helper 156 f.; Toponce 20; dirty cook Root and C 96; Borthwick 118, 128; Richardson 185, 493; Delano 224; verse Quiett 153; pies R and C 97; Kelly in *Stroll*; Delano 34; Greeley 144; Delano 41; Taylor II 9; Rickard *RM* 28; Beebe *U.S. West* 260; Wolle *Timberline* 14, 34; Drury 246; Twain 26; Haskins 50, 80; Shinn 215; Bowles 341-42; Colton 37; Borthwick 64-65; Bowles 325; *News* Oct. 8, 1864; Montanan Barsness 161; *World* Apr. 25, 1872; Denver Zamonski and K 98; Helper 136; Emrich *passim*; Boise Donaldson 140; Drury 54.

Chapter 9.   Coffroth, Dane 180; "If you want . . ." *Avalanche* Aug. 3, 1867; "Ye . . ." *News* Feb. 6, 1864; Jomes, *Avalanche* Nov. 16, 1867; *News* Oct. 20, 1863; Huntley I 200; Shirley Letter Sixth; *Avalanche* Nov. 25, 1865; Dimsdale 12; McClure 209 see also Bowles 367-68; *Avalanche* Mar. 3, 1867; July 6, 1867; Twain II 135; Colton 27; Helper 171-72; Richardson 479; *News* Nov. 14, 1863; Mar. 12, 1864; Kelly 126; Twain 32-34; *Statesman, Enterprise* Sept. 24, 1864; toad, Shuck *California Scrap-Book* 513; Bowles 72; Kelly 82; *Avalanche* Apr. 21, 1866; May 5, 1866; Oct. 6, 1866; *Statesman* Sept. 28, 1865; Mathews 112; Toponce 29-30; Bowles 423; *Avalanche* Nov. 18, 1865; tourist, in Paul 164; Kelly 134; Helper ch. 3; orator, Dane 187; M'Collum 160; McIlhany 104; Mathews 172; Richardson 504, 398, 446, 445; Donaldson 128; *News* Feb. 5, 1864; *Avalanche* Jan. 13, 1866; Wiggs, *Silver State Post* Mar. 13, 1941; *Avalanche* Feb. 10, 1866; Toponce 173-74; 52; *Avalanche* Jan. 27, 1866.

Chapter 10.   *World* Nov. 12, 1864; *Journal* Dec. 22, 1863; Donaldson 105; Wilde, Parkhill 34 tells the story with all the trimmings; Colton 143-44; Quiett 85; Borthwick 303-4; Stuart I 233; *Democrat* Sept. 23, 1868; hurdies, Trimble *MA* 150; Dimsdale 9 ff.; *Avalanche* Aug. 4, 1866; Oct. 7, 1865; Nov. 18, 1865; Apr. 14, 1866; July 28, 1866; *Statesman* Nov. 24, 1864; Tombstone, Myers 70-71; Drury 304-5; Goldfield, Glasscock *GH* on saloons; Beebe *US West* 263; Emrich 72; Drury 122; *News* Dec. 7, 1876; Drury 173; Borthwick 79; Greeley 170; Bowles 285; *News* Dec. 5, 1863; Colton 307; McIlhany 102; *News* Mar. 5, 1864; *World* Oct. 13, 1866; Greeley 36; Helper 71; Marryat 27; *Avalanche* July 6, 1867; Borthwick 76, 64; Woods vi; *Avalanche* Aug. 3, 1867, Feb. 28, 1868, Oct. 24, 1868; Meagher, Howard 43; Wharton, Parkhill 63-64; Richardson 186-87; Parkhill 64, 75; Beebe 68; Borthwick 69-70, 65-66; Buffum 165; Dyer,

Idaho Lore 98; Helper 71; Knower 96; Kelly 133; Dawson, Greever 372; Emrich 81 ff.; Quiett 23-24; Donaldson 43; Delano 350; Colton 302-3; Zamonski and K 37.

Chapter 11. Beebe in *Legends;* Otero 154; Klondike, O'-Connor 71; Parkhill xii, xiii; Hall 54; Otero 22-24; Bancroft *Madams* 8, 5, 6; Richardson 254; Stoll ch. 3; Belle, *Boise News* Oct. 22, 1864; *Tribune* Sept. 19, 1930; Emrich 118; Lewis *SK* 28-29; her crib, Zeke Daniels; Drury 36; Mathews 194; Davis 226; Stockings, Bancroft *Baby Doe* 10; Parkhill 54; Liz, Bancroft *Madams* 62-63; Mitchell, M. Adams in letter to author. The Charles Teeter journal was printed but never bound by the *Idaho World.*

Chapter 12. Colton 162; Driggs, Barsness 215; *Avalanche* Feb. 1, 1868; Denver, Zamonski and K 250 f.; Toponce 166-67; Alhambra, Drury 301-2; *News* Feb. 13, 1864; Ward and G, quoted J. Davis 214-15; Borthwick 336; Colton 216-18; Jackson in *AG;* Borthwick 276 ff.; Helper ch. 10.

Chapter 13. Mathews 167 f.; Father Judge, Quiett 357; Clum, *Ariz. Hist. Rev.* Jan. 1931; Casey 175 ff.; Otero; Parkhill 30; the opinions of Calamity Jane are in Sollid, by far the best of the books about her; Holbrook, Sollid 61.

Chapter 14. *Avalanche* Oct. 12, 1867; *News* Feb. 6, 1864; Bancroft *PT* I 578; Ferguson 153; Haskins 106; Twain *RI* II 134; Helper 70; Shinn 130; Dimsdale 8; Bryant 425; Mathews 81; O'Connor 75; McIlhany 88; Mathews 220; Marryat 229; Colton 174; Woods 75; hanging, *Midland Empire Farmer* Feb. 3, 1938; verse, J. Davis 31; Drury 305-6; Wolle 128; Donaldson 129-30; Rogers, Barsness 92; mail, *Idaho Lore* 71; Harlow *passim;* Knower 139; Woods 87; Page 316; Shaw 137; Zamonski and K 28; Richardson 386; Delano 291; Hogg 46 and *passim;* d'Easum 11-13; Toponce 49, 122; *World* Dec. 17, 1864; Twain 146-47; Bancroft *PT* I 127; *News* Sept. 29, 1863; *World* May 20, 1865; Teeter (unpaged); Borthwick 325 f.; Helper 142; Drury 120; Lazarus, *World* May 18, 1867; Tombstone, Myers 55; Ferguson 109; Buffum 128; Shaw 130; Page 305; Colton 156; *Statesman* May 17, 1866; Ferguson 201; Delano 284; Marryat 137; Donaldson *passim; World* Dec. 24, 1864; Jan. 28, 1865.

Chapter 15. *World* May 13, 1865; Dec. 1, 1866; *Avalanche* Apr. 14, 1866; Beidler 147-48; Davis 335; Shaw 37; Delano 305 ff.; Colton 68; Twain 132; *Avalanche* May 16, 1868; Bryant 272; Browne 201; Barnes 44; Ferguson 97; *Avalanche* Nov. 11, 1865; Stuart I 253; Chinese, M. Adams 53; Tassin, *Overland Monthly* XII 2nd Series 629; *Montanan* Jan. 30, 1873; on Chinese in Montana, see Davis; *Madisonian* Mar. 8, 1876; rime, *Montana Radiator* Jan. 27, 1866; Tonopah, Glasscock *GTH;* Tong wars, see Calif. histories; *World* Apr. 20, 1871; *Alta* Mar. 16, 1850; Bowles 67-70; *Avalanche* Aug. 31, 1867; Knower 74; Buffum 122-23; *Statesman* Mar. 10, 1868; Beebe *USW* 176; Quiett 160; *Statesman* Mar. 15, 1866; Bryant 363; pathos, Richardson 163; M'Collum 152; Greeley 119; Roosevelt in *WW;* De Smet quoted Davis 21; Richardson 367; Colton 30; Rickard 263; Marryat 263; Shinn 304-5, 308; gangs, Jackson *BC* 7; Twain II 105; Brace 212; Bowles 239-42; Bell 169; Donaldson 48 f. and *passim;* Chafee, M. Adams 46-47; Polly, essay in possession Caroline Bancroft.

Chapter 16. Gard 150; Shinn 142, 112, 111; Delano 254; Stoll 42; Borthwick 71; Jackson xi, xiii, 43; Bell 218-19; Bancroft *PT* I 151; Helena, Dimsdale 253-54; Buffum 107; Kip 38; Shaw 142-44; Taylor I 92; Springer,

Idaho Lore 46; Shinn 117-19; 169-70; Bancroft *PT* I 641; Shinn 209; Davis 234; *Avalanche* Apr. 28, 1866; Drury 73-74; Rickard *HM* 315; Richardson 509-10; Wells Idaho Bureau Mines Bull 22, 41-42; Glasscock *GTH* 31; Jackson *BC* xiv; Brace 200; Harlow *Waybills* 340; Dimsdale 23; *News* Dec. 9, 1879; Drury 151; Harlow 264; Beebe *USW* 291 f.

Chapter 17. Murder in Denver, *Boise News* Feb. 27, 1864; Colton 50, 192, 224, 227; Shinn, quoted Gard 256; Drury 21; Stoll 91 ff.; Gard 270; *News* May 21, 1864; Gard 260-61; Justice Barry, Shinn 193; Donaldson 151; *World* Apr. 22, 1865; Placerville, Gard 255; Kip 39; Bancroft *PT* I 600; Barsness 244; *Ariz. Review* July 1930; Otero 161 f.; Kelly 183; Bancroft *PT* 138; Lockwood, quoted Gard 258; Bean, West Texas Hist. Assoc. *Yearbook* XXXVII 92; C. L. Sonnischen *Handbook of Texas* I 129; Gard and Bell *passim;* Bancroft *PT* 332; Richardson 188; Collier and W 165; Reavis, Miller, *Arizona* 90-91; Shinn 192-93; Lewis *SK* 12; *World* Aug. 12, 1869; McKenzie, Miller, *Hist. of Alaska* 111; Parkhill xviii.

Chapter 18. Richardson 349; Shinn 114; Waters 290; Gard 154, 161-62; Burns 41; Terry, Buchanan 9; Houston, Horan *WW* 116; Bancroft *PT* I 679; Colorado, Z and K, 146, 215; Sanders, Beidler 53; *News* Mar. 9, 1864; Dimsdale 106-8; Gard 194; White 151; 174-76; Dimsdale 12; Marryat 35; *Avalanche* Jan. 25, 1868; Bancroft *PT* I 130; Twain *RI* II 56-57; *Avalanche* Oct. 14, 1865 (copied from *World*); *Herald* Sept. 29, 1867; *News* Mar. 12, 1864; Mike, Stoll 29-30.

Chapter 19. Twain *RI* II 58-59; Drury 168; Shaw 133; Berton *KF passim;* Drury 151 ff.; Bell, see Jackson *BC,* Harlow, many others. Apache Kid, Arizona histories; Soto, Jackson *BC;* ditto for Murieta; Sherman, *Calf. Hist. Soc. Quar.* Dec. 1944; *World* Nov. 9, 1867; Bart: Wells Fargo History Room, San Francisco, has huge piles of materials about him; see also Jackson, and Beebe 188. Slade: Twain in *RI;* Root and C 217 and *passim;* Toponce 148 ff.; Rivington *The Frontier* XIII; on Smith, Berton, Collier and W, and most books on Colorado and Alaska camps. Orchard is based on MacLane.

Chapter 20. Delano 359; Bancroft *PT* I 131 and *passim;* Trimble *MA* 153; Kelly 182; Helper 239; Richardson 487; Dimsdale 91; Huntley I 145; Otero 205-6; Borthwick 84; *Herald* quoted by Coblentz 44; Marryat 389; Bancroft 317; Otero 193; Beebe 265; *Avalanche* Feb. 3, 1866; Borthwick 219; Shirley Letter Eleventh; Shinn 217 ff.; the San Francisco work of the vigilance committees is set forth by Bancroft, White, Coblentz, and others; Magruder, see McConnell and Donaldson, and Sanders on Beidler; *Boise News* Feb. 13, 1865, June 25, 1865; Bancroft 675; Drury 159; Twain 62-63; *News* Mar. 19, 1864; Davis 46; *Statesman* June 6, 1865; July 22, 27; Aug. 8, 29; *Avalanche* Jan. 9, 1869; Harlow 278; *Avalanche* June 9, 1866; *World* Apr. 14, 1866, May 5; *Avalanche* Jan. 9, 1869; Donaldson *passim;* Bancroft 663 ff.; Gard 207 and *passim;* Osborne, Gard 210; Berton 262; *Idaho Lore* 46, 42, 39; Howard 44; Kid Hall, Peattie 120; Caughey 236; 188; Juanita, Jackson *AG* and Bancroft *PT* I 578 ff.; Borthwick 216; *Avalanche* Sept. 23, 1869; Beidler 116; Dimsdale *passim.*

Chapter 21. Johnny mine, Wolle 52; Rickard *RM* 324-25; Womack, see Sprague and Waters; Sprague 313; Eilley, Drury 28; Ralston, Drury 118-19; Twain *RI* II 34; Marshall, Sioli, *History of El Dorado County;* Pancake, Drury and others; Sam, Sprague 223 ff.; Comstock, Glasscock and many others; Waters, *Midas passim;* on Tabor, C. Bancroft and other Colorado historians.

Chapter 22.   Brace 189, 195; Birch, in Manning 74; Kearns, Utah histories and the booklet for visitors to Utah Hist. Soc.; Beebe *West* 226-27; Big Four, Lewis *SK*, Drury, many others; Drury 65; Mrs. Mackay, Manter, Lewis, others; Stratton, Waters is the standard; also Sprague, Lee, others.

Chapter 23.   Montez, Rourke, *Troopers*, Lewis, others; Baby Doe: Bancroft, *Doe* and *Matchless*; Hall; Colorado histories; Hoffman *Overland Monthly* VIII 2nd series 225; Sarah Hill: Wagstaff, Buchanan, Oscar Lewis but above all, Kroninger; Kate: Lucia chiefly.

Chapter 24.   Shaw 123; Richardson 485, 305; Haskins 156; Featherlegs, Parkhill 31; Quiett 167; Marryat 329; Smith, *Avalanche* Jan. 4, 1868; Otero 187; Knower 98; McConnell, see Donaldson; Colton 391-93; Stoll 13; Drury 218 f.; Spillman, Davis 28; Borthwick 301; Donaldson 126; Johnny, Gillis; Angel 345; Lyman 160; Gillis, *Gold Rush Days;* others; McCall, dozens of books and the Dist. Court archives Sioux Falls, S. D.; Patterson, Idaho histories, Donaldson 170, McConnell 278-79; Smith, Drago; Kansas Hist. Coll. XVII and XIX; and others; Brown, Emrich 137; Drury and De Quille; see also Morse, *Calif. Hist. Soc. Quar.* Dec. 1927; Jackson *BC* 284; Angel 344-45; Van Sickles *Nev. Hist. Soc. Papers* I, 192-93; on Packer, Colorado histories and Denver newspapers of the time; the Shores statement, Rockwell 343.

Chapter 25.   Bancroft *PT* I 122; Twain *Auto.* I 350; *News* Oct. 20, 1863; Marryat 354; Col. & Judge, *Statesman* July 26, 1866; Thorp 37, 55-56, 12; Mexican-Chilean, *Argonaut* II 1887; Riley, Zamonski 190-91; Editor, Miller, *Arizona* 92; Bell 226; Myers 94-95; Cronin, *Ariz. Hist. ev.* July 1930; Tucker, Banning 159 ff.; *World* Dec. 29, 1870; Broderick and Terry, Bancroft *PT* 320, Hittell *Overland Monthly* XIII 2nd Series 103, Jeremiah Lynch, *A Senator of the Fifties,* San Francisco, 1911, John Currey, *The Terry-Broderick Duel,* Wash., D.C., 1896, Oscar T. Shuck, *Bench and Bar in Calif.,* San Francisco, 1888; Denver *News* Aug. 26, 1877; also Buchanan, O'Meara, Hunt and Sanchez, Wagstaff, Kroninger; *World* June 18, 1870, July 28, Aug. 18; *Statesman* July 21, 1866; Twain, Gillis, Lyman, Paine, others; Mattie, Parkhill 207 ff., others. On firearms used in the West, Charles E. Chapel *Guns of the Old West* —thinks Cora used a genuine Philadelphia derringer when he shot Richardson; James L. Mitchell, *Colt,* James E. Serven *Colt Firearms:* "An astonishing number of variations are to be found in the Colt 1851 Navy pistols."

Chapter 26.   Emrich 282; King, Coblentz 132; Bagg, Miller 97; Drury 167, 313-14; d'Easum 119; Miller *Frontier* vi; Jernegan, Drury 183, 174-75; Marion etc. Miller 99 ff.; Langford, Barsness 123; Richardson 501; *World* Dec. 10, 1864; *Statesman* Jan. 12, 1865; Donaldson 129-30.

Chapter 27.   Silverburg, *Natural History,* March 1965; Drury 40; Watson *Calif. Hist. Soc. Quar.* Mar. 1932; Russell, Emrich 182-83; Twain, Quiett 20; Mount Pisgah, Sprague ch. 3; diamond hoax: the chief, almost the only, source is Harpending's book: "It would be difficult to estimate how many motion pictures, short stories, and novels have been inspired by this book." For hilarious reading see Edmund Pearson *Dime Novels* and Albert Johannsen *House of Beadle and Dime Novels.*

Chapter 28.   Petronius, *Idaho Lore* 117; *World* July 4, 1878; Emrich 279; Waters 200 ff.; Sophia, Waters 300 f.; Hogg 115; Coffman 169; Emrich 197; Beebe 161; Bell 138-39; Miller 249; Coffman 85 f.; Eberhart *Treasure Tales;* Jackson *BC*; Breen, Hult 184; Plummer, Davis 307. Some books on lost treasure: Philip A. Bailey *Golden Mirages;* H. D. Clark *Lost Mines of the Old West;* J. Frank Dobie *Apache Gold; Coronado's Children;* Ely Simms *Lost Dutchman Mine;* Harold O. Weight *Lost Mines of Death Valley.*

Illustrations

We learned that getting the photographs for a book as large and comprehensive as this one is a sizable task; besides journeying to the larger collections, and photographing many of the items ourselves in ghost towns and elsewhere, we wrote hundreds of letters over a period of more than two years. In gathering photos we encountered two unforeseen problems. One is that most authors give no source or an incomplete source for the photos they use. It isn't much help, for instance, to have only X. Y. Museum or Blank County Museum or John Doe, if the reference librarian in a big library can't for the life of her find out where the museum is or who John Doe is. To save time and emotional wear on those who may wish to use some of the photos appearing herein we give enough address for a letter to reach the source.

An even more exasperating problem has been the fact that many persons and even some agencies and institutions claim exclusive rights in photographs in which they have no such rights. When at the Bancroft we said to Dr. Tompkins, the curator, in regard to a certain photograph he put before us, that we knew of an institution and a person who laid exclusive claim to it he literally snorted with impatience. His lecture on that subject did a lot to enlighten us: in substance he said that there are hundreds of persons, institutions, societies, and agencies that claim a sole and exclusive right in photos in which they have no rights. It does not follow that no rights in photos exist; a copyrighted photo is private property and written permission is needed for use of it. Many great collections, such as the Rose, are private property; they have been bought and paid for. There are many uncopyrighted photos that belong to those who possess them. But in the public domain are tens of thousands of photos over which persons or institutions maintain a watchful custody, and claim the right to grant permission for use. The extremest instance of this that we encountered is a man who boasts that he has a staggering number of photos that belong exclusively to him. He has to our knowledge bluffed and intimidated authors and publishers and institutions; rather than risk suits or (in the case of public institutions) unfavorable publicity, they have yielded to his threats and paid his fees, which (if demands made on us are typical) are outrageously high.

Before each photo is its book page number; if more than one photo appears on a page, read from left to right and top to bottom.

451

A Selected Bibliography

The quality of these publications ranges from sound scholarship to unabashed re-telling of old legends. P, where used, denotes a primary source—records, that is, by persons who were on the scene (these also vary widely in quality); E denotes excellent, Pop, popular, and PE means an excellent primary source. The works of many nineteenth-century writers—Bancroft, Shinn, and others—are based on both primary and secondary sources.

The amount of time and labor demanded by such a project as this one is staggering. No more than index-research in all the volumes of *Overland Monthly, Harper's, Century, Pacific Monthly* and similar journals requires months, and a cursory searching of the publications and journals of state historical and similar societies would take years.

ABBEY, JAMES. *California. A Trip Across the Plains in the Spring of 1850.* New Albany, Ind., 1850. Almost the earliest of its kind.

ADAMS, RAMON F. *Burs Under the Saddle.* Norman, Oklahoma, 1964. Corrects thousands of errors in more than four-hundred books.

ADNEY, TAPPAN. *The Klondike Stampede.* New York, 1900. E, by one who was there.

ALLSOP, THOMAS. *California and Its Gold Mines.* London, 1853. Letters, chiefly about quartz mining and its problems.

ANDREWS, C. L. *The Story of Alaska.* Caldwell, Idaho, 1938.

ANGEL, MYRON. *History of Nevada.* Oakland, 1881.

AUGER, EDOUARD. *Voyage en Californie.* Paris, 1854. Thin.

AVERILL, CHARLES VOLNEY. *Placer Mining for Gold in California.* (Division of Mines, Bulletin 135.) San Francisco, 1946. Good picture of the whole thing.

BALL, NICHOLAS. *The Pioneers of '49.* Boston, 1891.

BANCROFT, CAROLINE. *Gulch of Gold.* Denver, 1958. The standard on Central City. Miss Bancroft has also published a number of booklets, on madams, the Tabors, etc.

BANCROFT, H. H. *Works.* (But chiefly *Popular Tribunals* and *California Inter Pocula.*)

BANKSON, RUSSELL A. *The Klondike Nugget.* Caldwell, Idaho, 1935. Based on the Dawson newspaper of that name.

BANNING, CAPT. WILLIAM AND GEORGE HUGH. *Six Horses.* New York, 1928. A standard.

BARNES, DEMAS. *From the Atlantic to the Pacific, Overland.* New York, 1866.

BARSNESS, LARRY. *Gold Camp; Alder Gulch and Virginia City, Montana.* New York, 1962.

BATES, MRS. D. B. *Incidents on Land and Water.* Boston, 1857. A woman's point of view.

BECKER, GEORGE F. *Geology of the Comstock Lode.* (USGS.) Washington, 1882.

BEEBE, LUCIUS, AND CLEGG, CHARLES. *U.S. West; the Saga of Wells Fargo.* New York, 1949.

BEIDLER, JOHN XAVIER. *X. Beidler: Vigilante*; ed. by Helen Fitzgerald Sanders. Norman: University of Oklahoma Press, 1957.

BELL, HORACE. *On the Old West Coast.* New York, 1930. An old-timer's account.

BENTON, JOSEPH A. *The California Pilgrim.* Sacramento and San Francisco, 1853. Lectures.

BERNEWITZ, VON, M. W. *Handbook for Prospectors and Operators of Small Mines.* 4th ed.; New York and London, 1943. A standard.

BERTON, PIERRE. *The Klondike Fever.* New York, 1958. E, by the editor of *Maclean's.*

BIRNEY, HOFFMAN. *Vigilantes.* Philadelphia, 1929. Pop.

BLACK, E., AND ROBERTSON, SIDNEY. *The Gold Rush Song Book.* San Francisco, 1940.

BORTHWICK, J. D. *Three Years in California.* London, 1857. PE.

BOWLES, SAMUEL. *Across the Continent.* New York, 1865. PE.

BRACE, CHARLES LORING. *The New West; or, California in 1867-1868.* New York, 1869. PE.

BREWER, WILLIAM H. *Up and Down California.* New ed.; New Haven and London, 1930. A big book but not much about mining.

BROWNE, J. ROSS. *Adventures in the Apache Country.* New York, 1869.

BRYANT, EDWIN. *What I Saw in California.* New York, 1849. PE.

BUCHANAN, A. RUSSELL. *David S. Terry of California, Dueling Judge.* San Marino, Calif.: Huntington Library, 1956.

BUCK, FRANKLIN A. *A Yankee Trader in the Gold Rush.* Boston, 1930. Letters.

BUFFUM, E. GOULD. *Six Months in the Gold Mines.* Philadelphia and London, 1850. PE.

BURLINGAME, MERRILL G., AND TOOLE, K. ROSS. *A History of Montana.* 3 vols.; New York, 1957.

BURNS, WALTER NOBLE. *Tombstone.* New York, 1929. Pop.

California Gold Rush. The Mother Lode Country. (Division of Mines, Bulletin 141.) San Francisco, 1948. Centennial ed. E.

CANFIELD, CHAUNCEY L. *Diary of a Forty-Niner.* Boston, 1920. The diary of A. T. Jackson.

CASEY, ROBERT J. *The Black Hills and their Incredible Characters.* Indianapolis, 1949. Pop.

CAUGHEY, JOHN W. *Gold Is the Cornerstone.* Berkeley, Calif.; University of California Press, 1948. An overall picture.

CHRISTMAN, ENOS. *One Man's Gold. The Letters and Journal of a Forty-Niner.* New York, 1930.

CLAPPE, LOUISE. *The Shirley Letters from the California Mines, 1851-1852.* New York, 1949. With an introduction by Carl I. Wheat. PE. Philosopher Royce thought them "the best account of an early mining camp that is known to me."

COBLENTZ, STANLEY A. *Villains and Vigilantes.* New York, 1936. Deals only with California.

COLLIER, WILLIAM ROSS, AND WESTRATE, EDWIN VICTOR. *The Reign of Soapy Smith.* New York, 1935. Pop.

COLTON, REV. WALTER. *The Land of Gold; or, Three Years in California.* New York, 1850. PE.

DANA, JULIAN. *The Sacramento.* New York, 1939. Pop.

DAVIS, JEAN. *Shallow Diggin's.* Caldwell, Idaho, 1962. Compiled from newspapers.

D'EASUM, DICK. *Fragments of Villainy.* Boise, Idaho, 1959. Pop.

DELANO, ALONZO. *Life on the Plains and Among the Diggings.* Auburn and Buffalo, 1854. PE.

DE QUILLE, DAN (William Wright). *The Big Bonanza.* A good edition of this well-known book is Knopf's, with introduction by Oscar Lewis.

DIMSDALE, THOMAS J. *The Vigilantes of Montana.* Norman, Oklahoma, 1953. PE.

DONALDSON, THOMAS. *Idaho of Yesterday.* Caldwell, Idaho, 1941. P.

DRAGO, HARRY SINCLAIR. *Wild, Woolly & Wicked.* New York, 1960. Debunks glamorizers.

EBERHART, PERRY. *Treasure Tales of the Rockies.* Denver, 1961. With maps.

ELSENSOHN, SISTER M. ALFREDA. *Pioneer Days in Idaho County.* 2 vols.; Caldwell, Idaho, 1947-1951.

EMMONS, WILLIAM HARVEY. *Gold Deposits of the World.* New York and London, 1937. E. Has a good but technical chapter on prospecting.

EMRICH, DUNCAN. *It's an Old Wild West Custom.* New York, 1949.

FARISH, THOMAS EDWIN. *The Gold Hunters of California.* Chicago, 1904. Pop.

FERGUSON, CHARLES D. *The Experiences of a Forty-Niner.* Cleveland, 1888. Very good on overland journey.

FINCK, J. W. *Prospecting for Gold Ores.* (Idaho Bureau of Mines.) Moscow, Idaho, 1932.

FINLAYSON, GEORGE. *Here They Dug the Gold.* New York, 1931. Pop.

FITZHUGH, EDWARD F., JR. *Treasures in the Earth.* Caldwell, Idaho, 1936. Excellent account by a geologist of how gold is formed and where found.

FOSSETT, FRANK. *Colorado.* Denver, 1876. A good early account.

GANTT, PAUL H. *Case of Alfred Packer, the Man-Eater.* Denver, 1952. Pop.

GARD, WAYNE. *Frontier Justice.* Norman, Oklahoma, 1949. A good overall picture.

GILLIS, WILLIAM R. *Gold Rush Days with Mark Twain.* There are many editions.

GLASSCOCK, C. B. *The Big Bonanza: The Story of the Comstock Lode.* Indianapolis, 1931. Pop.

————. *Gold in Them Hills.* Indianapolis, 1932. Pop.

GREELEY, HORACE. *An Overland Journey . . . in the Summer of 1859.* New York, 1964.

GRINNELL, JOSEPH. *Gold Hunting in Alaska.* Chicago, 1901. Pop.

HAFEN, LEROY R. AND ANN W. *To the Rockies and Oregon, 1839-1842.* Glendale, Calif., 1955. P.

————. (eds.). *Journals of Forty-Niners.* Glendale, Calif., 1954. P.

HALL, GORDON LANGLEY. *The Two Lives of Baby Doe.* New York, 1962. Pop.

HARPENDING, ASBURY. *The Great Diamond Hoax.* Norman, Oklahoma, 1958.

HASKINS, C. W. *The Argonauts of California.* New York, 1890. PE, by a Forty-niner; pages 360-501 list names of Forty-niners and other pioneers.

HELPER, HINTON R. *The Land of Gold.* Baltimore, 1855. PE. Written to correct what he thought was a dishonest picture of the gold regions.

HENDRICKS, GEORGE D. *Bad Man of the West.* San Antonio, Texas, 1942. Pop.

HITTELL, JOHN S. *History of California.* San Francisco, 1885. A standard.

HORAN, JAMES D., AND SANN, PAUL. *Pictorial History of the Wild West.* New York, 1954. Pop. Excellent for its photos.

HOWARD, JOSEPH KINSEY. *Montana High, Wide, and Handsome.* New Haven, Conn., 1943. Pop.

HOWE, OCTAVIUS T. *Argonauts of '49.* Cambridge, Mass., 1923. E on routes taken by Forty-niners.

HULBERT, A. B. *Forty-Niners; The Chronicle of the California Trail.* Boston, 1949. A standard.

HUNGERFORD, EDWARD. *Wells Fargo; Advancing the American Frontier.* New York, 1949. E.

HUNTLEY, SIR HENRY. *California: Its Gold and Its Inhabitants.* 2 vols.; London, 1856. (It is spelled "Huntley" in bibliographies but in ink is spelled "Huntly" on the Huntington Library copy.)

JACKSON, JOSEPH HENRY. *Anybody's Gold: The Story of California's Mining Towns.* New York, 1941. Pop but good.

————. *Bad Company.* New York, 1949.

JOHNSON, THEODORE. *Sights in the Gold Region.* New York, 1850. PE.

KELLY, WILLIAM. *An Excursion to California with a Stroll through the Diggings.* London, 1852. PE.

KNOWER, DR. DANIEL. *The Adventures of a Forty-Niner.* Albany, New York, 1894. P.

KRONINGER, ROBERT H. *Sarah and the Senator.* Berkeley, Calif., 1964. E, by a judge.

LANGFORD, NATHANIEL P. H. *Vigilante Days and Ways.* Chicago, 1912. A standard.

LAVENDER, DAVID. *Westward Vision; The Story of the Oregon Trail.* New York, 1963. See also *The Oregon Trail* in the American Guide Series.

LEE, MABEL BARBEE. *Cripple Creek Days.* New York, 1958. She lived there as a girl.

LEWIS, OSCAR. *Silver Kings.* New York, 1947. E, on the Big Four. His *Bonanza Inn,* with Carroll D. Hall, is also good.

LINCOLN, FRANCIS C. *Mining Districts and Mineral Resources of Nevada.* Reno, 1923.

LINDGREN, W. *Mineral Deposits.* 4th ed.; New York, 1933. A standard.

LUCIA, ELLIS. *Saga of Ben Holladay.* New York, 1959.

————. *Klondike Kate; The Life and Legend of Kitty Rockwell.* New York, 1962.

LYMAN, GEORGE D. *The Saga of the Comstock Lode.* New York, 1934. One of the best.

McCLURE, A. K. *Three Thousand Miles through the Rockies.* Phila., 1869. Good view of Montana in the sixties.

M'COLLUM, WILLIAM. *California as I Saw It.* Buffalo, 1850. PE. There is a 1960 ed. with excellent introduction by Dale Morgan.

McCONNELL, WILLIAM J. *Early History of Idaho.* Caldwell, Idaho, 1913. He was a vigilante captain.

McILHANY, EDWARD W. *Recollections of a '49er.* Kansas City, Mo., 1908. PE.

MACLANE, JOHN F. *A Sagebrush Lawyer.* New York, 1953. Reminiscences.

MANLY, WILLIAM LEWIS. *Death Valley in '49.* New York, 1929. E on the Death Valley journey.

MARRYAT, FRANK. *Mountains and Molehills.* London, 1855. PE.

MATHEWS, MRS. M. M. *Ten Years in Nevada.* Buffalo, N. Y., 1880. PE. Fine description of early Virginia City.

MATSON, R. C. *Prospecting and Exploration.* Mich. College of Mines, 1932.

MILLER, JOSEPH. *Arizona, the Last Frontier.* New York, 1956.

MORLEY, JAMES HENRY. His diary while in early Montana; transcript in possession of Montana State Historical Society.

MORGAN, DALE (ed.). *Overland in 1846*. Georgetown, Calif., 1963. P. From diaries and letters.

MORRELL, W. P. *The Gold Rushes*. New York, 1941. A good introduction to a big subject.

MYERS, JOHN M. *The Last Chance: Tombstone's Early Years*. New York, 1950. One of the more scholarly books on Tombstone.

O'CONNOR, RICHARD. *High Jinks on the Klondike*. Indianapolis, 1954. Pop.

————. *Wild Bill Hickok*. New York, 1959. Pop.

OTERO, MIGUEL A. *My Life on the Frontier, 1864-1882*. New York, 1935.

PAGE, ELIZABETH. *Wagons West; a Story of the Oregon Trail*. New York, 1930. PE.

PARKHILL, FORBES. *The Wildest of the West*. New York, 1951. Pop.

PATTERSON, LAWSON B. *Twelve Years in the Mines of California*. Cambridge, 1862. Gives an unvarnished account.

PAUL, RODMAN W. *California Gold; the Beginning of Mining in the Far West*. Cambridge, 1947. E.

————. *Mining Frontiers of the Far West*. New York, 1963. E.

PEARSON, EDMUND. *Dime Novels*. Boston, 1929. Very amusing.

PEATTIE, RODERICK (ed.). *The Black Hills*. New York, 1952. No bibliography and cites few sources.

PEFLEY, WYNN. *The A B C of Gold Hunting*. Boise, Idaho. By a mining engineer; it is chiefly about placers.

QUIETT, GLENN CHESNEY. *Pay Dirt; a Panorama of American Gold Rushes*. New York, 1936. One of the best overall pictures.

————. *They Built the West*. New York, 1934.

REVERE, JOSEPH WARREN. *A Tour of Duty in California*; ed. by Joseph N. Balestier. New York, 1849.

RICHARDSON, ALBERT D. *Beyond the Mississippi*. Chicago, 1869. PE, though with tendency to unqualified statement.

RICKARD, T. A. *Through the Yukon and Alaska*. San Francisco, 1909. E, by a geologist, with excellent photos.

————. *A History of American Mining*. New York, 1932. E.

————. *The Romance of Mining*. Toronto, 1945. E.

————. *The Bunker Hill Enterprise*. San Francisco, 1921. Good, on this later development.

ROBINSON, FAYETTE. *California and Its Gold Regions*. New York, 1849. Thin.

ROOT, FRANK A., AND CONNELLEY, WILLIAM E. *The Overland Stage to California*. Topeka, Kansas, 1901. Root was an agent and driver on the line.

ROSA, JOSEPH G. *They Called Him Wild Bill*. Norman, Oklahoma, 1964. Ramon Adams calls this the best of the books on Wild Bill Hickok so far.

RYAN, W. R. *Personal Adventures in Upper and Lower California*. London, 1850.

SAVAGE, E. M. *Prospecting for Gold and Silver*. New York, 1934.

SCHERER, JAMES A. B. *"The Lion of the Vigilantes," William T. Coleman and the Life of Old San Francisco*. Indianapolis, 1939.

SCHOLEFIELD, E. O. S., AND GOSNELL, R. E. *A History of British Columbia*. Vancouver, B. C. [1914]. On the gold rush.

SHAW, D. A. *Eldorado; or, California as Seen by a Pioneer, 1850-1900*. Los Angeles, Calif., 1900. PE.

SHINN, CHARLES HOWARD. *Mining Camps: A Study in American Frontier Government*. Published in 1884; latest ed., New York, 1948. The standard.

SHORES, CYRUS WELLS. *Memoirs of a Lawman*; ed. by Wilson Rockwell. Denver, 1962.

SHUCK, OTTO T. *The California Scrap-Book*. San Francisco, 1869. Interesting items from the newspapers.

SIOLI, PAOLO. *Historical Souvenir of El Dorado County, California*. Oakland, Calif., 1883. Has a section on barbarous crimes.

SMITH, GRANT H. *The History of the Comstock Lode, 1850-1920*. Reno, Nev., 1943.

SOLLID, ROBERTA BEED. *Calamity Jane*. Helena, Mont.: Historical Society of Montana, 1958. The best book on her.

SOULÉ, FRANK, AND OTHERS. *Annals of San Francisco*. New York, 1854. A valuable source.

SPRAGUE, MARSHALL. *Money Mountain; the Story of Cripple Creek Gold*. Boston, 1953.

STECKMESSER, KENT LADD. *The Western Hero in History and Legend*. Norman, Okla., 1965. E.

STELLMAN, LOUIS J. *Mother Lode*. San Francisco, 1934. Pop.

STOLL, WILLIAM T. *Silver Strike; the True Story of Silver Mining in the Coeur d'Alenes*. Boston, 1932. As told to H. W. Whicker.

STUART, GRANVILLE. *Forty Years on the Frontier*. 2 vols.; Cleveland, Ohio, 1925. A good primary source.

TALLENT, ANNIE E. *The Black Hills*. St. Louis, Mo., 1889. She was there.

TAYLOR, BAYARD. *Eldorado*. 2 vols.; New York and London, 1850. PE.

TEETER, CHARLES. His journal can be found in the Thirteenth Biennial Report of the Idaho Historical Society.

THORP, RAYMOND W. *Bowie Knife*. Albuquerque, New Mex., 1948. Corrects a lot of "history as written."

TOPONCE, ALEXANDER. *Reminiscences*. Ogden, Utah, 1923. Privately printed by his wife. Good, especially on freighting, but his memory was poor.

TRIMBLE, WILLIAM J. *The Mining Advance into the Inland Empire*. Madison, Wisc., 1914. E.

VISSCHER, WILLIAM L. *The Pony Express*. Chicago, 1908. He was on the scene.

WADDINGTON, ALFRED. *The Fraser River Mines Vindicated*. Victoria, B.C., 1861. He saw them.

WATERS, FRANK. *Midas of the Rockies; the Story of Stratton and Cripple Creek*. New York, 1937. E.

WEBB, TODD. *Gold Strikes and Ghost Towns*. New York, 1961. Pop.

WELLS, MERLE W. Bulletin 32, Idaho Bureau of Mines and Geology; possibly the best account of the Idaho City and Silver City areas.

WHITE, STEWART EDWARD. *The Forty-Niners*. New Haven, Conn., 1949. A popular account.

WIERZBICKI, DR. F. P. *California; a Guide to the Gold Region*. San Francisco, 1849. The first book printed in English in Calif.

WOLLE, MURIEL S. *Stampede to Timberline: The Ghost Towns and Mining Camps of Colorado*. Denver, 1949. Pop.

————. *Montana Pay Dirt; a Guide to the Mining Camps of the Treasure State*. Denver, 1963. Pop.

WOODS, REV. DANIEL B. *Sixteen Months at the Gold Diggings*. New York, 1851. PE. Perhaps the best one in presenting the hard labor and small returns of miners.

ZAMONSKI, STANLEY W., AND KELLER, TEDDY. *The Fifty-Niners*. Denver, 1962. Pop.

Index

459

McKenzie, Alexander, 294

MacLane, John F., 321-22

Madisonian, 262

Magnetometer, 114

Magruder, Lloyd, murder of, by
    Plummer gang, 332-33

Mahaffey, Mary, madam, 200-201

Mahoney, Mike, 82

Mail, a camp problem, 243-44

Mammy Pleasant, Negress madam,
    commitment of Sarah Hill, 390

Manuel, Moses, 21, 361

Maquereaux, 417

Marion, editor of Prescott *Miner,*
    feuds with Wasson of the Tucson
    Citizen and with Berry of the
    Yuma *Sentinel,* 421

Mariposa grant, 357

Marryat, Frank, 137, 190, 195, 241,
    253, 271, 301, 324, 395, 406

Marshall, James Wilson, 28-29, 360

Marysville, prices at (1850-51), 37-38

Mason, Gov. Richard B., 30, 42-43

Massachusetts Hill, gold vein, 19

Matchless mine (Colo.), 363-64, 385

Mathews, Mrs. M. M., 166, 171,
    215, 225, 241

Matson, John, marriage to Klondike
    Kate, 393

Mauk, George A., U.S. marshal, 227

May, Mollie, and Purple, Sallie,
    Denver madams, duel, 416

Meadows, Arizona Charlie, 90

Meagher, Gen. Thomas F., acting
    governor of Montana Territory,
    193, 348

Meiggs, Harry, 144, 145

Meissonier, M. Jean, 374

Meldrum, John, 73

Memphis *Appeal,* 323

Mexicans, prejudice against,
    271-72, 289

Milleian, Jean Marie A., murderer
    of Julia Bulette, 212, 214

Miller, D. G., gold hoax, 430-31

Miller, Joseph, 201, 202, 203,
    206, 417, 420

*Mineral Resources of the United
    States* (Geological Survey), *1907,*
    number of mining camps in
    U.S., 97

Mineralogical families, 18

Miners' earnings, 39-40; rich strikes,
    fate of, 360-61; banquet, 143

Mines, unusual names of, 40

"Mining Districts and Mining
    Resources of Nevada"
    (Lincoln), 358

Mining, how gold was taken
    out, 41-45

Mining Frontiers (Paul), 71

Missouri House hotel (Boise,
    Idaho), 153

Mizner, Wilson, a Klondike fast-buck
    man, 186, 196-97, 318 19

Monahan, Joe, male impersonator
    (Silver City), 109

Monk, Hank, 108, 124, 127-28

*Montana, The Magazine of Western
    History,* 1

Montana *Post,* 100, 171, 219-20,
    340-41, 348

Montanan, 146

Montez, Lola (Maria Dolores Eliza
    Gilbert): first marriage, 378; mis-
    tress to King Ludwig, 378; second
    marriage 378; arrival in San Fran-
    cisco, 379; performances in San
    Francisco, 379; third marriage, 379-
    80; Grass Valley, 380; return to
    Europe, 380; return to Grass Valley,
    380; journey to Australia, 380-81;
    death, 381

Moore, Jim, pony express rider, 131

Mormons, 63, 64

Morning Call, 412-13

Morrell, W. P., 33-34, 38, 145

Morro Velho lodes, 18

Morse, Sheriff Harry (San Francisco),
    capture of Juan Soto, 308-10

Mother Lode, 18, 19, 26

Mother Lode (Stellman), 349

Mount Davidson, 68

Mount Pisgah hoax, 430-31

Mulrooney, Irish Kate, Klondike rush
    fortune hunter, 390

Murieta, Joaquin, 2, 3, 7, 306, 310-12

Myers, John M., 408-9

— N —

Neagle, David: U.S. marshal, body-
    guard to Justice Field, 389;
    shooting of David S. Terry, 389-90

Negroes, prejudice against, 256-57

Nelson, Mrs., landlady, concerning
    Sophia Chellew's claim to W.
    S. Stratton estate, 438

Nevada City, Alder Gulch, 97

Nevada City, largest gold camp in
    1850, 38

Nevada (Reno) *Gazette,* 287

New West (Brace), 272

New York *Daily Tribune,* 39

New York Herald, 9, 30, 46, 384

New York *Herald,* 46, 384

New York *Sun,* 433

New York *Tribune,* 385

New York *World,* 320

Newspapers: feuds, 338-344, 419-27;
    names, 419; pro-Union and
    anti-Union factions, 421-27

Newsweek, Klondike Kate's death,
    393

Nez Perce Indians, 71, 270

Nichols, Col. George Ward, 9

Nickanora, king of Nevada stage
    robbers, 305

Nigger Hill mine, 357

Nome (Alaska), 20

Northwest Trail Blazers (Howard), 1

Noyes, Judge Arthur H., 294

— O —

O. K. Corral, gunfight at, 12

O'Brien, William S., 106, 367, 369

O'Connor, Richard, 9, 82, 219, 231,
    241, 417

O'Meara, James: 343; editor of
    Idaho City *World,* 426-27

O'Neil, Hugh, 219-20

Oddie, T. L., 358

Old Deadwood Days (Bennett), 235

Old Pancake, *See* Comstock, Henry
    Thomas Paige

*1001 Lost, Buried or Sunken
    Treasures* (Coffman), 439-40, 442

Ophir mine, 69

Orchard, Harry, 321-22

Oregon *Herald,* 255, 303

Orem, John C., 220

Oriental mine, 24

Oro, Colorado, gambling hall, 22

Oro Fino, 71

Otero, Miguel Antonio, 11, 199, 201,
    233, 289-90, 324, 325, 395

Ouray *Solid Muldoon,* 419

Overland journey, 46-65

Overland Monthly, 7, 326, 386

Overland Stage to California, The
    (Root and Connelley), 124

Owens, Sheriff, gunfight with Blevens
    family, 400-1

Owsley, William, told by Plummer
    of hidden gold cache, 444

Owyhee War, The, 282

— P —

Packer, Al: man-eater of the San
    Juans, case of, 403; parole of, 404

Page, Henry, 38, 51, 53, 60, 103,
    244, 251

Paget, Nellie, dance hall girl, 205

Palace Hotel, San Francisco, 366-67

Palmer, Dr., reminiscences of bowie
    knife duel between two women, 417

Palmer House hotel (Sheep Camp,
    Alaska), 90

Panic, financial, 1857, result of, 69

Panic, financial, 1875, 360

Panning for gold, 41-42

Pantages, Alexander: Klondike Kate's
    boyfriend, 391-92; scandal and
    court trial, 392

Paracelsus, 111-12

Parkhill, Forbes, 194, 205, 206, 207,
    233, 416, 417

Parkhurst, Charlie, 109, 129

Parsons, Judge, 290

Party of Law and Order, San
    Francisco, 332

Patterson, Ferd, killer of Sumner
    Pinkham, Idaho City (Idaho),
    398-99

Paul, Rodman W., 41, 68, 280, 367,
    370

Paxson, Frederic L., 10

Pearce, John, 24

Peg-Leg Annie, 109-10

Peg-Leg mine, lost treasure, 441

Pellet, Sarah, 170

People, types of, in mining camps,
    101-4; prostitutes, 104; Missourians,
    104; showoffs, 105; politicians, 105;